The Ecology of Animals

THE ECOLOGY OF ANIMALS

BY

N. P. Naumov

EDITED BY

Norman D. Levine

TRANSLATED BY

Frederick K. Plous, Jr.

University of Illinois Press
URBANA / CHICAGO / LONDON

Manufactured in the United States of America
Library of Congress Catalog Card No. 71–170965
ISBN 0-252-00219-9

Originally published as *Ekologiya zhivotnykh* by the Higher School State Publishers, Moscow, 1963.

Translated from the second edition with the aid of Grant CC–00037 from the National Communicable Disease Center, U.S. Public Health Service, Atlanta, Ga. Publication supported in part by Grant LM–00967 from the National Library of Medicine, National Institutes of Health, Department of Health, Education and Welfare.

EDITOR'S PREFACE

This book is a standard work on animal ecology in the Soviet Union. It was originally published in Russian in an edition of 8,000 copies by the Ministry of Higher and Secondary Special Education as a text for the state universities of the USSR. It is especially important because it illustrates the state of ecology in Russia and allows it to be compared with the same science in the western world. In addition, it is a rich source of Russian examples of ecological phenomena. Many of the references and examples which Naumov gives are unknown to western scientists.

Much of what Naumov says expresses standard viewpoints current throughout the world, but there are differences, and there are also curious and probably significant omissions. For instance, there is nothing in the book on energy dynamics, and the Odums' work is not referred to. The endocrine explanation of population crashes is mentioned, but with only one reference to the work of J. J. Christian and none to that of D. E. Davis.

Naumov denies, probably with justice, the occurrence of density-independent factors. The effects of even such nonbiological factors as weather changes are related to density; if a species' density is low, it uses only the most favorable shelters and is thus better protected from sudden or severe weather changes than if its density is high and it must use both highly favorable and relatively unfavorable shelters.

Naumov's distinction between the scope of ecology in the USSR and that in foreign countries is naive and should not be taken seriously. Perhaps in his country it is necessary to exalt the local and denigrate the foreign. Similarly, his acceptance of human ecology when it studies the effects on man of climatic, geochemical, and other landscape-geographic factors, but his rejection of it when it studies the effects on man of social and economic factors, is apparently due to the necessity of having to hew to the party line. He rejects the "tendency to biologize social phenomena," apparently equating this practice with interpreting social phenomena in favor of class interests. Here, again, his ideas are limited.

It is to be noted that Naumov attacks Malthus's ideas as "antiscientific, narrowly class-oriented 'fabrications.'" This is undoubtedly the party line, but it will undoubtedly be modified as the population of Russia increases further beyond the level at which all the people can be fed adequately. When I visited Leningrad in the summer of 1969, the people appeared well nourished, but potatoes were selling in the market for about $1.00 per pound (1 ruble per kg). This figure cannot, of course, be compared directly with the price of potatoes in the United States (about $0.12 per pound), since there is a considerable difference in personal income between the U.S. and Russia (in favor of the U.S.), and since the black-market rate of exchange between dollars and rubles is several times the official figure (in favor of the dollar).

A description of the method by which this translation was done is probably worthwhile. The book was sent to me by the late Dr. E. M. Cheissin of the Institute of Cytology, USSR Academy of Sciences, Leningrad. The translation was made by Mr. Plous and was then edited by me. The English meanings of a few Russian words (which we could not find in the dictionaries available to us) were supplied by Dr. Naumov. The illustrations were copied photographically by the University of Illinois Photographic Service. Mr. Plous translated the Russian words in them, and they were reprinted with all Russian words blocked out. The English equivalents were entered in place of the Russian words by Martha Dickinson of the College of Veterinary Medicine, University of Illinois, and the illustrations were copied again photographically. The whole manuscript, together with the illustrations, was then submitted for reading to Dr. S. Charles Kendeigh of the Department of Zoology, University of Illinois (one of the nation's leading ecologists and author of a standard American textbook, *Animal Ecology*). It was then given final editing before being published.

It is probably unnecessary to emphasize that this book is a translation and does not necessarily reflect the ideas of either the editor or the translator.

NORMAN D. LEVINE

Urbana, Illinois
3 May 1971

FOREWORD TO THE SECOND EDITION

In preparing the second edition the author has completely revised his book. Critical remarks from published reviews and from innumerable letters to the author have been taken into account. The necessity for a basic reworking of the text was dictated by the rapid development of ecological science; the new edition makes wide use of the latest Soviet and foreign ecological literature. The structure of the book has been greatly altered. Questions of intraspecies relationships and population ecology have been illuminated in detail; the chapter on the spatial orientation of animals has been newly written (with the active participation of G. N. Simkin).

The author is deeply grateful to all those who submitted criticisms, especially Professor S. M. Gilyarov, Professor G. A. Novikov, and his young co-workers V. S. Lobachev and G. N. Simkin, who provided invaluable aid in the revision of the book and the preparation of the text for publication.

THE AUTHOR

CONTENTS

PART III
THE ECOLOGY OF ASSOCIATIONS

The Ecology of Animals

INTRODUCTION

Ecology as a Biological Science

DEFINITION

Ecology studies those aspects of the relationship between the organism and the environment upon which the success of the organism's development, survival, and reproduction depend. Darwin (1859) referred to these aspects as "the struggle for existence," which he understood in the "broad and metaphorical sense, including the struggle for existence between individuals of the same species, between individuals of different species, or against the physical circumstances of life."[1] Darwin understood the struggle for existence not only in terms of predation, parasitism, the conquest of territory, refuges, food, moisture, light, and other vital resources but also in terms of the reactions of animals and plants to the physical (climatic) and chemical (gas and mineral resources, etc.) factors in the environment which the animal uses in order to adapt itself to existence under one or another set of concrete conditions.

The term "ecology" (*ecos* or *oikos*—dwelling, place of habitation; *logos*—knowledge; Gr.), as used to indicate a science of the relationships between organisms and the environment and their struggle for existence, was first used by E. Haeckel (1866, 1869) and was first used in its contemporary sense by the botanist Ye. Varming in his compendium *Oikological Geography of Plants* (1901).

Ecology studies not only the immediate relationships between organisms and the environment but also those things which have become historically formed because of them: (a) the adaptations of individuals, specific for each species, which furnish its various connections with the environment,

[1] From a Russian translation of *The origin of species*. See original for the English.

and the development and maturation of these connections; (b) the intra-species relationships and the population structure typical for each species; the latter determines the species' manner of life (solitary, colonial, herd), its method of utilizing vital resources, and its reproduction and settlement pattern, as a result of which its level of abundance and dynamic character are established; (c) the various associations of different species in various parts of the world which are connected by many mutual adaptations providing for a biogenic turnover of matter.

The theoretical basis of ecology is the concept of a dialectical, contradictory unity of organisms and environment. The idea of the historical unity of organisms and environment, first formulated by the Russian zoologist K. F. Rul'ye (1814–1858), was clearly expressed by the physiologist I. M. Sechenov (1861, p. 242): "An organism without an external environment to maintain its existence is impossible; therefore the environment influencing an organism should also enter into the scientific definition of the organism." The development and deepening of this idea is connected especially with the works of I. V. Michurin (1855–1935) on genetics and plant selection, I. M. Sechenov (1829–1905), I. P. Pavlov (1849–1936), and their pupils.

The interaction between organisms and the environment is studied by each biological science in its own area. Ecology has to do with only that aspect of the problem which determines the development, reproduction, and survival of individuals, the structure and dynamics of the populations they form, and, finally, the structure and dynamics of the associations of various species.

The basic problem of contemporary ecology is the abundance of living organisms (the biomass) and its dynamics. Knowledge of the principles of abundance fluctuation among the economically important species permits us to govern their numbers, which constitutes a practical goal of ecology. Its theoretical importance consists in the investigation of the interactions between organisms and the environment (the struggle for existence) and, by the same token, of the mechanisms of natural selection. No other branch of biology really illuminates these questions. Morphology studies the forms (construction) of organisms, clarifies their adaptive function, and establishes the principles of the evolutionary transformation of the organism's structure; physiology from this point of view studies the functions of organisms; embryology studies the development of forms and functions in the individual organism, while paleontology and the study of evolution (Darwinism) handle the historical aspect of their development; genetics studies the inherited basis of species differences and the principles according to which they change; finally, biogeography studies the geographic distribution of species. All of these sciences touch on ecology in one area or

another, but they do not concern themselves with its specific questions.

This definition of ecology is similar to that accepted by the Third All-Union Ecological Conference.[2] In foreign countries ecology is regarded as the science of "the interrelations between living organisms and the living and nonliving environment surrounding them, especially the intra- and interspecies relations" (Allee *et al.*, 1949) or as "the science of all the relationships of all organisms to everything which surrounds them" (Taylor, 1936). An interpretation as broad as this is incorrect, since it deprives ecology of specific problems, expands it to the limits of all biology, and leads, in essence, to its negation, to the recognition of the existence of a mere "ecological method" in all the branches of biology.

ECOLOGY: BRANCHES, OBJECTIVES, AND METHODS, AND RELATIONSHIP TO COMBINED SCIENCES

The long-existing division into the ecology of separate species (autoecology) and the ecology of associations (synecology or biocenology) is out of date. The ecology of animals now falls into three divisions, each of which is represented by its own object of study.

The first division is the ecology of individuals; it is devoted to the study of that interrelation of organisms (individuals) and environment which determines the possibilities of development, survival, and maturation. The relationships of individuals and the environment are different for different species; at the bottom of these relationships lie the morphophysiological reactions of organisms to external influences whose character is reflected in those animals. The study of such elementary reactions is the initial phase of every ecological investigation; it is primarily analytic in character, permitting the discovery of the specific relationships between species of individuals and their environment.

The dimensions of the body and of the individual organs and other in-

[2] A resolution of the Third Ecological Conference (Kiev, 1955) states that the basic task of ecology is "the study of the interconnections of organisms and the environment in the process of altering the abundance of species populations and the development of interacting groupings of species—biocenoses." The basic content of ecology should be regarded as (a) the study of the species adaptations of organisms and their historical determination as bases for an understanding of the variety of interrelations taking place between organisms and the environment; (b) the study of the principles of the formation and development of populations as forms of the existence of a species, their differentiation, and their dynamics of abundance; (c) the study of the principles of the formation and development of biocenoses as an expression of the interrelations between organisms and the environment in the concrete surroundings of a habitat.

ternal or external features, several biochemical indices, and the character of gaseous exchange, water exchange, and other physiological processes permit reliable estimates of the individual's condition to be made. Investigation of a sufficient number of individuals makes it possible for us to obtain a reliable picture of their condition, to compare the properties of the different age and sex groups, and to contrast them with similar individuals of other species (the comparative ecological method) inhabiting various biotopes, geographic regions, and zones (the ecologo geographic method), as well as to contrast their states at various periods (hours of the day, months, seasons, years). An analysis of such observations demands that we contrast these states with the climatic, geomorphological, and soil factors in the environment, for which we employ the data and methods of climatology, geomorphology, soil studies, and other branches of physical geography. Special attention is now being given to the influence of natural and artificial radioactivity—"radioecology" (Peredel'sky, 1957). Laboratory and field experiments permit the discovery of causal relationships.

The second branch—the ecology of populations—is devoted to the study of the circumstances surrounding the formation, structure, and dynamics of natural groupings of individuals of one species (populations) occupying parts of the species's range (geographic regions, biotopes) or various niches in an association.

The historically compiled structure and organization of species populations provide for the use of vital resources in the territory occupied, create the struggle for them between populations of different species (concurrents), and provide for the multiplication and dispersion of the animals. During the course of the population's adaptation to local conditions and to their periodic and nonperiodic changes, the level of abundance and the character of its fluctuations are established; parallel to this, the age, sexual, and genetic character of the population changes. Changes in abundance are accompanied by migrations and resettlements of animals, partly beyond the borders of the territory occupied by the population. In this way an exchange of individuals between populations is created, providing integrity and maintaining the existence of the species.

The object of population ecology, therefore, is the study of intraspecies relationships, the structure (organization) arising from them, and the biological significance created by these relationships. The most important problem of this branch is the dynamics of species establishment and the factors surrounding it.

The methods of population ecology are specific; using these methods, we evaluate the density of population and the distribution of individuals throughout the territory. Various methods of absolute and relative calculation of the abundance of animals, investigations of sexual and age divisions

in the population, of the multiplication, death, and dispersion of individuals—all are used to establish the level of abundance, the character of its dynamics, and the type of settlement typical of a given population. In studying the relationships and territorial associations of individuals, families, flocks, or herds, the animals are counted by banding, attaching tags and other markers, staining with permanent dyes, amputation of appendages (in amphibians, reptiles, and mammals), or introducing a radioactive tracer into the animal's food or into its marking device.

The third branch is called ecology of associations or biocenology; it connects animal ecology with the ecology of plants and microorganisms. It studies the associations of interrelated populations of various species of animals and plants inhabiting the same spot (biotope) and forming the association (biocenose).[3] The basic objects of its study are (a) interspecies relationships (plants and animals, predators and prey, parasites and hosts, concurrents, "hangers-on"—commensals, cohabitants, symbionts, etc.) and their origin and contemporary significance; (b) the structure (construction) of associations arising from these interspecies associations (hierarchical or stratified associations, sinuous or other types of species groupings) and the boundaries of associations, their hierarchies and relations with one another; (c) the relationship between associations and external (climatic, soil-and-ground, topographic, and other) conditions and their dynamics.

The interspecies connections, structure, and dynamics of associations are studied by the methods of population ecology (calculations of abundance, observations of activity, movement, and reproduction and death rates among the animals). These demand mathematical processing and interpretation, especially in the area of population dynamics. Population ecology and biocenology make wide use of field and laboratory experiments, including experiments with laboratory-raised animals and their artificial associations. Experiments under natural conditions aid us in evaluating the role of the animals in the formation and dynamics of phytocenoses; they also help us discover the importance of predators and parasites in controlling the abundance of their victims and hosts. Great results have been obtained from the study of methods for controlling destructive animals and pests and for protecting and increasing useful species, as well as from studies of forest exploitation, tree farming, etc., which are quite properly regarded as a broad, though for the time being still-to-be-analyzed, experiment in productivity.

Dividing ecology into three basic branches permits us to examine the principles of the interrelations of the animal world and the environment

[3] The term "biocenose" as an indicator of a natural aggregation (association) of interrelated species occupying a definite plot of territory was suggested by Möbius (1877) as a result of his study of an oyster bed.

on three different levels: individual organisms, populations, and associations. These levels are part of a hierarchy; i.e., populations consist of individuals of the same species, while associations consist of populations of different species. The mutual influence between individuals and the environment represents the primary, elementary reactions of these two categories. When individuals are combined into a population, there arises the possibility of a more active relationship to the environment, at the basis of which lie group reactions, which cannot be subsumed under the reactions of separate individuals. Just as life is impossible for an individual outside a population, so is the existence of a species unthinkable without the existence of other species serving as its food, creating its necessary microclimate and refuges, removing harmful products of metabolism, etc. Only a study of the relationships to the environment on all levels of organization of the plant and animal population permits us to make a complete analysis of relations with the environment and to reveal their historical unity.

In addition to this "horizontal" division of animal ecology, there is a "vertical" classification by objects which inhabit various environments: (a) the ecology of water animals, which is part of the discipline known as hydrobiology and occupies one of its leading positions; (b) the ecology of soil inhabitants, or pedobiology; (c) the ecology of terrestrial (dry-land) animals; and (d) the ecology of parasites, which occupies a leading position in parasitology (Table 1).

TABLE 1. *The branches of ecology.*

Ecology		*Of Water Animals (Hydrobionts)*	*Of Soil Animals (Pedobionts)*	*Of Terrestrial Animals (Aerobionts)*	*Of Parasites*
Of individuals	Autoecology	the study of the reactions of individuals to external factors and of their relations with the habitat environment (conditions of existence)			
Of populations	Autoecology	the study of the specific relationships of populations to the environment, of the conditions and the paths by which populations are formed, and of their structure and dynamics			
Of associations (biocenology)	Syn-ecology	the study of the formation, structure, and dynamics of natural associations (biocenoses) and of their interrelations with the environment			

Under the ecology of farm animals we usually include the study of their reactions to climatic factors, feeding conditions, confinement, and herd reflexes (Kashkarov, 1945; Ruttenburg, 1953; Slonim, 1961). The foreign ecological manuals also set aside divisions devoted to a so-called "human ecology," covering the influence on man of climatic, geochemical, and

other landscape-geographic factors, which is useful, necessary, and possesses great practical importance. But they also make an attempt to include under "human ecology" economic and social problems (Wells, Huxley, and Wells, 1939; G. L. Clarke, 1954; Bodenheimer, 1958).[4] Here there appears a tendency common among several foreign biologists and known since the times of social Darwinism: the tendency to biologize social phenomena and by the same token to express their meaning in favor of class interests. There is no doubt that natural factors exercise an influence on man's body, but, although for animals the interrelations with natural circumstances are the basic motive force of historical development (i.e., natural selection flows from them), for human society the motive forces are the productive social relations.

So far as domestic animals are concerned, the motivating factors will be those environmental conditions which man creates for them through selection and care, reworking the inherited nature of the animals in the direction he feels necessary. Therefore, the term "ecology of farm animals" must be understood in a limited sense, taking into account only the influence of natural-geographic (landscape) factors upon their status, life, and development. For wild animals the role of human activity is sometimes of basic importance in changing the natural conditions which surround them, just as man knowingly alters the situation of farm animals. This is coming to be of greater and greater importance, even though man does not knowingly create a new environment for wild animals, as he does for domestic ones.

Here it would be proper to recall the fact that ecology examines the relationship between organisms and the environment under natural conditions. "The discoveries of Darwin, the eminently useful naturalist," wrote the English ecologist C. S. Elton (1939), "forced the zoological world, which had worked for about 50 years inside buildings, to thrust their heads out into the open air." This study of the interrelations of animals and environment in natural associations opened the possibility of regulating the composition of the associations and the abundance of the species by means of a comparatively uncomplicated set of actions on their conditions of existence.

The ecology of animals, the ecology of plants, and the ecology of microorganisms are autonomous disciplines which enter simultaneously into the provinces of zoology, botany, and microbiology. The differences between animals, plants, and microorganisms are so great that this kind of division is completely logical. There are also many general ecological questions,

[4] G. L. Clarke (1954) calls ecology "the economy of nature," only a part of which comprises human society in his opinion.

but they are, in essence, general biological questions. For this reason even the discussions on general ecology (G. L. Clarke, 1954; Woodbury, 1954; Ioganzen, 1959) recommend short courses in general biology.

Adjustment of the content and limits of animal ecology aids us in examining its relationships with other sciences. Especially close connections exist between ecology and animal physiology, many of the methods of which are used widely in ecological research. There is a distinctly noticeable and rapidly developing border area of knowledge, ecological physiology (Bykov and Slonim, 1949; Skadovsky, 1955; Slonim, 1961).[5] It analyzes the reactions of animals to external factors. The ecology of individuals is actually a form of physiological ecology. But in this area the physiologist is primarily interested in the mechanism giving rise to the animal's reaction to the external factors, while the ecologist must of necessity evaluate its final result, the influence on the state of the organism. Thus ecological physiology represents the physiological basis of the ecology of individuals. So-called "population," or "mass," physiology studies changes in the basic functions of individuals (feeding, respiration, thermoregulation, behavior, reproduction, etc.) under the influence of changes in the density and structure of the population. The data from population physiology might be regarded as the physiological basis of population ecology.

Finally, the change in physiological functions under the influence of interspecies relationships, like the functional properties of close and distant species entering into an association and connected by tight interrelations (food and its consumers, concurrents, symbionts, etc.—their physiological coadaptations are mutual ones), represents an item of outstanding interest for our understanding of the interspecies connections and, by the same token, the structure of natural associations (biocenoses). Unfortunately, this question is still receiving little attention.

The connections between ecology and morphology-embryology are strong. Intermediate areas of knowledge also arise between them—ecological morphology and ecological embryology (Vasnetsov, 1938, 1947, 1948, 1953; Kryzhanovsky, 1939, 1953; Matveyev, 1953)—the object of which is to study changes in form and its ontogenesis under the influence of external factors and changes in the structure and density of populations.

Animal ecology serves as a basis for zoogeography, in which we usually recognize ecological geography, or the study of contemporary factors in the distribution of animals. The historic factors in establishment of the ranges of individual species and the means by which faunas were formed (historical zoogeography) have the same ecological basis, but in the past.

[5] See the anthology *Experimental studies of the regulation of physiological functions*, 1st ed., 1949; 2nd ed., 1953; 3rd ed., 1954; 4th ed., 1958.

Understanding zoogeography and phytogeography as sciences of the principles of the geographic distribution of animals and plants and the formation of faunas and floras, biogeography is sometimes seen to include the geography of associations, regarding it as a part of physical geography (A. G. Voronov, 1957).

Ecology is widely applied in systematics, where the significance of ecological and ecologo-physiological criteria of species is growing steadily. In this respect contemporary systematics (Huxley, 1940; Mayr, 1947; Mayr, Linsley, and Usinger, 1956) has returned to the general concept of species formed more than a hundred years ago by K. F. Rul'ye. The latter demanded not only investigation of the structural properties but also study of the ecology of the species and its position in the associations: "Only when we study the animal from all possible aspects do we have the right to speak with assurance about the relationship of its whole sphere to the closest other sphere or similar spheres of other animals, i.e., to classify its species" (Rul'ye, 1954, p. 24). Solution of the questions of intraspecies systematics and distinguishing of morphologically similar species (the so-called "doubles") are impossible without the use of ecological criteria (location preference, feeding, reproductive properties, and many others) and "ethological" features (persistent behavior traits), which change little under outside influences.[6] In studying one population or another, ecologists in turn must know their systematic positions precisely.

As a study of the struggle for existence and the mechanisms of natural selection, ecology is widely applied in population genetics, especially in the solution of problems of preservation, distribution, and fixing of genetic changes. In the study of phenotypes, genotypes, and their dynamics in populations, the interests of ecologists, systematists, and geneticists closely overlap. Inherited heterogeneity of populations must be taken into consideration by ecologists in their field and especially in their experimental investigations.

The achievements of ecology permit paleontologists studying the structure of fossil animals to evaluate their probable relationship to the environment; at the same time analysis of the position of fossil remnants in the primary deposits permits at least partial establishment of the type of associations existing at that time. These special problems of paleontology have grown out of a separate area of knowledge which has been designated "paleoecology" (Gekker, 1957). Its important branch is the general biological problem of the evolution of vital types (forms) and their relationship to the environment in connection with historical changes in the

[6] The term "ethology" was first used by J. S. Mill in his *Logic* to indicate the study of "characters of people." Later it was used by Geoffroy Saint-Hilaire (1859) to denote relationships of organisms in families, groups, and associations.

inorganic environment and the evolution of associations—a problem of ecogenesis.

Several problems of ecology are transitional between physical geography and biology. The study of physical factors of the environment in ecology is closely associated with meteorology and climatology, from which it draws its basic data. And climatology in turn, explaining the climatological differences of different territories, devotes a great deal of attention to the living earth cover, the climate-forming role of which is very great. This is emphasized even today by certain climatological classifications (Voyeykov, 1884; Dokuchayev, 1900; Berg, 1938) such as the "forest climate," the "pine climate," the "oak climate," etc.

The chemical properties of the atmosphere, hydrosphere, and lithosphere determine their living populations. The populations in turn change the chemistry of their habitats, actively participating in the biochemical processes taking place there, which take the form of a turnover of matter (biochemical cycles) such as is studied by biochemistry. The rise and development of the latter science is due especially to the labors of Academician V. I. Vernadsky (1926, 1939, 1940) and his pupils.

Geomorphology and soil studies are found in close contact with ecology. The structure of the earth's surface is determined as much by the action of nonliving forces as by the activities of animals, plants, and microorganisms. The living cover has changed the upper layer of the continental rock, creating a new natural body, the soil (Dokuchayev, 1886). For animals and plants it serves not only as a background, a substrate, but as a source of moisture, salts, and other substances.

The connections and relationships between ecology and the contiguous sciences can be represented as shown in Fig. 1.

BRIEF HISTORY OF ANIMAL ECOLOGY

Ecology arose and developed as an autonomous biological science under the demands of practice. It has served and continues to serve as a theoretical basis for practical work in controlling destructive wild animals and plants. The solution of problems of ecology became especially necessary in the nineteenth century during the development of world capitalistic production, with its intense, often predatory, use of natural resources which threatened even to exhaust them. Ecology as a separate science arose approximately in the middle of the nineteenth century, but its roots go back deep into antiquity. Even the primitive peoples, judging from kitchen remains, hunting weapons, cave and rock paintings, etc., already had

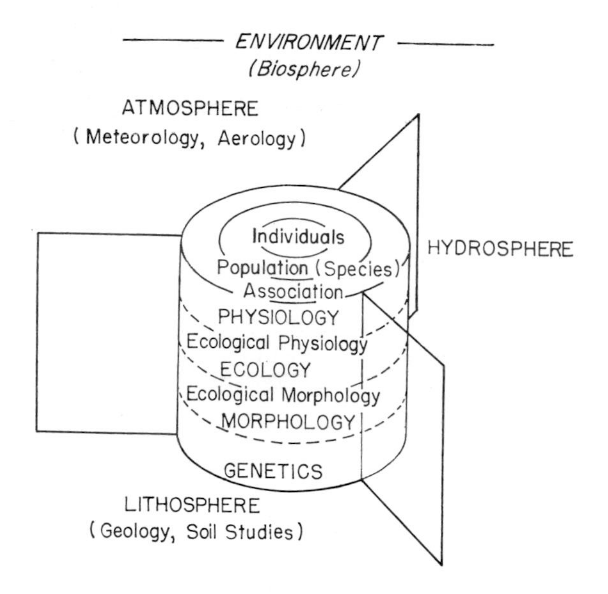

Fig. 1. Position of ecology and its relationship to the other sciences (original).

ideas about animals' habits and ways of life accurate enough to make them successful hunters, fishermen, and collectors of food.

The ancient Egyptian, Indian, Chinese, Tibetan, and European written sources contain much valuable information. The Indian epic poem *Ramayana* and the ancient *Mahabharata,* both assigned from oral tradition to the sixth to second centuries B.C., contain information on the habits and way of life of some 50 species of animals; there are recollections of changes in the abundance of animals and even of hunting regulations (the taking of females was considered a sin) (Nikol'sky, 1956). In the book *Bhagavata Parana* the role of wild animals in the spread of dangerous diseases is correctly evaluated. It recommends that the people abandon their homes "when the rat falls from the roof, staggers about the floor like a drunk, and falls dead—for then the plague is probably near." Similar information can be found in such other Indian sources as the Veddas and Manas as well as in early Chinese works.

The conceptions of an integral and constantly changing and developing world of nature held by the ancient philosophers were naive but in many ways correct. Empedocles (fifth century B.C.) wrote of the connections between plants and the environment and of their dependence on the surrounding world. Thoughts on the relationships between human illness and natural factors can be found in the "father of medicine," Hippocrates. "The father of many sciences," Aristotle (384–322 B.C.), distinguished ter-

restrial, amphibian, and water animals and classified them according to their way of life (solitary and social, sessile, migratory and wandering, diurnal and nocturnal), their means of movement (swimming, running, flying, slithering, wiggling), their feeding habits (carnivorous, herbivorous, omnivorous, specialized), and by "temperament" (kindly, sluggish, "ferocious," "timid," and "brave").

Many ecological observations were collected in the *Natural History* of Pliny the Elder and the writings of other ancient philosophers and writers. But zoology itself, much less animal ecology, did not exist as a separate science in those days. It was a time in which general biological facts were being accumulated and compared, leading to a sometimes true set of conclusions. The medieval period, with its halting of scientific work, yielded nothing for the advancement of either ecology or zoology in general, with the exception of a few isolated discoveries.[7]

The Renaissance period, with its great explorations, discoveries of new lands, and interest in ancient philosophy, was accompanied by development of the biological sciences as well, which was at first associated with the necessity for understanding the newly discovered exotic faunas and floras and for evaluating the possible agricultural importance of the new species. J. Ray (1628–1705) and C. Linnaeus (1707–1778) created the foundations of scientific systematics; the separate sciences of morphology and physiology arose. In the monographic descriptions of the flora and fauna of individual regions, such as L. Buffon's (1707–1788) thirteen-volume *Natural History*, much attention was devoted to the relations between organisms and the environment, and generalizations were sometimes made which have retained their importance to the present day (on the amount of heat necessary for the development of animals, on thermoregulation in the bee family, etc.).

Many new facts and ecological observations were collected by Russian naturalists, for example, P. S. Pallas (1741–1811), I. I. Lepekhin (1740–1802), S. P. Krasheninnikov (1711–1755), and others. Their explorations and investigations opened up to science the animal and plant world of Siberia, the Far East, the Urals, the Caucasus, Kazakhstan, and Central Asia. Their writings also contain broad generalizations, such as Pallas's idea on alterations of species or the establishment by Lepekhin of the relationship between the number of squirrels and the amount of conifer seeds. Still, even this period can only be called the second phase of the prehistory of ecology.

The separation of ecology into an autonomous branch of biology took

[7] For example, the discovery by Emperor Frederick Barbarossa of the rule that the dimensions of the bodies of warm-blooded animals increase toward the north (Dement'yev, 1935), which was later rediscovered by Bergmann (1847) and given his name.

place after the victory of the capitalistic means of production, when natural resources began to be exploited with growing intensity and the necessity for their profound study had arisen. Exhaustion of many natural resources had been observed at an early date and was quickly growing. Diminution of forests in Europe was being mentioned as far back as the twelfth and thirteenth centuries, while in America the forested acreage had been reduced by six times in the nineteenth. Pastures had been hacked out of forest land, and the washing or blowing away of soil layers had already begun.[8] The number of valuable game animals had dropped sharply, as had those of birds and fish. By 1917 the beaver and the desman appeared almost extinct in our country; the same fate threatened the sable, the otter, the wild hoofed animals, and other valuable beasts. As a result of unsystematic destruction of natural associations, mass destroyers of forests, meadows, and crops appeared. All this brought about the development of investigations in applied ecology, fish and game management, and pasture and forest culture and the necessity for studying crop pests, vectors of human diseases, and illnesses of domestic animals. The same thing was demanded by the advancing conquest of the deserts, the taiga, the tundra, and the tropical mountains, where human beings often encountered the danger of infection by "new" diseases of animals previously unknown.

Unsystematic use of lands and overtaxing of land facilities gave way to intense cultivation with perfected processes, fertilization of soil, and crop rotation. Studies of the influence of mineral fertilizer permitted the agrochemist J. Liebig (1840) to formulate a concept of limiting factors and a widely significant "law of the minimum." Information on the roles of temperature, moisture, the light regime, chemistry, and other aspects of the organism-environment relationship began to pile up rapidly. Much attention was devoted to properties of animal behavior and to the roles played in this behavior by congenital and acquired elements, taxis, and instinct.

The problem of animal abundance and its dynamics, which has occupied a leading position in ecology since that earlier period, has now been promoted to the primary position. It is attracting the attention of sociologists, economists, and statisticians, but many of these people are looking at the problem falsely, using the doctrines of Malthus (1798) on the limitation of vital resources and the halting of their growth because of an increase in the number of consumers. As we know, Malthus's antiscientific, narrowly class-oriented "fabrications" have been severely criticized by those who set the basic principles for Marxism. Similar metaphysical and mechanistic ideas are found in Spencer's (1852) work on the balance of nature, which asserts that an increase in the amount of energy expended on the mainte-

[8] In the United States erosion carries off some 1.5 billion tons of fertile topsoil yearly (Tugwell, 1935).

nance of life among individuals diminishes their fecundity and vice versa. Therefore, a rise in the organization and levels of life activity in the course of evolution is accompanied by a drop in fertility. Limitation of the growth of population was associated with the so-called "Farr's law" (1843), which attempted to show a connection between the increase in disease and death rates in England and the increase in population density. These and many other attempts followed along the same class lines as Malthus's tract. They are inapplicable to either human or biological situations, since they distort the character of the relationship between organisms and the environment, interrupting their historical unity.

The development of ecological research aided in determination of the basic problems of biology. The data which science had accumulated on the relationship between organisms and the environment permitted Lamarck in his *Philosophy of Zoology* (1809) to advance a theory of evolution. Lamarck's theory received no recognition then, even though it did contain many valuable thoughts on the interrelations between organisms and the environment.

The birth of ecology in Russia was connected with the spread of Lamarck's and Geoffroy Saint-Hilaire's ideas among Russian scientists. Its own paths of development were determined by the high level and practical diligence of Russian science and the glorious names of M. V. Lomonosov, P. S. Pallas, K. M. Ber, and later A. I. Voyeykov, I. I. Mechnikov, I. M. Sechenov, and many others. Russian science developed on the open fields of Russia, which permitted observation, comparison, and contrast of an immense variety of phenomena and tended to gain these phenomena acceptance in dynamics. It is not at all surprising that the study of soils and zonal geography of landscapes was born in our country (V. V. Dokuchayev).

The sources of Russian ecology are associated with K. F. Rul'ye, professor at Moscow University, evolutionist, and materialist-dialectician, whose views were close to the ideas of V. G. Belinsky and A. I. Herzen. Rul'ye wrote of the total integrity of nature, "A phenomenon existing alone, by itself, beyond any necessary relationship to others, is so unthinkable that man cannot imagine it." He expressed the unity of organisms and environment as follows: "No organic creature lives entirely of itself; each is called to life and lives only insofar as it finds itself in a mutual relationship with a world which is comparatively external to it. This is the *law of intercourse* or *dual origins*, showing that each living creature acquires the possibilities of life partially from itself, partially from outside." Rul'ye could not imagine an organism separated from the external world, since this "would mean imagining an animal which did not breathe, eat, move, which was not subject to the natural laws of gravity, pressure, evaporation, etc.; it would mean imagining not only the greatest, but even, in our opinion, an

impossible paradox." According to Rul'ye, who considered agriculture "the first of the primary sciences and the first of the primary arts," the goal of the study of natural phenomena should be the satisfaction of practical needs. He protested passionately against the lures of scholastic systematics: "The efforts of those who, at the cost of time, money, health, and sometimes even life, journey through distant countries merely for the purpose of describing dead fragments and never have a thought for whole, living people seem justifiably one-sided." He alone turned his attention to the connection between instincts, behavior, and way of life on the one hand and "bodily organization" on the other. He devoted much time and attention to the study of periodic phenomena in animals and their practical importance.

The ideas and traditions of K. F. Rul'ye, accepted and developed by an unbroken chain of his pupils and followers, connect his name and activity with the present. Among his pupils the first to be named should be the founder of Russian zoogeography, the zoologist and explorer N. A. Severtsov, and the physiologist I. M. Sechenov. N. A. Severtsov's master's dissertation, "Periodic Phenomena in the Life of the Mammals, Birds, and Reptiles of the Voronezh Guberniya" (1855), represents a mature ecological investigation. In studying periodic phenomena, Severtsov assumed that "attention should be directed primarily to the multiplicity of causes and the flowing formation or course of a phenomenon, which . . . depends: 1) on the body of the animal or plant; 2) on the whole complex of external conditions, climate, location, food, security; 3) on a whole series of previous vital phenomena shown by the animal or plant studied, a series in which each phenomenon depends on the preceding ones and determines the following ones" (N. A. Severtsov, 1950, pp. 22–23). To Severtsov belong the first attempts at classifying animals by biological types (life forms). He mentioned the connection between the characters of feeding, flocking, and migration in birds, and he turned his attention to the existence of an inverse relationship between length of life and the fecundity of animals. According to Severtsov, we see in birds a polarity between feeding—maintenance of the body—on the one hand and molting and reproduction on the other. Judging from the weight given to one pole or the other, the breed is seen to maintain itself either by rapid multiplication or by long life per individual, but never by both.

Severtsov's contemporaries, M. N. Bogdanov of Kazan' University and Academician A. F. Middendorf (1815–1894), enriched the science with many ecological reports which have still not lost their importance. In his *Siberian Travels* Middendorf gave many data showing the dependence of way of life, sessile behavior, migration, hibernation, distribution, and abundance of animals on local properties.

The publication of *The Origin of Species* by Darwin in 1859 and the victory of the theory of evolution in biology led the way for a new period in the history of ecology. It became the study of the interrelations of organisms with the environment in their struggle for existence. Chapter III of *The Origin of Species* presents the current concept of ecology and deals with the majority of its present problems. With the genius of his vision, Darwin in many cases pointed out the paths to their solution.

After Darwin the development of ecology picked up speed; it was stimulated by all the growing demands of farm, game, and forest management, medicine, and conservation. The theory of evolution provided a general orientation and opened broad perspectives to ecologists. And it was in this very period that accurate data began to be accumulated on the roles of temperature, moisture, and other external factors in the life and distribution of animals. Intraspecies relationships, population structures, and their mutual relations with the environment began to be studied. A great deal of attention was given to the habits and behavior of animals and to the changes in their numbers. And, finally, data on the structure and dynamics of the natural associations—the biocenoses—began to accumulate.

K. Möbius (1877) worked out and suggested the concept of the biocenose as a natural aggregation of species of plants and animals historically assembled under the influence of local landscape-geographic conditions and occupying a definite territory. Accurate data were accumulated on the abundance of species, their dependence on external conditions, and the importance of inter- and intraspecies relationships. A concept was formed concerning the "ecological tolerance" or "valence" of the different species toward temperature, moisture, and other environmental factors (Bachmetjev, 1901; Shelford, 1911, 1913). Changes in the abundance of species began to be regarded not as accidental fluctuations but as part of the regular dynamics of the population, covering the characteristic properties of various species (Turkin, 1898; Turkin and Satunin, 1901) and explainable in many cases by the interrelationships of species, especially the interrelationships of food and its consumers—predators and prey, parasites and hosts, etc. (Howard and Fiske, 1911; Seton, 1911). Along with theoretical works, the position of applied ecological efforts rose higher and higher to effect control of harmful species, the spreaders of diseases of human beings and domestic animals (Forbes, 1882, 1887, 1907; Shevyrev, 1892, 1893; Silant'yev, 1894; Kholodkovsky, 1909; and others). All this demanded the working out of accurate methods for observation and experimentation as well as of means for calculating the number of animals. Developing the latter was an indispensable condition for further development of ecology at this stage.

Ecological reviews, surveys, and special monographs on individual problems appeared in greater and greater numbers, especially those concerning the dynamics of abundance. Finally, in the second decade of the twentieth century, ecological scientific societies arose and began publishing special journals: *Ecology* since 1920, the *Journal of Ecology* since 1930, *Ecological Monographs* since 1931, *Voprosy ekologii i biotsenologii* [Questions of ecology and biocenology] since 1931, the *Journal of Animal Ecology* since 1932, and many others.

After the great October Revolution ecology became especially capable of broad development in our country. Its connection with practical demands stimulated goal-directed efforts, expanded the scope of the problems, and demanded solutions to the theoretical questions of ecology. The volume of research increased many dozens of times in a short period. Ecologists became indispensable participants in complex projects in agriculture, public health, and many branches of industry. An especially valuable role in the development of Soviet ecology and the propagation of its ideas and methods was played by D. N. Kashkarov, whose reports, "The Environment and the Association" (1933) and "Fundamentals of Animal Ecology" (1937, 1938, 1945), did much to aid in the growth of ecological research. Soviet ecologists achieved major successes in controlling agricultural pests (locusts, tortoises, grain flies, and many others), forest pests (army worms, bark beetles), and the vectors of such diseases as malaria, plague, and tularemia and in increasing the abundance of a number of valuable species (sable, beaver, elk, and others).

The development of ecology overseas was also accompanied by major practical successes at this time, especially in the control of agricultural pests, in conservation, in medicine, and in the management of game animals. In the Hawaiian Islands and on Fiji the number of imported harmful insects was decreased sharply by allowing their enemies and parasites to multiply and become acclimatized; major victories were achieved in the control of vectors of certain diseases of human beings and farm animals. Another and equally successful program concerned the use of plant-eating mites and insects (chinches, *Cactoblastis* butterflies, etc.) to control weeds (St. John's wort, *Opuntia* cacti, etc.) in Australia. But the contradictions of the capitalistic management methods, which are subordinate to the basic goal—maximization of profits—have in many cases rendered useless the achievements of ecology. The struggle against harmful animals in the capitalist countries has been slowed by the conditions of private enterprise. For example, extermination of rodents, the carriers of plague and other diseases (especially in the colonial countries), has not been carried out because of "low profits," while supplies of valuable animals continue

to run short. Thus the efforts of many great foreign scientists, quite valuable for the development of animal ecology, could not be completely utilized in a practical way.

The basic theoretical problems of ecology now attracting the attention of investigators both in our country and overseas are (a) regularities in the movement of population (population dynamics)—the theoretical foundation of regulating the abundance of useful and destructive species, and (b) the structure, organization, and dynamics of associations—the theoretical foundation of all practical measures for enriching and rationally using natural resources. The processing and solving of these problems takes place amid arguments, discussions, and struggles between the various ideologies.

The idealistic autogenetic conceptions associated with the Malthusian theory and the philosophical views of H. Spencer on the balance of nature take the form of a theory of "biotic potential," "environmental resistance" (Chapman, 1928, 1931), "automatic regulation" of abundance, and of a mathematical theory of "cycles" (Pearl, 1914, 1923, 1937; F. I. Baranov, 1918, 1928, 1947; C. S. Elton, 1924, 1925, 1927; Lotka, 1925; Volterra, 1926, 1928; Nicholson, 1933, 1947, 1958), or the form of a concept of the association as an organism or "microcosm" (Forbes, 1887; Petersen, 1918; Thienemann, 1925, 1931; and others).

The materialistic treatment of these problems is based on conceptions of the historical unity of organisms and environment—the dialectical unity—a contradictory unity which rests on a constantly developing mutual relationship between the organic and inorganic worlds. During the course of this mutual relationship morpho-physiological properties of organisms arise, and as a result of them connections and organization (structure) appear in the living cover of the earth (intraspecies organization and organization of associations) which furnish the biogenic turnover of matter. This point of view lies at the foundation of the works of the majority of the ecologists in our country.

A brief outline of the history of ecology permits us to demarcate on a scale of more than 2,000 years five basic stages in the development of the sciences. Only the last three are the history of ecology itself. Under the prehistory of ecology are:

1. Ancient period. Biology was undifferentiated, with inaccurate and often naive, but in many ways true, general representations of the organic world.

2. The Renaissance. Biology was divided into separate sciences, representations of the organic world and its variety were expanded, and new facts and accurate data were accumulated, including ecological data.

Under the rise and development of ecology as an autonomous science are:

3. The pre-Darwin period. Ecology was formulated as an autonomous

science with two basic branches—autoecology (ecology of species) and synecology or biocenology (the ecology of associations).

4. The Darwinist period. Ecology of animals became an autonomous science with its chief practical problem being the regulation of the abundance level of economically valuable species of animals and the changing of natural associations (biocenoses) so that they were directed toward man's benefit.

5. The contemporary period. Ecology developed intensively and on a broad scale, especially in the socialist countries. The basic problem was the governing of population dynamics and the productivity of associations, which made necessary the division of animal ecology into three branches—ecology of individuals, populations, and associations. Special organizations arose which were occupied with calculating the numbers of animals and observing their dynamics (the Anti-Epidemic Service for the Enumeration of Agricultural Pests of the Ministry of Agriculture of the USSR, the Ministry of Health Service of the USSR, the Biological Service of the Department of Agriculture of the United States, the Bureau of Animal Populations at Oxford University, the Locust Center in London, and many others).

BASIC DIRECTIONS AND TASKS OF SOVIET ECOLOGY

To each type of practical activity associated with the use of living natural resources there corresponds a branch of contemporary ecology. In animal husbandry there are many ecological questions associated with the regionalization of breeds of farm animals, since successful acclimatization of animals depends on the correspondence of the natural food and climatic conditions of the region and the demands of the animals brought in (Kashkarov, 1945). Ecology studies the influence of innumerable domestic and wild animals on the productivity of pastures. The efforts of domestic and foreign geobotanists and zoologists have clarified the principles of this influence.

The demand for protecting the yields of agricultural crops and fruit and forest crops and the necessity for protecting forest plantings are limited by immense losses still caused by wild animals. In years when the latter are undergoing massive multiplication, seedlings and plantings are often—sometimes entirely—wiped out. According to I. Ya. Shevyrev (1892), "The business of steppe agriculture has met a powerful obstacle head-on in the case of the insect." In Europe, Canada, and the United States massive de-

struction of forests, such as has been seen in the last hundred years, is being repeated, and its extent is increasing (Frideriks, 1932). Destruction of seeds by rodents and killing of seedlings and young trees are serious threats to natural renewal of forests and the growth of forest timber.

Field and experimental research on the ecology and physiology of insects and other plant destroyers has clarified the conditions under which they develop and multiply and has permitted us to establish the reasons for their abundance or lack of it at various seasons of the year and in various regions. On this basis methods have been worked out by which we can predict their numbers, surveillance has been established, and extermination programs are under way. Effective means of exterminating insects by poison, together with correct agrotechnology and forestry practices, have eliminated or cut the threat of mass swarms of locusts, chinch bugs, and many other pests. Great significance has been attached to the "biological method," or the use of a pest's natural enemies and parasites to control it. They become acclimatized to the region in which they were absent, or they multiply in laboratories or greenhouses and are let loose in the area when the pests are active (Rubtsov, 1948).

In fish farming the matter of the biological productivity of bodies of water has attracted attention. This calls for a study of food resources and their dependence on the physico-chemical and biological properties of the body of water. At the same time an intensive study should be made of the feeding, growth, multiplication, migration, abundance dynamics, and interrelations of all the game fish and invertebrates as well as of the influence which fishing has upon them. In the hunting and game industry prognosis of abundance levels is of great importance. Formulas for this kind of calculation have been worked out for squirrels, foxes, polar foxes, ermines, and certain other species.

In the preservation and increase of the supply of valuable game animals an important role is played by protection, re-establishing exterminated animals in regions where they formerly lived, and acclimatizing new species. The muskrat, the American mink, and the nutria have been deliberately introduced into our fauna. Many fur-bearing species (fox, marten, sable, beaver, nutria, polar fox, and certain others) as well as the elk are on the way toward domestication or are already domesticated and are bred in captivity. The question of raising sea animals is already being studied intensively. The production of works on the game ecology of the USSR is closely associated with the name of Professor B. M. Zhitkov, the organizer of the first game biology station in the country. The rise of works on the ecology of game should provide an answer to the growing demands of the country for fur-bearing animals and other products.

Medical and veterinary zoology (ecology) studies the animal agents,

harborers, and vectors of diseases and works out methods for controlling them. These tasks are resolved in close contact with microbiologists, epizootiologists, and epidemiologists, and they protect the laboring force and domestic animals from transmissible (transmitted by specific vectors) and parasitic diseases. These works are carried on by a network of special anti-epidemic stations and institutes. The biological basis of these efforts is Ye. N. Pavlovsky's (1946) study of the natural nidality of transmissible diseases. The successes of Soviet medicine in controlling such diseases are great; many epidemic diseases have disappeared from the USSR or have become rare (plague, tularemia, malaria, spotted typhus fevers, etc.). In old Russia they carried off tens of thousands each year.

Soviet ecology is closely associated with the practices of the national economy. Reconstruction of the fauna and flora of large regions, conquest through intensive agriculture and animal breeding in the virgin and marginal lands, introduction of new species of plants and animals, as well as tasks aimed at protecting the health of the labor force, all demand diligent deepening of the ecological study of useful and harmful species of wild animals and the working out of means for controlling their numbers.

PART I

THE ECOLOGY OF INDIVIDUALS

CHAPTER 1

General Principles of the Relationship between Organisms and the Environment

BIOSPHERE

"Biosphere" is the name given to the surface portions of the globe (atmosphere, hydrosphere, lithosphere) inhabited by living creatures which have changed its initial matter and state.[1] The biosphere includes the lower portion of the gaseous envelope—the troposphere—all of the water environment, and the upper portion of the solid envelope—the weathered crust. The biosphere was formed as a result of the appearance of life (living organisms), as a direct result of the general development of our planet. The existence of life on earth has been computed to cover 1.5–2 to 3–4 billion years (Vernadsky, 1926, 1928, 1934; Berg, 1951; Oparin, 1951, 1954, 1957; Zenkevich, 1951).

The water and the dry land are the actual possessors of life, since there are no purely aerial organisms unconnected with either the lithosphere or the hydrosphere. The air is occupied for a more or less lengthy period by actively or passively flying terrestrial (dry-land) and marine organisms (among them, especially, cysts, spores, etc.). But all of them need the lithosphere or the marine environment as a substrate or a place to feed. The upper limit of the biosphere is found at an altitude of approximately 10–15 km, where live organisms—bacteria, fungus spores, protozoa, and certain others—have been found (Fig. 2). The ozone layers, which filter out the ultraviolet and strong cosmic rays fatal to organisms, are the maxi-

[1] The term "biosphere" was suggested by Ye. Zyuss in 1885 and was widely used and redefined by Academician V. I. Vernadsky (1926, 1939, 1945).

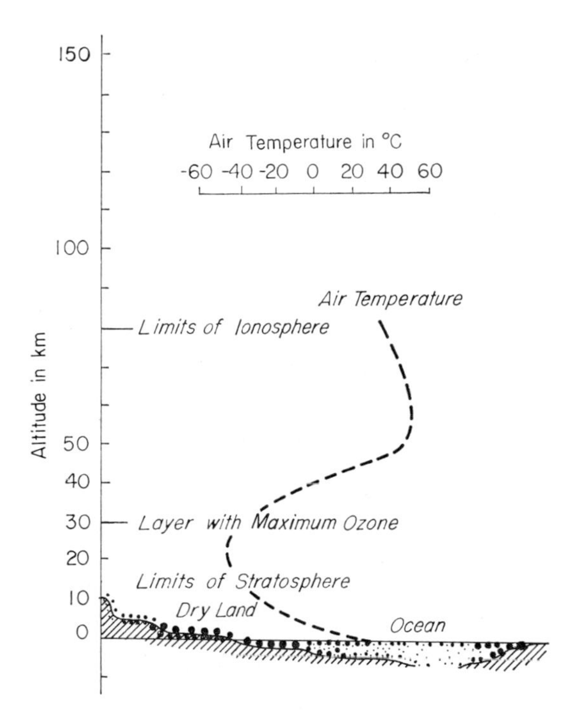

Fig. 2. Structure of the biosphere and the distribution of life within it (original). Size of dots shows the relative abundance of organisms.

mum upper boundary of the biosphere; these layers are found at an altitude approximately 20–50 km above sea level.

The basic mass of land-borne living creatures does not rise into the air more than 50–100 m above the surface of the earth. Above this height the number of organisms rapidly diminishes until it becomes insignificantly small above 1 km. Nevertheless, some birds fly at altitudes of 1–3 km, wind-blown insects have been found at 4–5 km, while in the high ranges, the mountains of Central Asia, a significant number of animal and plant species have become adapted to living at an altitude of 4–5 km above sea level. The mountain sheep *Ovis ammon* and the Tolay hare (*Lepus tibetanus*) have been observed at altitudes up to 5 and 5.5 km above sea level, while an expedition to Mount Everest observed alpine daws (*Pyrrhocorax pyrrhocorax*) at 8.2 km. The lower limit of the biosphere in the lithosphere is apparently at a depth of 2–3 km, where microorganisms have been found in the waters of petroleum-bearing strata. The roots of the most deeply penetrating trees reach 8–10 m, while the burrows of the deepest-living rodents (susliks and marmots) go to 6–7 m and of insects (termites) to 5–6

m. But the basic mass of organisms in the lithosphere is concentrated in the soil layer, which reaches an approximate depth of 1 m.

The hydrosphere is inhabited to a maximum depth of 11 km in the Tuscarora Deep of the Pacific Ocean. Here Soviet deep-water expeditions have found a unique and rather rich fauna (Zenkevich, 1951; Zenkevich and Birshteyn, 1956); a new type of abyssal animal (Brachiata)—the Pogonophora, belonging to the Deuterostomia—has been described (Ivanov, 1957, 1958). The deepest penetrators are carrion eaters, detritus consumers, and predators, while plants and plant-eating marine animals are confined to the upper layers down to a depth of 300–500 m.

Within the biosphere only the few oversalted bodies of water (the Dead Sea with 23 percent salinity, Lake Tuz-Gol in Armenia with 32 percent salinity) are completely free of animal organisms, as are places poisoned with hydrogen sulfide, carbon dioxide, or other toxic substances. But even here certain microorganisms (e.g., the *Microspira* bacteria) and quiescent stages (e.g., cysts, spores, etc.) may be found. The basic mass of organisms is concentrated in places where the three environments—water, surface horizons of the mountain rock, and air—are in contact, i.e., in the upper layers of the oceans and seas, in small marine waters and freshwater ponds, and in the soil. The latter represents the "daylight or surface horizons of the mountain rocks which have been naturally changed by the influence of water, air, and various types of organisms" (Dokuchayev, 1896).

TURNOVER OF MATTER AND ORGANISM VARIETY

The most characteristic property of the biosphere is the uninterrupted turnover of matter within it, the first link of which is composed of living organisms and their ability to exchange matter and energy with the environment. They represent "open systems" whose matter is constantly breaking down and re-establishing itself as a result of decay and synthesis, and they closely overlap one another. The persistence of living material and the ability of organisms to reproduce themselves rest on definite relationships between assimilation and dissimilation, representing a strictly coordinated and very complex system of chainlike biochemical processes which exist not only in a definite spatial association but in relationships which follow sequentially in time. This physico-chemical organization of living matter determines the basic characteristic of organisms—the ability to exchange matter and energy with the environment and as a result to grow and develop, multiply and inherit. This organization has developed

naturally on the basis of the survival of the best adapted (Oparin, 1951, 1958).

The "terrestrial" origin of organisms is confirmed by the fact that they consist only of elements found on our planet and are subject to its physicochemical laws. Of the 96 natural elements, 60 enter into the content of organisms. The rise of life on earth marked a definite stage in the evolution of the planet. According to A. I. Oparin (1951), this occurred in the waters of the primeval ocean, the chemistry and physical state (temperature, electrical properties) of which differed substantially from those of the present day. In aqueous solutions and in vapors saturated with electricity, conditions arose which led to the formation and polymerization of organic materials, including proteinlike high-molecular compounds. These formed the so-called "coacervate" systems, which displayed several of the functions of organisms and, in the opinion of Oparin (1951, 1957), represented the unique precursors of organic life ("proorganisms").

The formation of living matter capable of exchanging atoms with the environment and thus capable of growing and reproducing its own kind marked the appearance of a new form of matter in motion: life. Its substance—living (organized) protein matter—was distinguished by its unusually high chemical activity. With further development and the appearance of more and more varied forms of living organisms, there came successive changes in the earth's surface. Inorganic nature, having given birth to life at a definite stage of its development, had been irreversibly changed and continues to change under the influence of organisms.

Living matter, which comprises only 0.01 percent of the biosphere by weight, furnishes the free oxygen in the atmosphere; the turnover of carbon, nitrogen, and other "microelements" making up the basis of living material is associated with it. Finally, organisms cause the formation of the so-called causto-biolites—the coals, limestones, petroleum, sapropels, turfs, etc. Formation of living matter demands not only macroelements but also many micro- and ultraelements found in the bodies of animals and plants in very small amounts. With rare exceptions, all or almost all elements are involved in the turnover of matter; herein lies the explanation of the immense geophysical and geochemical role of living matter in the biosphere. It is so great that not only sedimentary and metamorphic rocks but the granites as well show evidence of former living organisms (A. P. Vinogradov, 1933, 1949; Vernadsky, 1934).

The physical state and climate of the atmosphere, the character of the weathering of the earth's crust, and the geomorphological processes were changed to their very roots by the appearance of animals, plants, and microorganisms. Life arose and was able to develop only as a special form

of the turnover of matter on the earth's surface. The character of the structures and functions of individual organisms and all forms of their organization (within single species or in associations of many species) are created by natural selection through mechanisms providing for an uninterrupted, ever-repeating, but still irreversible biogenic turnover of matter, which changes the face of the earth unceasingly, though relatively slowly.

This turnover can arise only in the presence of certain basic distinctions (varieties) in organic form. The necessity of variety proceeds from the very essence of the relationships between organisms and the environment. Each organism, ingesting food and other substances, returns only greatly changed and unusable products of metabolism: hence the necessity for organisms with different demands capable of using each other's products of metabolism. From the moment when the first living substances arose, they were already differentiated. Oparin (1957) assumes that certain coacervates had the ability to ingest others. Thus arose the biogenic turnover of matter, which had been created by one species eating another and the use by some species of the metabolic products of others. Originally such feeding must have taken place using organic substances which had been newly created in the oceans of that time; later it used the growth and multiplication of proorganisms and primeval organisms.

This prehistory of life apparently continued for quite a long time, longer than the whole succeeding period of life's existence. The biogenic turnover of matter acquired its contemporary character after the appearance of photosynthesizing plants. The conditions necessary for their appearance were a diminution in the spontaneous formation of organic matter in the primeval waters and a progressive decrease in temperature, along with a change in the chemistry and the electrical state of the waters of the primeval ocean. It became necessary for the organisms themselves to assemble organic substances out of materials from inorganic nature. "Certain living creatures actually succeeded in doing this. In the process of sequential development they worked out within themselves the ability to absorb the energy of the solar rays . . . and construct organic substances within their bodies. In this way . . . the proto-plants—the blue-green algae—arose. Other creatures preserved their previous methods of feeding, but now the algae began to serve as their food sources. So began the animal world in its primeval form" (Oparin, 1951).

The biogenic cycle of matter was completed when the world of saprophytic microorganisms appeared, organisms which could mineralize organic (primarily dead) matter and return it to its initial inorganic state. Three basic stages in the turnover of matter and energy had been assembled in the biosphere. On this basis the division of the organic world into three

"kingdoms" took place: the plants, producers of organic material; the animals, consumers of organic material; and the microorganisms, destroyers of organic material.

These three links lie at the basis of biochemical conversion in every part of the globe, and each complete association (biocenose) inevitably consists of plants, animals, and microorganisms. But the conditions of matter and energy turnover in the various portions of the biosphere are not identical, owing to differences in climate, chemical makeup of the substrate, and other local features. The character and speed of matter and energy turnover differ. The degree of climatic favorability and the richness of the hydromineral resources of a given area control the abundance and variety of its plants, and these in turn determine the number and makeup of animals and microorganisms. The growth of the mass and the distribution of living organisms over the surface of the earth have increased the variety of species in all three "kingdoms," complicating the biogenic turnover of matter and attracting ever greater amounts of inorganic matter and energy into it. From this point of view the evolution of the organic world has been the development and complication of the turnover of matter. The total turnover of matter expresses well the unity of the organic world with the inorganic environment, with the active element of this unity undoubtedly being living organisms, since they are associated with the basic changes in nature after the appearance of life on earth.

The biogenic turnover of matter at any point on the earth's surface might be represented as an aggregation of atoms and associated energy; they are autonomous, but they flow in one direction, like tributary streams which combine with one another in the complex system of tributaries and branches found in the delta of a river, now merging, now splitting apart. The present species assortment of organisms shown in Table 2 might give some idea of the variety of biochemical cycles throughout the earth's surface.

The variety and richness of the organic world in certain areas reflects the intensity and breadth of the biogenic turnover of matter, especially its seizure of the basic environments (atmosphere, hydrosphere, and lithosphere) and the different portions of the earth's surface. This variety arose first of all out of nutritional relationships and nutritional specialization (plants, animals, saprophytic microorganisms; herbivorous animals, predators, parasites). Adaptations to local conditions were also important—the environments of life (water, terrestrial, and soil), climate (inhabitants of deserts, clayey soils, rocks, screes, etc.), and others. Concurrent demands for food, light, and space led to expansion of animals and plants into unsettled or little-settled areas. This was accompanied by adaptation to new circumstances of existence and complication of the construction of associa-

TABLE 2. *Species assortment of contemporary organisms.*

	Number of Species
Animals[a]	1,400,000–1,500,000
Invertebrates	810,000–1,250,000
Protozoa	15,000–30,000
Coelenterates	5,000–9,000
Flatworms	6,000
Roundworms	5,000–10,000
Annelids	5,000–7,000
Arthropods	1,100,000–1,300,000
Mollusks	80,000–100,000
Echinoderms	6,000
Chordates	40,000–71,600
Fish	20,000
Amphibians and reptiles	6,000
Birds	8,590
Mammals	3,200
Plants[b]	500,000
Algae	25,000
Bacteria, fungi, etc.	100,000
Lichens	18,000
Mosses	20,000
Lycopods	800
Filices	6,000
Gymnosperms	600
Angiosperms	200,000–300,000

[a] After Zenkevich (1937) and Geptner (1956).
[b] After P. A. Baranov (1955).

tions; in particular, it led to the appearance of stratification (hierarchy).

The structures and functions of individuals of various species, and the form and structure of populations and associations, display their adaptations to one another and to the local physico-chemical conditions. These adaptations provide for the organism's exchange of matter and for the turnover of matter in nature which arises out of it and changes its initial status. In such a relationship between initial inorganic nature and living, productive matter we see the clearest expression of the historical unity of organisms and the environment.

THE ENVIRONMENT AND FACTORS OF EXISTENCE

We refer to environment as all that which surrounds organisms and directly or indirectly influences their state, development, survival possibilities,

and reproduction. The organism is influenced not only by the chemical content of the environment but by its physical state (temperature, pressure, radiation conditions, motility, particles, ionization of solutions, electrical state, etc.). The content of a given species's environment also includes all other organisms of its own or other species which directly or indirectly come into contact with it. From the environment the organism obtains its necessary matter, and to the environment it yields up the products of its metabolism. Taking and returning of matter and energy should be at least relatively balanced; otherwise the "equality of the organism with its environment"[2] is greatly disturbed and its existence is threatened.

Permanent or temporary environments of organisms are the air (troposphere), the water, and the weathered crust of the lithosphere. In the last the heavily settled portion is the soil, which represents a mechanical conglomerate of three phases—gas, liquid, and solid. The environment of parasites is the body of the host, while for saprophytes it may be dead organic matter: some nematodes live successfully in a bite wound, while larvae of the *Psilopa* fly live in petroleum, feeding on the microorganisms which thrive there.

Different species, even those living in the same place, use the environment in various ways, consuming different food and differing in their manner of gas, water, and mineral exchange and in their varied relationships to temperature, moisture, and other conditions. In this sense it has been said that each species possesses "its own environment" (Nikol'sky, 1955; Bodenheimer, 1958). Thus birds dwelling in the hollows of trees in our middle-latitude forests feed in different ways. The striped woodpecker feeds on seeds in tree bark and the larvae of xylophagous insects living in tree trunks. The tomtit collects insects, their larvae, and their eggs from the surface of the trunk and the larger branches (the great tomtit *Parus major*) or from the small branches and shrubs (*Poëcile borealis* and the long-tailed tomtit *Acredula caudata*); they also eat small tree and grass seeds. Flycatchers primarily snare flying insects (butterflies, flies). Among steppe and desert rodents the moles (*Spalax, Myospalax, Ellobius*) feed on roots, tubers, onions, and other subterranean plant organs. Susliks live on onions and green leafy parts of plants and willingly eat insects. An even greater amount of insects forms the food of jerboas and hamsters. Relationship to climate is likewise varied, since many species live in natural or artificial refuges (burrows, tree hollows, etc.) and others live in the open; some are active by day, some by night, still others throughout the whole 24 hours. All of them, though they may live in one place, actually exist in the most varied climatic surroundings.

[2] I. P. Pavlov's term.

Not everything which surrounds the organism is necessary for its existence or influences it. However, those elements of the environment to which a particular species is indifferent may play a role in the life of its neighbors and may thus influence the given species indirectly. Everything which acts upon the organism, regardless of the character of the influence and unlike the indifferent elements, is called an environmental factor. These must in turn be divided into those determining the possibility and success of development, growth, survival, and reproduction; they are usually called the conditions of existence or conditions of life. They include (a) supplies of matter needed for metabolism which are available to the organism; (b) favorable radiation conditions and temperature for the formation of proteins and nucleic acids and for the maintenance of the biochemical reactions pertaining to them; the range of such conditions is not large as compared with their known fluctuations on earth; (c) the environmental density and pressure necessary to maintain the forms of the organism and the possibility of its movement; (d) removal of harmful products of metabolism by environmental transport or else neutralization by other species and absorption of life-killing rays less than 2,860 Å in length by the already-mentioned ozone screen.[3]

Factors in the environment never act on the organism in isolation but always in an uninterrupted chain, which in no way precludes the known autonomy and specific action of each of them. The totality of mutually acting factors (their constellation) forms, in particular, such well-known phenomena as droughts, drying winds, rains, frosts, snow storms, and many others which may play an immense role in the ecology of individual species.

The tasks of ecological research should be the establishment of (a) which factors in the environment, in what combinations and in what quantities, are necessary for the establishment of a given basic process or life cycle as a whole; i.e., which ones determine not only the quantitative but also the qualitative aspects of life and are therefore the leading factors; (b) which combinations of factors and their quantitative significance primarily determine the intensity of the process, i.e., its quantitative aspect; (c) which factors merely make up the neutral background on which the phenomenon is played out, i.e., those factors which are indifferent to the phenomenon (Monchadsky, 1949, 1958).

Division of environmental factors into most significant, less significant, and indifferent ones is absolutely necessary, but it must be kept in mind that this division is relative. Most physico-chemical factors (temperature, radiation, moisture (environmental humidity), movement of environmental particles, and many others) are favorable for organisms within a definite

[3] One angstrom (Å) equals 1×10^{-8} mm or 0.0001 micron.

range, but beyond this range they first impede and then preclude the possibility of existence. The role of individual factors changes depending on their relationships with others. Thus the daily rhythm of mosquito, horsefly, and housefly activity is determined by the light regime in the warm southern regions, while in the north it is determined by changes in temperature, since temperature here is close to the threshold of viability for these animals (Monchadsky, 1956; Berezina, 1957).

Factors are distinguished by their nature as inorganic or abiotic, organic or biotic, and anthropogenic. The first category includes the chemistry of the environment (gas content of the air, chemistry of bodies of water, chemistry of the basic rocks and their products, the soils), its physical state and climate (temperature, radiation conditions, electrical state, density, pressure, motion), its mechanical content (aggregation), and its geomorphological properties (structure of the surface, etc.). The second category includes the connections and interrelations between individuals of one species (intraspecies relations) and between different species (interspecies relations, including nutritional ones). The third category—anthropogenic factors—includes the effects on the organic world as a whole of the activities of man (agricultural conquest of land, industry, transport, conservation of nature, exploitation of useful species and extermination of harmful ones, etc.). The significance of these factors undeniably grows according to the speed with which man more and more fully conquers nature and makes it subordinate to himself.

The most important factors for animals, in addition to the physicochemical ones which furnish metabolism, include food—an interspecies biotic factor, since animals feed on other organisms. Further forms of inter- and intraspecies relationships (predators and prey, parasites and hosts, concurrence, cohabitation, symbiosis) may play an important and even a pivotal role in the life of a species or its individual populations but may not have any obligatory significance for individuals taken singly, since they may not be encountered in the life of the latter. This once again emphasizes the relativity of the division of factors into primary and secondary ones. They acquire one meaning or the other only in relation to a concrete situation.

CHANGEABILITY OF THE ENVIRONMENT

The inorganic and organic environment in which animals exist does not remain constant. At the basis of its changes lie cosmic factors which determine the seasonal, lunar, and daily cycles as well as the nonperiodic

and, in the majority of cases, irreversible changes in living and nonliving nature which accompany its development (evolution). In other words, cyclic and nonperiodic fluctuations, together with irreversible changes, give us a complex picture of the dynamics of almost all elements of the environment. The rise and development of life on earth have taken place against a background of an uninterruptedly changing environment, reflected in the rhythms of biological phenomena: alternating periods of rest and activity, rise and fall of intensity in the life processes, periodicity of growth and reproduction, and, finally, fluctuations in numbers.

Based on the character of the factors, A. S. Monchadsky (1958) has suggested that we distinguish:

1. Stable factors, which do not change over the course of long periods of time and for this reason do not give rise to changes in numbers and geographic distribution of animals. Such factors are the force of gravity, the solar constant, the makeup and properties of the atmosphere, hydrosphere, and lithosphere, relief, etc.

2. Changing factors: (a) the regular-periodic (daily, seasonal, and other changes, depending on the movement of the planets in the solar system) factors determine daily, seasonal, and other biological cycles, dynamics of abundance, and range boundaries and have little apparent effect on the perennial course of abundance changes because of their true cyclic character; (b) those without regular periodicity (temperature, wind, precipitation, moisture, food, disease, parasites, predators, etc.) have little effect on seasonal and biological cycles but influence changes in population abundance in different years and in the distribution of animals within the confines of a range.

Even when there is strict periodicity of daily and seasonal cycles, they do not always occur as complete repetitions of one another; i.e., they differ in the state of the weather and in a series of other phenomena. There are also zonal properties of these cycles, which are expressed in different ways in high and low latitudes.

The longer the periodic fluctuation, the less regular it is. Fluctuations are associated with changes in climate and other phenomena lasting over long periods of time and stemming from cosmic causes, tectonic phenomena (orogenesis), etc. More or less regular alternations of warm and cold, dry and moist periods lasting 3–4, 6–7, 11, 35, and 80–100 years have been established. The short periods (the 3–7 year cycles) have not been distinctly expressed, and their meaning has not been examined intensively, although fluctuations in the numbers of small rodents and certain other biological phenomena are repeated at this frequency in some areas (Elton, 1942; Kalela, 1944, 1950, 1951, 1957; Siivonen, 1948, 1952, 1954, 1956, 1957; Siivonen and Koskimies, 1955). The 11-year cycle of sun-

spots is better known. Diminution in their number due to amalgamation, accompanied by growth in the area of the sun's active territory, strengthens solar radiation. When this happens, the oxygen molecules in the upper layers of the atmosphere, being transformed into ozone in greater numbers, withhold the heat-producing rays of the earth, which raises the temperature of the latter's surface. Increase in solar activity is accompanied by a rise in baric contrasts and an increase in atmospheric circulation, causing heating in the high latitudes and cooling in the low ones (Fig. 3). The centenary fluctuations in climate are repeated approximately

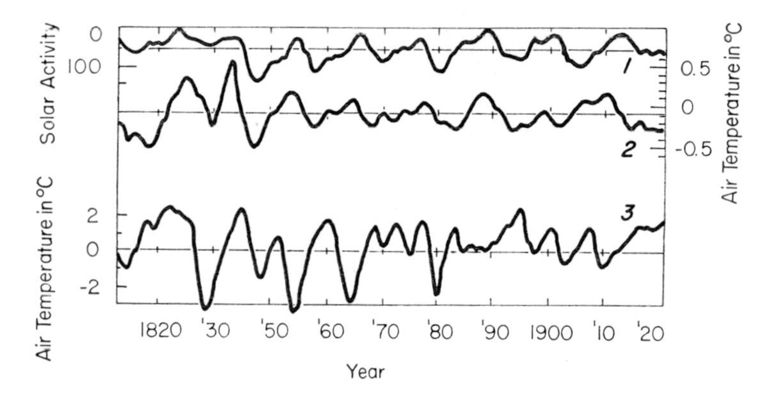

Fig. 3. Solar activity (Wolf's figures, 1) and deviations of temperature in the tropics (2) and in the temperate latitudes (3) (after Eygenson, 1948).

every 80–100 years. A great warming was observed in the middle of the nineteenth century and has been repeated in approximately the middle of our own century. It was especially well expressed in the Arctic and in the North Atlantic and was associated with an increase in the flow of the Gulf Stream, the pressure of which grew with the general warming and strengthening of water circulation (Fig. 4).

Fluctuations in the level of the Caspian Sea (Fig. 5), which depend on the amount of precipitation and drainage in the Volga basin, show an average period of wet-dry fluctuation lasting 17–18 years (the so-called 35-year Brückner cycles). These fluctuations repeat themselves in reverse in the Aral Sea, the level of which changes in relation to the melting of the snows in the Tyan'-Shan' and Pamir-Alay, which increases during hot, dry periods. There is a close correlation between the fluctuations in the level of the West Siberian and North Kazakhstan lakes, the biological effects of which have been studied by A. N. Formozov and his pupils (Formozov, 1934, 1937).

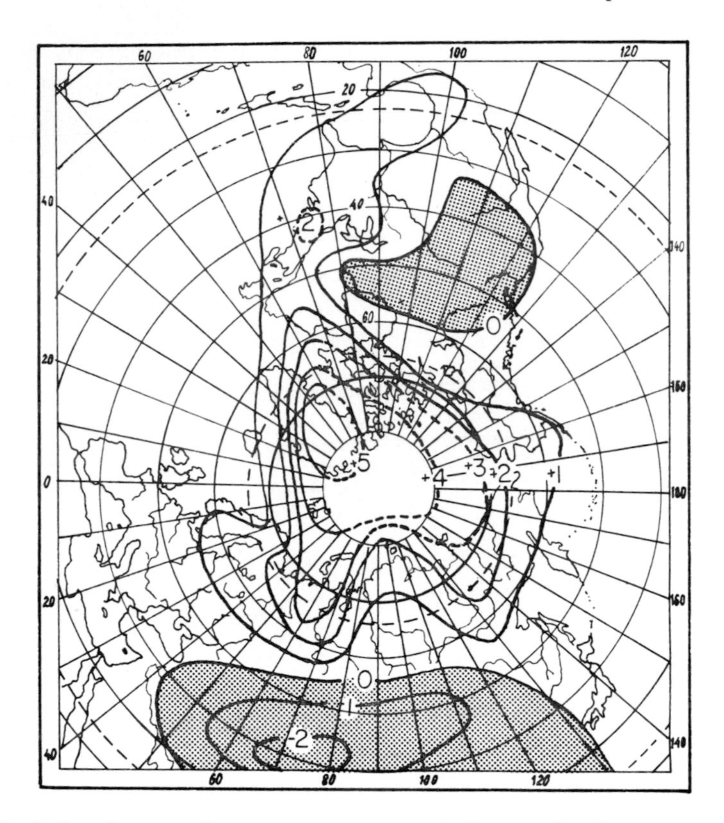

Fig. 4. Deviations from yearly average temperatures in January for the period 1929–1938 (after Rubinstein, from Markov, 1951).

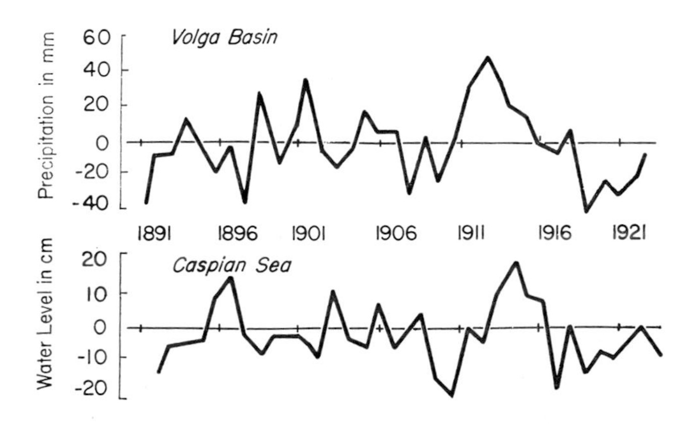

Fig. 5. Fluctuations in winter precipitation in the Volga basin and in the level of the Caspian Sea (after Kaminsky).

Attempts to observe a strict cyclicity in similar fluctuations and to establish a direct correlation with such complex biological phenomena as changes in the abundance and status of animals, harvests of agricultural crops, and growth of wild vegetation have not proved successful (MacLulich, 1937; Elton, 1942) (Fig. 6). It has been found that the length of these periods fluctuated within broad limits and sometimes lost its periodicity altogether (Berg, 1938). All these biological phenomena are the result of the interrelationships of organisms with the whole complex totality of environmental factors and, as a rule, show no precise correlation with the dynamics of any meteorological factors. The influence of the latter is great, but it is filtered through a complex system of organism-environment connections.

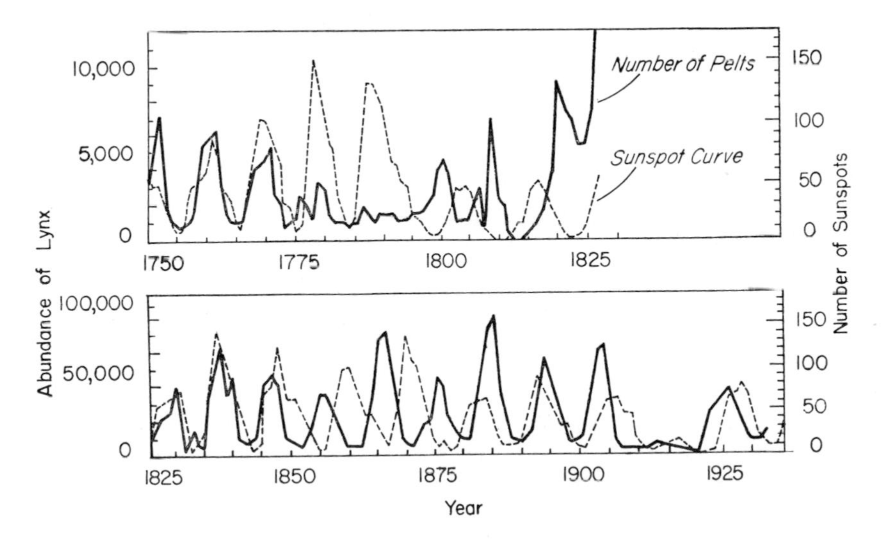

Fig. 6. Changes in the number of lynx pelts prepared by commercial firms in Canada compared with the sunspot curve (after MacLulich, 1937).

A series of animal adaptations is associated with the changeability of the environment. The first example would be the periodic daily physiological processes with which we associate the alternation of periods of rest and activity as well as migration, change of habitation, etc. The circle of adaptations is wider and more varied with regard to seasonal changes, which are expressed to different degrees in the various zones. One example is the seasonality of development and life activity, which is especially widespread among the animals and plants of the temperate and high latitudes, where seasonal contrasts in meteorological conditions are very strongly expressed. Seasonal morphological adaptations include poly-

morphism, expressed in different ways in different animal groups; cyclomorphosis, or seasonal alternation of generations and changes in their form; seasonal molting of skins; and several others. Seasonal physiological adaptations include changes in an animal's metabolism and energy status; these lie at the base of reproductive periodicity, fat accumulation, hibernation, and many other phenomena. An animal's behavior, including its complex forms, changes, and as a result food is stored away, migrations arise, and intraspecies relationships and ways of life change. Solitary existence may be turned into flock or herd-type existence, a sessile way of life into a nomadic one, etc. Ultimately the number of individual species grows or diminishes, their relationships to one another change, and the structure and content of the entire association are altered. Finally, the inherited makeup of the population changes at various seasons, its genetic variety being maintained by the seasonal variety of conditions of existence and by the deaths at different seasons of individual inherited lines associated with it.

One reflection of the perennial and centennial changes in climate are changes in the abundance and content of populations as well as in the content and structure of associations. The latter acquire a warmth-loving, drought-loving, or moisture-loving character, depending on the prevalence at various periods of one or another type of weather (climatic aspects of fauna and flora). Thus in the forests of the European USSR, especially on the border between the European and Siberian faunas, the warm and moist years are characterized by domination of the red vole (*Clethrionomys glareolus*) and the common vole (*Microtus arvalis*), but after dry, frosty winters they yield to the Siberian species *C. rutilus* (Formozov, 1948). On the steppes of the southern Ukraine the abundance index of the forest mouse *(Apodemus sylvaticus)* drops after dry summers, as does that of the common vole, and the dominant position is occupied by the drought lovers—the steppe lemming (*Lagurus lagurus*) and the society vole (*M. socialis*). A multitude of similar examples can be given for other animals.

Centenary fluctuations of climate are accompanied by even more profound alternations, as a result of which the ranges of some species are expanded while those of others are curtailed. Zoogeography and floristics have accumulated many data of this type (Geptner, 1936; Bobrinsky, Zenkevich, and Birshteyn, 1946). Finally, geological changes in climate and in other environmental factors, also known to be repetitive to a certain degree (Markov, 1951), are accompanied by replacement of flora and fauna. But evolution of the organic world, accompanied by increase in the variety of animals and plants and by greater complexity in their construction and functions, is a development which may not be associated only with the influence of the external environment. So far as we know, the

sun has never changed in the character of its radiation throughout the entire immense span of life on earth (Fesenkov, 1950); the cyclic character of orogenesis has in general been preserved (Markov, 1951). But changes in the biosphere during this same time are of a distinctly progressive character; i.e., rapid development of the contradictory organism-environment unity, in which the leading role of living organic matter is seen quite distinctly, goes on rapidly (Vernadsky, 1926, 1944).

VIABILITY OF ORGANISMS AND FACTORS OF THE ENVIRONMENT

The substances entering the body and the energy it absorbs are used in the work of the organs, and of the organism as a whole, for repairing destroyed tissues, for increase in mass, and for creating new structures (growth and development). The use of matter and energy in the body's work is known as the energetic aspect of metabolism; the body's growth and development are handled by the constructive aspect of metabolism. The relationship between these conditionally distinguished aspects of the total exchange of matter is different in young and adult animals; the former are bent chiefly on construction, the latter on energy exchange.

In species with a high level of activity, such as the warm-blooded animals (birds and mammals), the relationship of energy exchange to construction exchange is higher (30–10:1) than in fish and insects (8–3:1). Assimilation of matter and use of energy resist their collapse (dissimilation) and loss of energy (dissipation into the external environment). The necessity for a definite balancing of both aspects of metabolism is obvious, since in catabolism toxic breakdown products would build up, and the tolerable maximum and minimum limits of body temperature, water supplies, etc. would be exceeded.

Since the speed and degree of matter and energy acquisition by the body are controlled by environmental factors, it is with this speed that the degree of metabolic balance is associated, and this balance in turn determines the state of the body, i.e., its possibilities for growth, development, and reproduction. The introduction of exchange of matter corresponding to the state of the environment and its individual factors was called "equalization of the body with its environment" by I. P. Pavlov. Departure from it is accompanied by destruction of the balance of body-environment relations, slowing down of tempo, slowing or halting of development, and, in the extreme case, death of the organism. N. I. Kalabukhov (1946, 1950, 1951) has suggested "energy balance" instead of

"equalization" and has included the same contents in its definition. This balancing is accomplished by mechanisms for regulating metabolism; in animals it is done by the nervous and endocrine systems.

Still, complete equalization is never achieved even under optimal conditions because of the constant changeability of the environment and the time lag in the body's reactions to external changes. Therefore, destruction of the energy balance is common and explains many developmental anomalies, failures of reproduction, and reduction of animal survival. The indices of degree of equalization and of the state of the organism dependent on it are weight and size, fatness of the individual, state of the gonads (during the mating season), amount of reserve matter (especially fat) in the animal's body, relative sizes of the most important organs (heart, lungs, liver, adrenals, thyroid, etc.), and a series of biochemical (blood sugar level, ascorbic acid level, etc.) and physiological (level of oxygen consumption, respiratory coefficient, etc.) indicators. All this taken together permits detailed and multifaceted characterization of the viability of the individuals examined (Shvarts, 1956, 1958, 1959).

HOW ENVIRONMENTAL FACTORS ACT ON THE BODY

Environmental factors act on the animal body in several ways. One should distinguish between direct or reflective (signal) action and oblique or indirect action. Direct perception is the immediate action of a factor at the basis of which lie the general physico-chemical laws: the body receives external actions just as any nonliving body would. Such reactions are simple and widespread. Thus a rise in environmental temperature raises the temperature of the body surface and then the temperature of the body as a whole, as long as the complex thermoregulating processes in the organism do not interfere; an increase in the surrounding humidity reduces evaporation from the body; absorption of solar radiation by the animal's surface changes not only its temperature but also the speed and often the direction of its biochemical reactions. All of this may take place without the participation of the nervous system. In other words, in these cases the action of external factors does not depend on the structure of the perceiving body, and its reactions are of a nonspecific, passive character.

This method of reaction, or others similar to it, has an important significance for plants and lower animals. One example is the fluctuation in the amount of matter exchange (consumption of oxygen—Fig. 7), body temperature, and motility of cold-blooded animals—almost directly pro-

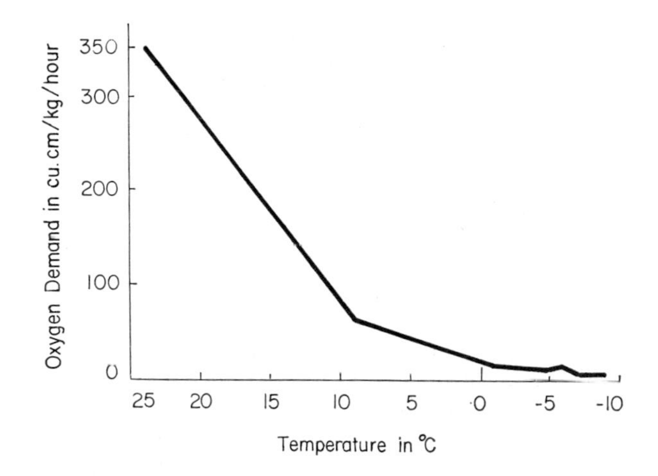

Fig. 7. Consumption of oxygen by lizards of the genus *Eremias* at various environmental temperatures (after Rodionov, 1938, original).

portional to a rise or fall in the external temperature (Fig. 8). In many of the so-called "cold-blooded"[4] aquatic animals the saline contents of the body fluids also change in response to changes in the salinity of the external environment (Fig. 9). The passive character of the perception of external influence, expressed in the proportionality between changes in the environment and changes in the body, depends on the absence of special regulating mechanisms to furnish known resistance by the basic vital processes.

The growth of metabolic intensity in a historical series of animals has brought about the development of thermoregulation, osmoregulation, and other adaptations which diminish the body's passive dependence on the environment and which change the character of the body-environment relationship. The nervous system, aided by receptors which perceive external changes, analyze them in special centers, and transform the observed disturbances into movement or other physiological acts, began to play a pivotal role in relations with the environment. These acts serve as responses to external influences. In this way the organism's relations with the environment became not only more complicated but more complete. The organism became more active and found itself in a more profitable position.

The reflexive character of relations with the environment appeared in

[4] Incapable of actively regulating the salinity or osmotic pressure of the body fluids during environmental salinity fluctuations.

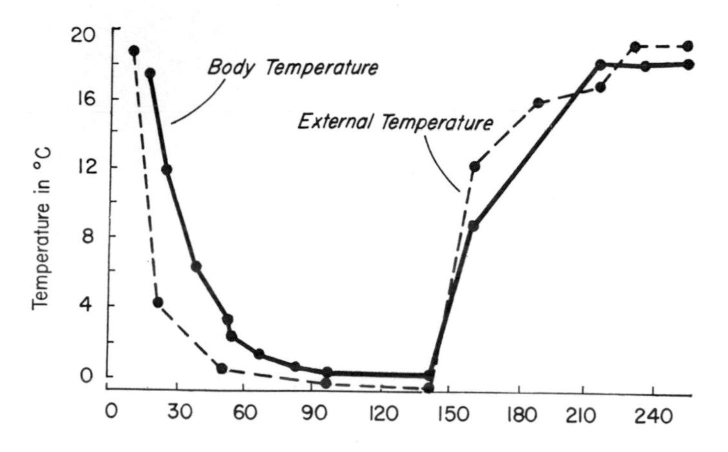

Fig. 8. Changes in body temperature of the grass frog under the influence of external temperature (after Eisentraut, 1933).

the early stages of development of the animal world, and the above-mentioned examples of "direct" influence of the environment on the body actually represent reflexive acts, since in no case do we observe a complete parallelism between external and internal changes (see Figs. 7–9). They do, however, preserve their character as primarily passive reactions associated with lack of special regulating mechanisms in lower

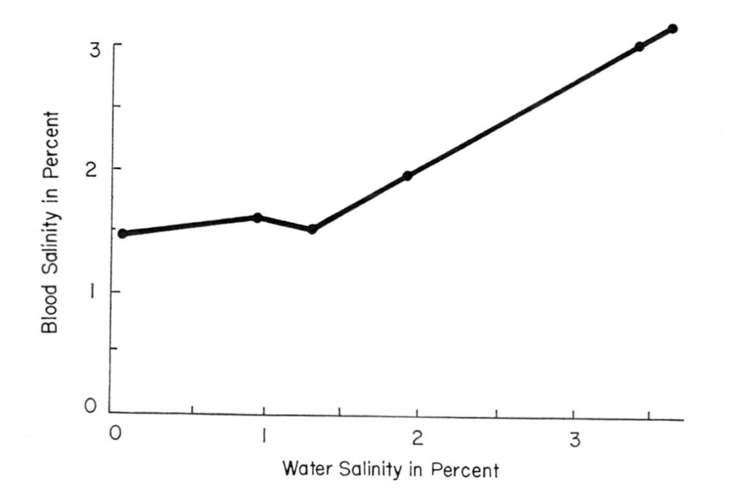

Fig. 9. Dependence of salinity fluctuations in the crab *Carcinus maenas* on salinity of the environment (from Zernov, 1949).

animals. Active regulation of functions in response to external influences increases in proportion to the complexity of the nervous system, perfection of the organs of blood circulation, motility, and formation of protective coverings; it is associated with total intensification of the functions. This process was perfected with the appearance of homoiothermic "warm-blooded" animals and their higher nervous activity. The formation, on a basis of nonconditioned reflexes, of temporary (conditioned-reflex) relations with the environment (as also proved possible with animals having noncomplex nervous systems) provided quick and adequate equalization of internal processes with changes in the external environment and greatly expanded the ability to use the environment.

The irritability of living matter made possible the formation of reflexive connections with any element of the environment, including those which did not play a direct role in the organism's existence but which were preparatory to the onset of changes important to it. Such changes could become fixed by natural selection, since biologically they were entirely useful because they prepared the organism to meet new conditions. This kind of preliminary adaptation (preadaptation)[5] places the organism in unusually profitable circumstances; the ability to react in this way rests on complex morpho-physiological mechanisms (conditioned-reflex reactions) which have developed under the influence of natural selection. Such connections deserve to be called signal ones, for they operate as factor-signals, which in themselves cannot act on the organism directly but which must of necessity precede any changes in the environment.

The seasonal biological phenomena in plants and animals (accumulation of reserve matter, supplying of food, migration, hibernation, reproduction, change from solitary to group life and vice versa, etc.) are good examples of preliminary adaptation and are established according to this scheme. External factor-signals, such as seasonal changes in the amount of daylight, in the temperature, in the appearance and disappearance of food, etc., are perceived by the animal's receptor analyzers (sensory organs). The obtained information enters the nerve centers, is processed, stimulates changes in the endocrine system, and brings about a general reaction on the part of the body. As a result, matter exchange is altered, and consequently reserve matter is either accumulated or expended, the seasonal change in body covering takes place, reproduction begins or ceases, and

[5] But not in the idealistic sense in which this term was defined by L. Cuénot (1925, 1936), L. S. Berg (1923), and other vitalists. They assumed a primordial ability of living organisms to make desirable reactions to conditions which were new to them. We are speaking of the materialistic conception, which subsumes under preadaptation the preliminary preparation for meeting new conditions as provided by the historically formed mechanisms of a signal-reflex character.

migration, hibernation, etc. are begun or ended. In vertebrate animals the signal connections are set in motion according to the diagram below.

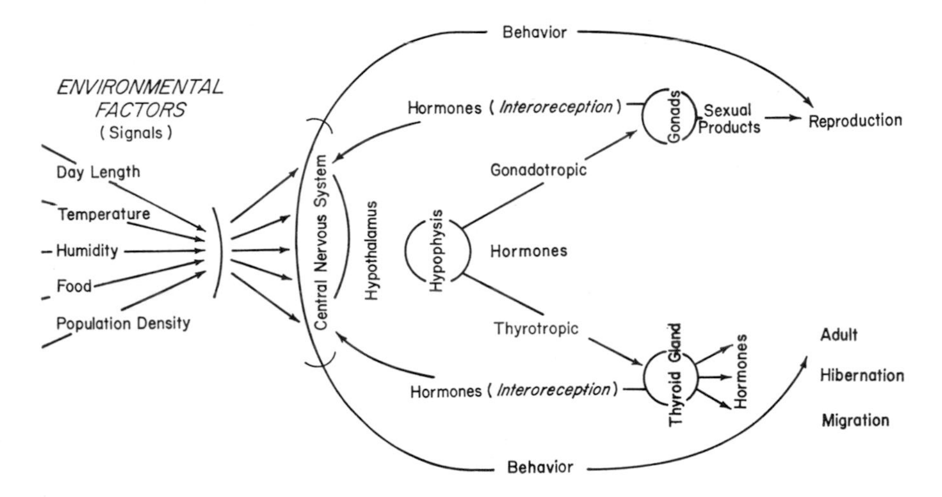

In order to acquire signal significance, the external factor need not be connected directly to the natural seasonal phenomena to which the body is going to adapt; it is necessary only that it commonly precede it. Thus accumulation of reserve material is an adaptation to the approaching insufficiency of food; hibernation precedes another period of unfavorable temperature or shortage of moisture or food; the period when the young are born tends to coincide with a period of abundant food supplies and favorable weather. Despite the variety of connotations, such adaptations as are possessed by molting, hibernation, fat accumulation, reproduction, etc. all arise similarly and are caused by one or several factor-signals, most often by a change in day-length (Rowan, 1926, 1938; Bissonette and Wadlund, 1932; Bissonette, 1936; Promptov, 1941; Larionov, 1945; Belyayev and Utkin, 1949; D. K. Belyayev, 1950; Bullough, 1951; and others).

Signal connections which provide for such complex phenomena in the lives of animals as feeding, reproduction, migration, and resettlement are instinct or chain reactions, at whose basis lie the elementary nonconditioned and conditioned reactions associated with them. The stimulus may be any external event which forms a permanent connection in time with those natural phenomena to which the animal must adapt. A natural phenomenon (a factor, a condition of existence) to which a given chain-reflex reaction or instinct is an adaptation, G. R. Backer (1938) proposed calling an initial primary or ecological factor, while the factor-signal which aids in the periodic introduction of the action mechanism he called the final, secondary, or physiological factor.

QUANTITATIVE PRINCIPLES OF FACTOR ACTION

The influence of a factor on the body depends not only on the character (specifics) of its action but also on the quantity (dosage) in which the factor is perceived by the body. A decrease or increase in the dosage beyond the favorable range works to an identical degree to lower the body's level of activity, and the attainment of certain known magnitudes (the maximum or minimum) precludes the possibility of its existence. The magnitude of the optimal zone and the entire range of tolerable fluctuations characterize the tolerance of the species (or individual) with regard to a given factor; this ability to exist under large or small fluctuations of a given factor or aggregate of factors is called ecological valence or species plasticity. It is different even in closely related species, sometimes between subspecies, and as a rule differs with age (Fig. 10).

Differentiation is made between the highly tolerant species, the eurobionts (*euro*—broad, all-embracing; Gr.), which tolerate a wide range of fluctuations in a given factor, and the stenobionts (*steno*—narrow, limited;

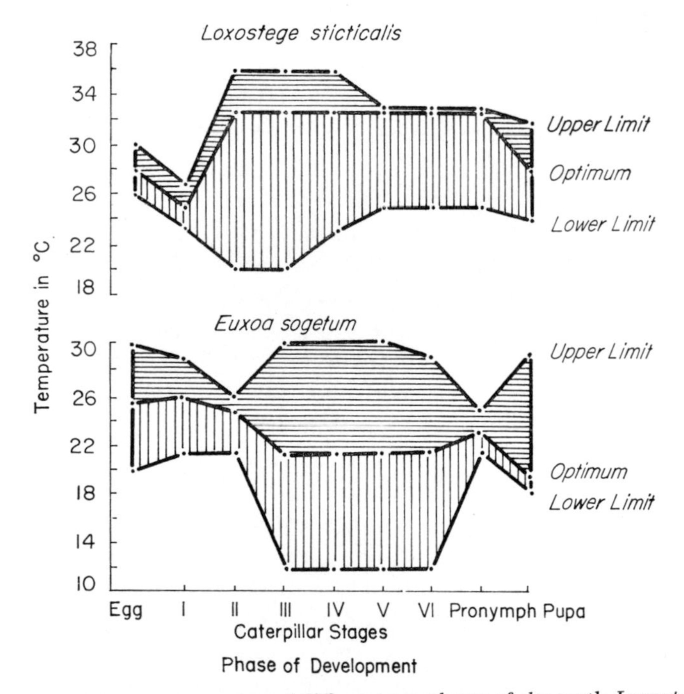

Fig. 10. Relationship to temperature of different age phases of the moth *Loxostege stricticalis* and the winter owl *Euxoa sogetum* (after Kozhanchikov, 1937).

Gr.), which are able to exist only under definite, narrowly limited, and rarely changing conditions. Depending on the kind of factor we have in mind, we speak of species which are eurythermic, tolerating large fluctuations in temperature, and correspondingly of stenothermic animals; steno- and euryphotic animals, which are differentiated by their ability to tolerate light; and steno- and euryhalinic species, capable of tolerating greater or lesser fluctuations in the amount of salt in a body of water. In addition to tolerance of fluctuations, species may differ in their preference for one or another range of a given factor. Thus we have thermophiles and cryophiles, warmth lovers and cold lovers; xerophiles and hygrophiles, drought lovers and moisture lovers; and photophiles and photophobes, lovers of light and lovers of shadow.

A factor whose quantitative expression or pressure is higher than the optimum limits the possibility of growth, development, and reproduction and reduces the chances for a species's numbers to increase; i.e., it becomes a "limiting factor." If we speak of a shortage of substances necessary for metabolism, we are speaking of the law of the minimum, first formulated by Liebig (1840). This scientist demonstrated the dependence of crop size on the amounts of those necessary elements in the soil (potassium, sodium, phosphorus, nitrogen) which supply the demands of the plants to the least degree. Later the law of the minimum began to be used to express the dependence of body development or population abundance on any of the environmental factors which might have quantitative characteristics. Thus, according to A. Thienemann (1925), "that factor of the surrounding environment which acts on the stage with the least ecological valence in quantity or with an intensity most distant from the optimum is the one which determines the population density of a given species of living creatures." Shelford's (1911) and Taylor's (1934) "law of tolerance" is close to this definition. The latter emphasized that the abundance of organisms is limited by the necessary factor (condition of existence) which occurs in the least quantity, as measured against demand, during the critical period or during an especially unfavorable year.

These formulations of the law of the minimum, which sunder the unity between the organism and the environment and set them in opposition to one another, are not only inaccurate but also untrue, since they oversimplify and distort complex ecological phenomena. On the one hand the organism actively selects its necessary conditions of life in the environment, while on the other it reconstructs and adapts itself and its exchange of matter in compliance with the concrete circumstances surrounding it. Finally, the law of the minimum does not take into consideration the interaction of factors perceived by the organism as an inseparable whole in which not only the individual parts but also the relations between

them are important. This latter correction was introduced by E. A. Mitscherlich (1921), who showed the influence on harvests not only of the substances which are in minimal supply but also of all others. Liebig did not consider the fact that a change in quantitative relationships is accompanied by qualitative changes, that when the character of each factor's action changes, the position and dimensions of the optimal zone change, etc.

It is also true that the concept of ecological valence and the law of the minimum are not proper or are completely inapplicable to many biotic factors in which the concepts of optimum, maximum, and minimum have no meaning. It is impossible to operate with them even in the study of interspecies relationships (predation, parasitism, etc.). Finally, the law of the minimum, resting as it does on physico-chemical principles, does not deal with the complex mechanism of the organism-environment relationship. In living creatures responses to external actions are screened through a perceiving and transforming system which changes irreversibly their character, strength, and sometimes their time. This is basically the phenomenon which makes life unique as a special form of matter in motion. It is for this reason that only a few living phenomena have a more or less satisfactory quantitative correlation with the external disturbance (see Fig. 8). More often it simply does not exist, owing to a unique "biological inertia," for example, when the appearance of abundant food or other favorable factors does not have an immediate influence but may be expressed significantly later, sometimes several years afterward.

Thus there is no possibility of summing up quantitatively the state of organisms, their abundance, and their changes in one result stemming from a mere set of factors (conditions of existence). In addition, it is necessary to recognize that ecological valence and the law of the minimum are concepts applicable only to a description of the quantitative influences of primarily abiotic factors. Even in this case the relationship of the animal to temperature, humidity, pressure, movement, and the chemistry of the inorganic environment is of a complex reflex character, all of these relationships depending on one another, on intrapopulational connections, and on interrelations of species within their associations. In this case the mutual relations between environmental factors are especially distinct in their action.

Academician T. D. Lysenko (1952) has worked out a theory of stadial development which separates ontogenesis into a series of stages in each of which various factors in the environment (temperature, light, moisture) have pivotal importance for the success of development. He has shown for plants the resolving importance of developmental conditions in one stage for the success of the succeeding ones.

The interrelationship of factors is expressed in the fact that they are

perceived by the organism not only as individual elements of the environment but also as a total aggregation, a totality, in which the character and strength of one factor's action is determined by the states of the others. The interrelationship of climatic factors is easily shown in the case of temperature and humidity. Thus the tolerable fluctuation in humidity for the moth *Ephestia kühniella* is broad under optimal temperature conditions but narrows quickly as the thermal maximum or minimum is reached. The same is true for the reverse situation (the action of temperature when humidity is reduced).

In nature not only pairs but also more complex systems of factors are at work, including the biotic conditions of intra- and interspecies relationships. The importance of the relationship between the character and effect of physico-chemical factors and changes in population density has long been known for certain species, and we have long known the extent to which the abundance and content of species in the biocenose depend on these factors. Likewise, we know the extent to which weather, water supply, gas supply, availability of minerals, etc. influence the population of certain species, the character of intrapopulational connections, migrations, and reproduction, and the relations to populations of other species.

This has forced many foreign ecologists to distinguish between environmental factors which are "dependent" on or "independent" of the population density. Such a division is incorrect, for it ignores environmental factors which form a single and united system. The status, possibilities for reproduction, and survival of the various species, even of those living in one place and under the same set of external alterations, change in different ways. It is in this very manner that the environmental specificity of each species is expressed (Nikol'sky, 1955). The factors limiting the possibility of existence and the numbers of the different species in one place are also different for the same reason.

Limiting factors may replace one another periodically, from season to season or in different years. In certain periods (seasons or years) the possibilities for development and reproduction of individuals are limited, while mortality increases. When such a time comes, we call it a critical or sensitive period. V. I. Tikhvinsky (1938) compared changes in weather factors with fluctuations in the numbers of several species of animals (hares, ermines, martens, foxes, and others). He discovered how to show the existence of such critical periods and learned to note the factors whose fluctuations correspond to movements in the number of animals. He was the first to suggest using the discovered relationships to predict abundance. The winter is especially important for the common squirrel (*Sciurus vulgaris*), for it is then that these animals eat the seeds and nuts of trees (fir, cedar, larch, pine, oak). In years of poor harvest they starve, survive

the winter poorly, multiply little, and suffer marked losses in numbers during the following year (N. P. Naumov, 1930; Formozov, 1935; Kiris, 1947, 1948, 1956). Great multiplication followed by a rise in numbers among many species of voles and mice accompanies a warm autumn, when large amounts of seeds are abundant, and the animals' reproduction continues into November and December, often without interruption into winter and spring. In the north the critical factor is the total amount of heat during the growing season (Bashenina, 1949). In southern regions reproduction of rodents is curtailed or ended by cold weather or by a dry summer. Successful reproduction and survival of young heath grouse on the Kola Peninsula and in Finland depend on the weather during the laying and setting periods, especially on the temperature in June (O. I. Semënov-Tyan-Shansky, 1938, 1959; Siivonen, 1948, 1957; Siivonen and Koskimies, 1955).

INTERRELATIONS OF ORGANISMS AND ENVIRONMENT AND THEIR HISTORICAL UNITY

Organisms and the environment in which they dwell are in constant interrelation. This arises out of the exchange of matter and energy carried on between living creatures and inorganic nature. The interrelationship itself is expressed in the fact that the status of the organisms—the possibility and success of their development, survival, and multiplication—depends in the most direct way on the totality of environmental factors. At the same time the biological exchange of matter and the transformations of energy associated with it continue to change the environment markedly, just as they have irreversibly changed it in the past. In this interrelationship between initial inorganic nature and productive, living, organic matter the historical unity of organisms and environment is expressed in the most obvious and patent manner. The active, guiding element in this unity is undoubtedly the living population of the biosphere, since it alone determines those basic changes which have taken place in the biosphere since the moment when life appeared on the earth. At the same time, being the leading element of unity, organisms, especially if we take each one individually, depend completely on the surrounding environment, without which their existence is impossible. *The state and viability of individuals,* no matter to which species they belong, depend on the *conditions of matter exchange.* Thus the unity of organisms and environment is, *au fond,* contradictory.

The scale of exchanges in the biosphere is immense. None of the earth's envelopes (atmosphere, hydrosphere, and lithosphere) can avoid some part of it. The chemistry and physical state of the atmosphere have changed; large amounts of free oxygen have been produced to create the ozone barrier. According to V. R. Vil'yams (1948), the modern climate in a natural zone is the ruler of vegetation formation.

In the hydrosphere both physical properties and chemistry have been changed as a result of the ability of living creatures to attract, change, mix, and dispose of various substances partially withdrawn from the biogenic turnover by accumulation of dead organic remains in causto-biolites (sapropels, turfs, coals, silts, limestones, petroleum, gases, etc.). The transformation of the lithosphere is every bit as vital; its upper horizons have been turned by plants, animals, and microorganisms into a special natural body, the soil. Many dozens and hundreds of meters down this influence continues on a smaller scale.

The leading role of organisms in the biosphere is manifested only through their associations, i.e., systems of species of various biological types connected with one another (plants, animals, lower fungi, bacteria, and viruses) whose metabolisms represent different lines in the turnover of matter. However, relations between the environment and individual organisms or populations of one species are distinctly contradictory, since the ingestion of nutritional substances impoverishes and the excretion of metabolic products contaminates the surrounding environment. This necessitates the reverse-directed activity of organisms of another sort, so that life can exist here uninterrupted.

The environment determines completely the possibility of an individual's existence and development in a given place at a given time; its subjection to the environment is great and multifaceted. Natural selection of animals also occurs on this basis. During the process the population adapts to local conditions of existence. In the population (a group of individuals of one species) the organisms acquire certain, but limited, possibilities of resisting unfavorable factors and effectively utilizing their environment. Only in associations of various species does this possibility display itself in full measure, and only under these circumstances do we discover the active, transforming role of the living population. Here, too, we see the contradiction of the unity of organisms and environment.

The closest mutual connections between organisms and the environment do not preclude their autonomy and independence from one another. At each given moment any organism and the environment surrounding it resist one another as separate systems. This autonomy is displayed in the fact that different species may arise out of the same initial material, in one place, and under generally identical conditions. The entire evolution

of the organic world, with its increasing variety of life forms (species) and its complication of their organization, illustrates the correctness of this Darwinian principle of divergence. On the other hand, many changes in the environment are independent of organisms and are determined by tectonic, cosmic, or other causes. All this explains the relative adaptability of species to the environment; it is constantly being upset from both sides, both by changes in inorganic nature and by the activities of organisms which exhaust and contaminate it.

Consequently, the unity between organisms and the environment is of a historical character and cannot be treated as a balanced interrelation. In associations of animals, plants, and microorganisms which have developed under the influence of natural selection and which represent organizations which produce biogenic turnover of matter, the mutual supplementation and neutralization of the results of one-sided metabolism create the possibility of a resistant existence of a living population on a given plot of ground. But this resistance rests not on balance but on a very complex combination of historically correlated, but in many ways contradictory, processes which are not always balanced. This explains the dynamism and changeability of the organic world. It may be seen not only in the example of evolution of species or alteration of faunas and floras over vast stretches of time but also in those reconstructions of biocenoses, changes in their content, dispersion of their species, or expansions of their ranges which are the consequences of all sorts of alternations in nature and which occur in short spaces of time that we can comprehend.

The unity of organisms and the environment is the expression of a historically assembled dependence between groups of interconnected species (associations) and the aggregate factors of nonliving nature. Only in such a system and as a result of the contradictory interactions of its parts can automobility (development), the evolution of species, the changes in their associations, and the alternation of landscapes ultimately arise.

CHAPTER 2

Feeding

BASIC TYPES OF FEEDING AND ADAPTATIONS ASSOCIATED WITH THEM

Feeding is one of the most basic forms of activity and is the most important form of connection with the environment. It provides for consumption of the majority of necessary substances. Autotrophic organisms are capable of using the energy of solar radiation (chlorophyll-bearing green plants) or the energy of chemical transformations (chemosynthetic bacteria).[1] They construct their bodies by consuming inorganic matter, creating high-molecular compounds with a large supply of potential energy. Such organisms are called producents, or producers, of organic matter.

Heterotrophic organisms use prepared organic matter and from its reduction obtain the compounds and energy they need. They are called consumers of prepared organic matter. Academician A. I. Oparin (1951) has suggested that this, the most ancient type of nutrition (in a very primitive form, of course), was typical of the primeval organisms. From their food and drink the heterotrophic organisms also obtain inorganic compounds—electrolytes which provide the osmotic state necessary for their internal aqueous environment. Among the heterotrophic organisms animals feed primarily on living food (plants, animals), while microorganisms (saprophytes) live basically at the expense of dead bodies and refuse; they mineralize dead organic matter, returning it to its initial inorganic state. They are thus called reducers, or destroyers, of organic matter. Each of

[1] They include a not-too-numerous group of bacteria of the nitrogen-fixing type, sulfur, hydrogen, and certain other microorganisms. Nitrogen-fixing bacteria oxidize ammonia, the sulfobacteria oxidize hydrogen sulfide, the hydrogen ones oxidize hydrogen, and the ferrobacteria oxidize ferrous oxide.

these stages in the biological turnover of matter consists of one or several successive links.

Among the animals we distinguish primary, secondary, and tertiary consumers. The first are the plant eaters, or phytophages, which feed on phytoplankton, the vegetative parts of the higher plants (leaves, sprouts, roots, root tips, tubers, bulbs), or their seeds, fruits, or juices. The second, the zoophages, live on the phytophages. Among them the "peaceful" species (insects, many crustaceans, plankton eaters, benthon-eating fish, the majority of amphibians and reptiles, insectivorous birds and mammals, etc.) feed on tiny species of food;[2] predators (carnivores) live on relatively large prey; external or internal parasites use the flesh or juice of the host's body for food. Finally, the tertiary consumers—the saprophages—eat dead organic matter; they include the detritophagous animals (many worms and snails) or necrophages and the excrement-eating coprophages. These "clean" types of feeding are widespread. Many phytophages are found among the insects, roundworms (herbivorous nematodes), and other groups of animals. But the common members of these groups are zoophages (the parasitic nematodes, cestodes, and trematodes, predatory insects, parasites, etc.).

Sometimes one encounters a feeding pattern which is mixed to one degree or another, especially in the higher animals (mammals, birds, etc.). Even such carnivorous mammals as the sable, the marten, and certain others, whose basic foods are animal, sometimes eat fruits, seeds (cedar "nuts"), and berries. Those ducks (Anatinae) which feed on the bottom growth of small bodies of water in summer eat a great number of plant seeds in fall and winter. Because of the construction of their dental systems and intestinal tracts, rodents are typical phytophages; however, many species, and even genera (mice, jerboas, hamsters, dormice, and others), eat insects in significant amounts. This food is not accidental for them and is an important element of their food supply; the dormouse *Selevinia betpackdalensis* feeds only on tiny invertebrates. The so-called grain-eating birds (bunting, sparrow, finch, linnet, and many others) eat small invertebrate animals in large numbers during the summer, and their young are fed primarily on the same food. The "insectivorous" blue titmouse and small shrews transfer to the eating of tree and grass seeds in wintertime. At the same time the seemingly typical phytophages—horses, cattle, goats, and sheep—eat small invertebrates with their grass while in pasture, and these small animals are not without value in satisfying the larger animals'

2 Plankton are tiny, actively or passively swimming organisms which spend their entire life in the water itself (out of contact with the ground). Benthic animals live either on the surface of the bottom of a body of water or in the bottom itself.

demand for proteins. The northern reindeer eat fish rather willingly and even catch small animals.

Among the phytophages the zooplankton—small and very tiny water animals (protozoa, lower crustaceans, etc.)—and the larvae of fish feed on lower aquatic plants (algae and, to some extent, fungi). The lower fungi serve as an important component of the food of ixodid ticks and other arthropods, including certain bloodsuckers. They are eaten by larval and adult animals living in dung, nests, and burrows, usually referred to as detritus eaters. The reproductive bodies of the higher fungi are used by a wide circle of invertebrates (terrestrial snails, insect larvae, etc.), some of which are their parasites, since they live in the bodies of the fungi. Reindeer, squirrels, hares, and other mammals willingly eat the reproductive portions of fungi, although fungi usually constitute only a small, sometimes seasonal, part of the diet of these animals.

Green plants serve as food for many animals and are used in various ways. Herbivorous bugs, worms, aphids, and ticks suck plant juices. Many butterflies (adults), Hymenoptera (bees, bumblebees), and, among the vertebrates, brush-tongued parrots, hummingbirds, and certain marsupials (*Tarsipes*) feed on nectar; they usually serve as plant pollinators. Many animals feed on the supraterrestrial vegetative parts of plants, while some insects (especially their larvae), some nematodes, certain mammals, and even some birds feed to a greater or lesser extent on roots, bulbs, tubers, and other subterrestrial plant parts. Some animals destroy the plant completely (rodents and certain insects), others only gnaw or strip it (mollusks, gnawing insects, many vertebrates), a third group penetrates the tissues or leaves or drills into the woody portions, bark, or other parts of the plant (some of the Diptera, butterflies, sawflies, bark eaters, weevils, certain termites). An attack by certain nematodes, acarines, or insects against a plant leads to swelling of the tissues and scar formation. Eating of fruits, seeds, generative bodies, and plant spores is widespread; in some cases it may actually limit spermatic reproduction. Many insects destroy pollen, eat flowers, and harm fruits and seeds.

Among the zoophages the so-called "peaceful" species feed primarily on tiny, widespread animal foods which do not demand special effort in order to be captured (insects, plankton, crustaceans, worms, aquatic mollusks, and many others). Predators live on rather large prey (mainly vertebrates) which are usually more rare and therefore harder to catch. Finally, many parasites use their host not only for food but also as an environmental habitat and as a means of locomotion. Among the tertiary consumers, earthworms and many terrestrial and aquatic mollusks feed primarily or exclusively on dead plant remains, mainly on those microorganisms which

decompose these remains (detritophages). The coprophages use excrement, and the necrophages use dead bodies (remains) of animals.

Cannibalism, or the eating of one's own kind, is common among the predatory zoophages. For example, for the adult Balkash perch, the young of the same species actually prevail over other kinds of food as the main source of nutrition. In waters where only perch or pike live, the young may make up 100 percent of the food. Here cannibalism permits the predator population to use plankton, which would otherwise be unavailable to the adult fish. Cannibalism is regularly found among predatory birds whose young differ in body size. When food supplies are insufficient, the larger ones claw the smaller ones to pieces and eat them; during times of severe starvation the parents even eat their offspring. When water supplies are low and abnormally small supplies of food exist, some beasts also eat their own young. Cannibalism in dipterous larvae is expressed more and more strongly as the creatures grow more motile and their prey-seizing reflexes become developed. The highly motile pigmented larvae of the mosquito *Cryophyla lapponica* eat one another more often than do the larvae of *Chaborus cristallinus,* which are not very motile and are weakly pigmented. Cannibalism among dipterous larvae is distinctly connected with population density and with temperature and other conditions upon which motility depends (Monchadsky and Berezina, 1959).

Various types of feeding have historically determined the basic features of the structure and function of the animal body, making it possible for the animal to live and catch its food in a definite environment (aqueous, terrestrial, etc.). Sessile or motile way of life, solitary or group existence, and many other properties of animals depend on the means by which food is taken. This is hardly surprising, for feeding, as was stated above, is the basis for association with the environment, and consequently it is the basic function of the organism. Obviously, the morpho-physiological and ecological properties of a species cannot be connected only with feeding, but all of them bear the distinct stamp of nutritional specialization. In many cases this specialization has been the leading determinant in the historical formation of these properties.

The character of feeding is connected with the properties of fermentation. In the phytophages there is a quantitative prevalence of amylase with high fermentational activity, while in the typical zoophages their place is taken by protease, and the enzymes associated with reduction of carbohydrates are less strongly represented. Cellulase, hemicellulase, and lichenase, which digest plant cell walls, are absent in the majority of animals and are observed only in certain invertebrates. The latter includes mollusks (grape snails, sea snails of the genus *Aplysia,* and certain others

which attach themselves to the wooden parts of ships and marine installations) and certain crustaceans and protozoa (amoebae). Assimilation of these cellular components into the bodies of most animals, including the vertebrates, takes place with the aid of symbionts (protozoa and bacteria).

Highly specialized feeding patterns correspond to unique enzymes. In the adult locust, which feeds exclusively on carbohydrates, only the following carbohydrases are present among the enzymes: cellulase, amylase, anulinase, saccharase, and maltase (Nenyukov and Tareyev, 1937), while in nectar-sucking butterflies only invertase is present. Keratinase is found in the mallophagous insects (clothes moths, hair-eating mammal parasites, and feather-eating bird parasites); it breaks down the poorly soluble scleroproteins of the horny epithelium, skin, hair, and feathers. It is interesting that this same enzyme is found in the crop of certain predatory birds (*Accipiter gentilis, Aegipius monachus*). In the oral glands of many mollusks acids of high concentration are produced which participate in breaking down the carboniferous exoskeletons of food objects. Bloodsucking parasites (leeches, fleas, lice, ticks, bloodsucking Diptera) and vampire bats are abundantly supplied with proteolytic enzymes and substances which impede the coagulation of the swallowed blood (anticoagulants). The choice of enzymes is poorer in the parasites (cestodes and nematodes), and in *Trypanosoma evansi* (the agent of "surra" disease) even the usual lipase, amylase, maltase, lactase, pepsin, and trypsin, characteristic of the majority of animals, are missing.

A surprising property of many parasites and predators feeding on relatively rare and hard-to-obtain food, and also of some herbivores (like snails), is their ability to go hungry for long periods of time.

The eating of hard-to-digest matter, especially that rich in cell-wall structures, leads in all animals to a lengthening of the digestive canal and is accompanied by the development of supplementary compartments in it. The digestive tract of the hoofed animals is complicated; an important part of its work is accomplished by symbionts (protozoa and bacteria). In the predatory fish the length of the intestine slightly exceeds that of the body, while in the Trans-Caspian *Variocorhynus herafensis* (which lives on periphytons and detritus) the intestinal tract is seven times longer than the body (Fig. 11).[3] The posterior portions of the intestines of small grass-eating mammals are rather large, since it is in them that protozoal and microbic breakdown of cellulose is accomplished (Table 3).

[3] Periphytons are vegetative and sessile animals which accumulate on objects present in water.

TABLE 3. *Relative length of intestinal divisions in small rodents, depending on character of feeding (after N. P. Naumov, 1948).*

Species	*Small Intestine*	*Large Intestine*	*Cecum*	*Character of Food*
House mouse	74.3%	18.6%	7.1%	plant seeds
Yellow-throated mouse	65.5	25.3	9.2	plant seeds
Red vole	58.8	28.8	12.4	seeds and green parts of plants
Steppe lemming (*Lagurus lagurus*)	38.2	34.7	27.1	primarily green parts of plants

METHODS OF OBTAINING FOOD

The assortment of foods available to a species is closely associated with the species' methods of obtaining it, but these methods are determined by the structure of the body and the properties of the sense organs and nervous system.

Insects use all their sense organs but primarily the olfactory ones. The searching activities of insects and other invertebrates are represented primarily by nonconditioned-reflex acts. Female butterflies which drink nectar lay their eggs on food plants, apparently because they have retained from

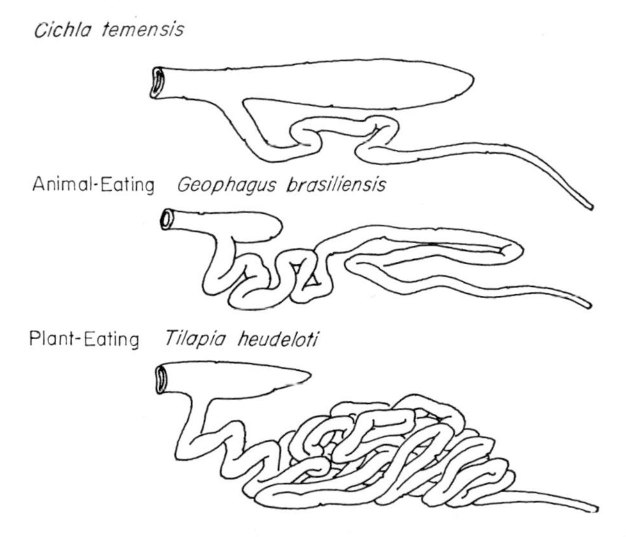

Fig. 11. Intestine of fishes of the family Cichlidae with different types of feeding (after Pellegrin, 1929).

caterpillar age a "memory" of the smell of these plants. The same thing is true of female wasps when they search for food for their young. Some part in the capturing of food by insects is played by conditioned reflexes. Thus dragonflies, which follow every flying object, quickly abandon it if it is not fit to eat; the initial reflexes are impeded, consequently, by controls from the higher centers. Bees trained on honey in bowls with colored linings fly toward those colors even when there is no honey. Even the extent of cannibalism among mosquito larvae depends on the temporary associations with the available food at the time. Thus the springtime larvae of the *Maclonyx* mosquito, which feed in nature on larvae of the *Aëdes* mosquito, in a laboratory experiment destroyed 63 percent of their own larvae, which are externally similar in appearance, while larvae of the second and third hatches of *Maclonyx*, which feed primarily on *Daphnia* in nature, destroyed only 4–8 percent of their own larvae because of this new food reflex (Monchadsky and Berezina, 1959).

In food searches made by aquatic animals the olfactory organs and the organs which detect mechanical fluctuations in the environment (the lateral line of fish), as well as the organs of sight, are of the greatest importance. Sound and touch are of lesser importance, although perception of ultrafrequency sounds plays an important role in the life of aquatic animals. In amphibians, reptiles, and birds vision is of the greatest importance. In birds it is developed especially strongly and is associated with the structure of the eyes (the presence of a "comb"). In L. R. Dice's experiments (1945–1947) an owl distinguished and quickly seized a model of an immobile bird but did not touch a model of a mouse unless it moved its feet. This can apparently be explained by the fact that most birds sleep at night and are immobile, while small mammals are active.

Smell, as we know, is poorly developed in birds. But sound, especially at night, is widely employed. The structure of the aural labyrinth of the bird is very closely associated with its methods of moving and catching its food. In good flyers and climbers the bony labyrinth of the inner ear is distinguished by its delicate walls and large semicircular canals. In climbing birds the longest semicircular canal is the lateral one, while in good flyers it is the upper one. In birds which catch their food on the ground the semicircular canals are short and thick (Levin, 1955). Perception of sound depends markedly on the structure and position of the feathers around the outer ear aperture, and this feathering serves to some degree as an analog of the outer ear of mammals (Il'yichev, 1960). The sense of smell is especially well developed in snipes, ducks, flamingos, and other species which search for food in the soil, silt, or similar locations. Their sense of smell enables them to distinguish edible from inedible material.

The classic works of I. P. Pavlov showed that mammals recognize food

through all their sense organs but that vision and smell are of special importance to dogs. Any phenomenon corresponding in time and space with the appearance of food may serve as a food "signal." The formative temporal association (conditioned reflex) can be maintained if it is periodically reinforced. Mammals are characterized by quick formation and persistence of conditioned food reflexes as well as by a general complexity of food-catching activity. They make use of sight, smell, and other senses, depending on the circumstances associated with catching the food. In the case of easily noticed food and definite reference points, vision is of dominant importance. When seeking hidden food, rodents orient themselves primarily by the sense of smell; in experiments they have found seeds buried 8–10 cm deep and down to 25–30 cm (Yershova and Fal'kenshteyn, 1948; Sviridenko, 1951, 1954, 1957; Larina, 1952; Yershova, 1952; Slonim, 1954). Attempts to protect planted seeds by smearing them with repellents have led to the reverse effect, since rodents use the smell to seek out the seeds and eat them. Among the insectivores, which are oriented primarily by sight, scent is not very keen.

The great variety of feeding and food-catching methods in the animal world can be summed up in three basic types:

1. Passive feeding is characteristic of species with a sessile or little-motile way of life (sponges, coelenterates, sessile or little-motile crustaceans, certain worms, echinoderms, tunicates, *Amphioxus*, etc.). A low level of matter exchange, small food demands, and few metabolic wastes are typical of all of them.

Pütter's (1909) theory of exclusively or primarily osmotic feeding by many aqueous animals (by penetration through the body surface of water-soluble organic substances) has proved incorrect, although some animals (the protozoon *Chilomonas*, larvae of the *Aëdes* mosquito and of the oyster, and others) have been shown to obtain and absorb significant amounts of food in this manner (Skadovsky, 1955). Basically, passive feeding means seizure of food brought by natural or often artificial currents of water which are created by palpi, twitching epithelial filaments, or other adaptations. These adaptations are supplemented by "trapping nets" of varying structure in the individual organism (Fig. 12).

Animals with very small-meshed nets can even filter colloidal substances.[4] The intensity of filtration is very great in some species. Larvae of the *Anopheles* mosquito filter an amount of water in 24 hours which exceeds their body volume by hundreds of thousands of times, and the *Midia* mollusk, which is 30–40 mm long, filters 1,000 ml of water per hour (Voskresensky, 1948). These filterers create powerful currents of water and

[4] In *Urechis caupo* the size of the mesh is 0.004 micron.

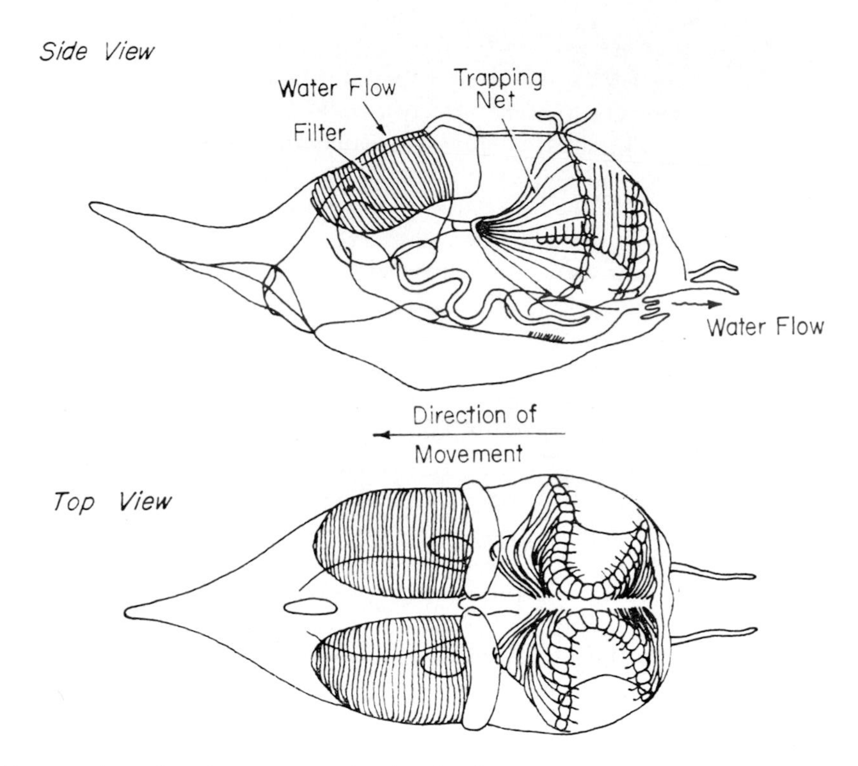

Fig. 12. Adaptations for water filtration in *Oicopleura*.

play an important role in the biological purification of such bodies of water (Skadovsky, 1955). Even some large animals approach filtration of water in their feeding habits. One such is the sperm whale, but its method of feeding is different; it actually deserves to be called a "herdsman" (see below).

2. Parasitic feeding in many respects approaches the preceding type but demands preliminary penetration of the host organism or an ability to maintain life on its surface. In the organs and tissues of the host or on its surface the parasite expends small effort in obtaining and even ingesting its own food, which makes possible the rapid growth and great reproduction rate necessary to compensate for the high mortality rate of the earlier developmental stages of the parasites. Simplicity of catching, lack of processing, and sometimes lack of digestion of food are the basic reasons for the simplification and often reduction of digestive organs in parasites.

3. Active feeding, characteristic of the majority of animals, is typical of the higher levels of consumption; it demands special efforts in searching for and catching food. The level and character of activity differ in the

various species and are usually related to the abundance and availability of food. Four basic forms of active feeding may be distinguished:

1. Pasturing (collection) is characteristic of the phytophages and "peaceful" zoophages living on the numerous immobile or little-moving and easily caught forms of food (woody and grassy vegetation, plankton and benthon, insects and other terrestrial invertebrates). In pastoralism only part of the food supplies in a given area are disturbed, making possible their quick re-establishment, while the searching out and catching of food amount to simple collection. In animals which have no permanent abode the so-called nomadic way of life arises from this situation. In this form of life the animals (usually groups of them—flocks or herds) wander over a large territory, changing pastures (hoofed mammals, sperm whales, many fish during the fattening period, wandering birds, etc.). Migrations, flocking (herding), and the absence of a permanent refuge have determined the speed of movement (trotting, flying, swimming) of such species as well as their well-known caution and group protection from predators.

2. Complete eating differs from pastoralism in that a relatively available, evenly distributed, and abundant food supply is used on the spot completely or almost completely, which leads to local destruction (eating up) of food supplies. This makes it necessary to change the place of feeding and to return only after the food supply has re-established itself. This method of feeding is characteristic of species whose individuals are attached to permanent dwellings (burrows, lairs, nests, etc.) and collect food on adjacent plots, which are often protected from penetration by other individuals of the same or other ecologically close species. Such is the behavior of many birds during the nesting period, of rodents, predatory and insectivorous mammals, lizards and snakes, solitary-living fish, certain crustaceans, insects, mollusks, etc. When food is abundant, there is no necessity for dividing it or for protecting feeding grounds (insects).

3. Watch-and-guard (ambush) feeding is typical of predators which attack their prey unexpectedly, usually from cover (among fish, the pike, *zherekh* [*Aspius* sp., a cyprinid fish—Ed.], and sheatfish; among birds, hawks and certain owls; among mammals, the cats). In the cats (Felidae), which use primarily their senses of sight and sound when hunting, the eyes and ears are especially large, while in dogs the nasal region and the olfactory organs are more developed. Unlike dogs, cats do not begin to salivate upon the appearance of food but only after it has been seized, since cats must sometimes wait and observe their food for hours before catching it (Ugolev, 1950, 1953).

4. Stalking, usually combined with searching, is the most active and complex form of food acquisition. It is widespread among birds and mammals, is usually distinguished by great changeability in its external forms,

and is often combined with such other forms of hunting as ambush and sometimes pastoralism.

The means and methods of obtaining food are more constant ("routine") in the lower animals, since individual experience plays a smaller role in their activity. In birds and mammals they are more changeable, but particular species properties exist in them too. The place and time of feeding depend on the methods of getting food. Thus the noble falcons (peregrine falcon, gerfalcon) hit birds in flight, striving to attack them from above like a fighter plane and striking with the posterior claw of the foot, which is kept solidly pressed against the body while on the attack. This kind of hunting can be carried out only in open spaces, which explains the presence of these falcons mainly in the tundra, the steppe, and the desert.

The hawks hunt by ambush and mainly in the forest. Hiding in branches, they attack the approaching prey unexpectedly. Stalking, if the ambush has been unsuccessful, is carried out over short distances. A long tail and short wings and well-developed steering feathers provide the necessary maneuverability in flight. The nocturnal predatory owls hunt by noiseless, furtive flight or by long observation of the prey. The invisibility of the bird's body combines with broad vision, made possible by the ability to turn the head 180° or more. Observation of prey during relatively cold weather is made easier by the heat-retaining properties of a dense and fluffy feather pattern.

Among piscivorous birds the kingfisher and the heron watch their prey, the former from an observation point (a stone, a bush) on the bank near the deep parts of the body of water and the latter standing on its long legs in shallow water. The stalking terns, with their long, sharp wings and forked tails, inspect their moving prey from the air while maneuvering. Seagulls fly slower and catch larger and less motile, often sick and decaying, fish; this results in transmission of fish worms to their definitive hosts, the birds (see Chapter 17). Cormorants and loons usually catch their food rather far beneath the surface, and the specific gravity of these birds' bodies is relatively great. On the other hand, it is small in the pelicans; these birds float easily on the water and catch small fish near the surface as if with a dip net, using their feet with a web of skin between the claws.

The shoveler ducks feed in shallow waters, getting their food from the silt. Their claws are supplied with a unique straining apparatus in the form of little ribs along the top and bottom of the feet. Vision plays no part in their search for food; thus they can feed while on the alert even at night, when the danger of predators is less. The *Nyroca* ducks feed on the benthos of the deeper waters, dipping 8–10 m for their food and using their vision to spot it. For this reason they feed in the daytime. Thus

herons feed on sandbanks, where there is no place for kingfishers; *Nyroca* needs a definite depth; and the birds which hunt over the water need an open surface. All of the above also determines the places where each species will hunt.

SPECIALIZATION OF ANIMAL FEEDING

The specialization of animal feeding is quantitatively characterized by the number of food species capable of being used by the animals. We distinguish monophagy, living on one species of food; oligophagy, living on several species of one biological type (seeds of woody plants, their vegetative parts, grassy plants, bulbs, insects, small birds, etc.); polyphagy, feeding on many foods of various biological groups; and, finally, pantophagy (omnivorousness), using all or almost all species of existing foods. There are clear-cut boundaries between the groups mentioned. Therefore, we prefer to speak of limited (or specialized) feeding, or stenophagy, and nonspecialized feeding with a large choice of foods, or euryphagy.

Monophagy is widespread among the invertebrates, especially the insects, and is almost completely absent among the vertebrates; it is often found among abundant species of genera, families, and orders. Thus among members of the subfamily Halticinae, which is composed of 125 species, 40 species are true monophages, feeding on only one plant; 30 species use several species of plants of one genus; 40 species feed on several species of plants of various genera; and 15 species live on a large number of different food plants. Monophagy is characteristic of the so-called biological races and closely related species; they are often indistinguishable morphologically but can be distinguished on physiological grounds because they feed on different foods. Such are the habits of *Carpocapsa pomonella*, which lives on apples and walnuts; of the plant nematodes, which live in onion, garlic, and other bulbous plants; and of the collective "species" *Anopheles maculipennis*, forms of which are known to attack primarily man or animals.

Monophagy is common among the endoparasites. The nematode *Cystoopsis acipenseri* infects only the sterlet [*Acipenser ruthenus*—Ed.], forming subcutaneous swellings on the sides of the fish's body. The branchiate fluke *Ancyrocephalus siluri* is found only in the sheatfish, while the parasitic gastropod mollusk *Parenteroxenus dogieli* is found only in the Far Eastern *Cucumaria japonica*. The paddle-footed crayfish *Xenoceloma brumpti* parasitizes only the polychaete *Polycircus albicans*. Specific par-

asitization (monozoid) is also observed in the Epicariidae.[5] It is very characteristic of gregarines and other parasitic protozoa.

Among the ectoparasites monophagy is encountered less often; most species of lice parasitize a definite host, and the fur and feather mites are also highly specific (Dubinin, 1956), as are the fleas. Fleas' specificity is associated with the structure of their oral parts, their sense organs, and the specific anticoagulants in their saliva, which are designed to prevent blood clotting only in certain hosts. Bloodsucking is quickly broken off if the flea feeds on the wrong host. In nature only specialized "burrowing" species of our fauna are monozoid: the fleas of the cliff swallow (*Ceratophyllus strix*), the mole rat (*Ctenophthalmus spalacis*), Prometeyev's mouse (*Ctenophthalmus inoratus*), and certain others (Ioff, 1941). But they are found in small numbers on predators, and in experiments they may also bite warm-blooded animals and man. Consequently, specificity in this case is determined not so much by physiological characteristics as by ecological conditions—a host living an isolated life.

Many species of mites of the suborders Trombidiidae (family Demodicidae and others) and Sarcoptiformes (family Sarcoptidae and others) are highly specific. The mange mites (*Sarcoptes*) may still move to another species of host. Among the ixodid ticks bi- and tri-hostal species predominate. But the tick *Boophilus calcaratus* parasitizes cattle in all its stages of development. Among vertebrates the palm eagle (*Gypohierax angolensis*) is found in Africa only where the palm tree *Elalis guineensis* grows; it lives on the fruit of this tree. Many hummingbirds and bristle-tongued parrots suck nectar only from flowers of a definite species of plant.

Oligophagy occurs more often than monophagy. Flukes, acanthocephalans, nematodes, cestodes, and many ticks change their hosts as they develop and are capable of using several species of hosts in one stage. The mosquitoes *Anopheles maculipennis* and *A. superpictus* feed on domestic cattle and cluster in large masses near settlements, ignoring the spaces between them. Many species of the family Culicidae (the majority of the *Ochleroratus*) and *Anopheles hyrcanus* suck primarily the blood of small mammals and birds; they are distributed uniformly and do not form clusters around herds of hoofed animals, although they attack them willingly (Beklemishev, 1944). Many Hemiptera, Hymenoptera, and Lepidoptera feed on the juices, nectar, and pollen of a limited number of plants.

Among marine invertebrates some live on plankton organisms and others feed on inhabitants of the bottom or on detritus with the bacteria and fungi living in it (chironomid larvae, Tubificidae, and mollusks). The

[5] A monozoid organism feeds only on one species of host.

rather large planktonal euphausid crustaceans, which play an important role in the feeding habits of game fish and the sperm whale, are distinguished by a definite ability to select their food; they take mainly copepods, which are tiny crustaceans (Fig. 13).

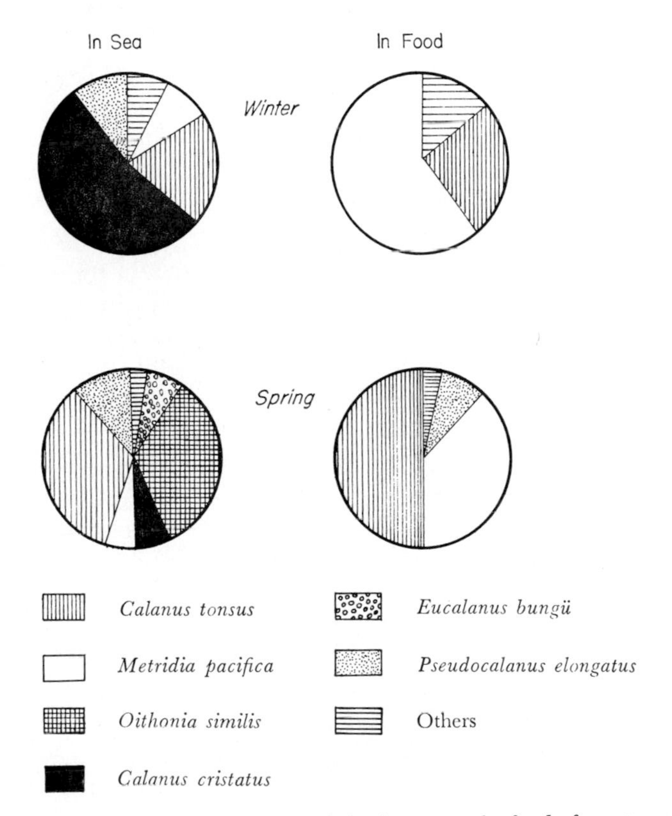

Fig. 13. Relationship of number of copepods in the sea to the food of crustacean euphausids of the Sea of Japan living at the 0-100-m level (after Ponomareva, 1955).

Among the fish of the northern Caspian, A. A. Shorygin (1952) has distinguished worm eaters (sterlets, bullheads, and long-tailed bullheads), in which worms comprise 96, 88, and 44 percent of the diet respectively; mollusk eaters (Caspian roach, *Beuthophilus ctenolepidus*, the Gobiidae), whose food consists of the dreissenid Adacna, Didacna, and Monodacna in amounts of 82, 100, and 52–54 percent respectively; crab eaters (*Abramis brama, Neogobius fluviatilis, Hyrcanogobius bergi*), in which the diet is made up of 54, 71, 69, and 91 percent crabs respectively; and predators (sevruga sturgeon, shad, sturgeon, zander, beluga, and others), with a diet consisting primarily of fish.

Among the amphibians *Rana aesopus* feeds almost exclusively on vertebrates (especially the toads *Bufo lentiginosus* and *B. fowleri*); *R. adspersa*, in addition to vertebrates, also eats earthworms. The majority of reptiles are oligophages. Among the snakes there are species and even genera known to feed on small mammals, birds, or reptiles (the Viperidae, *Ancistrodon*, and *Eryxmilioris*), insects, or other invertebrates (the family Typhlopidae and the broad-headed varieties of the family Amblycephalidae). The food of the African *Dasypeltis scabra* and snakes of the genus *Elachistodon* (India) is birds' eggs almost exclusively. The steppe tortoise (*Testudo horsefieldi*) eats only the juicy green parts of plants, which in Central Asia limits its active period to 2.4–3.0 months of the year.

Among passerine birds the feeding habits of the warblers (Sylviidae: *Acrocephalus, Locustella, Lusciniola, Cettia*, etc.) are the most specialized (primarily *Ephemera* and chironomids). The genus *Loxia (mariae, curvirostra, altaiensis, himalayensis)* feeds on the seeds of conifers. The beaks of these species intersect like the blades of a scissors and are specially adapted for getting the seed out of the cone. In different geographic regions, depending on the prevalence of pine, fir, or silver-fir seeds, the heaviness and dimensions of the beak are different (Fig. 14).

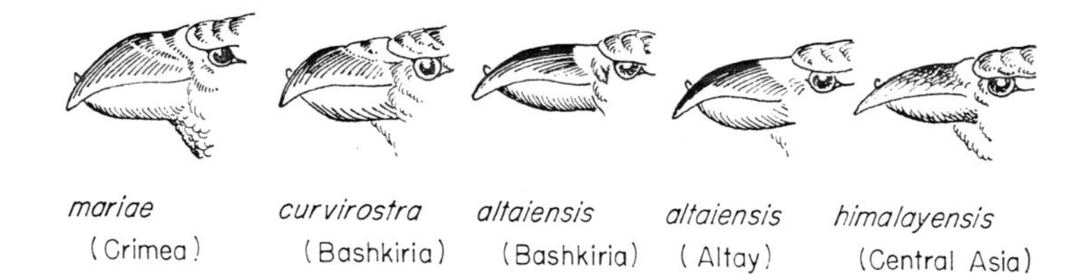

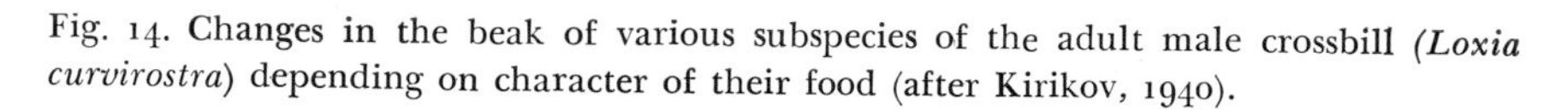

Fig. 14. Changes in the beak of various subspecies of the adult male crossbill *(Loxia curvirostra)* depending on character of their food (after Kirikov, 1940).

Oligophagy is seldom encountered among mammals. Bats, which feed on flying nocturnal insects, have a limited season of activity in our latitudes. Single-type feeding limits the active period of marmots, susliks, and dormice, which all go into hibernation when their food has disappeared. The abundance and geographic distribution of the yellow-throated mouse are limited by the abundance and distribution of three basic species of food plants—the oak, the lime, and the filbert. The Siberian lemming is affected in the same way by several species of *Carex*.

Euryphagy (polyphagy and pantophagy), or feeding on the maximum number of species possible, is often encountered among the animals of temperate and high latitudes. The corn moth (*Pyrausta nubilalis*) feeds on 160 species of plants. The list of foods for the meadow moth (*Loxostege*

sticticalis) and many other species runs to the same length. The food of young fish of the species *Dorosoma cepedianum* includes 140 species of animals and plants from the Myxophyceae, Euglenidae, Peridinea, Desmidiaceae, diatoms, protococcal and filamentous algae, branchiated and paddle-footed crayfish, rotifers, and protozoa. More than 40 species of animals and plants have been found in 1 cu. mm of the intestinal tract of one individual of the genus *Styela*.

Omnivorous or at least multivorous species are found among the frogs, while among the reptiles the water turtle (*Emys orbicularis*) and certain species of lizards are included in this category. Hawks of the genus *Buteo*, *Milvus korschun*, *Falco vespertinus*, *F. tinnunculus*, certain owls, and other little-specialized birds eat more than a hundred species of food: mammals, birds, amphibians, reptiles, fish, insects, mollusks, worms, and other invertebrates and even vegetable matter. The euryphages are most of the sessile or nonwandering birds (crows, daws, rooks).

Predatory mammals (martens, canines, bears) eat both animal and vegetable matter. In certain seasons vegetable food (fruits and berries) even prevails among them (Fig. 15). In the Siberian taiga an important food of many beasts, including carnivores, is the cedar seed (or "nut") (Table 4). In years when the cedar harvest is good (1925/26 and 1927/28) or when mouselike rodents are abundant (1926/27), the sable population is well provided for. An insufficient amount of both of these basic food species (1923/24 and 1929/30) leads to an increase in the trapping of sables and a drop in their numbers, since during a famine the sables are more mobile and catching them is easier. Several intersecting food groups

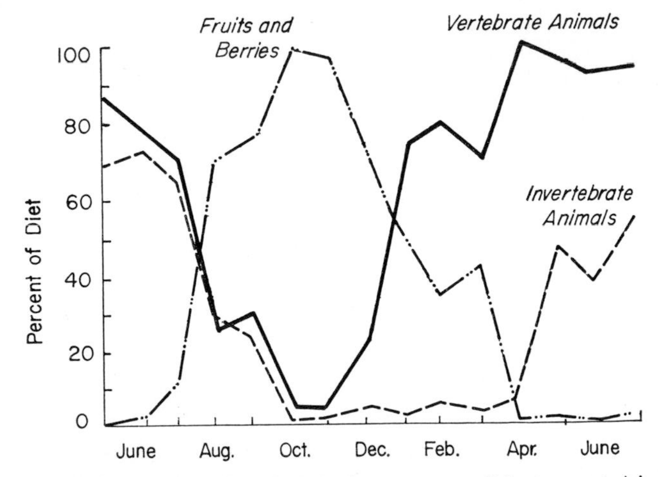

Fig. 15. Frequency of encounter of food of the forest marten (*Martes martes*) in the Caucasus National Park (after Donaurov, Teplov, and Shikina, 1938).

are also characteristic of the badger (Table 5) and the crow *Corvus corax* (Fig. 16).

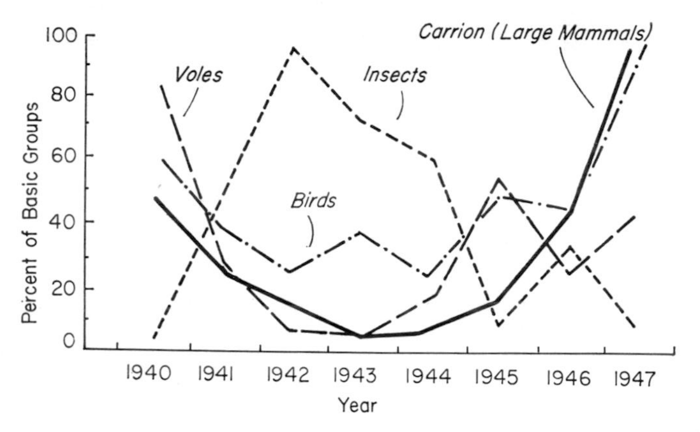

Fig. 16. Presence of basic groups in the food of the crow *Corvus corax* at Tul'sky Notch at various seasons of the year (after Likhachev, 1951).

TABLE 4. *Presence of basic food groups in the stomachs of sables* (Martes zibellina) *in the Barguza National Park in winter (after Folitarek, 1947).*

	YEAR						
Basic Food Group	*1923/24*	*1924/25*	*1925/26*	*1926/27*	*1927/28*	*1928/29*	*1929/30*
Small rodents	60.0%	75.0%	37.5%	78.6%	12.5%	15.4%	34.4%
Squirrels	10.0	25.0	25.0	2.4	33.3	23.1	15.6
Chipmunks	20.0	25.0	6.2	19.0	8.3	15.4	9.4
Belyak hares	10.0	0	12.5	7.1	12.5	30.8	3.1
Moles	10.0	0	0	21.4	8.3	15.4	9.4
Birds	10.0	0	18.7	14.3	12.5	15.4	18.8
Insects	0	0	6.2	7.1	25.0	15.4	6.2
Cedar seeds	20.0	50.0	62.5	35.7	79.2	53.8	68.8
Berries	0	0	6.2	9.5	4.2	7.7	3.1
Average number of sables taken by hunter	1.3	0.8	0.9	0.9	0.5	0.6	0.8

TABLE 5. *Presence of basic food groups in the stomachs of badgers* (Meles meles) *in the Tul'sky oak groves (after Likhachev).*

	SUMMER SEASON			
Basic Food Group	*1936*	*1937*	*1938*	*1939*
Mouselike rodents	59.4%	75.0%	85.0%	31.5%
Birds	46.9	16.6	6.4	68.4
Insects	85.0	65.0	50.0	95.0
Evaluation of abundance of mouselike rodents in oak groves	poor	average	large	very poor

Mutually replaceable full-valued foods, or vetches, are an important adaptation to existence in places with a changeable (nonresistant) food base. The possibility of such substitution explains the persistence of sable and badger abundance. As a rule, where a species has a long list of possible foods, only a few of them are full-valued; these are the basic foods. The others, casual or surrogate foods, can only maintain existence; they do not provide for growth or reproduction. In the multivorous fox as well as the Arctic fox the basic food is only a group of tiny rodents. When they are abundant, the foxes multiply intensively, their offspring survive well, and they increase in numbers. When rodents are absent, sterility rises sharply in the females and the size of the litters diminishes (Formozov, 1935; Chirkova, 1947, 1948, 1951, 1953).

The basic food of the common vole (*Microtus arvalis*) consists of the juicy parts of broad-leafed cereal grasses and bean plants. The steppe lemming (*Lagurus lagurus*) feeds primarily on feather grass and narrow-leafed cereals. Seeds and subterranean plant parts are of special importance in the food regime of the society vole (*M. socialis*), which is a dry-steppe and semidesert dweller.

Pantophagy may change depending on conditions. The common squirrel (*Sciurus vulgaris*), which eats more than a hundred species of foods during hot weather, confines itself to the seeds of one or several trees when snow is on the ground. The summer pantophagy of the Arctic fox is replaced in winter by feeding primarily on lemmings, and it sharply increases the level of casual (surrogate) foods in the diet. In years of hunger the euryphages typically have a more varied diet (see Tables 4 and 5). The list of foods consumed by a species throughout its entire range is always greater than the food list for an individual population. The common vole feeds on almost 500 species of plants, while the food list for individual populations rarely comprises more than 80–100 species.

The presence or absence of steno- and euryphagy is different. In stenophagy the digestive process is specialized, since it must digest one or a few species of food, and it is this specialization which leads to high effectiveness of food use. Digestion is more complex in the euryphages, especially when food is alternated. But stenophagy is possible only when a single or a small number of food species is in constant supply, as often occurs in the tropics with their steady climate and absence of sharply expressed seasonality. Stenophagy is more common among representatives of the rich and complex fauna of the tropical forests and is less often encountered in the temperate and high latitudes. Thus the average number of basic food components among the fish of the Aral Sea is eight, while in Lake Victoria in the tropics it is only three (Nikol'sky, 1947).

In zones with a less steady food base, a drop in feeding specificity is observed, while sometimes there is even an increase in specific foods. The common vole in the forest steppe and on floodplain meadows lives on 80–100 species of plants; the steppe lemming in the dry steppe and the complex semidesert feeds on 100–110 species of plants; the society vole in the semidesert uses 155 species (N. P. Naumov, 1948). An average of 20 species of higher plants has been found for 1 sq. m of meadow, while 18 species have been found for the meadow steppe, 11 species for the dry steppe, and only 8–9 for the semidesert. The ability to eat many foods thus makes existence possible in zones with an impoverished and unsteady food base, while stenophages cannot live in regions where their basic food is abundant only temporarily. Stenophagy testifies to the existence of sufficiently abundant and steady supplies of the basic food.

Food specialization has historically been subsumed under the influence of food concurrence. In studying feeding patterns and food concurrence among the fish of the northern Caspian, A. A. Shorygin (1952) differentiated the *amount* of concurrence, expressed by the similarity of foods in concurring species, from its *tension* (the relationship of concurrent consumers to the presence of that food which causes them to concur). The placing of volume of concurrence against tension of concurrence yields the strength of concurrence; the latter index serves as its most objective expression. Concurrence grows stronger as the number of claimants to the food increases, as their food lists coincide, and as the supply of food drops, as a result of which the consumers concur. Even among the same sets of species it may change in different years and at various times during the year (Fig. 17).

Coincidence of foods is usually higher in species of the same genus than among different genera. In the fish genus *Neogobius* coincidence among the individual species is 47 percent on the average, in the genus *Bentophilus* it is 39.5 percent, but between these two genera it reaches only 16 percent. The strength of food concurrence is especially great among closely related species.

A decrease in the strength of concurrence is achieved historically either by demarcation of pastures (spatial separation) or by changing the assortment of foods (nutritional specialization). Both forms of adaptation are characteristic of groups rich in species, such as the insects, and are especially frequent in complex biocenoses, such as the tropical forest, where one observes the highest percentage and the most surprising cases of narrow food specialization. Reverse examples also exist; narrow specialization of feeding occurs among certain insects of the taiga and tundra which live in associations poor in species.

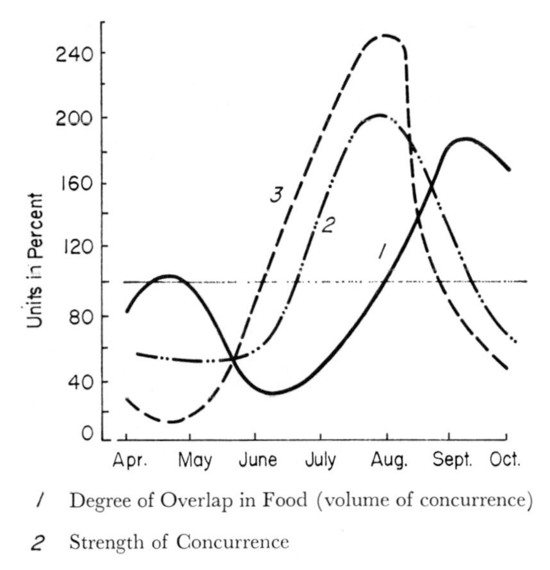

Fig. 17. Seasonal changes in the food relationships of the bream and the Caspian roach in the Caspian Sea (after Shorygin, 1952).

FOOD SUPPLY AND VIABILITY, SURVIVAL, AND REPRODUCTION OF INDIVIDUALS

Demands for food are not constant among animals. They change with age, with the season of the year, and even with the hour of the day. Not only the amount demanded but its content may change. At certain seasons, especially in autumn and spring, organic and mineral curative and tonic substances or growth and development stimulants may acquire great importance. Demands change also in relation to the circumstances of the animal's metabolism. The necessity of balancing intake and expenditure of matter and energy causes a raised demand for food in all cases where a decrease in surrounding temperature, a decrease in humidity, a rise in wind velocity, or many other phenomena causes an increase in the loss of heat, water, or other substances. The state of the animal, and consequently its chances to survive and reproduce, depend on satisfaction of the constantly changing demands for food, drink, and mineral matter.

Animals must have protein, carbohydrates, vitamins, and other substances which provide for growth and development. Certain mineral sub-

stances are also required in insignificantly small amounts (microelements), but even they are indispensable, since they enter into the skin pigments and participate in many important biological reactions. Finally, animals obtain a significant part and sometimes all of their water in their food.

A shortage of food may be the result of a "poor harvest," nonavailability, or growth in the number of consumers. A general lack of food is called caloric hunger. When widespread, it causes depression of the population, movement of individuals, and ultimately death and a decrease in abundance. The quality of the food—its vitamin content, the relationships of its various organic and mineral substances, and its ability to respond to the demands of the species—plays a great role. Thus the survival time and development period of the moth *Ephestia elutella* are best when it feeds on the embryos of wheat grains and poorest when it feeds on beans (Table 6). *Ephestia elutella* does not develop unless its diet contains riboflavin, while a closely related species, *E. kühniella,* also lives on food without this vitamin but grows more slowly (Fraenkel and Blewett, 1946).

TABLE 6. *Length of larval development and survival of larvae of* Ephestia elutella *when fed on various foods (25° C and 70 percent relative humidity) (after Waloff, 1948).*

Food	*Percent of Larvae Surviving*	*Length of Caterpillar and Chrysalis Phases (excluding diapause) in days*
Wheat	87	50
Wheat embryos	100	42
Tobacco	10	120
Figs	10	134
Cocoa beans		
Beans	6	diapause never completed
White flour	25	137

Fecundity and length of life of the Australian *Thrips imaginis* are good when the animal can feed on plant pollen and poor when it cannot (Table 7).

The Meloidae and the Cerambicidae in the adult stages need plant pollen (Brues, 1946). The blowfly *Lucilia, Calliphora* flies, and *Musca domestica* normally multiply only if they can eat animal protein and do not lay eggs at all if maintained on cane sugar and water. The bloodsucking Diptera (mosquitoes) behave the same way. The aphid *Brevicoryne brassice* lays an average of 445 eggs per female when raised on cabbage planted in bright light and rich in protein, while those fed on cabbage raised in shadow and poor in protein lay an average of only 160 eggs in the same 24-day period (Evans, 1938). All insects need vitamins of the B group

TABLE 7. *Influence of various foods on fecundity and life span of the adult* Thrips imaginis *(23° C) (after Andrewartha and Birch, 1954).*

	NUMBER OF INDIVIDUALS		
Food	*At Start*	*Remaining Alive after 20 Days*	*Average Number of Eggs Laid by 1 Female in 20 Days*
Leaves of *Trifolium repens*	22	2	0.8
Leaves of *Trifolium* + pollen	19	15	158.0
Leaves of *Plantago lanceolatum*	19	3	0.2
Leaves of *Plantago* + pollen	16	9	156.0
Leaves of *Antirrhinum*	9	9	0.3
Leaves of *Antirrhinum* + pollen	7	7	120.0

except *Lasiodermys, Stegobius,* and *Oryzaephilus,* which obtain these vitamins from symbiotic microorganisms living in their bodies. *Dermestes,* incapable of using plant sterols, can live only on food of animal origin.

The caterpillars of the Siberian bombyx *Dendrolimus sibiricus* grow well on larch cones, less well on silver-fir cones, moderately well on cedar cones, and poorly on fir or pine cones, regardless of which cedar, larch, or fir population is selected for the experiment. On larch cones the caterpillars attain their greatest dimensions and spin the heaviest cocoons (up to 6 g), and the resulting moths are the most fecund (an average of 826 eggs per moth). Those fed on cedar cones are smaller in size, and the moths are only half as fertile (400 eggs). On fir or pine cones the caterpillars do not develop, the majority die, and the resulting moths are distinguished by their very low fecundity (Boldaruyev, 1958). The development and survival of the unmated bombyx also depend on its food (Table 8).

TABLE 8. *Development time and survival span of caterpillars of the unmated bombyx (after Getsova and Lozina-Lozinsky, 1955).*

	MAY		JUNE	
Food Plant	*Average Length of Development in days*	*Percent Pupated*	*Average Length of Development in days*	*Percent Pupated*
Oak	28	46.6	43–60	46
Bird cherry	30	30.0	died in stage II on 14th day	–
Birch	33	34.0	46–70	18
Willow	died in stage III on 20th day	–	48–60	10
Lime tree	died in stage II on 13th day	–	44–64	18
Alder	–	–	48–60	16

Substances which hinder or slow down development may also be very important. Development of the larvae of *Melanophila californica* depends, apparently, on the state of some subtle properties of the chemistry in the trunk of the ponderosa pine, in the bark of which the insects develop. In a sick or weakened tree the whole developmental cycle ends in the course of one summer, while in a healthy tree it slows down, and the larvae go almost unfed and live in the passages near the cambial layer for up to 4 years. If the tree remains healthy, the larvae die in the first phase of development; but if it weakens, the larvae, regardless of age, began to feed more actively and by autumn finish their development (West, 1947). The larvae of the beetle *Tribolium confusum*, which feed on wheat flour rich in carbohydrates, end their development in approximately 28 days, but they stretch it out to 32–50 days if they are cultured on a substrate of yeasty or meaty flour which is poor in carbohydrates. On the other hand, the larvae of *Ptinus tectus* develop more slowly when there is a carbohydrate surplus in their food; development speeds up by 20 percent if yeast is added to the flour. When well fed at high temperatures, *Daphnia* live an average of 30.9 days and lay 200–300 eggs in their lifetime; but when hungry, their life span stretches to 38.6 days, and the number of eggs drops to 74.5. *Daphnia* which have gone hungry through the twelfth and even the eighteenth stages reach normal growth when their food is improved and give birth to the same number of young, but hunger which follows proper nutrition leads to a shortening of life and a drop in fecundity (Skadovsky, 1955).

The mouselike rodents (mice, voles, hamsters) grow quickly and reach sexual maturity, sometimes within a month, as long as they feed on such full-valued foods as plant seeds. In this case they bear large litters, which follow one another at intervals of 35–40 days. But these same creatures grow slowly and multiply much less, or not at all, if their food is in poor supply or is not full-valued (lack of seeds, insufficient moisture in the food). In such years even the spring litters do not attain sexual maturity by autumn, and they begin to reproduce only the following spring at the age of 10–12 months; and even in these litters there are fewer offspring. An autumnal decrease in abundance and a deterioration in quality of the food supply have the same effect on late-summer and autumnal litters of small rodents; they multiply poorly and enter hibernation without having reached sexual maturity. But in a warm autumn, with massive ripening of new seeds, feeding on these valuable and vitamin-rich foods accelerates the maturity of the young, even those from late litters, and they reproduce in the autumn of the same year and often continue doing so throughout the winter (Formozov, 1937; N. P. Naumov, 1948). Regardless of weather conditions, the multiplication of small rodents continues all winter in

haystacks, straw piles, and piles of unthreshed grain. The body temperature of common voles living in haystacks during the winter proved to be 1.0–1.5° C higher than that of creatures living in fields and burrows (Strel'nikov, 1940; Kucheruk and Rubina, 1953).

Depending on the seed "harvest" from woody plants, the fecundity of squirrels, crossbills, nutcrackers, and jays may be affected (Formozov, Naumov, and Kiris, 1934; Kirikov, 1940; Kiris, 1944, 1947, 1956, 1958; Siivonen, 1948, 1957; Kalela, 1954; Lek, 1957; and others). The intensity of multiplication and the success and length of development in young red and Arctic foxes also depend on the populations being supplied with full-valued food (Chirkova, 1948, 1951, 1953; Shibanov, 1952).

Reproduction of animals may be stimulated by the appearance of special substances in their food. M. Kh. Friedman (1939, 1941) isolated a substance from cereal grains which had the ability to stimulate the sexual organs and thus the multiplication of rodents in the manner of the gonadotropic hormones of the anterior lobe of the pituitary. But I. T. Broadbury (1944) has shown that this substance has nothing in common with hormones and is actually a neurotoxin which stimulates the activity of the pituitary. It appears in nature periodically, for short periods of time, and may be the cause of outbreaks of multiplication in small field rodents. The appearance in moldy lucerne hay of substances from the dicoumarin group which hinder clotting of the blood (anticoagulants) may be the cause of mass murrains among animals which feed on such hay (Lik, 1937).

Numerous ichthyological studies have shown a clear relationship between growth, maturing, and fecundity of fish and supplies of their food groups. The direct or tangential connection between these phenomena and many other factors (temperature, salinity, pH, population density, etc.) shown by the same investigators does not take away the prime importance of food. The immense field of raising domestic animals also recognizes the importance of proper diet in the attainment of maximum productivity. Not only a general insufficiency of food or a qualitative deficit in the food but even overfeeding, especially of food which is not full-valued and has not been supplemented and tested, leads to a breakdown in the organism's metabolism and development, a drop in its productivity, and a decrease in fecundity, perhaps even sterility.

In changing the metabolic and physiological state of the animal, various diets also cause changes in its behavior—the character of its daily activity, preferred temperature, and other phenomena. This indicates the profound and many-faceted ecological importance of food supplies. Such knowledge is very important when animal mobility increases during famines and when mass outflows of animals occur from regions and areas suffering from lack of food. Finally, hunger, and especially the avitaminoses, are accom-

panied by lack of resistance to infections (decreased immunity) and the spread of illness.

All of these phenomena, including a drop in fecundity and, in extreme cases, cessation of reproduction, cannot be regarded as mere consequences of insufficient food. They are not merely a direct and passive reaction on the part of the animals to a shortage or surplus in the food supply. These phenomena also have an adaptive meaning, since they are underlain by biological mechanisms which have developed out of the process of natural selection. The biological importance of decreased fecundity, decreased growth, diminution of the size of individuals, and increased mobility when food supplies are low is completely clear, since they decrease the danger of mass starvation, and in many cases the value of one individual's death is the survival of the others. As a rule, all these phenomena arise and develop before the existing supply of food is used, since their basis is to be found in those signal connections with the environment which were described earlier. But because of this, the natural food supply is never or only rarely used up (C. S. Elton, 1938).

The degree to which animals use the food supply in nature is determined first of all by the abundance and searching ability of the species (Fig. 18). But it also depends on the availability of the food and the concrete conditions under which it is caught. That which is often called the "biological supply" of food (unusable leftovers) is, in reality, the part which either exceeds the demands of the few consumers or is unavailable because of the difficulty of getting it.

Effective use of food is often made more difficult by the distances of the

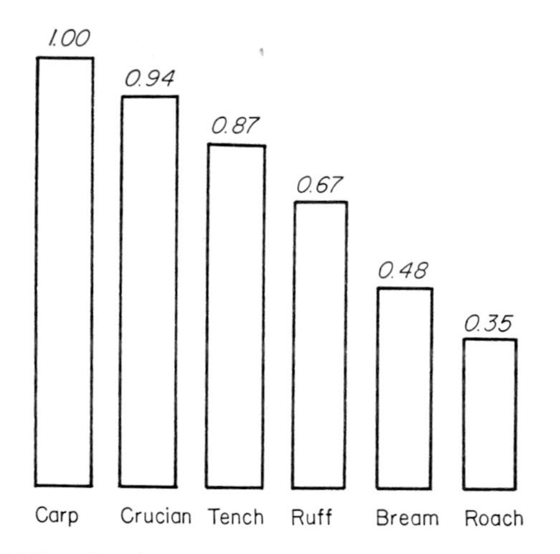

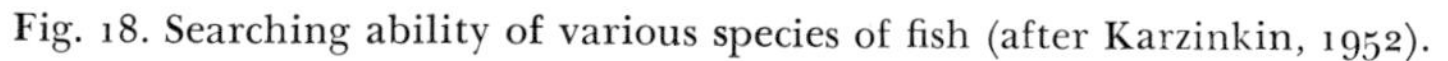
Fig. 18. Searching ability of various species of fish (after Karzinkin, 1952).

territory where it grows, so that expenditures in searching are not rewarded by results. The hymenopterous hyperparasite *Lygocerus* lays its eggs in the larvae of hymenopterous *Anorophus sydneensis,* which parasitizes the caterpillar of the butterfly *Pseudococcus longispinus.* When the abundance level of *Pseudococcus* is up and its caterpillars are not heavily infected with *Anorophus* larvae, *Lygocerus* must inspect a large number of caterpillars before it finds infected ones in which to lay its eggs. This hinders egg-laying and, in the final account, limits the role of *Anorophus* in aiding the spread of *Lygocerus* (DeBach, 1949).

The Australian *Thrips imaginis,* which feeds on flower pollen, multiplies intensively in spring when there are abundant flowering plants; up to 190 insects have been found on one flower. In summer during a drought, or in winter when flowering is poor and the plants in flower are far removed from one another, the chances of landing in the flowers are sharply diminished, and the pollen, contrary to what one might expect, is used to a lesser degree than when it is readily available. The result is that an average of only five thrips is found on a flower (Andrewartha and Birch, 1954). Sporadic distribution of food (flowering plants) hinders its use in this case. The same situation obtains for squirrels, crossbills, and other seed eaters when the trees are suffering a poor harvest and only a few of them are producing seeds. In this case the small number of seeds is just about equal to none at all. The effectiveness of the food in bringing about growth is lost in energy expended on searching for it (Fig. 19).

Assimilation of food is also determined by physical conditions and the

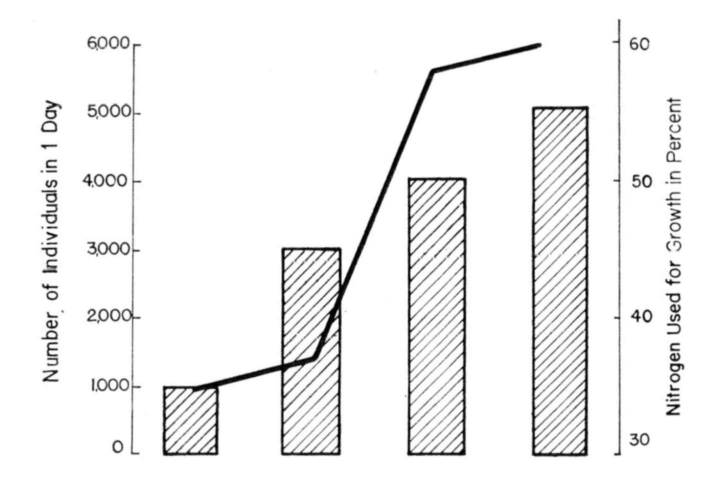

Fig. 19. Use of nitrogen in food for growth (curve) in the young sevruga sturgeon, depending on concentration of food organisms (columns) (after Karzinkin, 1952).

state of the organism. The Caspian roach's "assimilation" of the dreissenid mollusk in the spring-summer period at a water temperature of 20° C is approximately 31 percent, at 18° C it is 25 percent, and in winter, at a temperature of 10° C, it assimilates 29 percent. But nitrogen is assimilated at corresponding rates of 85, 84, and 80 percent (Bokova, 1939). Naturally, consumption of food intended for increasing the body's mass is not identical in the different species (Fig. 20).

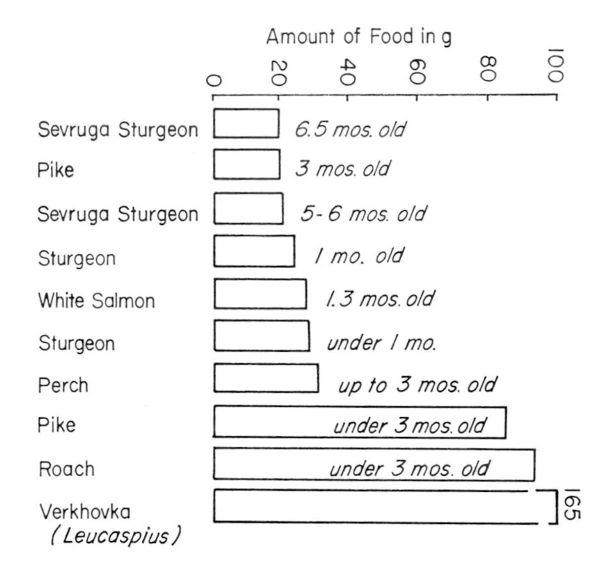

Fig. 20. Amount of food consumed for producing fish offspring per g of "raw protein" (after Karzinkin, 1952).

In other cases an abundance of food may hinder its acquisition. Rotifers feed intensively on phytoplankton when there are 4,000–10,000 cenobia or alga cells per ml. Under conditions of greater density their motility drops, and general depression sets in (Fig. 21). The temperature optimum for feeding is equally distinct for the rotifers (Erman, 1956) (Fig. 22). The density and character of the food supply's spacing are especially important to animals with limited mobility. The leaf-gnawing beetle *Chrysomela gemellata,* which feeds on the leaves of *Hypericum perforatum,* destroys the entire food supply in the spot where the colony is situated, after which the beetles migrate. But because of their poor mobility, they cannot go farther than 60 m. If no food is found at this distance, the colony perishes, while supplies of sporadically distributed food, though they be abundant, remain unusable (Clark, 1953).

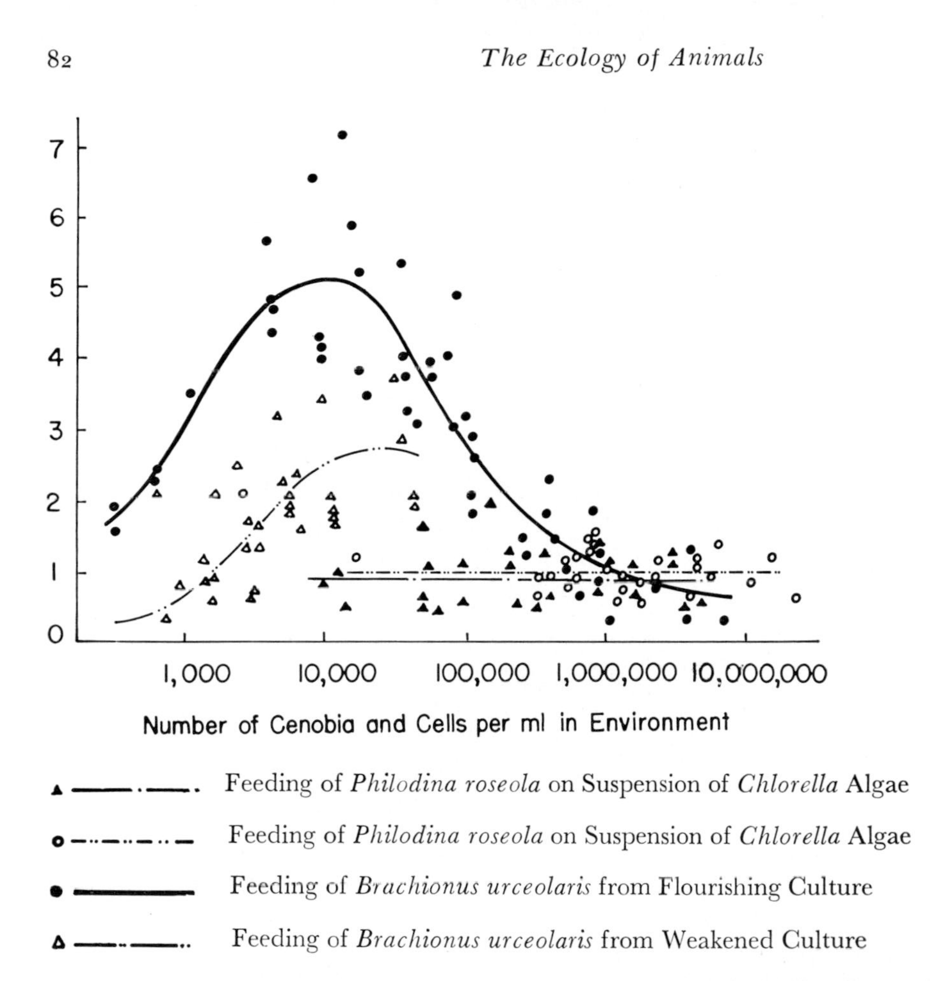

Fig. 21. Intensity of feeding in rotifers in varying concentrations of food (after Erman, 1956). Scale along horizontal axis is logarithmic.

The stereotyped behavior and the food connections based upon it (such as the above-mentioned example of *Maclonyx* mosquito larvae) exert a great effect on the degree to which available food will be used. Predatory birds and mammals usually get the most available food, concentrating their hunting activities in places where it is abundant and almost abandoning their activities in other territories. Thus other kinds of food are sometimes ignored completely. When the supply of one food drops and another one appears, the switchover does not occur at once, and for some time the species feeds on its accustomed food, continuing to search for it in the familiar surroundings. This so-called "selectivity" of feeding in many cases has the character of "feeding inertia" such as is found in the

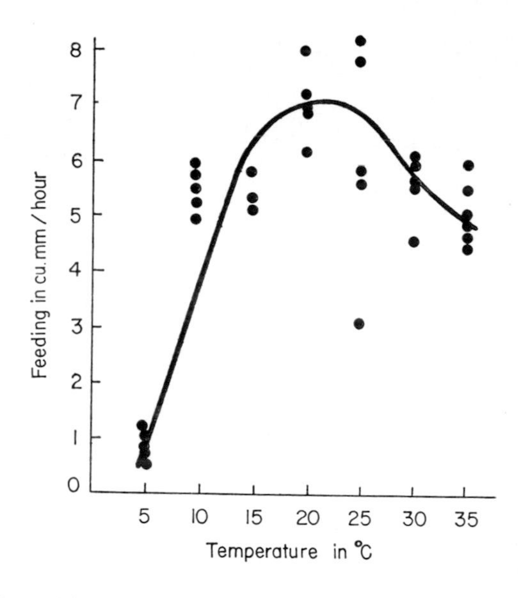

Fig. 22. Feeding of *Brachionus* at various temperatures of the environment (after Erman, 1956).

plant-eating species and which is associated with the preserved stereotyped behavior (Grigor'yev and Teplov, 1939; Osmolovskaya, 1948, 1949).

Availability of food changes under the influence of weather. In temperate and high latitudes the seasonal drop in temperature and the falling of snow are of great importance (Formozov, 1946). Unfavorable weather, which suppresses the activity of many species, curtails the seeking and obtaining of food. The relationship between intensity of feeding and the surrounding conditions is especially strong in poikilothermic animals, whose activity is determined by the temperature and humidity of the environment, and in water animals, in which the intensity of food absorption is governed directly by the pH of the environment (Fig. 23).

Thus the state of the animals and their chances for existence and reproduction are governed not by the absolute food supplies existing at a given time in nature but by that part which proves available and can be obtained by the animals. This part depends on the abundance and character of the food's distribution, the degree to which it is hidden, the means of searching it out and obtaining it employed by the different species, the conditions under which the species hunts its food, and, finally, those temporary associations which arise among the animals and control the local features of their food-getting activity.

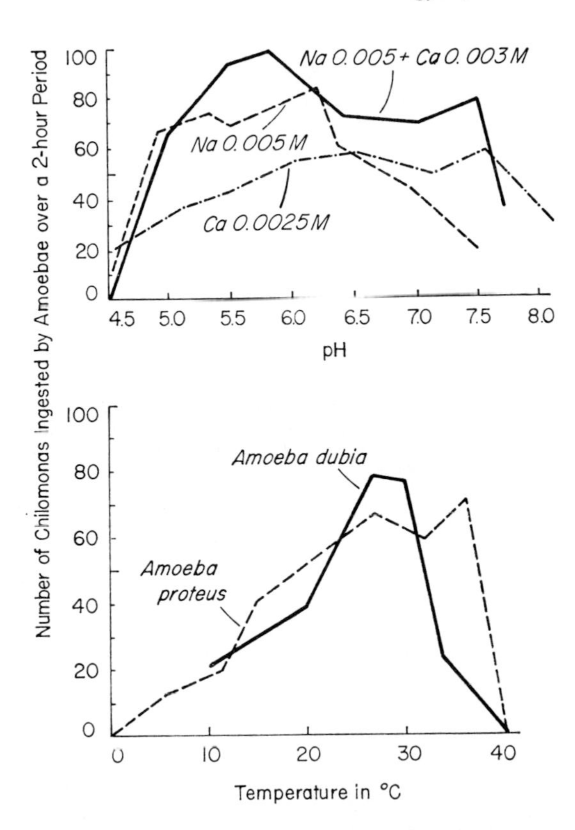

Fig. 23. Phagocytosis in amoebae in relation to pH and to temperature of the environment (after Mast, from Skadovsky, 1955). At various pH's and with different contents in the environment.

The ultimate result of feeding—successful development of the animals and their preparation for reproduction—depends also on the degree to which they assimilate the absorbed food. This changes with the qualitative content and the relationship of the organic components in the diet (especially vitamins), the content of moisture and mineral salts, and the external conditions under which digestion occurs (temperature of the air and other environmental factors). All this testifies to the complex and varied influence of food supplies, the most important condition of life, on the state of separate individuals and populations and underlines the dependence of feeding (successful use of food supplies) on many environmental factors.

CHAPTER 3

Water-Salt Exchange and Mineral Nutrition

GENERAL IMPORTANCE OF WATER

Living creatures first arose in an aquatic environment. Its physico-chemical properties govern the exchange of matter in hydrobionts, all the features of construction and biology of which are adapted to life in the water.[1] Water serves as a means of distribution (flow) for many species. The water in an organism is the basic medium in which the biochemical processes of both aquatic and terrestrial animals take place. It forms colloidal systems—the protoplasm. The amount of water and the amount of salts dissolved in it govern osmotic pressure and the status of the other body fluids (blood, cavity fluids), determining inter- and intracellular exchanges. Nutritive substances and products of dissimilation are transported in the form of aqueous solutions. Only the water envelope can give rise to absorption of oxygen and excretion of carbon dioxide in respiration; the sense of smell works the same way. Finally, evaporation of water from the surface of animals, which demands 530 calories per g, effectively lowers the animal's temperature.

The water content in the body fluctuates between 46 and 92 percent in insects, between 80 and 92 percent in mollusks; it is 84 percent in the sponge *Chondrosia reniformis*, 77 percent in the river crayfish, 87 percent

[1] Hydrobionts are animals and plants living in water and wholly adapted to existence in this medium. Usually they are distinguished from the aerobionts, inhabitants of dry land, and the pedobionts, inhabitants of the soil. Amphibionts are those species which lead part of their lives in water and part on land.

in the lancelet, 93 percent in the frog tadpole, and as high as 95 percent in the jellyfish *Rhizostoma cuvieri.*

A constant exchange of water and substances dissolved in it takes place between the body of the animal and the external environment, accompanied by a change in the amounts of each in the body of the animal. Decrease of the water content below a known limit leads to breakdown of the animal's activity and even to death. Water losses are more serious than hunger. A starving human being can, without threat to his life, lose up to 40 percent of his body weight, including half his proteins and almost all his carbohydrates and fats, but he suffers disturbance of his functions if he loses as much as 10 percent of the water in his body and will die if he loses 20 percent. The acquisition and loss of water and mineral substances are accompanied by changes in the level of hydration of the colloids and of the osmotic pressure of the body fluids. These two basic phenomena, along with changes in the ion state of the electrolytes, change the conditions of intra- and intercellular matter exchange by altering the permeability of the biological membranes and the electrical state of the cellular structures. Large changes in these intimately related phenomena disturb growth, development, and the entire life activity of the organism, which explains the necessity for a definite maintenance of the state of the internal water environment. This is also bound up with the essence of water-salt exchange, which, as is clear from the above, is an inseparable part of matter exchange.

The sources from which water and salt are obtained are penetration through the circumambient medium (in an aquatic environment),[2] absorption of moisture from the air (in amphibians, insects, mollusks, worms, and many others), drinking water, water and minerals in the food, and metabolic water (which is formed during the oxidation of compounds, especially fats, in the body). The body gives up water and salts by evaporation from the surface of the respiratory organs or the skin, by excretion from the sweat glands and other glands, and through the special excretory organs.

The necessity for acquiring, dispensing, and maintaining water in the body, like the necessity for regulating the amount and content of dissolved mineral substances, is identical for terrestrial animals, soil inhabitants, and hydrobionts, although the character and mechanisms of water-salt exchange are different for these three groups of animals. Physiological dehydration may arise even in bodies of water, while harmful surpluses of moisture can occur on dry land.

2 So-called "polar electrolytes" penetrate through the biological membranes only with difficulty and very selectively.

WATER-SALT EXCHANGE IN HYDROBIONTS

Water contains all the things necessary for construction of the bodies of plants and animals. They are extracted from solution and, with the aid of photosynthesis, are directly assimilated by aquatic plants. Animals use them through the medium of plants, but they also acquire part of them through osmosis. Certain substances are accumulated in the bodies of animals and plants. Compared with seawater, the body of an aquatic organism contains approximately a thousand times more phosphorus, silicon, and zinc, a hundred times more sulfur, iron, copper, and iodine, and ten times more potassium, boron, and fluorine. The silicon compounds are found in great amounts in diatomic water plants (up to 19–20 percent of the raw weight of *Chaetoceras*), while among animals the radiolarians and the sponges (which have a silica skeleton) contain up to 88 percent. Calcium, especially lime compounds, is accumulated in the body of certain kinds of bacteria (*Bacterium calcis*), algae, rhizopods, many mollusks, brachiopods, echinoderms, anemones, and corals. Calcium oxide makes up 53 percent of the body weight of the *Madrepora* coral. We know of certain "iron" organisms which contain as much as 20 percent iron. Finally, certain sponges, the *Gorgonia* corals, and many algae are rich in iodine and are used industrially to obtain it.

Selective extraction of substances from the water changes the water content and leads to the accumulation of organogenic precipitates in the form of silt and sapropels (limes, flints, etc.). The salt regime of a body of water is determined by the content and amount of substances dissolved in it at a place where they are capable of being brought in by natural flow and by the activity of its living population. The relative stability of the salt content of many bodies of water is associated with the latter.[3]

In natural bodies of water the primary salts are those of the nitrogen, phosphorus, sodium, carbon, and other acids. Pure solutions of potassium, sodium, calcium, and magnesium are toxic to organisms, but when mixed in definite proportions, their cations mutually exclude the poisonous effects of one another. The water in most natural bodies of water is such a balanced "buffer" mixture. At the same time the level of metal-ion equalization depends on the relationship in solution of salts of the mono-

[3] According to L. S. Berg's (1908) calculations, the amount of salt in the waters of the Aral Sea could have been put there by the Syr Darya and the Amu Darya in a mere 300–400 years. Actually, however, it has existed as a landlocked basin for no less than 2 or 3 millennia. Deposits of crystalline salts in estuaries and their removal by the winds, on the one hand, and the use of salts by the living inhabitants of the basin, on the other, explain the absence of rapid salinization, which is unavoidable in the hot and dry climate of the Central Asian deserts.

valent (potassium, sodium) and bivalent metals (calcium, magnesium). This relationship between antagonistic ions (the ion coefficient; Leb, 1926) changes during fluctuations in the general salinity. In waters with little salinity it is 1.5 times less than in ocean waters (3.5–3.7 percent salinity), which have more effective cation buffering combinations. It is important to note that the saline content and the body-fluid ion coefficient of animals are close to those of seawater.

The maintenance of the osmotic pressure necessary for matter exchange and ion state of solutions in the bodies of animals is provided by regulating water-salt exchange with the environment. Two types of water animals are distinguished on the basis of the character of their salt exchange. In the majority of marine invertebrates (the poikilo-osmotic animals) there is no special mechanism for regulating water-salt exchange. In the others (the homo-osmotic animals) this regulation is well developed. Until recently the existence of osmoregulation in poikilo-osmotic animals was completely denied (Zernov, 1949). However, D. M. Belyayev (1951) showed that this is not completely true (Table 9). In normal seawater the majority of marine invertebrates are slightly hypertonic, while the salt content in

TABLE 9. *Degree of anisotonicity (lack of agreement between osmotic state of body fluids and state of external environment) in marine invertebrates (after D. K. Belyayev, 1950).*

	BELYAYEV'S DATA		OTHER AUTHOR'S DATA	
Animal Group	*Number of Species Studied*	*Average Degree of Hypo- (−) or Hypertonicity (+)*	*Number of Species Studied*	*Average Degree of Hypo- (−) or Hypertonicity (+)*
Coelenterata	–	–	5	0
Priapulida	1	+0.12	–	–
Annelida	6	+0.09	3	+0.080
Mollusca	15	+0.03	8	+0.060
Brachiopoda	1	+0.03	–	–
Cirripedia	3	+0.13	–	–
Amphipoda	13	+0.11	2	+0.040
Isopoda	2	+0.11	1	+0.100
Mysidacea	1	−0.25	–	–
Shrimp	–	–	3	−0.510
Most grapsoid crabs	–	–	11	−0.240
Other Decapoda	5	+0.08	18	+0.030
Xiphosura	–	–	1	+0.020
Echinodermata	–	–	8	0
Enteropneusta	1	+0.10	–	–
Tunicata	–	–	2	+0.004

Note: The degree of anisotonicity is evaluated by the magnitude of deviation from the freezing point, depending on the salinity of the body fluids as compared with seawater.

the body fluids of certain crustaceans (mysids, the shrimp *Palaemonia*, most of the grapsoid crabs) has proved to be lower than that in the environment (hypotonicity). Only coelenterates and echinoderms are isotonic with seawater.

The incompleteness of osmoregulation in the poikilo-osmotic species is expressed by the fact that raising the external salinity lowers the weight of the body by causing it to give off water, while decreasing the salinity of the water brings about an increase in body weight owing to inflow of water (Fig. 24). There is a parallel change in the salinity of the body fluids and a drop in freezing-point depression (Δ°), which reflects it. Most of the

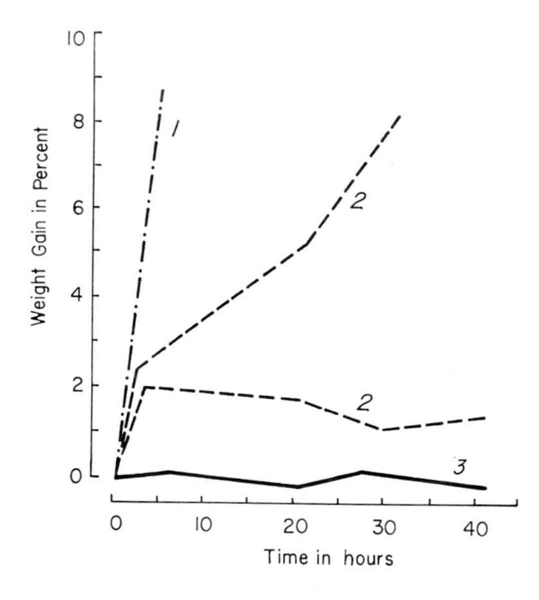

Fig. 24. Changes in the weight of *Hyas araneus* (1), *Cancer pagurus* (2), and *Carcinus maenas* (3) when placed in fresh water (from 32 to 20 percent) (after Skadovsky, 1955).

marine invertebrates react in this way (Fig. 25). It is natural that this type of salt exchange is possible only in bodies of water in which the salinity of the water is similar to that of the body fluids and fluctuates within narrow limits (Table 10). Many of the seas, and especially the oceans, are of this type.

In bodies of water with highly changeable salinity or with one that differs markedly from that of the body fluids, only species with osmoregulatory adaptations which maintain the internal osmotic state at the necessary level can live. In such homo-osmotic animals the salinity and the ion state of the mineral substances in the body fluids are not, as a rule, even

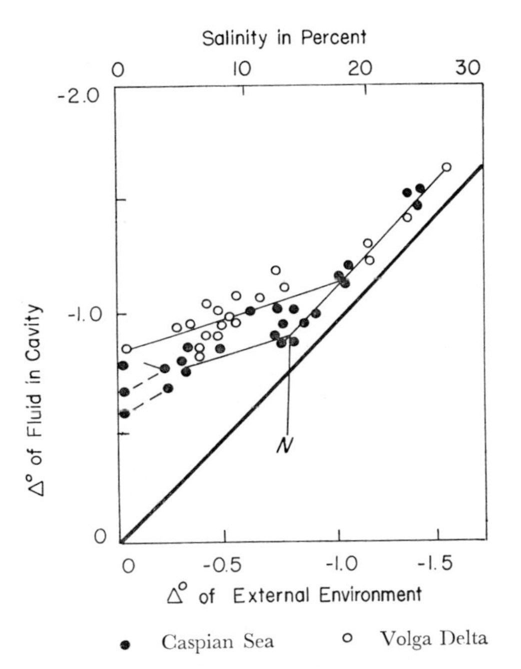

Fig. 25. Freezing-point depression of the blood of *Dikerogammarus haemobaphes* after 3 days in waters of varying salinity (after Skadovsky, 1955).

with those of the surrounding environment. This includes the freshwater and brackish-water invertebrates, inhabitants of "oversalted" waters, and all the aquatic vertebrates. They are divided into two groups: (a) hypertonic animals, inhabitants of fresh water whose body fluids have a higher osmotic pressure than the surrounding environment; (b) hypotonic animals, inhabitants of saline water with a smaller amount of salts in the body fluid than in the environment.

In the freshwater species maintenance of the normal osmotic pressure

TABLE 10. *Correspondence between freezing-point depression of seawater in different bodies of water and freezing-point depression of body fluids of animals living in them (from Zernov, 1949).*

Species	*Character of Fluid*	*Freezing-Point Depression in °C*	*Freezing-Point Depression of Seawater in °C*
Sea stars (*Asterias*)	coelomic fluid	2.312	2.290
Common lobster (*Homarus vulgaris*)	blood	2.292	2.290
Sea urchin (*Echinus esculentus*)	coelomic fluid	1.860	1.900
Scallop (*Pecten maximus*)	blood	1.905	1.910
American lobster (*Homarus americanus*)	serous fluid	1.780	1.760

is usually achieved by organs which excrete surplus water entering the organism. The frequency of the contraction of the contractile vacuole of the freshwater amoeba at 20° C, with different concentrations of sodium chloride in the water, was as follows (Kherfs, 1922):

Percent of NaCl in water	0	0.25	0.50	0.75	1.00
Contraction period in seconds	6.20	9.30	18.40	24.80	163.00
Amount of excretion per hour in body volume	4.80	2.82	1.38	1.08	0.16

Organisms with this adaptation cannot live in bodies of water with a salinity higher than that of their body fluid, since diffusion of water from the organism leads to dehydration of the cells and tissues. The absence of amphibians from marine waters is apparently explained by this phenomenon.

Maintenance of the salts necessary to the organism is also aided by the excretory system. In the river crab *Potamobius astacus* the excretory organs (the green, or antennal, glands) are more complex and better developed than in the marine species (Fig. 26). The urine formed in the coelomic cavity is enriched in the labyrinth by organic substances, while in the nephridial canal the reverse operation—absorption of salts—takes place. As a result, the urine contains 20 times less chlorides and 10 times less potassium but only 4 times less calcium and only 2.5 times less magnesium than the blood (Florken, 1947). The salt content of the urine is 9 times

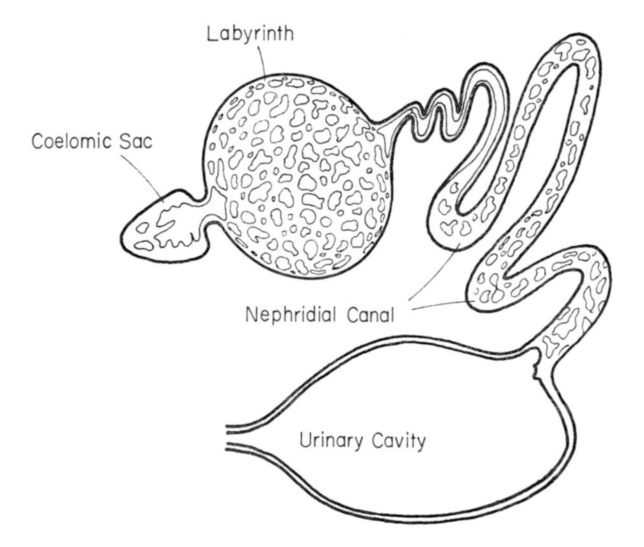

Fig. 26. Antennal gland of the river crab *Potamobius astacus* (after Schlipper, from Skadovsky, 1955).

less than that of the blood. As is apparent, the organism is especially active in retaining those ions which are few in fresh water. In seawater the work of the river crab's excretory system is sharply decreased, leading to the animal's death from poisoning by metabolic products.

The water-salt exchange of the various fishes is unique. Teleost freshwater fish remove surplus water entering their bodies by increasing the work of the excretory system. Their kidneys contain a large number of malphigian glomeruli, while their urine is abundant and, compared to the blood, strongly hypotonic. Such teleost fish can absorb salts through their gills (Fig. 27), and absorption occurs against the saline gradient. Changes in the osmotic pressure of the blood occur within limits in an environment with a varying salt content (Fig. 28), but they are accompanied by changes in the water content of the tissues and in the intensity of respiration (Fig. 29), which indicates the mutual dependence of the various aspects of metabolism.

Teleost marine fishes, which are threatened by dehydration because of the permeability of their skins in salt water, live essentially under conditions of high physiological drought. Their tissues struggle with the danger of dehydration by using the gills to excrete surplus salts encountered in the water they drink. The kidneys of marine fish are distinguished by their small number of glomeruli; they excrete little urine, and it is only slightly hypotonic compared to the blood. The entire mechanism of

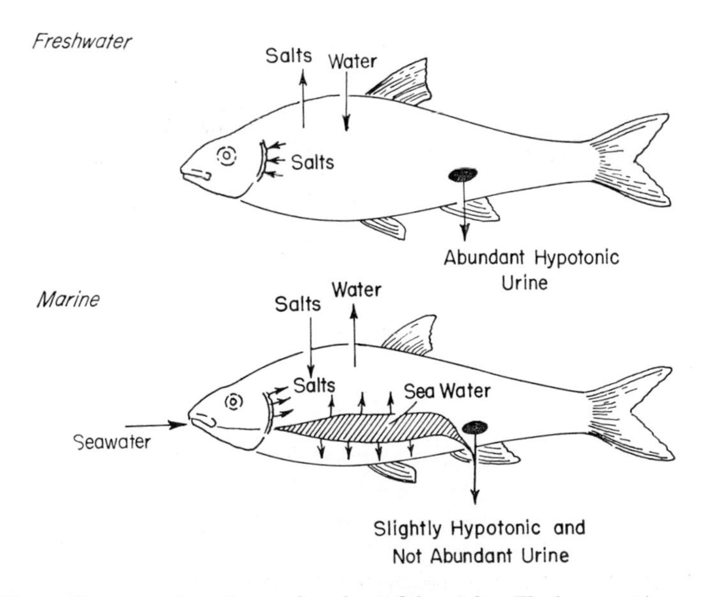

Fig. 27. Types of water-salt exchange in teleost fishes (after Florken, 1947).

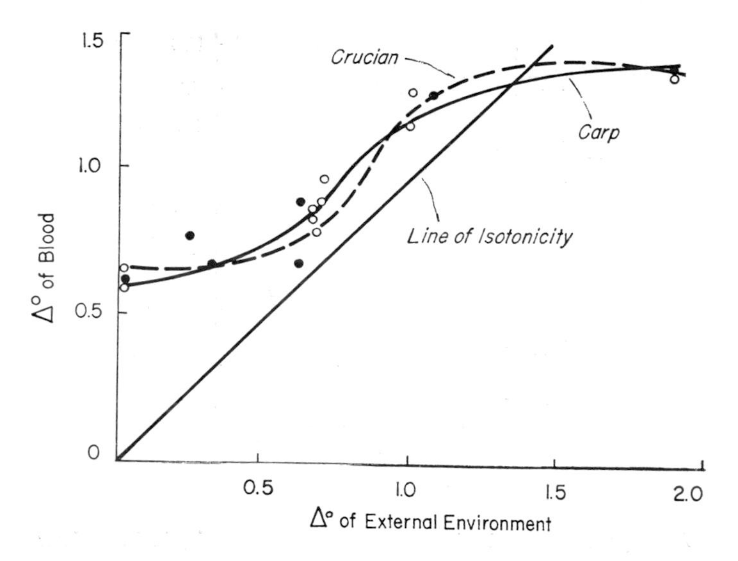

Fig. 28. Relationship between osmotic pressure in the blood of the carp and the crucian and salinity of the external environment (after Veselov, from Skadovsky, 1955).

excretion in these fish is directed toward retaining water in the body and removing surplus salt (see Fig. 27).

The mechanism of osmoregulation is unique in the sharks; in these fish a state of isotonicity with the environment is attained by retaining urea in the blood (Fig. 30); although this is uremia and is fatal to other groups of animals, it has no harmful effects on sharks.

In many animals homo-osmoticity is strengthened by skins made impermeable by chitin or horny substances (in the larvae and eggs of freshwater insects, crustaceans, adult invertebrates, and freshwater vertebrates), by mucus (the eggs of freshwater mollusks, fish, amphibians, many Diptera, etc.), and by the complex skin coverings of the higher vertebrates. These adaptations have been acquired by few groups of animals, which explains the qualitative poverty of freshwater fauna and especially of the inhabitants of oversalted bodies of water.

Regulation of osmotic pressure permits animals to enter bodies of water with an unfavorable salinity level. However, its changes may influence the size, development, and reproduction of all hydrobionts. The larger animals are marine. Among the cephalopods the squid *Architeutis princeps* reaches a total length (with tentacles extended) of 18 m, and tridacnon mollusks have been known to have a diameter of 2 m and a weight of 200 kg. The freshwater representatives of this group are small animals. Among the arthropods the Japanese crab *Kaempfferia kaempfferi*, with appendages 3 m long, is the biggest, while the largest species of fish is the giant

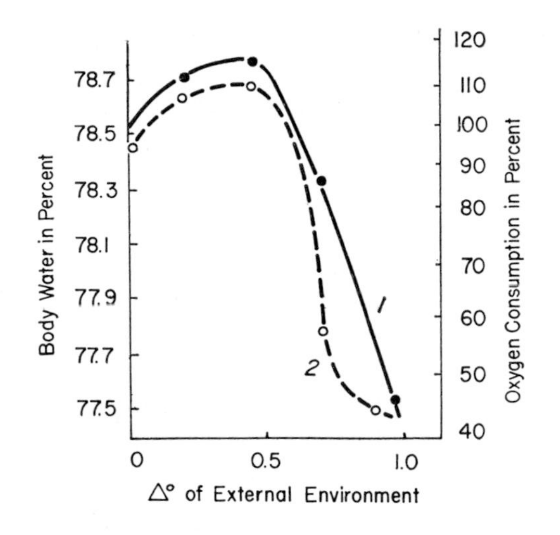

1 Oxygen Consumption in Percent of Average Breathing in Fresh Water

2 Water Content in Tissues

Fig. 29. Relationship between salinity of the external environment, rate of respiration, and degree of hydration of the tissues of the gudgeon (after Veselov, from Skadovsky, 1955).

shark *Cethorinus maximus,* which is 15 m long. Finally, the largest of contemporary mammals is the great blue whale (*Balaenoptera musculus*), with a length of up to 33 m and a weight of 120 tons.

The size of animal bodies has been determined historically not only by the influence of water-salt exchange but by all the properties and aspects of biology. The immense sizes of aquatic animals are associated both with the displacement possibilities of large bodies and with the increased heat dissipation possible in an aquatic environment. They also serve as protection from predators; the large and very large contemporary animals are basically peaceful species. Finally, the dimensions of these animals depend on the dimensions of the bodies of water, the abundance of the food supply in them, the oxygen content, the favorability of the temperature regime, and many other factors. Of course, the direct importance of the conditions of water-salt exchange is still unquestionable. In waters with varying salinity the measurements of individuals of the same species are not the same. The edible mussel *Mytilus edulis* is 110 mm long in Kiel Bay, which has a salinity of 1.5 percent, but reaches only 27 mm in the Gulf of Finland, which has a salinity of 0.2–0.5 percent. The transitional form of the shad *Alosa finta* is 45 cm long, while *A. f. lacustris,* the permanent inhabitant of the northern Italian lakes, is only 25 cm long. The tempo

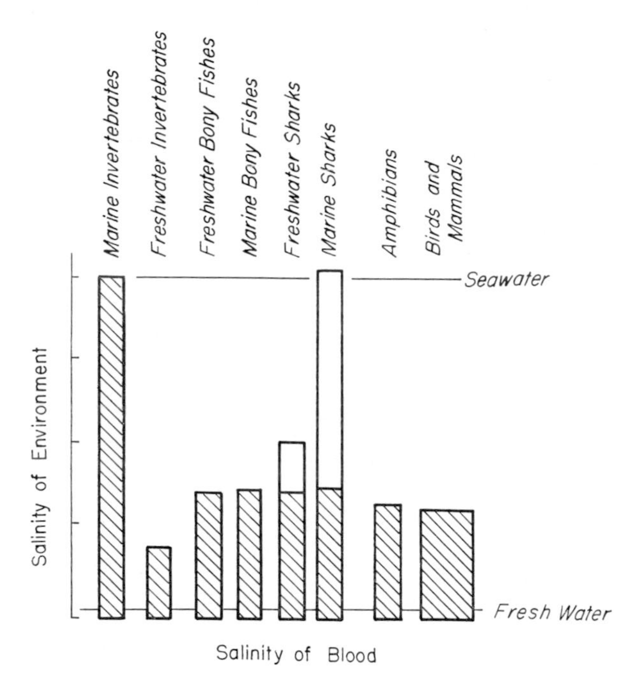

Fig. 30. Salinity of blood in sharks, marine and freshwater invertebrates, and teleost fishes as contrasted with salinity of their habitats (after Florken, 1947).

of growth also differs between fresh- and saltwater fish (Fig. 31). In fresh water the various salmon species mature almost identically, but in the sea the various salmons increase their growth markedly, while the freshwater trout remain as before.

When salinity changes, the morphological properties of water animals change too, e.g., the construction of the abdomen in the brine shrimp *Artemia salina*. In fishes (herring, cod, etc.) living in waters with lower temperatures and higher salinity, the number of vertebrae in the spine increases; this serves as an adaptation for movement in a denser environment (Jourdan's law).

Usually marine animals are more fecund, but they lay much smaller eggs; freshwater fish lay fewer eggs, but they are larger and have larger amounts of nutritional substances. Marine fishes in less salty bodies of water are less fecund or have a shortened reproduction period (the marine *Gobius* and the sea needles *Nerophis* and *Syphonostoma* in the eastern part of the Baltic). The brackish-water hydroid *Cordylophora lacustris* diminishes in size in fresh water, and its gonophora bear 3–6 eggs each instead of the usual 6–12 each.

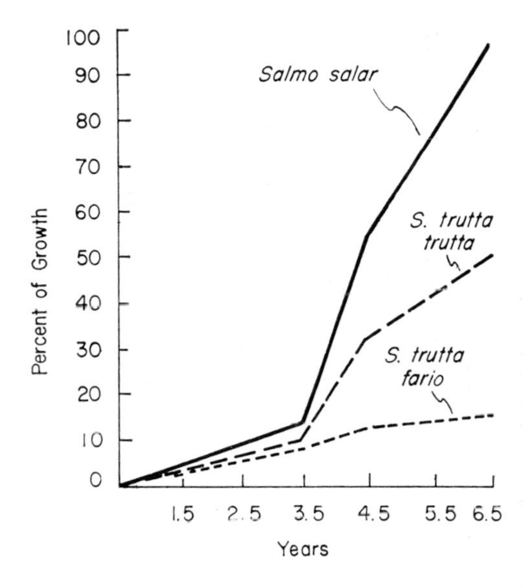

Fig. 31. Curves of growth in salmon (after Zernov, 1949).

When the environment changes, reproduction, survival, and abundance undergo substantial fluctuations. The speed of division in plankton cells, which is at a maximum at a salinity of 2.0 percent, is curtailed by two times at 1.0 percent and at 3.0 percent (Braarud, 1951). Oysters live only in brackish seawater, but their larvae may get into water with a salinity of 0.56 percent for a short time, although normal reproduction and development take place only in estuaries (freshened portions of seas near the mouths of rivers) with a salinity of 1.6–1.8 percent. The *Bankio* mollusk, which attacks wooden underwater marine installations and ship hulls, has never caused any trouble in San Francisco Bay; it cannot live in coastal waters because it cannot withstand a salinity lower than 1.0 percent. But in 1913 another wood-attacking mollusk, *Teredo,* was imported there; it lives well at a salinity as low as 0.6 percent, and in several years it had spread over the bay region, causing $25 million worth of damage (G. L. Clarke, 1954).

With regard to the chemistry of the aquatic environment and its fluctuations, we distinguish high-demand stenohaline species, tolerating only small changes in salinity, and the euryhaline species, capable of tolerating significant fluctuations in the salt content of water. Among the former any deviation from the optimum salinity causes suppression of respiration, increased mortality, and an acute reduction in abundance (of the biomass) (Fig. 32A, B, C), all of which is strengthened or weakened by the influence

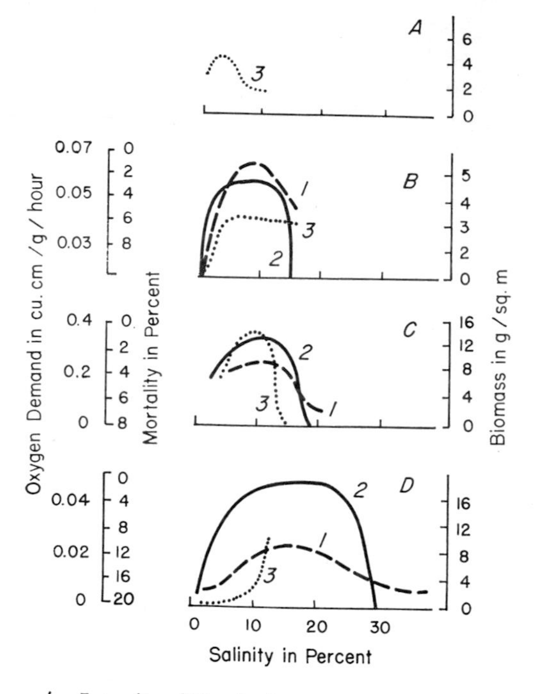

1 Intensity of Respiration

2 Average Daily Percent of Mortality

3 Biomass in the Sea

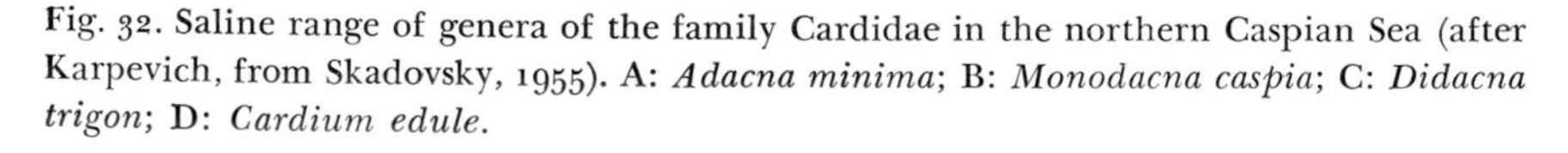

Fig. 32. Saline range of genera of the family Cardidae in the northern Caspian Sea (after Karpevich, from Skadovsky, 1955). A: *Adacna minima*; B: *Monodacna caspia*; C: *Didacna trigon*; D: *Cardium edule.*

of other factors (Fig. 33). Among the eurybionts the negative action of salinity fluctuations is expressed less strongly (Fig. 32D).

The active acidic or alkaline reaction is closely associated with the salinity and the ion state of the aquatic environment of free ions of water (H and OH ions) and is closely related to the calcium content and the carbonate system in the body of water. It is denoted by the pH index.[4] The biochemical activity of H and OH ions is a thousand times higher than the action of the ions of the physiologically important salts, which explains their immense importance for organisms. The active reaction of the environment influences the pH of the organism's internal environment, changes the permeability of the skin and the water-salt exchange as a

[4] The pH index refers to a neutral reaction when H and OH ions are in equal supply. A drop in the pH index from 6.9 to 1 characterizes a rise in acidity; its increase from 7.1 to 13 indicates increase in alkalinity.

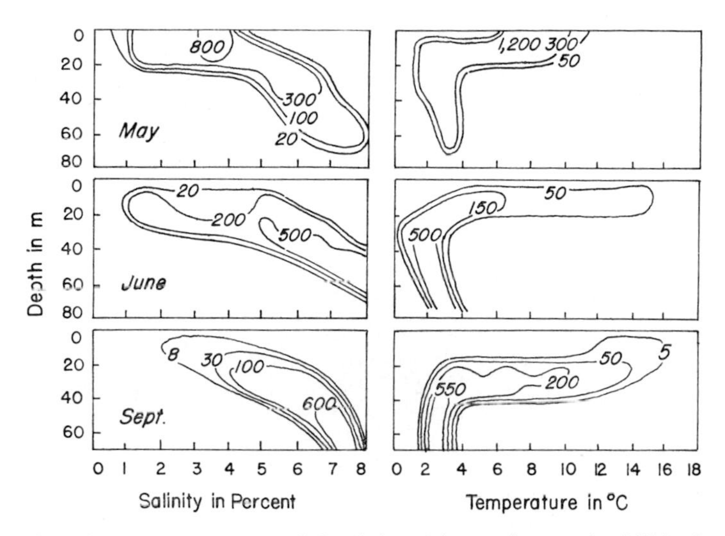

Fig. 33. Salt and temperature range of the shrimp *Limnocalanus grimaldii* in the Gulf of Finland (after Bityukov, 1960). Figures represent abundance of crayfish in cu. m.

whole, has a profound effect on respiration and absorption and release of water and salts, and, in the same way, influences all the basic vital phenomena of hydrobionts. The mechanism through which the pH of the environment operates is apparently connected with excitation or depression of the nervous system.

The optimum pH is different for different species (Fig. 34). Only a few

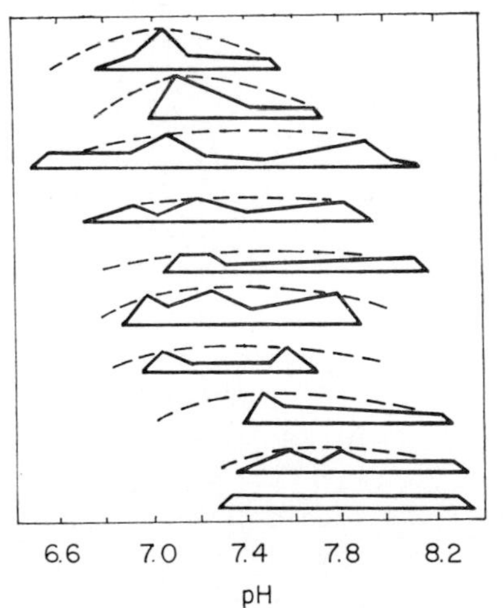

Fig. 34. Distribution of various species of fishes in relation to pH of the environment (from Zernov, 1949).

species tolerate wide variations in this index. The larvae of the midge *Chironomus plumosus* can tolerate an acid reaction down to pH 2–3 and will die at pH 1 only after several hours. They can also live in an alkaline environment. The protozoon *Euglena mutabilis* sustains pH fluctuations from 1.8 to 7.9, and the tapeworms have a range of 4 to 11. On the other hand, the ciliate *Stentor coeruleus* is able to exist only at a pH of 7.7–8.0. A special, small group of stenobiotic animals prefers to live in an acid medium. It occurs in lakes among sphagnum swamps with a pH of 3.8, and it includes some flagellated protozoa (*Carteria obtusa*), rotifers, and certain other animals.

The highly motile pelagic marine forms (*Scomber scombrus, Clupea harengus*) are most sensitive to changes in pH. Among freshwater species (the carp and the crucian [*Carassius*—Ed.]) maximum growth occurs at a pH of 5.5–6.5 (a slightly acid medium) (Fig. 35), while the copepod multiplies best at pH 5 and 9 (Fig. 36). The brine shrimp *Artemia salina* tolerates an alkaline medium less well; in experiments with a pH raised to 7.48–8.24, its population died off over a span of 13 months, having yielded only 2 generations, while at a lower pH a total of 12 completely viable generations was raised. Phagocytosis proceeds normally in the freshwater *Vorticella* only in a neutral environment, where the pH is close to 7; when acidity or alkalinity is raised, phagocytosis stops, although the animals themselves remain alive. In bodies of water with a varying pH

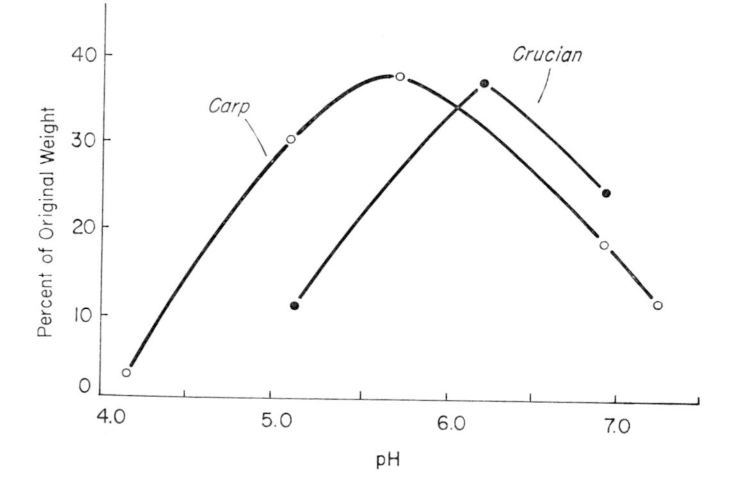

Fig. 35. Growth of the crucian and the carp (in percentage of original weight) over a period of 2 months, depending on pH of the environment (after Bryukhatova, from Skadovsky, 1955).

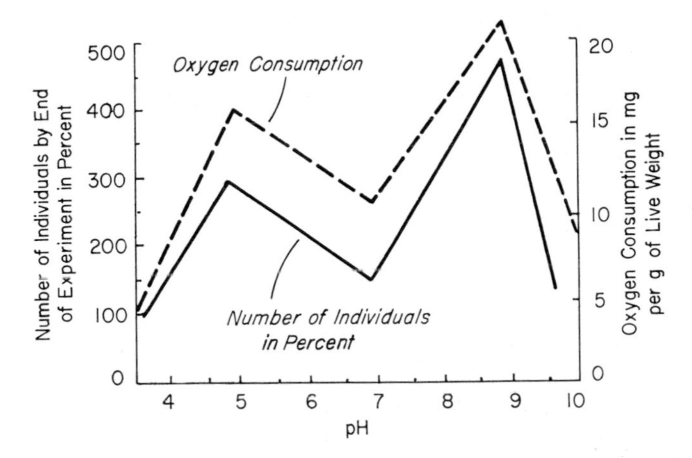

Fig. 36. Effect of pH on consumption of oxygen and on reproduction in the crayfish *Chydorus ovalis* (after Yatsenko, from Skadovsky, 1955).

even the forms of the species living there are changed (Fig. 37). Animals flee waters with an unfavorable pH by means of "hydro-ionotaxis."[5]

Since pH depends on the relationship of oxygen and carbonic acid (changes in the carbonate system) in the water, it rises and falls regularly with the daily and seasonal decrease and increase of photosynthesis. The vertical daily and seasonal migrations of water animals are associated with these changes. It is supposed that pH may be the guiding factor in the return of the Pacific salmon to its spawning streams after its fattening and maturation in the seas (Shelford, 1925).

The salinity and reaction of a body of water are closely connected through the carbonate system. The chemistry of a body of water influences the organisms living in it as a closed system of interacting factors. An example of such complex interaction is the influence of the "contamination" of bodies of water with dead organic matter (suspensions, colloids, and solutions). With the aid of bacteria this matter is broken down and mineralized, which changes the salt and gas content of the water, its ion state, and consequently its pH. Therefore, organic contamination of a body of water is an important and powerfully acting factor of the environment. We distinguish (a) polysaprobic (highly contaminated) waters, where only the primary phases of the decomposition of organic compounds can take place; (b) mesosaprobic (moderately contaminated) waters, where

[5] A taxis is an involuntary (compulsory) movement of the animal toward, or away from, the source of a stimulus (toward light, heat, a chemical agent, etc.). Their bases are, of course, different in different animals, but they are always associated with the nerve elements or with the nervous system. In many of the animals, especially the higher ones, taxes are unconditioned reflexes.

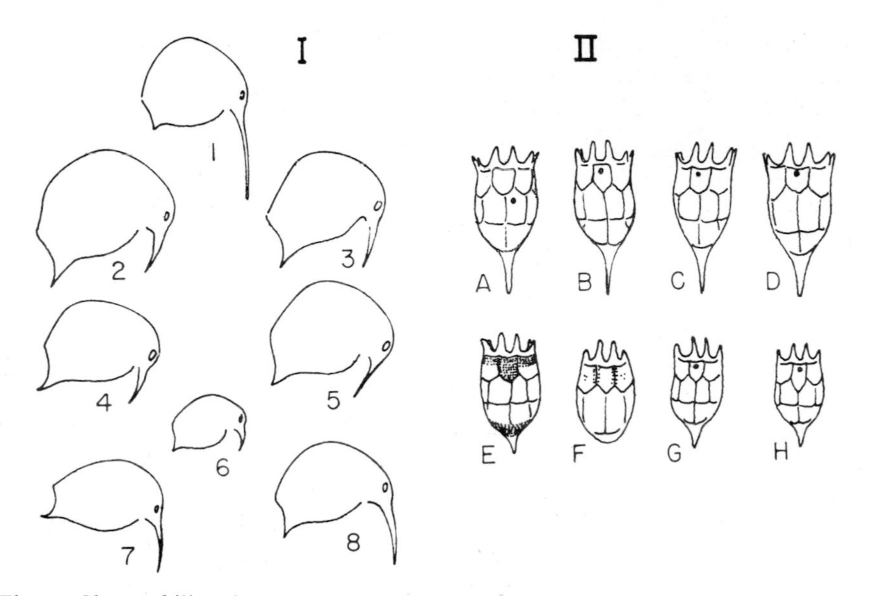

Fig. 37. Changeability of *Bosmina coregoni* (I) in lakes with a pH of 7.32 (1), 7.1–6.7 (2–5), 4.75 (6), and 4.54 (7–8), and changeability of *Keratella cochlearis* (II) in lakes with a pH of 4.5 (A–D), 7.3 (E–F), and 6.75 (G–H) (after Skadovsky, 1955).

oxidation goes further; and (c) oligosaprobic (slightly contaminated) waters, with complete mineralization of organic matter. The degree of contamination can be judged from the typical inhabitants of the body of water. Polysaprobic waters are characterized by infusorians like *Paramecium putrinum,* the tube worm *Tubifex tubifex,* and the rat-tailed maggot (*Eristalis tenax*). Mesosaprobic waters are inhabited by rotifers and the larvae of the midges *Tendipes* and *Chironomus plumosus.* Oligosaprobic bodies of water contain the rotifer *Notholca longispina,* the cladoceran *Daphnia longispina,* and also such fish as trouts, sterlets, and minnows. There are similar biological indicators which can be used to establish the degree of contamination of a body of water.

The whole aggregation of chemical factors in a body of water is one of the most important conditions determining the species assortment of its population, the number of individual species, and the stability and fluctuation of the abundance index. The ecological importance of fluctuations in salinity, ion state, and pH of the water environment is especially great in places with great fluctuations in these factors, as in continental lakes, with their changing levels depending on the inflow of rain-, ground-, and river waters, and in estuaries. It is in such places that the euryhaline forms are abundant.

The most important factor—salinity of natural bodies of water—fluctu-

ates from 0.001 percent (pools on the surface of sphagnum swamps) to 280 g of common salt (Lakes El'ton and Baskunchak) and even 347 g of Glauber's salt per l (Lake Tambukan). The world ocean, with its predominance of chlorides and stable salinity of 3.2–3.8 percent (3.5 percent on the average), and the seas connected with it (salinity from 1.6 to 4.7 percent) have a pH of 8 and contain 2–4 mg per l of organic matter. These polyhaline bodies of water differ in their complex fauna, the species abundance of which decreases as salinity drops. Among the marine animals stenohaline poikilo-osmotic species prevail, which explains the importance of the stable chemistry of seawaters. When rain falls, certain species of oceanic plankton actually abandon the surface and hide in the depths, where the fresh rainwater has no effect.

The brackish basins (mesohaline) (the Azov and Baltic seas and the large continental lakes–the Caspian and the Aral), with a salinity of 0.05–1.60 percent, a pH of about 8, and a large amount of organic matter in the water, differ in their seasonal changeability of the ion-salt regime, which is associated with fluctuations in rainfall. They are inhabited by freshwater forms resistant to a certain amount of change in salinity, by marine forms which tolerate lowered salinity, and by specific brackish-

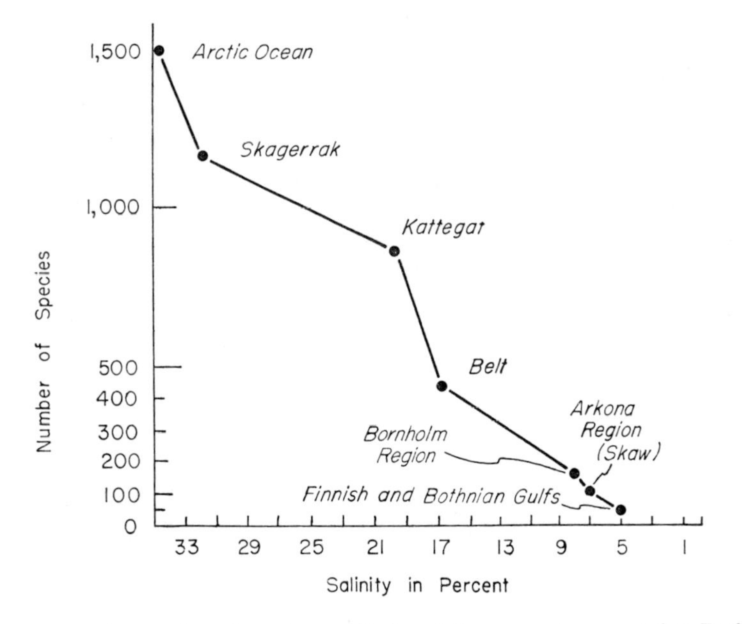

Fig. 38. Decrease in number of species as salinity of the ocean decreases (after Zenkevich, 1947).

water inhabitants. As in marine waters, the species assortment of fauna in the brackish basins diminishes if there is an increase in salinity changes or if the salinity as a whole is decreased (Fig. 38). The overwhelming majority of the inhabitants of brackish basins are euryhaline (homo-osmotic). Some of the Gobiidae living there can tolerate fluctuations in salinity of 0.1–0.6 percent.

Freshwater (oligohaline) bodies of water with a reliably low salinity (0.02–0.05 percent) are poor in sodium, potassium, and chlorides and are relatively rich in calcium, magnesium carbonates, and sulfates (Fig. 39). We distinguish bodies of water of the carbonate type (alkaline) and humus-rich subacid types. Fresh waters are inhabited by a unique fauna, but one which is rather impoverished compared with that of the sea. Typically, freshwater animals are stenohaline and hypertonic. The bivalve mollusks *Unio* and *Anodonta* cannot tolerate a salinity higher than 0.2 percent. The river *Hydra*, when transferred to seawater, dies in approximately 1 minute, while the majority of freshwater crayfish can tolerate only a short period in salt water. The continental salt lakes differ in their various types of salinity, fluctuating between 0.05 and 34.70 percent. The animal population in lakes with little salt is similar to that in fresh water. As salinity rises, it grows poorer, and in high concentrations of salt (over 10.0 percent) there are either no animals left or only isolated species (larvae of the *Ephydra* fly, the brine shrimp *Artemia salina*, and certain others).

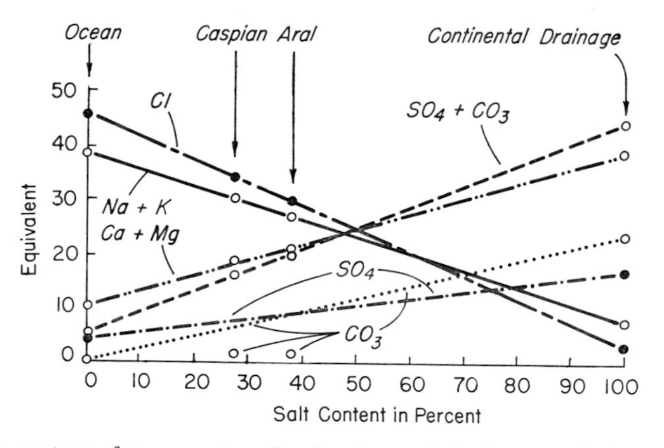

Fig. 39. Salt content of ocean water, the Caspian and Aral seas, and river water (after Blinov, 1947). Straight lines running at angles show theoretical changes in waters of continental origin; concentration of sulfate observed in "intermediate" basins is higher, while carbonate is lower, than the theoretical points; dotted lines correspond to the sum of the equivalents.

WATER EXCHANGE IN TERRESTRIAL ANIMALS

Water exchange and mineral nutrition are just as important to terrestrial animals as they are to the aquatic ones, but the ecologo-physiological mechanisms providing the necessary stability of the salts and ions in these animals' body fluids are not the same as those in the hydrobionts. On dry land the danger of a large amount of water invading the body and destroying the osmotic pressure of the internal aqueous environment does not arise, but on land there is always the possibility of losing water from the body by evaporation and the danger of water shortage in the external environment. Therefore, the most important things for terrestrial animals are the regulation of water and salt excretion in association with air and soil moisture, the water content of food, the presence of drinking water, and the availability of mineral substances.

The circumstances of water exchange in terrestrial animals are relatively independent of the conditions governing salt exchange and are determined by climatic moisture—its aridity or humidity, i.e., the amount and type of precipitation—and temperature. These factors also influence the water balance of organisms.

Absorption of water by the skin plays an important role in the life of amphibiotic and soil animals. Frogs, toads, and earthworms have skins which water can penetrate; they behave in water and moist soil like poikilo-osmotic hydrobionts, excreting surplus water in the form of abundant hypotonic urine. The entrance and elimination of water are subject to neurohumoral regulation, in which the posterior lobe of the pituitary plays an important role with a hormone it secretes (Fig. 40). Absorption and evaporation of water are especially important among the inhabitants of dry areas.

Insects and ticks are also capable of absorbing water. Orthopteran eggs absorb water even before they begin to develop. The penetration of water into orthopteran eggs is not subject to simple physical (osmotic) laws, but it is regulated by special "hydrophilic" cells situated on the posterior part of the egg, where they form a "hydrophilic" patch on the cuticle. Water can pass into the egg only through this patch (Fig. 41). At the same time absorption proceeds against the osmotic gradient even from such media as a glucose solution with an osmotic pressure of 14.6 atmospheres. Similar adaptations are found in the eggs of *Tachycines* crickets (Krause, 1936) and in *Pteronarcys* (Plecoptera) (Miller, 1940) as well as in the semicoleopteran *Notostira* (Johnson, 1934, 1937).

Unengorged females of the tick *Ixodes ricinus* absorb water from air when it has a relative humidity of 92 percent. This absorption against

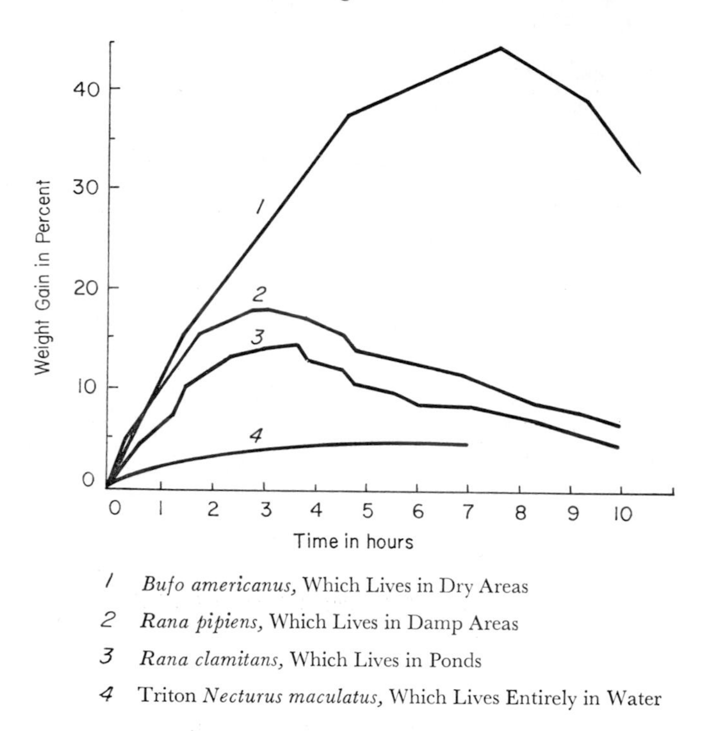

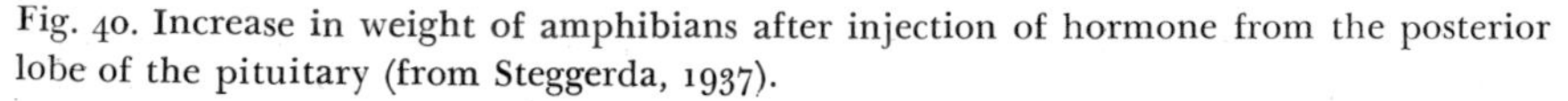

Fig. 40. Increase in weight of amphibians after injection of hormone from the posterior lobe of the pituitary (from Steggerda, 1937).

the gradient of water-vapor pressure can be accomplished only with the aid of special secretory systems similar to those in the hydrophilic cells of orthopteran eggs. Absorption occurs after a preliminary "drying out" of the tick and continues until normal osmotic pressure and a normal ionic state have been re-established in the hemolymph, after which water absorption is curtailed (Lees, 1946). Bedbugs, *Leptinotarsa* beetles, and the larvae of the beetle *Tenebrio molitor* can absorb moisture from the air if the relative humidity is 88 percent (Wigglesworth, 1931). The grasshopper *Chortophaga viridifasciata* is able to absorb moisture even from air with a relative humidity of 82 percent, while the flea *Xenopsylla cheopis* can absorb water from air with a relative humidity of 50 percent. Absorption of moisture from the soil or atmosphere is of vital significance for the development of the eggs of dry-land reptiles. On the other hand, inhabitants of areas with permanently high humidity—terrestrial isopods like *Armadillidium* and *Legia*—absorb water only at a relative humidity of 98 percent (Edney, 1951). The ability to absorb water vapor permits the species to exist in dry habitats.

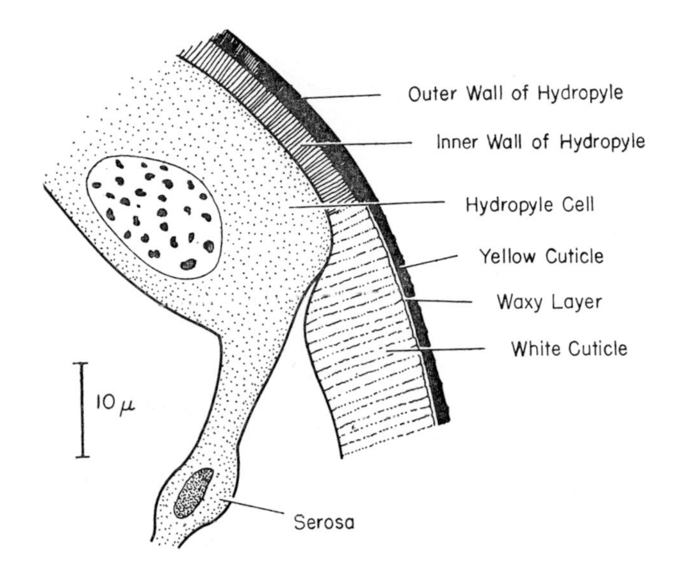

Fig. 41. Structure of the hydropyle of the egg of *Locusta pardalina* (after Matte, from Andrewartha and Birch, 1954).

Water in food and in drinkable form is the most important means that dry-land animals have for obtaining moisture. When the air grows dryer, the animals switch over to moister foods and thus tolerate the lower air humidity better (caterpillars of the meadow moth (*Loxostege sticticalis*); Kozhanchikov, 1935, 1937). Many animals require drinking water. There are few insects that drink regularly, but the beetle *Ptinus tectus* must have drinking water added to its food constantly. Water is also necessary for culturing houseflies and other flies in the laboratory. Insects which drink nectar or blood actually obtain a surplus of water and salt with their food (especially chlorides); these are eliminated by increased work on the part of the excretory organs. The majority of reptiles drink willingly; this includes the desert species. Birds also drink potable water. But the desert skylark, the sasquavl jay, and many of the chats manage without drinking. Desert grouse and hoofed mammals cannot live long without water and make regular migrations to bodies of it. During droughts carnivores, bats, and birds wander in search of water.

The majority of insects, many birds, and the rodents can manage without water, satisfying themselves with the moisture in their food. When the temperature rises from 10 to 25° C, the common vole needs twice as much water, while the society vole, an inhabitant of the desert, needs only 1.4 times as much (I. Ya. Polyakov, 1952).

An important item for desert animals is metabolic water, which forms

taining air saturated with moisture; the trachea also opens into the cavity, preventing loss of moisture during respiration (Dizer, 1955). The inhabitants of very dry areas have a comparatively thick cuticle which allows very little moisture to escape. They lose water only by respiration and excretion, while in wet regions the cuticle is delicate and water can be easily evaporated through it (Gilyarov, 1949; Gilyarov and Semenova, 1957; Semenova, 1958).

Most mollusks quickly lose water in dry air and often need drinkable water or moist food, but snails, snug in their shells, stay dormant for a month or more when the air gets too dry. Amphibians which live in moist habitats have skins which are more permeable to water than those of amphibians living in dry areas. This permeability is so great in the grass frog that the weight loss during a dry period is practically identical among individuals with undamaged skins and is not significant, while toads, which live in dryer areas, are distinctly different (Fig. 43). Their skin permeability falls sharply when the mucus covering it dries out. Skin freed of mucus

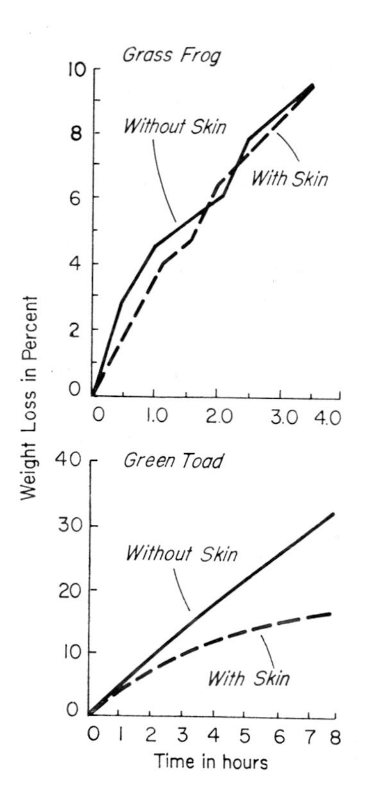

Fig. 43. Losses in weight during dehydration (after Dinesman, 1948).

admitted 151.0 cu. mm of water through 1 sq. cm of surface every hour, untreated skin (natural skin with uncoagulated mucus) admitted 30.5 cu. mm, while the skin of a toad left in the sun before the experiment admitted only 19.6 cu. mm (Dincsman, 1948).

Evaporation through the skin differs markedly between amphibians and reptiles; among the former it decreases the body temperature noticeably, while among the latter it has little influence upon it (Table 12). Among the birds water is evaporated chiefly by the respiratory system (trachea, lungs), the oral cavity, etc., while in the mammals the sweat glands are of great importance in eliminating water. The work of these glands is governed by the brain's heat center.

The elimination of water also depends on the temperature of the air (Table 13). Among the inhabitants of the wetter areas (the large gerbil, the crested gerbil, and the gray hamster) skin and pulmonary losses of water are two to three times higher than they are among the inhabitants of arid biotopes. Both species of gerbils live in the desert zone, but the

TABLE 12. *Difference between temperatures of the air and of the body in reptiles and amphibians under conditions of differing humidity but constant air temperature (20° C) (after Hall and Root, 1930).*

	RELATIVE HUMIDITY IN PERCENT				
Species	*7*	*25*	*50*	*75*	*95–100*
Triton (*Plethodon glutinosus*)	−9.20° C	−6.34° C	−4.62° C	−2.54° C	−0.29° C
Frog (*Rana pipiens*)	−8.60	−6.75	−4.63	−3.01	−0.13
Toad (*Bufo fowleri*)	−7.33	−5.31	−3.98	−2.48	∓0.74
Lizard (*Sceloporus undulatus*)	−0.73	−0.70	−1.15	+0.30	+0.64
Horned lizard (*Phrynosoma*)	−0.77	−0.02	+0.11	+0.90	+0.38
Painted turtle (*Chrysemys marginata*)	−0.72	+0.57	+0.52	+0.41	−0.12
Land turtle (*Testudo major*)	−0.34	−0.23	−0.11	0	+0.15
Alligator (*Alligator mississippiensis*)	−0.30	−0.26	--0.15	−0.08	+0.18

TABLE 13. *Skin and pleural losses of water at different air temperatures (after Felatova, 1944; Shcheglova, 1949).*

	WATER LOSS IN CM PER KG OF WEIGHT PER HOUR		
Species	*At 10° C*	*At 20° C*	*At 30° C*
White rat (*Rattus norvegicus*)	80	87	95
Large gerbil (*Rhombomys opimus*)	49	52	56
Yellow suslik (*Citellus fulvus*)	–	54	–
Severtsov's jerboa (*Allactaga severtzovi*)	–	60	–
Gray hamster (*Cricetulus migratorius*)	–	110	–
Crested gerbil (*Meriones tamariscinus*)	–	170	–

crested variety is closely associated with the relatively moist areas, feeding primarily on seeds and leading a strictly nocturnal life. The large gerbil, which is active by day and which feeds on juicy plant parts, is an inhabitant of dry areas.

Excretion of metabolic wastes must of necessity be accompanied by elimination of water and mineral salts. The marine animals have already seen the necessity of holding water within the body and have accomplished it by curtailing the excretion of urine. This same necessity confronts the terrestrial animals with even greater acuteness. The means of resolving the problem are similar. The kidneys of present-day terrestrial vertebrates (amniotes) are distinguished by a comparatively small number of glomeruli and by the presence of long, thin canals (Fig. 44), in parts of which (Henle's loops) reverse absorption of water occurs, making the urine of vertebrates hypertonic with relation to the blood. In birds, reptiles, and many insects uric acid is excreted; this acid does not need water to be transported out of the body. Among the amphibians, some mammals, and all the true water vertebrates urea is excreted in solution (Florken, 1947). In the terrestrial invertebrates there are adaptations with similar ecologo-physiological importance.

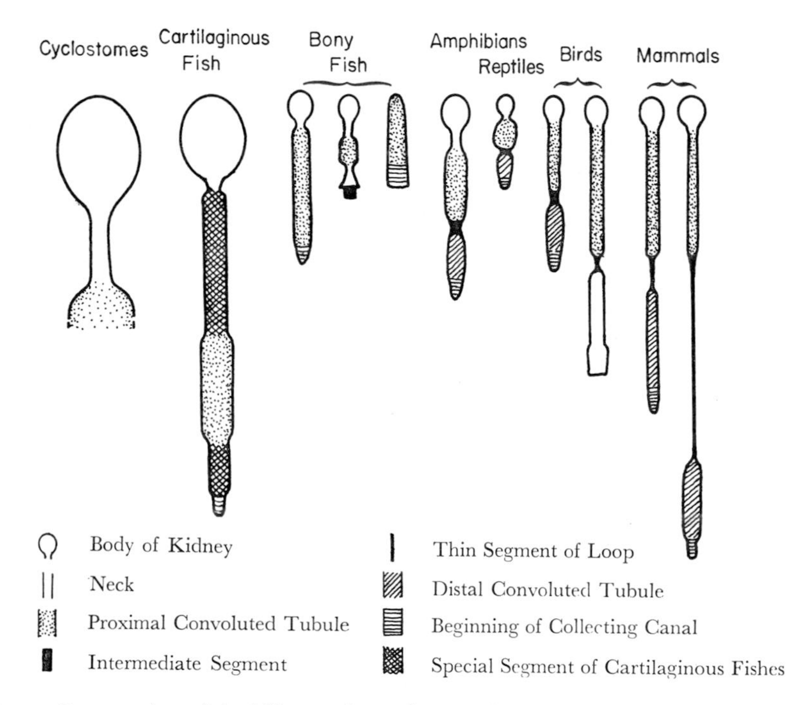

Fig. 44. Construction of the kidneys of vertebrates (after Koshtoyants, 1950).

With regard to the humidity of the air and the amount of water demanded in food, terrestrial animals are divided into moisture-loving (hygrophilic) and dryness-loving (xerophilic) types. There is an intermediate group, the mesophiles. Among the hygrophiles the above-described mechanisms for storing and holding water in the body are poorly expressed or absent entirely, but they are present in the xerophiles. We also distinguish stenohygrobiontic species, capable of existing only under definite and limited moisture conditions, and euryhygrobiontic species, which tolerate a broad range of fluctuations in the humidity of the air or in the amount of moisture in the food. Moisture-craving is usually associated with a high level of water exchange and a lack of special adaptation for its regulation. The degree of moisture-craving or dryness-craving, varying in different species, usually corresponds to each species's conditions of existence (Fig. 45).

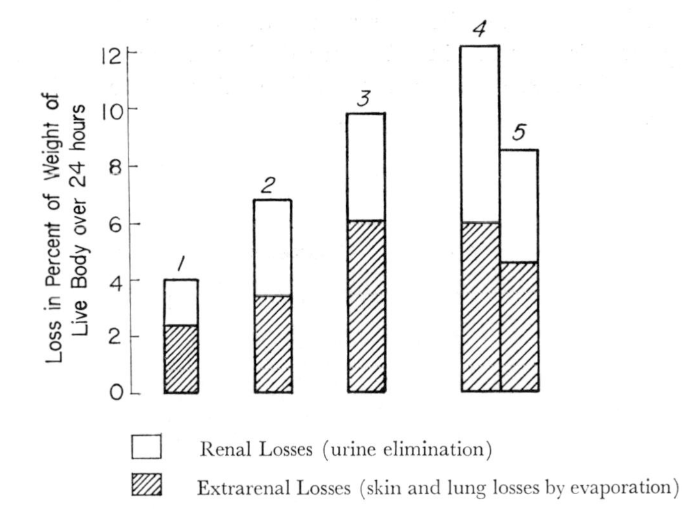

Fig. 45. Daily losses of water in thin-fingered (1), yellow (2), and minor (3) susliks in March and April (after Shcheglova, 1952) and in the muskrat in June (4) and September (5) (after Shcheglova, 1953).

The conditions of water exchange play an important role in the development and multiplication of terrestrial animals. Most important are air humidity, water content of food, and drinking water. Ability to tolerate water losses is different for each species (Tables 14, 15).[7]

[7] These tables give only a relative representation, since they do not consider length of life after the water loss has occurred and do not take into account physiological differences between the individuals, which are sometimes quite significant.

The conditions of water exchange influence fecundity, which decreases sharply in hygrophiles and mesophiles when there is a shortage of moisture in the air or the food source (Figs. 46, 47). A shortage of water in the food

TABLE 14. *Amount of water an animal may lose and still remain alive.*

Species	*Loss in Body Weight in percent*	*Water Loss in percent*	*Author*
Mollusks	50–60	–	Howes and Wells (1934)
Various snails and *Limax tenulus*	80	–	Künkel (1916)
Crustaceans	50	–	Edney (1951)
Various annelids:			
Allobophora foetida	61	73	Schmidt (1918)
A. chlorotiscus	70	83	Hall (1922)
Lumbricus terrestris	43	–	Jackson (1926)
Insects:			
Tenebrio (larvae)	53	100	Hall (1922)
Austroicetes cruciata (eggs—resistant diapause)	70	87	Andrewartha and Birch (1948)
Amphibians:			
Ambystoma punctata	47	–	Hall (1922)
Rana pipiens	31	45	Thorson and Svihla (1943)
Reptiles:			
Chrysemys marginata	33	–	Hall (1922)
Sceloporus spinosus	48	–	Hall (1922)
Mammals:			
Peromyscus leucopus	31	–	Hall (1922)
Mus musculus	24	–	Hall (1922)

TABLE 15. *Minimum loss of water leading to death in amphibians (after Thorson and Svihla, 1943).*

	MINIMUM WATER LOSS CAUSING DEATH		
Species	*Percent of Body Weight*	*Percent of Water in Body*	*Living Conditions*
Scaphiopus halbrooki	48	60	burrows in ground
S. hammondi	48	60	burrows in dry ground
Bufo boreas	45	56	terrestrial
B. terrestris	43	55	terrestrial
Hyla regilla	40	50	terrestrial, but lives in moister areas
H. cinerea	39	49	terrestrial–boreal
Rana pipiens	36	45	terrestrial–semi-aquatic
R. aurora	34	43	semi-aquatic
R. grylio	30	38	aquatic

suppresses or stops the activity of the gonads in rodents (Alikina, 1959). When rodents ate food containing only half the normal amount of water, all of them suffered a decrease in growth tempo (Table 16). In the com-

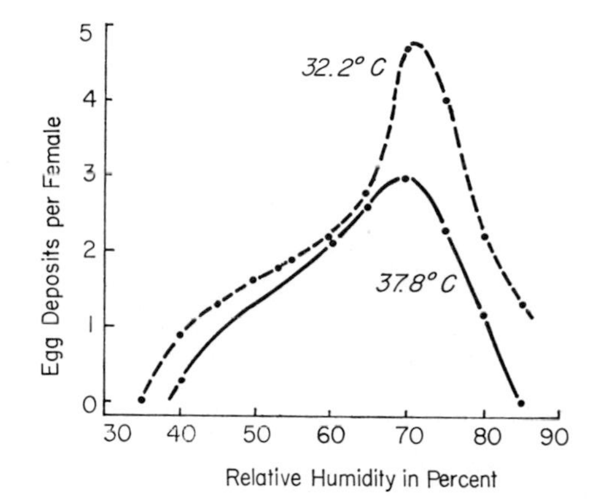

Fig. 46. Degree of egg-laying by *Locusta migratoria migratorioides* maintained at various levels of relative humidity (after Hamilton, from Andrewartha and Birch, 1954).

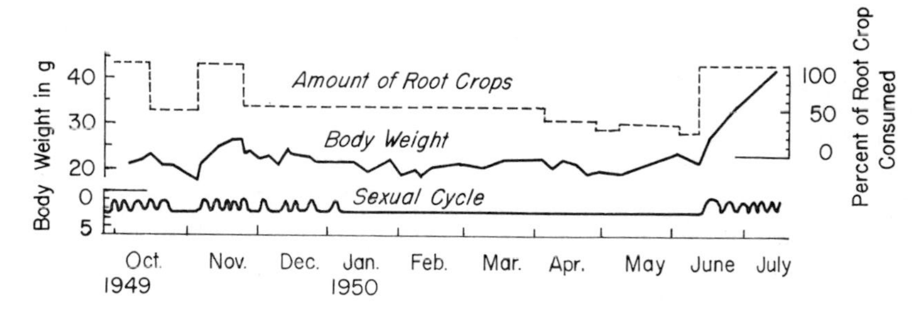

Fig. 47. Alternation of phases in the sexual cycle of the common vole under differing water regimes (after Alikina, 1959).

TABLE 16. *Dependence of growth and reproduction of small rodents upon moisture content of food (after N. P. Naumov, 1948).*

	CONTENT ON DRY FOOD			CONTENT ON MOIST FOOD		
Species	*Number of Animals in Experiment*	*Final Weight as Percent of Initial Weight*	*Young Born to 1 Female*	*Number of Animals in Experiment*	*Final Weight as Percent of Initial Weight*	*Young Born to 1 Female*
Common vole	29	96.3	0	15	152.5	3.9
Steppe lemming	11	100.1	5	12	101.5	9.5
Society vole	4	101.2	0	4	102.7	0
Forest mouse	10	137.3	0	10	143.8	0

mon vole growth stopped, and only those females which were given sufficient water bore offspring. Steppe lemmings (*Lagurus lagurus*) produced litters in both groups, but the amount of reproduction in the group with the dry diet was half of that in the other group.

The same dependence exists for speed of development, life span, and mortality (Fig. 48). Atrophy of cocoon contents at room temperature (25° C) among miller moths, which live normally at rather high levels, rises to 75 percent at a humidity of 10–20 percent and drops to the minimum at 45–60 percent humidity; it rises again to 65 percent mortality when the relative humidity rises to 90 percent. The survival rate of rat fleas is maximum at 89 percent relative humidity. Their average life span at 32° C and a relative humidity of 80–90 percent is 152 hours, while at the same temperature but with a relative humidity of 27 percent they live only 27 hours (Bacot and Martin, 1924).

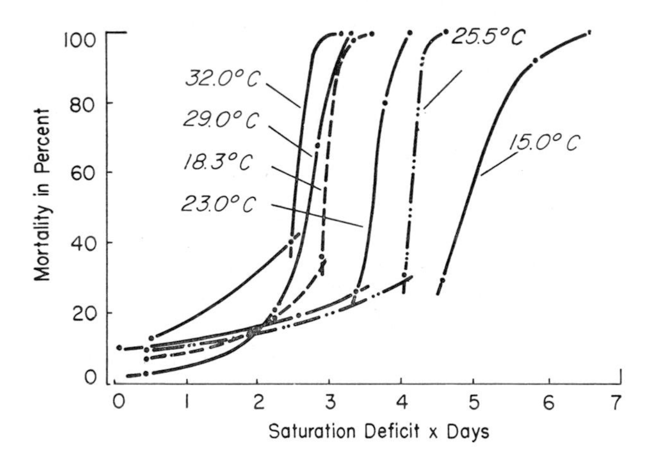

Fig. 48. Influence of saturation deficit on death of eggs of *Calandra oryzae* at different temperatures (after Birch, 1944).

Reproduction of the desert locust proceeds normally if it is sufficiently dry at a definite stage of development.[8] Under these conditions diapause takes place, gonad development stops, the individuals' activity decreases, and their pigment fades. Having passed through diapause, the individuals quickly mature as the humidity rises and produce abundant and fertile batches of eggs. Among those which did not observe diapause, reproduction is either absent or suppressed. Fig. 49 is a diagram of the influence of water exchange on the most important vital phenomena in two types of species. Fig. 49A shows a hygrophile which did not experience any sup-

[8] In nature this usually coincides with the time of the summer drought.

pressive influence from the humidity of the air. Fig. 49B shows a species of xerophile for which the zone of possible existence is narrower; i.e., conditions of increased humidity are precluded.

The normal conditions of water exchange are often furnished by adaptive behavior. While studying the caterpillar *Choristoneura fumiferana* in a device with a simultaneous temperature and humidity gradient, W. G. Wellington (1949) found a distinct preference on their part for conditions of identical intensity of evaporation (Fig. 50). This preference, however, changed with the state of the insects, since hunger and dehydration were of great importance.

Preference for a definite humidity is typical of most animals. In a device with a humidity gradient (at constant temperature), moisture-loving insects (the wood louse *Porcellio scaber*, the larvae of *Agriotes*, the beetle *Tribolium confusum*, and many others) are more active at its dry end and less active at a humidity of 90–100 percent, so they collect at the "moist" end of the machine. The inhabitants of dry areas behave in just the opposite manner. The same thing happens in nature, leading to a concentration of animals in the areas which they prefer.

At the same time the movements of most arthropods are undirected. This type of reaction is called "orthokinesis" (Fraenkel and Gunn, 1940).

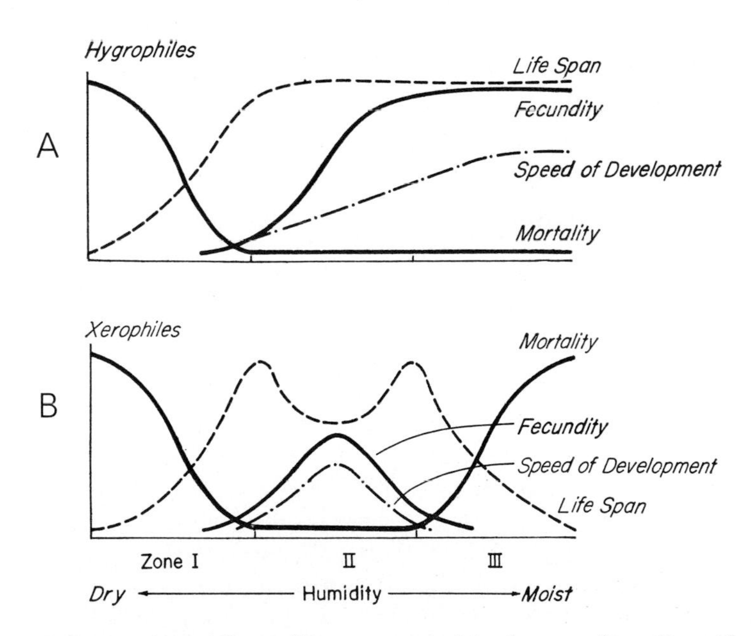

Fig. 49. Influence of humidity on life span, speed of development, fecundity, and mortality (after Andrewartha and Birch, 1954).

In the clothing louse *Pediculus humanus* accidental motions are of the feedback type (circular), but even these grow stronger in an unpleasant zone, permitting the animal to avoid it. This type of reaction is called "clinokinesis" (Wigglesworth, 1941). Finally, the larvae of *Agriotes* move in a directed manner—away from dryness—which deserves the name "clinotaxis." The latter is typical of vertebrate animals and can be traced easily in devices with a gradient in air or substrate humidity (Kalabukhov, 1951).

Daily fluctuations in humidity may be the cause of migrations among animals. *Anopheles homonculus* and *A. bellator*, which live in the moist tropical forests of Trinidad, remain in different tiers at different times of day, seeking out places with optimal and approximately identical humidity. At the same time *A. homonculus* always stays below *A. bellator*. Among the amphibians the way of life and the distributional pattern bear

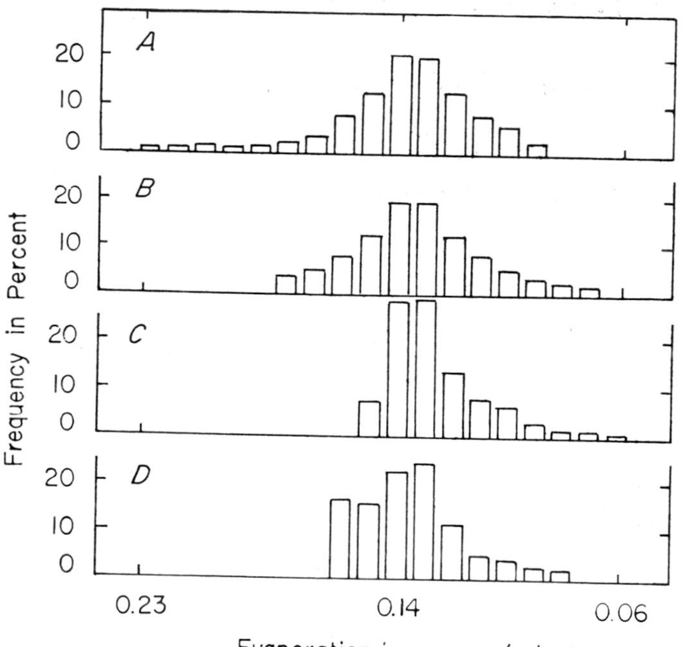

Fig. 50. Distribution of caterpillars of *Choristoneura fumiferana* in a gradient device with differing evaporation rates (after Wellington, 1949). A: at gradient temperature in dry air (zone with a temperature of 36° C higher than the zone with an evaporation rate of 0.23 cu. mm per minute); B: the same in moist air (zone with a temperature of 36° C corresponding to an evaporation rate of 0.18 cu. mm per minute); C: the same in a very moist atmosphere (36° C corresponding to 0.15 cu. mm per minute); D: at a constant temperature of 20.6° C.

the distinct stamp of dependence on air humidity (Table 17, Fig. 51). L. G. Dinesman (1948) has divided these animals into three groups:

1. Inhabitants of bodies of water. In summer these do not move more than a few meters from the water. They spend a significant part of the 24-hour period in the water (green and lake frogs).

2. Inhabitants of moist places. These live in meadows, moist forests, gardens, etc. (grass and sharp-faced frogs, gray and spadefoot toads).

3. Inhabitants of dry places (steppes, fields, dry forests, etc.). These wander several kilometers from bodies of water (green toad, little chorus frog).

Species in the first group are active in the daytime, since their being in water or near it eliminates the danger of overheating and losing large

TABLE 17. *Encounters with* Pelobates fuscus *in relation to humidity and temperature of the air (after Dinesman, 1948).*[a]

	AIR HUMIDITY IN PERCENT			
Air Temperature in °C	*70*	*78–89*	*90–95*	*100*
10–15	no data	4.4	13.5	19.2
15.1–20.5	4.0	no data	15.3	23.6

[a] Number of animals computed over a 1-km route.

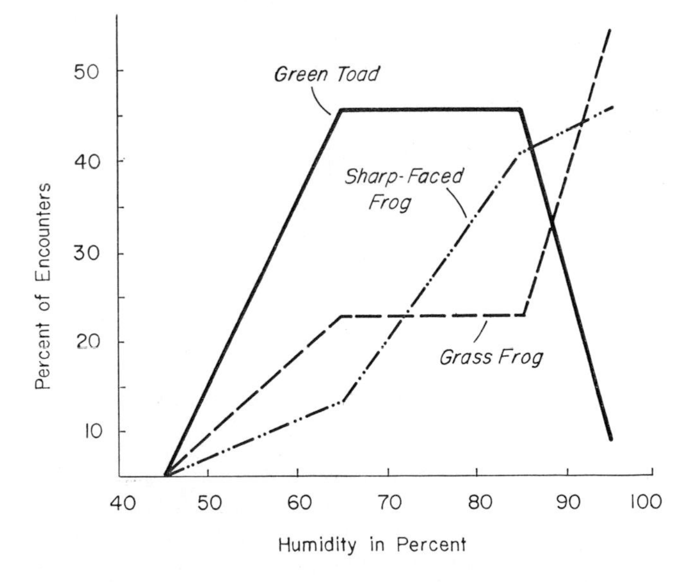

Fig. 51. Frequency of encounters with different amphibians at differing humidities (in percent of total number of encounters with each species) (after Dinesman, 1948).

amounts of water. Representatives of the other two groups are primarily nocturnal animals, leaving their refuges in the daytime only in the event of cloudy weather.

Amphibians react to variations in the humidity not only of the air but also of the soil. The Australian frogs *Lymnodynastes ornatus* and *Helioporus pictus*, which live in the central deserts, dig temporary or permanent burrows in the beds of intermittent streams and successfully find moist soil there. They spend the day in these burrows whenever the weather is dry. The burrowing form of life and the nocturnal activity pattern of amphibians are adaptations against excessive water losses. The grass mites (Oribatei), carriers of monieziosis (a disease of domestic cattle), survive very well in moderately moist soils and die off immediately in both very moist and dry soils (Fig. 52). Their distribution depends entirely on the humidity of the habitat (Moskacheva, 1960).

Those invertebrate animals which are demanding so far as moisture conditions are concerned are active primarily in the night hours, but in areas with a small deficit they may be active in the daytime as well. For example, hungry female malarial mosquitoes may attack their targets in broad daylight in spring and autumn when the temperature is rela-

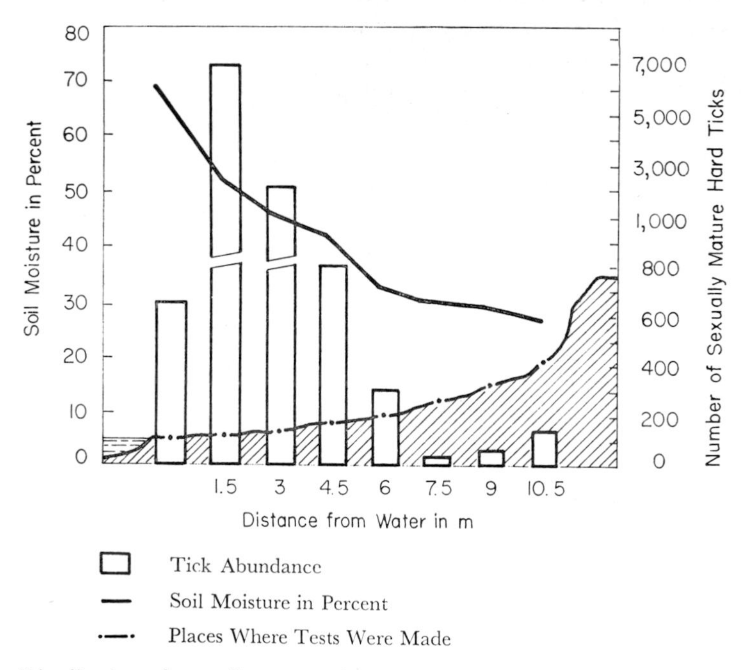

Fig. 52. Distribution of sexually mature hard ticks in relation to humidity of the soil in a floodplain (after Moskacheva, 1960).

tively low. They are active around the clock in shady and moist biotopes (the forest) (Beklemishev, 1944). Fleas leave the burrows and nests of their hosts more at night.

A unique form of adaptation to insufficient moisture is torpor or hibernation. These phenomena may take place either during acute drops in the humidity of the air (especially among the small invertebrate animals) or as a result of a decrease in the amount of moisture in the food. The inhabitants of mosses and lichens growing on trees and rocks may go into torpor many times during a single season; in the European forests we find 70 species of nematodes, 19 of tardigrades, 19 of rotifers, and 11 of protozoa living in such places. When shallow bodies of water or ponds dry up, their inhabitants fall into torpor. Earthworms become encapsulated in dry soils. Wood lice may continue to exist up to 60 days in a state of torpor. Certain species of them (*Hemilepistes*) are common inhabitants of the desert and are encountered en masse on the clayey soils of the Ust-Urt plateau, where slightly more than 100 mm of precipitation fall per year. Terrestrial mollusks dig themselves into the litter or the soil when drought sets in, or else they hide under stones, covering the openings of their shells with a film of dried mucus and then going into torpor (Tsvetkov, 1939). Narrow-footed crayfish and the sea acorn *Chtamalus stellatus* dry out, cover their shells, and go into torpor when the sea is calm and the waves cannot reach the stones to which they have fastened their shells. When the waves grow stronger, the crayfish become active and begin to feed. When drought hits the African fish *Protopterus,* it forms a burrow-capsule on the bottom out of silt and a secretion of its own skin; there the fish falls into a summer hibernation until the body of water is refilled. The South American *Lepidosiren,* another lungfish, survives droughts in the same way. The loach *Misgurnis fossilis* hibernates in the silt of the dried-up tributaries in the Danube delta.

The steppe tortoise (*Testudo horsefieldi*) wakes up in February or March and goes back to sleep at the beginning of June (it is active only 2.5–3 months) in the flat deserts of Uzbekistan. In the foothills of the Pamir-Alay and the Kopet Dagh it is active from March through August.[9] Yellow susliks begin their hibernation in summertime, when their food plants dry up (Kashkarov and Leyn, 1927). In the deserts of Uzbekistan this occurs at the end of May or in June, while in the Volga-Ural sands it takes place at the end of July or in August.

Atmospheric precipitation represents a complex and involved factor. The humidity of the air depends on it. It may have a direct influence on the behavior of animals or it may cause their death. In the form of snow this precipitation serves as a background for the animals, hindering their

[9] The mechanisms of summer and winter hibernation are very similar.

movement but at the same time covering the soil and preventing it from freezing. Therefore, many biological phenomena are regulated by changes in the amount of precipitation.

The amount of precipitation falling in the southern Ukrainian steppes in August governs the growth of weeds and grains and, hence, the reproduction and fluctuations in abundance of the kurgan mouse (*Mus musculus hortulanus*) (Fig. 53). Changes in the abundance of locusts have the same indirect dependence on the amount of precipitation. Their appearance en masse usually comes in years with a small amount of precipitation, although individual fecundity is actually lower in such years because of a decrease in the number of ovarian tubules among the females (Rubtsov, 1933, 1935, 1936, 1938). This reverse dependence of locust abundance on the amount of precipitation has been explained in different ways in different cases. The great increases in the abundance of the Asiatic *Locusta migratoria,* which take place about every 20 years, correspond with the periodicity of droughts, which partially dry up the deltas of the southern rivers (their nesting territory) and thus increase the surface area suitable for egg-laying. Food in the form of reed and cane shoots is always abundant. When abundance increases, the locusts disperse and form new nidi. The increase in abundance is usually interrupted by high water in a moist year (Zakharov, 1946, 1950). Among the Siberian locusts, although high humidity increases fecundity, it favors to an even greater degree the reproduction of parasites and the development of fungous and other diseases of these insects (Rubtsov, 1936, 1938). The American and Australian locusts are likewise drought-loving, and the limit of the zone favorable to them correlates very well with isopleth 1 for September in eastern Australia

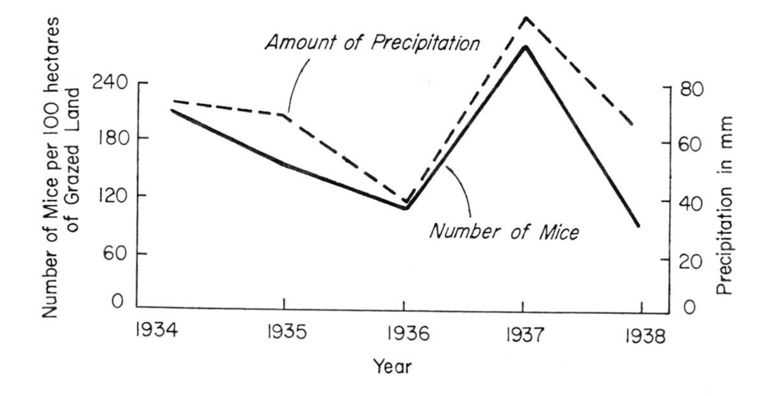

Fig. 53. Changes in the amount of autumn precipitation and in the abundance of kurgan mice in the steppes of the southern Ukraine (after N. P. Naumov, 1948).

and for August in western Australia (Andrewartha and Birch, 1934). The multiplication cycles of the desert locust, on the other hand, come during periods of increased moisture, also corresponding to a 10–20-year pattern (Predtechensky, 1928, 1935; Shcherbinovsky, 1952).

In dry years (during spring droughts with less than 100 mm of precipitation) the moth *Porosagrotis orthogonia,* an important destroyer of newly sown cereal grains, appears en masse. The same behavior is known for the destructive Canadian forest moth *Choristoneura fumiferana.* It has not been established just how a dry period affects these species favorably. It seems probable that the effect is complex, largely indirect, and apparently not the same for all species. Of special importance are fungi, viruses, and other disease agents which spread during moist periods and in damp locations.

High moisture levels may aid in decreasing the abundance of certain animals such as silkworms by limiting their fecundity. A moist habitat and heavy precipitation aid the development of helminth infections among hares. Precipitation may also be a direct cause of lowered abundance. Such is the influence of cloudbursts on small rodents; they often perish en masse, the different species being affected to different extents (Fig. 54). The

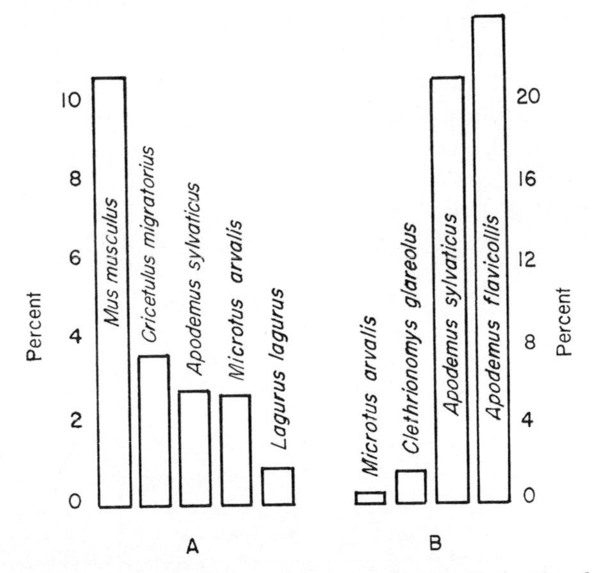

Fig. 54. Influence of cloudbursts and ice storms on changes in the abundance of certain species of small rodents (after N. P. Naumov, 1948). A: growth in the number of animals from June to September (in percent), withheld by July cloudbursts (southern Ukrainian steppes); B: percentage of animals surviving the ice storm in the winter of 1938/39 in the Tul'skaya oblast' forest.

relatively nonmobile voles, which stay quite close to their burrows, are killed off to a relatively great extent during cloudbursts and ice storms. Mice often escape such a fate because they move around more and are not so closely associated with subterranean refuges. Long and heavy rains cause heavy deaths among small birds and many invertebrate animals, including the soil-dwelling species.

Periodic floods and high water have proven to be a factor in the life of river floodplains and certain lakes. They control the times when periodic phenomena occur among floodplain inhabitants. Reproduction of the water vole (*Arvicola terrestris*) in the Volga delta is interrupted at high water (Fig. 55). Farther to the north, however, it goes on without interruption. Multiplication of other inhabitants is governed by the time when the necessary places, refuges, and available food are free of water. Soil animals pass the flood period in a state of torpor; as a result, long floods tend to impoverish the populations of river floodplains. The success with which animals will survive a flood depends on their mobility and the presence of refuges. Large animals (hoofed animals, carnivores) usually leave the floodplain for this period. Small animals collect on "islands of salvation" (unflooded patches, bushes and trees, flotsam and jetsam, etc.).

On the floodplain of the middle course of the Dnieper, high water has a different effect on insect life during certain years. Soil insects survive high water well in their wintering places if the water is cold and the insects have not yet emerged from their state of winter torpor, but a flood of high temperatures leads to mass death. The surface and litter inhabitants which

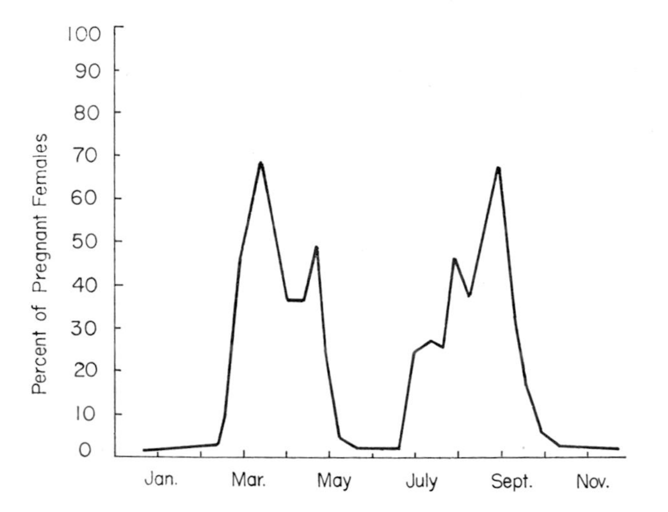

Fig. 55. Change in the percentage of pregnant females in the water vole (*Arvicola terrestris*) population in the Volga delta (after Kondrashkin, 1946).

have emerged from their winter torpor migrate actively to high ground at the approach of the water and save themselves more effectively than in cold weather, when the flood finds them in the torpid state. A short flood usually causes insects to leave their wintering places together and then multiply (Kryshtal', 1955).

Among small rodents success in flood survival is determined by climbing or swimming ability (Table 18). These abilities are typical of red voles,

TABLE 18. *Deaths of small rodents owing to predators during a flood on the Mologi River (after N. P. Naumov, 1948; Adol'f, 1951).*

	SHARE OF POPULATION ACCORDING TO DATA FROM REPORTS IN PERCENT			FREQUENCY OF ENCOUNTER WITH FEATHERED PREDATORS IN PERCENT		
Species	*In Field and Meadow*	*In Woods*	*Average*	*Owls*	*Carrion Crows*	*Average*
Common vole	79.0	14.0	46.5	46.2	79.0	65.3
Economist vole	0	16.0	8.0	3.8	7.9	6.7
Red vole	15.0	67.0	41.0	19.2	7.9	12.0
Field mouse	6.0	3.0	4.5	30.8	5.2	16.0

which are saved by climbing up trees, and economist voles, which dive and swim well. The common vole has been exterminated almost completely, since it clusters on high and dry islands where it is easily available to stalkers. Settlement of the river valley by this species took place slowly as they came down from the banks. The red vole and the economist vole quickly resettled the entire forest floodplain simply by jumping back down from the trees.

For the majority of floodplain inhabitants a low level of standing water in the fields and a short flooding period are the most favorable conditions, while a high water level over a long period of time causes a decrease in abundance (Fig. 56). The latter is caused either by the death of the animals directly from flooding (ermine) or by the outbreak of an epizootic among the animals clustered together on high ground. However, on floodplains which are not flooded yearly (the so-called high-level plains) only a high flood, filling the intermittent lakes and other bodies of water of the floodplain, favors the reproduction of the amphibionts associated with it (water voles) (Fig. 57).

The destroyers of floodplain forests might serve as an example of the complex influence of floods. On elevated, seldom-flooded ground in the Dnieper valley a greater number of parasites and predators of the tree-threatening moths are preserved. The leaves of the willow and the oak in these places contain more tannin and are less favorable for the feeding

of gypsy-moth caterpillars. The chrysalis of this species weighs 11 percent less here than it does in forests flooded in the lowlands. So in elevated areas reproduction of harmful forest moths seldom occurs and does not usually last long. But in flooded areas the chemistry of the tree leaves is more

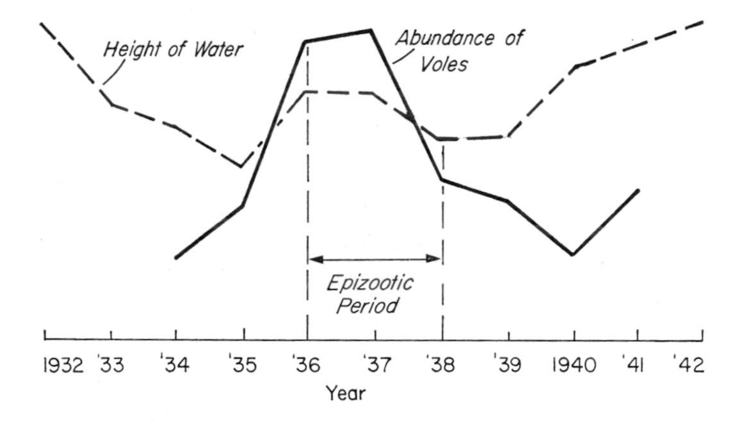

Fig. 56. Height of river water, abundance (from records of skins taken) of water voles, and development of tularemia epizootic among them in the Azov–Black Sea area (after Formozov, 1946).

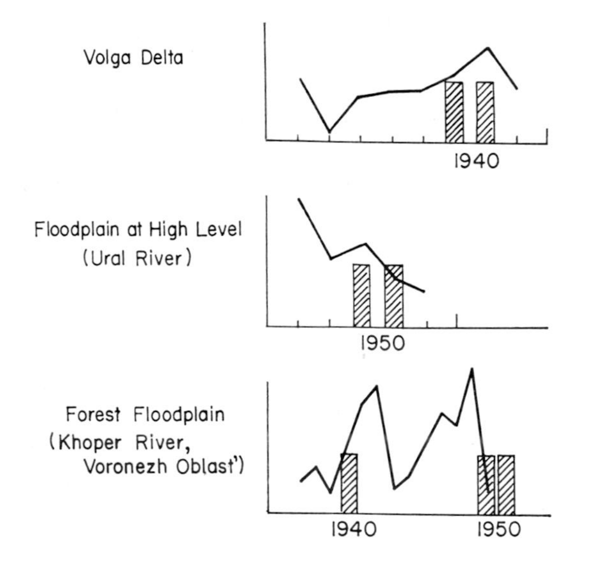

Fig. 57. Relationship between flood size and abundance of water voles (after Shilov, 1954). Curves show level of floods (within given boundaries); columns represent years of high vole abundance.

favorable to the *Bombyx,* and the pests are more fecund there. At the same time predators and parasites survive poorly here, since entomophages die off quickly in floods. This is why the larger and more resistant nidi of harmful forest Lepidoptera (gypsy moths, annulated silkworms, tussock moths, etc.) are found in the flooded forests and the lowland woods (Lozinsky, 1960).

Rising rivers govern the periods and the success of fish fattening. Thus the ide in the Ob' River fattens primarily on "litter," the grain-filled waters which flood low-lying cereal-growing fields. In years of high water the flood stays for long periods, sometimes until autumn, the period of fattening is lengthened, and the growth of the fish is accelerated; in years of low water the fecundity of the ide may drop from 20 to 40 percent.

Fluctuations in lake levels are more or less periodic in character and are typical of drought-afflicted areas of the Old and New Worlds. They are accompanied not only by changes in the amount of water—the disappearance or re-appearance of watering places necessary for many animals—but also by changes in salinity, contamination, and other properties. Therefore, such fluctuations affect the entire circle of animals connected in one way or another with the watercourse by directly or indirectly altering their fecundity, mobility (migration), abundance, and distribution. Among our people these fluctuations are well known in southern Siberia and North Kazakhstan, where they are repeated approximately once every 15–20 years (Formozov, 1937, 1953).

MINERAL NUTRITION OF TERRESTRIAL ANIMALS

Most terrestrial animals, especially the carnivores, easily find the necessary quantities of salt in their food and drinking water. But in some places these demands, especially among the grass eaters (ruminants, rodents, etc.), are not completely satisfied, and the necessity arises for finding supplementary mineral foods in salt licks, etc. Many animals are included here: hoofed animals, carnivores, and certain others. Salt shortage is usually seasonal, which explains the seasonal visits the animals make to the salt runs. Hoofed animals and carnivores usually visit them in the greatest numbers during the summer.

In addition to salts, many mammals need clay, which aids in the formation of dry masses of dung; this is especially important for the highly mobile hoofed animals. Since a change of diet is usually accompanied by a disturbance of digestion, the demand for soil becomes especially great

in spring (during the transition from winter to summer food) and autumn. In both periods animals are frequently observed eating earth on the salt licks. In addition to these animals, wolves, foxes, and others are frequently seen to visit salt runs and eat earth at this time (Ivanov, 1953; Smirin, 1955).

In recent years we have greatly increased our knowledge of the animals' demands for substances encountered in very tiny amounts in nature (the microelements). Previously considered to have turned up accidentally in the body, these elements have proven to be indispensable for many biochemical reactions. A lack of microelements depresses or disturbs many functions, and their absence causes improper development (malformation), illnesses, and premature death. The composition of a biologically active substance such as vitamin B_{12} includes cobalt, thyroxin requires iodine, insulin and the enzyme carboanhydrase need zinc, the enzyme xanthine oxidase needs molybdenum, trypsin needs chromium, thyroxinase and the respiratory pigment hemoporphyrin need copper, arginase needs maganese, and the vanadocytes need vanadium. Naturally, a surplus or shortage of these elements in the external environment (the soil, bodies of water) affects the exchange of matter, the vital resistance, and the form of an organism and aids in the working out of adaptations to compensate for improper chemistry in the environment. In areas of the globe distinguished for their lack or surplus of certain biologically important chemical elements—in certain "biogeochemical provinces"—plants and animals may differ in their forms and physiological properties, and among them one may encounter specific diseases caused by the geochemical features of the area.

The study of the interactions of organisms and the geochemical environment in concrete biogeochemical provinces makes up the discipline known as geochemical ecology. Its tasks include the discovery of the biochemical, physiological, and structural adaptations to definite amounts of biologically important chemical elements in an environment and the study of the distribution and causes of the so-called endemic diseases and other anomalies caused by geochemical factors (Vinogradov, 1936; Koval'sky, 1957, 1958).

A study of the non-chernozem forest-belt region, which is poor in cobalt (its content there varies from 3.7×10^{-5} to 25×10^{-5} percent), has shown that the cobalt insufficiency has complicated the synthesis of vitamin B_{12} by the microflora in the animals' intestines. Some 15–20 percent of the local animals were observed to suffer from hypo- and even avitaminosis of B_{12} (acobaltosis). In the Azerbaydzhan SSR, where the cobalt content of the soil is high (3×10^{-3} percent), the livers of sheep have an increased amount of vitamin B_{12} ($5–6 \times 10^{-5}$ percent). But the highest level of B_{12} in the liver

($7–8 \times 10^{-5}$ percent) was found in sheep in regions where the amount of cobalt in the soil was 1×10^{-3} percent. The relationships of organisms to a shortage or surplus of cobalt in the environment are not identical in animals from the various biogeochemical provinces. A shortage is tolerated better by inhabitants of poor provinces, while a surplus is withstood best by inhabitants of areas rich in this element.

Non-chernozem soils are also characteristic of regions poor in copper and, to a certain extent, iodine. Under conditions of copper hunger (insufficient copper in plant foods) the growth of young farm animals slows down, their coats grow shabby and rough, and their bones become light and brittle. The meat and milk production of adults falls off, and anemia similar to that caused by iron deficiency develops, since copper takes part in the body's assimilation of iron. Endemic atoxia, characterized by early impairment of coordination, is associated with copper deficiency; apparently this is caused by decrease in activity of the oxidation-inducing enzymes of the central nervous system.

In a region rich in boron (northwestern Kazakhstan) about 3–5 percent of the sheep suffer from "boron enteritis," which may be associated with a decrease (of two times) in the activity of the proteolytic enzymes of the intestine. (In this area enteritides of children are also frequent.) As a result of iodine deficiency in the non-chernozem belt, and especially in the mountains, abnormal enlargement of the thyroid gland, endemic goiter, and prevention of thyroxin synthesis are observed. An underactive thyroid lowers the body's gas-exchange level, oxidation processes, and heat production. Among farm mammals this is accompanied by stunted growth, low fecundity, and decreased milk and wool production; in birds, it is accompanied by decrease in egg production.

In the mountains of Armenia there is an area where the plants contain a great deal of molybdenum. In 24 hours a sheep here will obtain an average of 8.0 mg of molybdenum per kg of live weight, as opposed to 0.8 mg outside the area. An increase of 1.5 to 2 times in the amount of molybdenum in the blood of animals or man encourages the synthesis of xanthine oxidase, which accelerates the transformation of xanthin and hypoxanthin into uric acid (up to 6–10 mg per 100 ml in the blood, as opposed to the normal 2–4 mg per 100 ml). Gout is widespread among the people of this region. In areas with a surplus of nickel (North Caucasus) the mineral is abundantly stored in epidermal formations, in particular in the corneas of the eyes of lambs and calves, which leads to the so-called "nickel blindness" in 15–30 percent of the animals. An increase in the nickel level in bodies of water where *Bufo viridis* undergoes metamorphosis speeds its development but slows down growth.

The so-called "Urov" biogeochemical regions are situated in the moun-

tain-taiga area of the eastern Trans-Baykal and the Amur oblast'. Here animals and man are affected by the endemic Urov disease (after the Urov River). It is expressed by disturbance of the epiphyses and the cartilage; in animals (and in man) it affects the joints, dissolves the joint cartilage, interrupts growth, and deforms the bones. Bone fractures are frequent here and are even observed in wild animals (roe, fox); the animals are often dwarfs, with low productivity. These regions have very little calcium and large amounts of strontium and barium in the soil.

An increased amount of fluoride leads to the development of "fluorosoma" in animals. Because of favorable micro- and macroelements, endemic diseases are not observed in the chernozem zone and are seldom seen in regions with gray forest soils (Koval'sky, 1957, 1958).

All this reveals the extraordinarily important, but still poorly studied, influence of geochemical properties in the environment on the lives of animals and plants. Undoubted connections have been established among specific changes in the exchange of matter, the structural properties of organisms, and the geochemical factors. These connections could not help but affect the evolution of animals and plants. Actually, the animal populations inhabiting a definite biogeochemical province are adapted to its chemistry. This is proven by the low percentage (usually no more than 10–20 percent) of illness caused in the animals by the specific deficiency, while 80–90 percent of the local animals are resistant to the disease. Individuals from other provinces are, as a rule, always affected.

CHAPTER 4

Gas Exchange

GENERAL SIGNIFICANCE OF GAS EXCHANGE

Animals are aerobes or oxybionts; i.e., they need free oxygen for respiration. Only a small number of protozoa, certain worms dwelling in silt (the Tubificidae), the dipterous larvae of *Corethra* and *Chironomus*, mollusks of the genus *Pisidium*, and crustaceans of the genus *Candona* are capable of existing for any appreciable length of time under anaerobic (nonoxygenated) conditions.

Absorption of oxygen from the external environment is accomplished either by the entire surface of the animal's body or occurs in special respiratory organs (lungs, gills, tracheae, etc.), but it must take place through a film of water. The acquisition, transport, and delivery of oxygen to the individual organs, tissues, and cells is performed by the blood with the aid of the blood pigments. In most animals the work is done by iron in the form of hemoglobin, in arthropods and mollusks iron is mixed with copper (hemocyanin), and in the tunicates it is mixed with vanadium. The blood pigments may be concentrated in special structures—erythrocytes—or dissolved in the blood or other internal fluids. One characteristic of the blood pigments is their ability to absorb (fix) free oxygen when it is abundant, forming unstable compounds (the oxyhemoglobins), and deliver it into the surrounding environment whenever an oxygen deficiency exists. This ability of the pigments is different in the various species of animals and to a significant degree explains the fact of their existence in places having different quantities of oxygen.

In the body of an animal oxygen maintains the exothermic oxidizing processes in which matter and energy are liberated to be used in the development and activity of the animal. This includes the metabolic water so important in the water balance of animals living in dry areas. In this

total process of metabolism the basic exchange includes all matter and energy liberated for use in the basic functions of the body: respiration, blood-building, excretion, neurohumoral regulation, and growth. By calculating the magnitude of basal metabolism, which is measured by the consumption of oxygen registered while the animal is in a resting state, one can judge the total level of vital activity. Movement and other forms of activity demand a supplementary expenditure of matter and energy, expressed in an increased demand for oxygen.

The amount of energy set free by oxidation depends on the substances to be oxidized. Oxidation of carbohydrates, in which the greatest amount of energy is liberated, is characterized by the relationship between oxygen absorbed and carbon dioxide released. This relationship is called the respiratory coefficient and is equal to 1 when normal. When proteins or fats are oxidized, the respiratory coefficient usually drops to 0.8–0.7. But if it is raised and exceeds 1, this means that a significant amount of anaerobic fermentation is occurring, a phenomenon which is less profitable from the standpoint of energy and which is characteristic of animals when there is a deficiency or lack of oxygen.[1] In this case the glycogen reserves are used up ten times faster than under aerobic conditions, while the blood fills up with carbon dioxide. This is observed not only in hydrobionts living where there is little or no oxygen (Fig. 58) but also in mollusks inhabiting

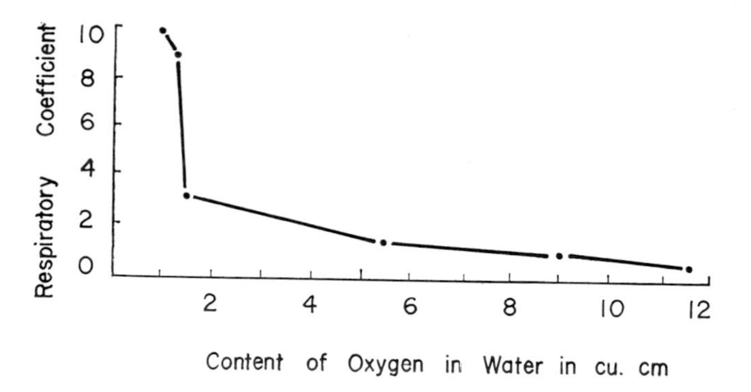

Fig. 58. Change in respiratory coefficient in *Sipunculus* in relation to oxygen content of the environment (after Skadovsky, 1955).

littorals, which cover themselves completely with shell when the tide goes out. After 30 hours in the air *Mytilus edulis* has 3.5 times more carbon dioxide in its blood than when in water.

[1] When 180 g of sugar are fermented into alcohol or lactic acid, 26.0–34.2 calories are released, but 720 calories are released if it is "burned."

Heat energy obtained from the oxidation process is used to heat the body and is lost if heat is released into the external environment. The size of such losses depends on the relationship between the body's mass and its surface area. In small bodies the relative surface area per unit of body mass is greater, and the relative degree of heat release is less among the large animals. In connection with this, some animals have a tendency to reduce their consumption of oxygen as their body dimensions grow larger. This principle, known in physiology as Richet and Rubner's law (Richet, 1889; Rubner, 1927), is generally applicable to the cold-blooded animals as well (Table 19 and Fig. 59).

TABLE 19. *Absorption of oxygen by freshwater crustaceans at 20° C (after Shcherbakov, 1950).*

Species	*Weight of 1,000 Individuals in g*	*Absorption of Oxygen in mg per g of Weight in 24 Hours*	*Respiration per Unit of Surface*
Cyclops bucarti:			
Male	8.2	50.0	10.1
Female	30.1	49.3	17.3
Diaptomus gracilloides:			
Male	55.0	34.4	13.1
Female	99.3	26.1	12.1
Daphnia longispina:			
Female	280.0	19.1	12.5
Cyclops stensus:			
Female	411.5	16.9	11.4

GAS EXCHANGE IN AQUATIC ANIMALS

Aquatic animals differ from one another in their demands for oxygen. The highly demanding ones live in bodies of water containing not less than 7–11 ml of oxygen per cu. l; these are the oxyphilic stenoxybionts. Less demanding species can be satisfied with a total of 0.5–4.0 ml of oxygen per cu. l. Widespread among these are the euryoxybionts, but there are also oxyphobic stenoxybionts, inhabitants of places with poor oxygen supplies.

Among the fishes the oxyphiles include the forms living in swift-flowing and cold streams and rivers (brown and aurora trouts, *Charius,* chub, gudgeon). The euryoxybionts, some of which are oxyphobes and flee high concentrations of oxygen, are represented by the inhabitants of ponds and by bottom fishes in slow-flowing rivers, backwaters, and lakes (ruff,

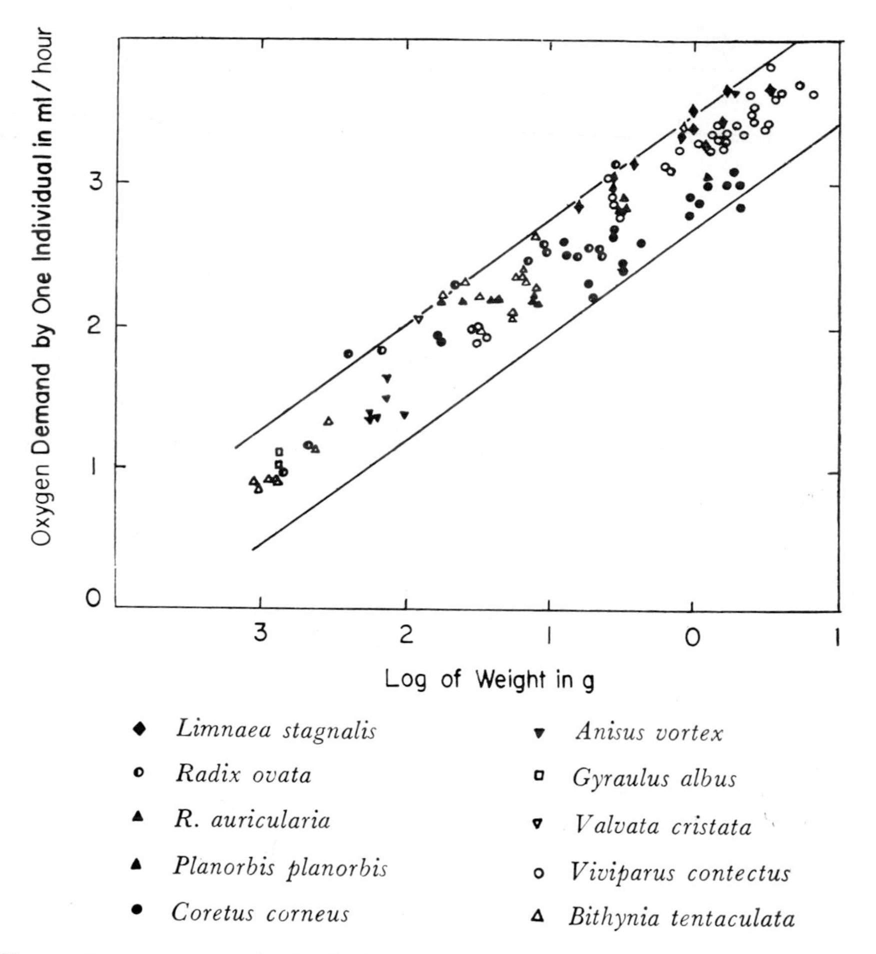

Fig. 59. Oxygen consumption by freshwater gastropod mollusks at 20° C (after Vinberg and Belyatskaya, 1959).

perch, carp, bream, crucian). The oxyphobes include the majority of those fish inhabiting the bottom, especially those which live in the mud. Many protozoa, worms, crustaceans, and mollusks inhabiting the silt may live for a long time with almost no oxygen at all. When oxygen is completely absent, most of them are inactive; they stop moving, swell up, and fail to react to external stimuli (Vinberg, 1950; Skadovsky, 1955). Among this group—for example, among the larvae of *Chironomus plumosus*—absence of oxygen in the environment even brings a halt to exchange, which normally adapts these animals to a state of "oxygen debt" (a low oxygen content in the tissues).

Consumption of oxygen differs among the hydrobiont species depending on their activity and general level of vital functioning. In Table 20 one may see the sequential growth of the level of matter exchange in the phylogenetically younger and more highly organized groups, as compared with the primitive ancients.

TABLE 20. *Consumption of oxygen and respiratory coefficient of aquatic animals (from Zernov, 1949).*

Species	*Oxygen Demand in ml per kg of Weight per hour*	*Carbon Dioxide/ Oxygen*
Sponge (*Suberites*)	6.50	—
Ctenophore (*Cestus*)	2.60	0.79
Ctenophore (*Beroe*)	5.00	0.84
Medusa (*Rhizostoma*)	7.20	0.90
Starfish (*Asteracanthion*)	32.00	0.79
Holothurian (*Cucumaria*)	13.30	3.00
Leech (*Hirudo*)	22.98	0.69
Oyster (*Ostrea*)	13.40	0.79
Mussel (*Mytilus*)	12.20	0.76
Octopus	44.10	0.86
Crab (*Astacus*)	38.00	0.86
Lobster (*Homarus*)	68.00	0.80
Shrimp (*Palaemon*)	125.00	0.83
Gammarus	132.00	0.72
Salpa	8.10	1.12
Lancet "fish" (*Amphioxus*)	35.70	0.96
Carp (*Cyprinus*) at 8° C	25.00	0.72
Carp (*Cyprinus*) at 24° C	104.50	0.83
Flounder (*Pleuronectes*)	80.00	0.60
Trigla	94.50	0.71
Phoxinus	140.00	0.86
Mullus	171.00	0.86
Trout (*Salmo*) at 10° C	100.00	—
Trout (*Salmo*) at 15° C	220.00	—

The oxyphiles, usually stenoxybionts, make up the so-called rheophilic fauna, which populates swift-flowing streams and rivers with a constantly high level of oxygen. Their gills are, as a rule, reduced as compared with those of the oxyphobes; their hemoglobin level is high, corresponding to their high blood-sugar level. They have no biochemical adaptations to remove high concentrations of carbon dioxide from the blood, as a result of which ions of alkaline-earth metals predominate in their blood, while their nerve centers quickly become depressed when carbonates build up in the bloodstream.

The opposing group—oxyphobes—has a more effective respiratory system. Oxyphobes have a hemoglobin level capable of absorbing oxygen

even when it is present in the environment in low concentrations, and even when potassium and sodium prevail over calcium and magnesium in the mineral portion of the blood, raising the system's ability to buffer high concentrations of carbon dioxide. Finally, the oxyphobes differ in the ability of their nerve centers to tolerate high concentrations of carbon dioxide in the blood (Florken, 1947).

Within the boundaries of the oxygen optimum, when the amount of oxygen in the environment satisfies the demands of the animal and does not depress it, respiration is not changed. From the point of view of energy, the consumption of oxygen here is at its most effective level, and the respiratory coefficient is especially low. In zones of oxygen adaptation, when the concentration of oxygen changes (rises or falls) beyond the optimum and when the level of carbon dioxide changes correspondingly, there is also a change in the frequency of respiration and in the effectiveness of oxygen use, i.e., in the respiratory coefficient (Lozinov, 1950).

Among the inhabitants of places with a frequent deficit or a generally low level of oxygen, the consumption of oxygen maintains a steady level despite wide fluctuations in availability and drops only after a definite oxygen threshold is reached in the external environment (Fig. 60). Among the oxyphiles the consumption of oxygen quickly drops when the gas is in short supply; the respiratory coefficient increases, and ultimately a general disturbance of exchange sets in.

In addition to oxygen, such gases as carbon dioxide, hydrogen sulfide,

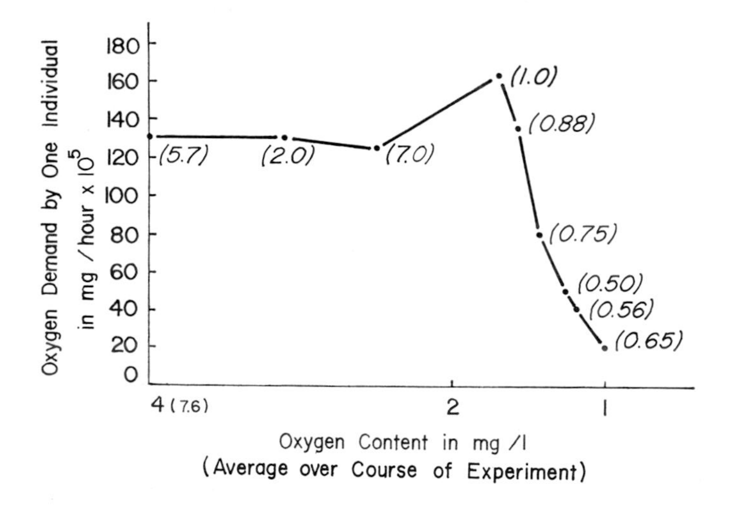

Fig. 60. Consumption of oxygen by *Daphnia magna* in relation to amount of gas in water at 20° C (after Skadovsky, 1955).

nitrogen, methane, etc. are important to animals living in water. Oxygen and nitrogen are absorbed by the surface water layers and are transported to the depths by turbulence and convection currents. Maintenance of these gases depends on the temperature and salinity of the body of water; their level diminishes when temperature and salinity rise. The amount of both gases in the water differs from the amount in the air because of the better solubility of oxygen, which is 1:2 instead of 1:4. The sources of carbon dioxide and other gases in bodies of water are the respiration of the inhabitants of the water and the action of microbes which break down organic matter.

The amount of oxygen in the various basins or portions thereof fluctuates between broad limits. In marine waters oxygen penetrates everywhere, which explains the fact that animals live even in the deepest parts of the ocean. But their abundance in the seas does not always correspond to the maximum supply of oxygen, which is explained by the ability of the blood pigments (hemoglobin and hemocyanin) to supply the demands of the organism by energetically fixing oxygen even when there is a comparatively small amount of oxygen in the environment.

In rivers and streams the amount of dissolved oxygen depends on aeration of the water, which is best brought about by complex longitudinal, transverse, and corkscrew-shaped movements of the water. It is thus best in mountain streams and rivulets and less noticeable in rivers with a slow flow. The supply of oxygen in lake basins depends on the size of their populations, the amount of decaying matter in them, and the depths of the bodies of water. In mineral-rich and densely inhabited "eutrophic" lakes the amount of oxygen quickly diminishes with depth and is practically nil at the bottom; there is usually none in the silt, where the restorative processes occur. In less densely inhabited, mineral-poor, and usually deep "oligotrophic" lakes oxygen is abundant and is distributed almost equally. In "dystrophic" waters situated in swamps and in almost uninhabited lakes there is a great deal of phosphorus, nitrogen, and calcium, but organic matter is also abundant; pH is low, and free oxygen is almost entirely absent.

A drop in the amount of oxygen in the water corresponds to a rise in the amount of carbon dioxide, while in certain bodies of water it corresponds to a rise in hydrogen sulfide and methane. The absence of vertical circulation leads to anaerobic conditions and accumulation of substances on the bottom, accompanied by the development of hydrogen sulfide fermentation. In the Black Sea this takes place with the participation of sulfate-reducing bacteria of the genus *Microspira* and bacteria which break down proteins and give off hydrogen sulfide; compared to the role of *Microspira*, the role of the latter bacteria is not great. As a result of both,

the depths of the Black Sea below 100–200 m are poisoned by hydrogen sulfide and are bare of animal life. Hydrogen sulfide fermentation is also known in certain fiords of Norway, in the Caspian Sea, and in other areas.

The availability of oxygen undergoes regular seasonal fluctuations, usually reaching its maximum in our latitudes in the winter. In abundantly populated waters or in those containing a greater amount of organic matter which is being broken down, the free oxygen under the ice may be completely used up, causing mass death among fish and other animals. These kills are of grandiose dimensions in the Ob' River below Tomsk. Usually beginning some 200 km above this city, the kill spreads at the rate of 30–40 km per day, passing down the course of the river and covering 1,800 km in a month and a half. In February and March it appears in the lower course of the river. The cause of the kill is the huge amount of humus material and iron oxide compounds carried into the Ob' by its tributaries flowing from the swampy West Siberian depression. These compounds absorb the free oxygen supplies, which cannot be replenished in winter because of the ice (Fig. 61). In the Vasyugan River free oxygen is almost totally absent during this period, while in the swift-flowing Yenisei, which flows over a stony bottom, the deficit is lower. The whitefish and white salmon withstand the kill poorly and die in large numbers, while the black fishes (ide and roach), gudgeon, and crucian suffer little from the kill. The decreased fish productivity of the Ob' is associated with these kills, and supplies of food are not used by the fish.

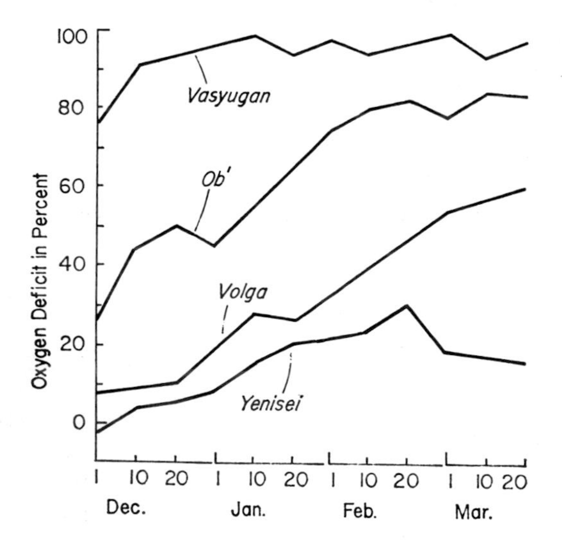

Fig. 61. Winter deficit of oxygen in Siberian rivers and in the Volga River (from Zernov, 1949).

Kills also take place in winter in ponds and lakes when the body of water may be covered with ice. The kills are more significant in eutrophic lakes, while inhabitants of oligotrophic waters do not usually experience a winter deficit of oxygen. The oxygen deficit is felt distinctly in the deep layers of eutrophic lakes in summer (Fig. 62). Diminution in the amount of oxygen is associated not only with its consumption by animals, plants, and decaying organic matter but also with the decreased photosynthetic activity of plants, which is accompanied by an increase in active environmental acidity, i.e., by a drop in pH.

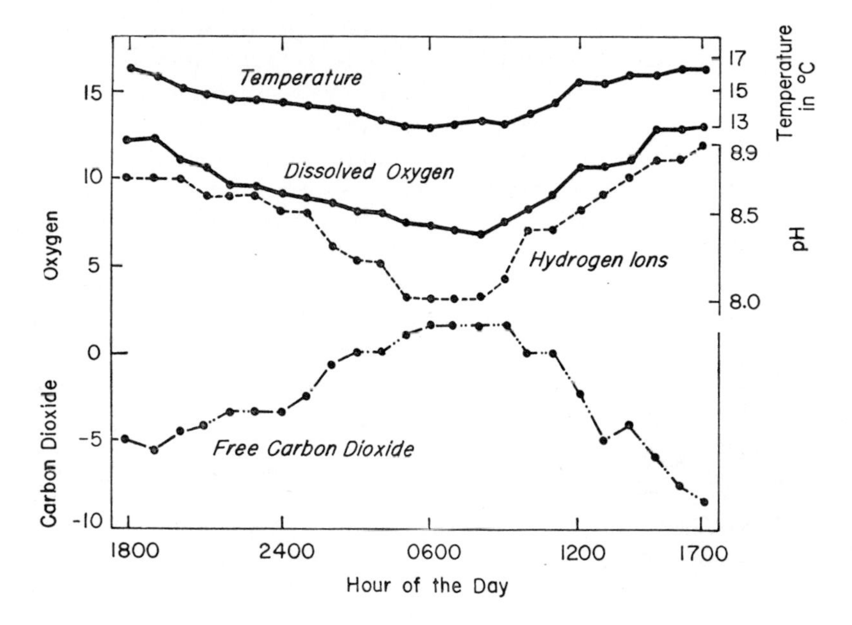

Fig. 62. Daily fluctuations in temperature, oxygen content, acid reaction (pH), and carbon dioxide content in the upper layers of a lake at the end of summer (from G. L. Clarke, 1954). A negative reading for carbon dioxide indicates carbonates dissolved in water necessary for neutralization of phenolphthalein.

The kills in the Azov Sea and in certain southern rivers are unique. They occur in summer when there are long periods of calm during which the surface and the upper layers of the water do not mix with one another. The breakdown of the abundant microbe-infested matter on the bottom leads to rapid exhaustion of the oxygen supply, a spread of the restorative zone down toward the bottom, and the production there of hydrogen sulfide fermentation caused by *Microspira aestuaria*. As a result, the benthos dies off completely in such areas, and a part of the fish population

also perishes. The first gale, however, eliminates this condition. When fish die in this manner in estuaries and lakes, conditions sometimes arise which lead to the development of anaerobic bacterial decay in their carcasses, or botulism. The eating of botulism-poisoned fish by seagulls leads to mass deaths among these birds and sometimes wipes out entire populations and colonies. On the banks of a salt lake in Manych Sound, Bakeyev, Skalon, and Chugunov (1957) counted 353 seagulls of the species *Larus gelastes* which had been wiped out in one stroke by botulism. These birds included 257 young and 96 adults—almost the entire colony.

GAS EXCHANGE IN TERRESTRIAL ANIMALS

In the transition from an aquatic environment to an open-air habitat expenditures of energy for movement (overcoming the force of gravity) are increased, which calls for an increase in oxidation and hence in respiration. When amphibians change from an aquatic (tadpole) to a terrestrial (adult form) way of life, the amount of hemoglobin (per unit of body mass) increases several times, while the cardiac index rises three to four times (Korzhuyev, Akatova, and Zubina, 1959).

The chemical content of the atmosphere is distinguished by its constancy. It deviates substantially from the norm only in caves, crevices, burrows, and a few other places where carbon dioxide, sulfurous fumes, and other gases from the earth's core accumulate, such as in the famous "Valley of Death" on the Yava. In cities and industrial centers where coal, petroleum, and other fuels are burned, the air becomes polluted with the products of combustion. Similar pollution may be observed near active volcanoes. With these exceptions the atmosphere below 1,000 m above sea level fully satisfies the gas demands of animals. Oxygen deficits associated with decreased partial pressure become important only at high altitudes. In the soil the amount of oxygen decreases with increasing depth at a rate depending on the soil structure, its mechanical composition, and its hydrothermic regime as well as on the abundance of its organic debris and the speed of bacterial decay.

An oxygen deficit or a concentration of carbon dioxide higher than 0.03 percent (the norm) will disturb the gas-exchange scheme in terrestrial animals, increase the frequency of their respiration, impede development and growth, reduce fecundity, and, in species which hibernate, will bring on hibernation sooner. Terrestrial animals differ in their sensitivity to changes in oxygen levels. Birds which migrate at high altitudes are more resistant to oxygen deficit. Domestic ducks have tolerated an "altitude"

up to 6,000 m in a barochamber; carrion crows and swamp owls still felt well up to 8,000 m but died at 11,000. Adult rooks and pigeons died at 8,500 m. The greatest "altitude" toleration was shown by daws, magpies, starlings, and windhovers (9,000–10,000 m). These birds died at "altitudes" of 10,000–11,800 m. The fledglings of these birds sustained even higher "climbs" and died at "altitudes" of 11,000–11,400 m (Vasil'yev, 1949; Stroganova, 1950).

Death owing to lack of oxygen is associated with a decrease in the level of metabolism, a phenomenon especially easy to observe at temperatures which deviate far from the optimum (Table 21). The inability of most animals to maintain the oxidation process and conserve heat during a period of oxygen debt hinders them from penetrating to high altitudes.

TABLE 21. *Consumption of oxygen in cu. mm per minute by rats, depending on atmospheric pressure (after Dzhaya and Dzhelineo, 1935).*

Pressure in mm of Mercury	ENVIRONMENTAL TEMPERATURE IN °C	
	16.0	*33.5*
759–750	4.119	2.741
742	3.934	2.607
580–572	3.594	–
460–430	2.289	2.622

Terrestrial animals which often encounter oxygen deficits (mammals and birds which stay under water for some time and the high-altitude species) have a greater oxygen capacity in their blood (Korzhuyev, 1952). This increased capacity is associated with a rise in the amount of hemoglobin in the blood and an increase in the number of erythrocytes—a phenomenon observed in many species. Of the five species of rodents studied by N. I. Kalabukhov (1935, 1937, 1950), the forest mouse (*Apodemus sylvaticus*) and the yellow-throated mouse (*A. flavicollis*) from the Pre-Caucasus flatlands increased their hemoglobin content from 9 to 27 percent when raised in a barochamber to altitudes of 1,500 and 3,000 m and when actually taken to that altitude in the mountains. Unlike forest mice, yellow-throated mice died at an altitude of 3,000 m, even at a temperature of 11° C, but they survived at a temperature of 20–25° C. This explains why they can live at such altitudes in the winter only in human dwellings (Markova, 1948). In the field mouse (*Apodemus agrarius*) and the red vole (*Clethrionomys glareolus*), as well as in gray hamsters, the hemoglobin content rose insignificantly and even dropped 5–10 percent (Fig. 63). In nature the field mouse does not actually go more than 600–700 m up in the mountains of the Caucasus (Sviridenko, 1936, 1944) and 1,600

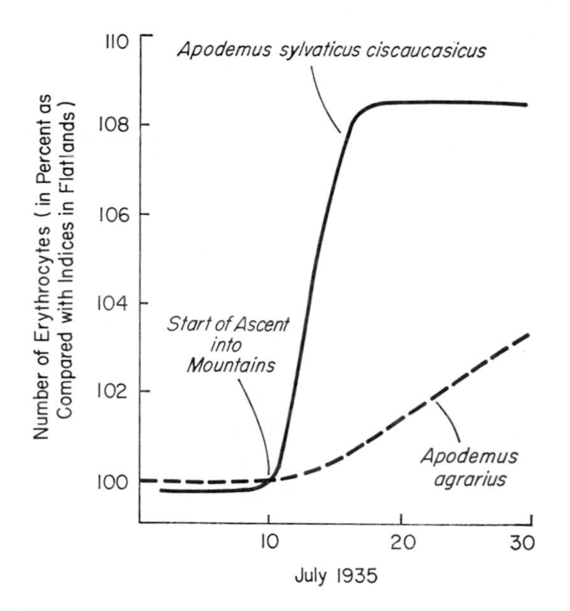

Fig. 63. Change in the number of erythrocytes when a forest mouse (*Apodemus sylvaticus ciscaucasicus*) and a field mouse (*Apodemus agrarius*) from the flatlands of the Pre-Caucasus were elevated to 1,500 m above sea level (after Kalabukhov and Rodionov, 1936).

m above sea level in the Tyan'-Shan' (Shnitnikov, 1936). The red vole never goes very high, nor does the gray hamster of the Caucasus.

Curtailment of tissue demands for oxygen also serves as an adaptation to life at high altitudes. This is observed in domestic animals which have long been accustomed to high-altitude life (local breeds of sheep—Barbashova and Ginetsiansky, 1942; Ol'nyanskaya and Slonim, 1947; horses—Ponugayeva and Slonim, 1949; and the minor mountain susliks—Bondarenko, 1950; Kalabukhov, 1950). Among them, as in the yellow susliks, crested and Asia Minor gerbils, and long-eared and European hedgehogs, the blood may be far from saturated with oxygen at 1,000–2,000 m above sea level and may contain large amounts of carbon dioxide for long periods of time as a result of suppression of the respiratory center. This lowers the level of metabolism but does not preclude the activity of such animals (Filatova, 1946, 1949; Shcheglova, 1949; Slonim, 1952).

The necessity of special adaptations for living at high altitudes unites the mountain fauna, the specific outlines of which are well known. The jerboa *Allactaga sibirica,* which penetrates to the high levels of the Tyan'-Shan', is free of its specific fleas there, although they are widespread among jerboas on the plains. At high altitudes these fleas are replaced by those of the high-mountain vole and the narrow-skulled vole (Ioff, 1949).

Amphibiotic animals (like the mountain species) have a series of adaptations for gas exchange which permit them to stay under water for long periods of time. In whales there is not only an increase in the hemoglobin content of the blood but also a feature which enables the alveoli in the lungs to "lock in" large amounts of air even at great depths. In the sperm whale *Physeter* the right nostril has been transformed into a unique air reservoir. In all the amphibionts the bufferability of the blood for carbon dioxide and the ability of the respiratory center to withstand a concentration of carbon dioxide in the blood are very high (Korzhuyev, 1952). These same physiological properties distinguish the diving ducks from other types (cormorants and others).

Generally, the birds, which are highly motile as compared with other vertebrates, have a hemoglobin content which is raised to 10 percent. In connection with this, the organs which form red blood cells (tubular bones) are distinguished by their greater development.

CHAPTER 5

The Significance of Radiant Energy

GENERAL SIGNIFICANCE OF EXCHANGE OF ENERGY WITH THE ENVIRONMENT

The exchange of matter in the organism must of necessity be accompanied by the absorption, production, transformation, and elimination of various types of energy, i.e., by its exchange with the environment. Both aspects of metabolism (exchange of matter and exchange of energy) are inextricably associated and mutually determined. One item which is especially important for organisms is exchange of heat and the absorption and elimination of other types of light energy. Kinetic energy, which is associated with the pressure and movement of the aquatic environment and the atmosphere, occupies a prominent place; so do the force of gravity, sound waves, mechanical alternations, etc. Therefore, the energy state of the environment includes its movements, mechanical fluctuations, temperature, electromagnetic field, and amounts of visible infrared, ultraviolet, and shortwave penetrating light waves, all of which influence the course of matter exchange and, by the same token, the state of the organism.

The energetic regime of the environment has a direct influence on processes within the organism, accelerating or decelerating them as a result of the body's having absorbed energy or, on the other hand, having dissipated it into the environment. This sort of influence is subject to the usual physical laws. But these same external phenomena, when picked up by receptors (the sense organs of the animal), are accepted as information about the environment and processed by the nerve centers, causing responsive reflex reactions which may not be applicable to simple physical laws and which rest upon historically formed, complex reflex mechanisms. With

the aid of these mechanisms, the animal, "predicting" oncoming changes in the environment, rearranges its metabolism and meets the changes in a state of preparedness. From this point of view we can see immense biological importance in factors of the environment which are of little power and have no substantial direct influence on the processes of exchange and life activity in the animal but which acquire significance as signals, since they regularly tip the animal off to daily, seasonal, and sometimes other changes in the environment. They include mechanical fluctuations, especially sounds and ultrahigh-frequency sounds, certain types of light energy, etc.

The different forms of radiation are of great importance in the exchange of energy with the environment, but the total significance of radiation in the life of animals cannot be regarded as well studied. The five basic types of light energy are electromagnetic light, with a wavelength of 1 m or more (radio waves); thermal infrared rays (wavelength of 4 mm to 7,700 Å); visible light (wavelength of 7,700 to 3,900 Å); ultraviolet light (wavelength of 3,900.0 Å to 40.3 Å); and the penetrating gamma and X rays (40.30–0.01 Å) (Fig. 64). Visible light and the infrared and ultraviolet rays adjacent to it are the best-studied forms of light. In the meantime, facts have been accumulating in the last few years showing the important and multifaceted significance of a greatly varied range of light energy for animal life.

It is known that in areas where powerful radio stations are operative, carrier pigeons lose their orientation and cannot find their way. Moreover, Academician A. F. Middendorf has expressed some considerations on the significance of the earth's magnetism in the orientation of birds during migration. Although today this orientation is treated as a complex instinctive process in which the most important role is played by the bird's experience in the postnidal period (hatching) of life (Promptov, 1941), the

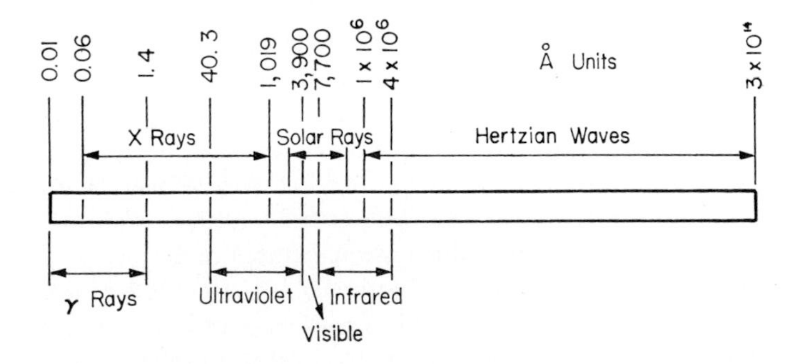

Fig. 64. The electromagnetic spectrum.

question about the importance of physical factors in orientation has not disappeared.

The means by which infrared rays act on animals and the importance of these rays within the range of 7,700–4,000,000 Å are set forth in Chapter 6. The information obtained by the sense organs enters into the thermal centers of the nervous system and serves to regulate not only the oxidation processes but also the motile reactions. The latter are established either by the type of thermokinesis (nonoriented movements usually strengthened when the temperature rises) or by thermotaxis (oriented movements in the direction of a given preferred temperature).

Visible rays (range 3,900–7,700 Å) have a mixed effect. Heat effects are concentrated in the red part of the spectrum, while the blue and violet end have mainly a chemical effect which is expressed in changes in the speed and direction of various reactions. The indirect (thermal) effects of visible rays on the life of certain species of animals are just as important, since they are associated with changes in the temperature and humidity of the environment, photosynthesis in green plants, and the level of activity displayed by the animals. Finally, visible light is especially important because of its signal significance, since daily and seasonal changes in illumination cause daily and seasonal biological cycles, the bases of animal life (see Chapter 9).

The ultraviolet part of the spectrum is distinguished by a very high level of chemical activity (Fig. 65). In the range from 2,000 to 3,000 Å the rays have a powerful bactericidal effect, destroying microorganisms in water, air, and on the surface of objects. In the range from 2,500 to 3,000 Å they cause sterols to form the antirachitic vitamin D. And at wavelengths from 2,000 to 4,000 Å they cause sunburn on the skins of human beings. The erythemic sensitivity of the skin (burning) is maximal at 3,000 Å, and the protective pigment in the skin is formed primarily under the influence of waves measuring 3,250 Å (Frank, 1950). Ultraviolet waves, combined with other environmental factors, affect even the most generalized ecological phenomena. It is interesting from this point of view to consider the correlation observed by Shelford (1951) between changes in the population abundance of certain animals and fluctuations in ultraviolet illumination.

The perception of visible and ultraviolet radiation by the receptors provides the animal with immediate orientation in the environment and enables it to find favorable conditions through the use of photokinesis and phototaxis.[1] The reflex reactions of metabolism caused by changes in the light regime prepare the organism for the approaching changes in the

[1] Photokinesis is a nonoriented augmentation of the animal's movements when illumination is intensified; phototaxis is movement oriented toward or away from a source of light.

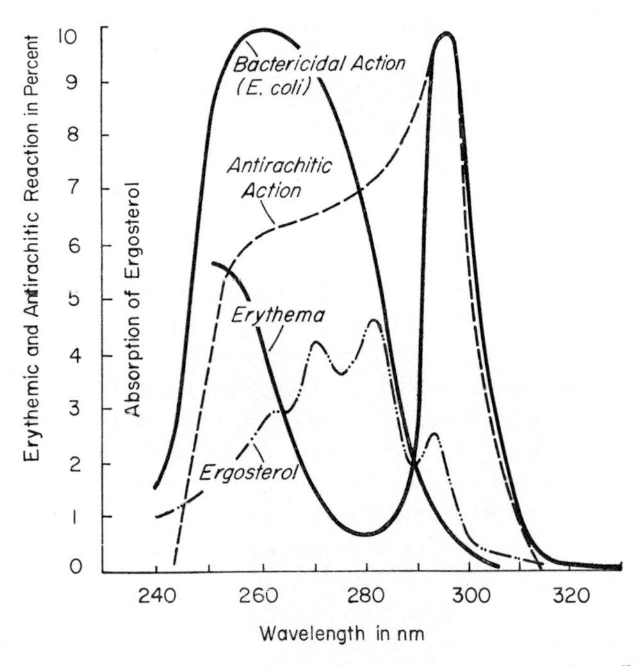

Fig. 65. The bactericidal, antirachitic, and burning (erythemo-formative) effects of ultraviolet rays with varying wavelength and absorption of ergosterol (after Koblentz, 1939).

environment and serve as a physiological basis for such phenomena as fat storage, hibernation, molting of skin or feathers, reproduction, and seasonal migration. The receptors of light energy in this range may be either the entire surface of the body (in the lower animals) or special organs (vision, sensation).

In the majority of cases penetrating shortwave radiation is fatal for animals. Even a short exposure leads to irreversible chain reactions associated with the formation of peroxide compounds within the cells; this destroys metabolism and, in the final reckoning, brings the animal to the point of death (Tarusov, 1957). The growing amount of radioactive dust in the atmosphere and the charged particles in water do much to augment the significance of this factor in the life of plants, animals, and man. "The radioactive background" grows and in many places becomes greater than should be permitted. However, the factor of penetrating radiation is not a new one on this planet. The overwhelming majority of cosmic rays are absorbed by the upper layers of the atmosphere, but a small part of them still manage to reach the surface of the earth. Radioactive decay within the earth's crust is also an important source of ionizing radiation. Being associated with definite parts of the earth, it has determined the existence

of zones with different levels of radioactivity. Investigations have shown that resistance to penetrating rays among plants and animals in zones with a higher natural radioactivity is significantly higher than in populations living in other regions. This means that ecologists now have the task of investigating the practically unstudied role of this growing environmental factor (Peredel'sky, 1957).

SIGNIFICANCE OF VISIBLE LIGHT AND ULTRAVIOLET RAYS

All animals need light of one intensity or another. The exceptions are the permanent inhabitants of deep caves (troglobionts), the soil, and the marine deeps below 1,500 m, since light can penetrate no lower even in translucent water. These species develop special adaptations for collecting minute amounts of light energy (greater size and complexity of the eyes) and have highly sensitive organs of hearing, sensing, and smelling. Some have special organs of illumination, while in others the eyes are wholly or partly atrophied. Lack of light is usually accompanied by a reduction in coloration.

When light has a direct influence (absorption), its ecologo-physiological result depends on the amount (dosage) of absorbed rays. Various species of animals are able to tolerate illumination of a definite spectral content, duration, and strength; exceeding these limits suppresses their viability and may lead to death. Using these features, we distinguish the light lovers (photophiles) and the shade lovers (photophobes), the euryphotics, which tolerate a wide range of changes in illumination, and the stenophotics, which can withstand only a narrowly limited set of illumination conditions. The deep-sea, cave-dwelling, nocturnal, and soil-dwelling animals are good examples of the last, while the diurnal species are usually more tolerant of light fluctuations. Nocturnal and twilight mammals (voles and mice) withstand direct illumination by the sun for only 5–30 minutes, while the diurnal forms (susliks) can take it for several hours (Babenyshev, 1938). However, in bright sunlight even the desert lizards *Eremias grammica, E. velox,* and *Phrynocephalus mystaceus* cannot withstand direct illumination long, since the temperature of the body rises to 50–56° C in 5–10 minutes and the animal dies (Strel'nikov, 1934; Kashkarov, 1945).

Illumination of the eggs of the sea urchin and of many insects accelerates their development, but only if illumination does not go beyond a known degree of magnitude which is different for different species. After this limit has been exceeded, development is impeded or stopped altogether,

especially if large doses of ultraviolet light are received. The larval period of the fruit fly *Drosophila* is shortened by illumination up to 2,500 metro-candlepower; anything over this figure causes development to slow down. The development of trout eggs (which are usually buried in the ground) stretches to 4–5 days in the light, while the eggs of the flounder, the Caspian sturgeon, and the herring (which normally mature in the light) develop a day or two later if kept in the shade. Any reaction to illumination will depend on the phylogenetically based adaptations of the animal to conditions of illumination and will correspond to the animal's way of life and the properties of its habitat.

The pigmented coverings of certain organs are usually regarded as a protection against excessive doses of light energy, but they may have a significance in heat exchange as well. The eggs of aquatic animals (fish, amphibians, etc.) developing in bodies of translucent water or swimming in the surface layers are usually pigmented, which aids their heat maintenance. The pelagic larvae of the herring *Spratella* and the cod have a dark-pigmented cover on the nerve trunk. Many reptiles, especially the southern species, have a black covering on the abdominal cavity. Such a covering is found again above the brain of the thin-fingered suslik (*Spermatophylopsis leptodactylus*), which lives in the deserts of Central Asia (V. Ye. Sokolov). In all these cases it is the more important organs which are covered: the nervous system, the organs of reproduction and blood formation, and others. Among the desert-dwelling *Phrynocephalus* lizards the pigmented spot on the peritoneum is largest in the southern populations and smallest in the northern ones. The jumping lizard *Lacerta exigua*, which lives even farther to the north, has a very small spot (Fig. 66). The intensity of pigmentation differs also. The *Phrynocephalus* lizards from the desert (Aral'sk) pass 13 percent of the light energy through their pigmented peritoneums, those from the southern semidesert (the sands of Irgiz-Kum) pass 18 percent, and those from the northern semidesert (the sands of To-Sum) allow 21 percent of the light to pass. Figures for the lizard *Eremias velox* are similar: 14.8 percent in the desert and 32.3 percent in the semidesert (Dinesman, 1949).

LIGHT AND ANIMAL BEHAVIOR

Animals avoid extremes of light, taking refuge in hiding places, cowering in the shade, or digging into the ground. The characteristic behavior of *Agama* lizards—the tendency to crawl onto bushes during the hot part of the day—usually regarded as a measure to prevent overheating, is actually

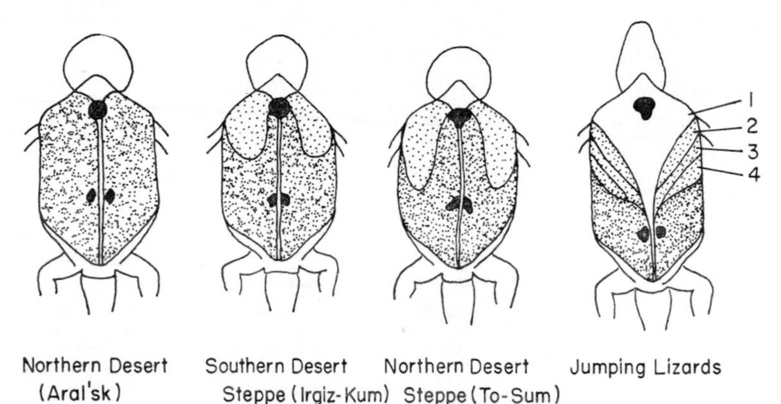

Fig. 66. Pigmentation of the peritoneum in long-eared *Phrynocephalus mystaceus kruglogolovki* from different areas and in jumping lizards (after Dinesman, 1949). Frequency of dots denotes intensity of pigmentation; in the jumping lizard the lines show the borders of the pigmented zone in individuals from Irgiz (1), Turgay (2), Naurzum (3), and near Ves'yegonsk (4).

typical only of the males and is associated with hunting for females and protection of nesting areas (Shibanov, 1940). Daily and seasonal changes in illumination determine not only changes in level of activity (periods of rest and activity) but also such cyclic biological phenomena as reproduction, migration, molting, etc. The photoregime acts here as a leading signal-factor, the mechanism of which is described in Chapter 10.

The ability to see is a varying one among the animals. Many vertebrates respond to light much as does man, distinguishing color, form, and number of objects; others have only black-and-white vision and do not sense the colors of objects. Nocturnal animals, as a rule, can see clearly in weaker light than diurnal animals. Some animals can sense polarized light. Illumination sensed as a signal-factor determines many features of behavior and spatial distribution of animals. Larval and adult mosquitoes react to definite degrees of light. During the daylight hours malarial mosquitoes prefer weakly illuminated refuges (0.15–1.50 lux) and avoid both complete darkness and strong illumination. The invertebrate soil dwellers are almost all photophobes. For them light is a signal warning of unfavorable conditions (dryness, high temperatures, etc.). The appearance of nocturnal insects and the disappearance of diurnal ones in the morning or evening take place under degrees of light which are definite for each but different for the different species of insects. The marble beetle *Polyphylla olivieri* usually appears 5–6 minutes after sunset (Prints, 1932). The waking hour of songbirds changes with the seasons and occurs earlier on clear days

(Fig. 67). The strength of illumination necessary to awaken the birds is called "waking brightness" by Promptov (1940).

Differing relationships to the degree of illumination are often observed even in closely related species which may live side by side (Fig. 68). The yellow-throated mouse, living in broad-leafed forests, prefers a low light intensity; the widespread forest mouse has no preference for any definite intensity of illumination. Two closely related species of susliks, the speckled *Citellus suslica* and the minor *C. pygmaeus*, living in different zones (steppe and semidesert) but in the same kind of biotope, both behave in the same way with regard to light (Kalabukhov, 1950).

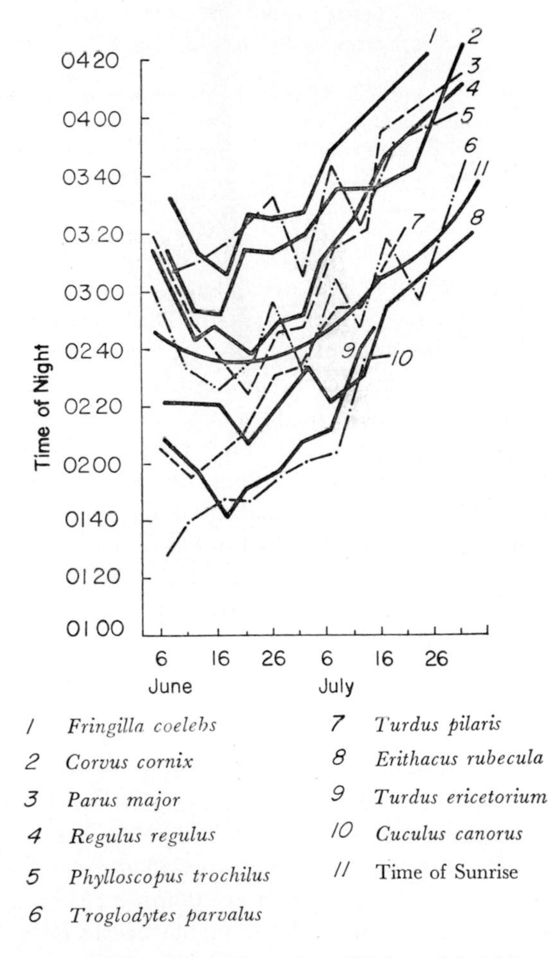

Fig. 67. Correlation between waking time of songbirds and brightness of illumination (after Promptov, 1940).

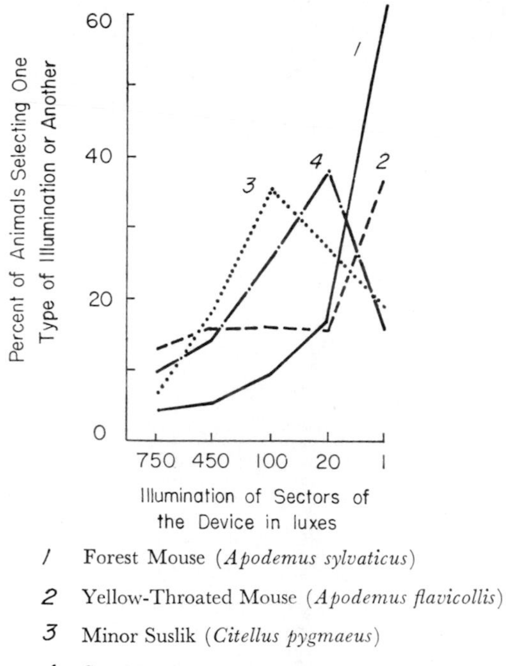

Fig. 68. Reaction to light in two closely related species of mice and two susliks (after Kalabukhov, 1938).

Depending on the degree of illumination and the temperature, the hunting grounds of birds may change. The flycatchers, chiffchaffs, and other small birds living on the edge of the forest will catch their food in the depths of the forest in the daytime but will hunt in the open spots at dusk and dawn. The same is true of woodpeckers, tomtits, and nuthatches. An example of the indirect influence of the light regime on animal movement is the changing pattern of catches associated with the mackerel (*Scomber scombrus*) as this fish approaches the shores of the British Isles (Fig. 69). An increase in the number of daylight hours in February and March aids in the development of phyto- and later zooplankton. The latter's abundance attracts mackerel to the shore, which explains the correlation between the catch curve in May and the sunlight curve for February and March in a given year.

The photoregime limits the geographic distribution of certain animals. Uninterrupted sunshine during the summer months attracts countless and varied bird populations, and sometimes mammals, to the higher latitudes. The overwhelming majority of them wander south in the autumn, since

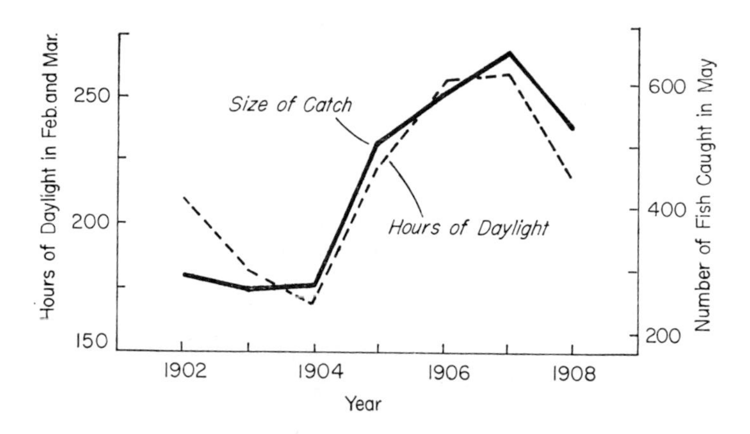

Fig. 69. Correlation between size of the May mackerel catch near the British Isles and number of hours of daylight in February and March (after Allen, 1909).

the short northern day is insufficient for many species to obtain enough food. Small birds with an especially high level of metabolism experience the limiting effect of short daylight to the greatest degree. This is expressed in their wintertime migration. In order to obtain a sufficient amount of nutritive matter, small birds must fill their stomachs five or six times a day. When prey is small (insects, larvae, eggs, seeds, etc.), food searches and selection require more time than the short northern day permits. This makes the existence of such species impossible there and serves as the basic cause of their flights to the south. Linnets, waxwings, pine finches, various species of tomtits, kinglets, woodpeckers, nuthatches, and certain others behave this way. They are driven south in the winter not by low temperatures (which their feathers protect them from very well) nor by a lack of food but by the sheer impossibility of getting enough food during the short northern day. On the other hand, the uninterrupted Arctic day in the summertime, with its increased possibilities for feeding, permits the young to grow quickly and cuts down the time needed for reproduction. A huge number of snipe, passerine birds, gulls, waterfowl, and predatory birds successfully mate and raise their young in a violent and quick process during the Arctic summer. The abundance of light in the high latitudes in summer provides that abundance and intensity of animal life which have so surprised visitors to the tundra, even though the climate remains severe during this period.

The light regime has an opposite effect on the distribution of nocturnal birds and mammals. In the north they are rare, while in the south they sometimes prevail even over the diurnal species. The number of goatsucker species, for example, diminishes toward the north, so that in our middle

latitudes we have only one, *Caprimulgus europaeus*. The females of this species lay two eggs, while the southern species lays three or four. The reverse relationship here is clear, since the period of activity for these species occurs in the night, and the possibilities for feeding the young depend on its length.

CHAPTER 6

Heat Exchange and the Temperature of the Environment

SIGNIFICANCE OF TEMPERATURE OF THE ENVIRONMENT

Directly or indirectly, the temperature of the environment affects the development, status, survival, and multiplication and hence the abundance and distribution of animals. The direct effects are associated with absorption of heat by the animal's body or dissipation of heat into the external surroundings. These phenomena depend on the temperature of the environment and determine the thermic state of the organism. This state is in turn associated with not only the intensity but also the character of metabolism, i.e., development and life activity. The speed of the biochemical processes in the organism, like that of all chemical reactions, increases two or three times when the temperature is raised 10° C (the Van't Hoff rule). This coefficient of acceleration (Q_{10}) for biological processes is operative only within certain temperature limits which permit vital activity. Reactions do not accelerate above this temperature zone; they decelerate or halt altogether. Below a definite threshold (the threshold of development) they cease.

But even within the boundaries of the favorable zone temperature does not always determine the speed of biochemical processes. There is no direct correlation between temperature and the speed with which eggs develop in many reptiles. The hatching of chicks cannot be accelerated by raising the temperature; they develop normally only over a period of 21 days at 40–41° C. An even hyperbola is the best expression of the correlation between temperature of the environment and development, activity

level, and even abundance of animals. But even the hyperbola expresses this correlation only approximately, since on the one hand each process has its own lower, optimum, maximum, and upper temperature limits, while on the other hand any of these processes may depend not only on the temperature of the environment but also on many other factors. Finally, the effect of external temperature on metabolism, development, and activity may be completely changed by absorption of solar radiation.

The indirect effects of environmental temperature on a given species consist of changes in conditions of development, reproduction, survival, activity, abundance, and distribution of animals and plants. "Temperature information," obtained by special sensing organs (thermoreceptors), serves as a means of orientation in the environment and is accompanied by active searching for favorable temperature conditions. Environmental temperature has important significance as a signal and as a regulator of the time and direction of movement (taxes, migrations); it affects the distribution and reproduction of animals.

Heat exchange in animals is closely associated with the whole energy scheme of the body. The acquisition, production, and expenditure of thermal energy is regulated with the aid of the reflex activities of the nervous system. Two basic types of thermal exchange exist: poikilothermy, or coldbloodedness, and homoiothermy, or warm-bloodedness. Poikilothermic animals are distinguished by their irregular level of matter exchange, varying body temperature, and almost complete lack of mechanisms for regulating the latter. Body temperature differs little from the temperature of the environment and changes right along with it (see Fig. 8); this phenomenon is closely associated with the activity of such animals (Fig. 70). Homoiothermic animals are distinguished by a higher (average) and more regular level of matter exchange, which acts as a basis for regulation of heat production and dissipation and provides a relatively constant body temperature. An intermediate group is composed of the heterothermic mammals and birds. During winter hibernation (susliks, dormice, hedgehogs, jerboas, certain birds) or deep sleep (bats, badgers, bears, hummingbirds, swifts) the matter-exchange level falls and the body temperature is only slightly above that of the environment. In their active state they are homoiothermic; i.e., they have a more or less constant body temperature.

POIKILOTHERMIC ANIMALS

At present an enormous number of poikilothermic animals live comfortably not only in bodies of water, where the temperature fluctuates

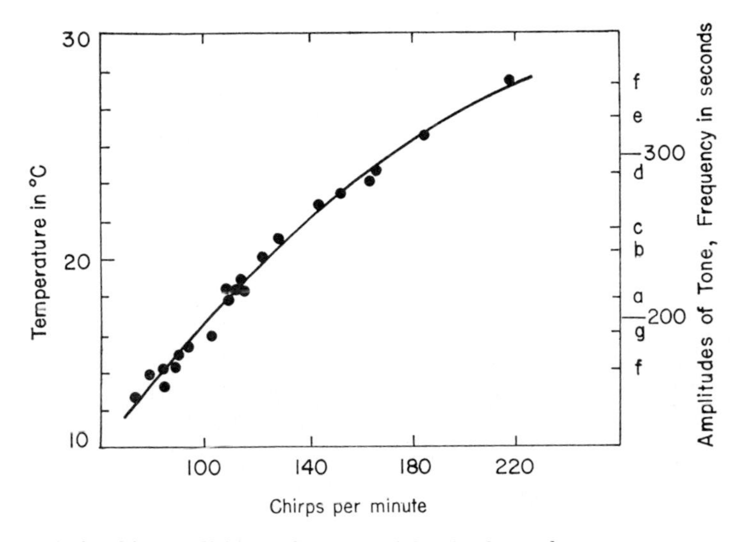

Fig. 70. Relationship of clicking of the wood beetle *Oacanthus* to temperature (after Matthews, 1942).

relatively little, but also on dry land in the temperate and even Arctic latitudes, with their sharp seasonal and daily changes in temperature.[1] This means that the heat exchange of such animals is adapted to a severe temperature regime; when unfavorable temperatures occur, heat exchange is associated with suppression of the body's activity level. This leaves the poikilothermic animals in a less profitable position as compared to the homoiothermic ones, especially in the higher latitudes.

The state of the poikilothermic organism changes regularly with fluctuations in the external temperature (Fig. 71). In such animals a zone of thermal suppression lies just below the upper temperature limit. Activity is still impossible within it. Below this area is the activity zone, in which development and reproduction are carried on intensively. The thermal optimum, with the best conditions for development and maximum fecundity, lies approximately in the middle of this zone. Below the temperature of the active state lies the zone of frigid torpor, which passes into the stage of supercooling, extending to the point where the body fluids freeze. When the body fluids freeze, the liberation of latent heat of fusion raises the temperature of the body for a short time, but afterward a new drop in temperature sets in, ending in final freezing of the body fluids and death of the animal. P. I. Bakhmet'yev calls the supercool temperature the "critical point."

[1] The number of terrestrial animals in the Arctic is thousands of times lower than in the tropics.

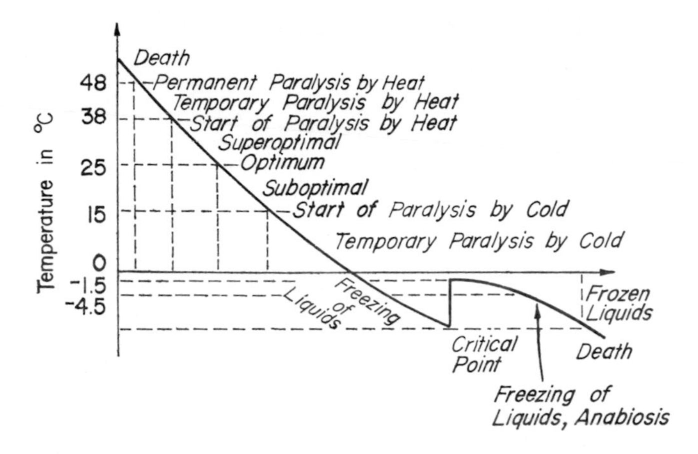

Fig. 71. Changes in body temperature and state of poikilothermic animal in connection with a decrease in environmental temperature (after Bakhmet'yev, 1902, with changes).

The upper limits of temperature are different for the various species and their stages (phases) of development. Above these limits existence is impossible because of the lack of harmony between vital processes, which accelerate at different speeds when the temperature is raised, and because of the inability of the respiratory organs to keep pace with the increase in heat (oxygen deficit). Excretory functions fall behind and paralyze the nervous system with the poisonous products of runaway metabolism.

The thermal maxima of the different species and groups of animals bear the stamp of their evolution under definite circumstances. The heat resistance of the tissues of the various species corresponds to these maxima (Ushakov, 1959). The majority of marine invertebrates can tolerate heating of the body only up to 30–32° C, rarely up to 38° C. Freshwater animals can usually withstand heating up to 41–44° C (green frogs, *Limnaea truncatula*). But larvae of flies of the family Ephydridae have remained alive at a water temperature of 55 and even 65° C, and rhizopods have survived at 58° C. In fishes the upper boundary is lower. The development of eggs in the common sturgeon, Caspian sturgeon, and herring stops at a temperature higher than 20° C, that of the king salmon at 10.6° C, and that of the cod and haddock at 12–14° C (Derzhavin, 1922). The adults of these species can withstand higher temperatures before they die. In the hot spring waters of California *Cyprinodon macularis* can successfully withstand temperatures of 52° C and higher. However, an extended stay in such temperatures is impossible, and permanent inhabitants are found only in springs with a temperature no higher than 45° C.

Dry-land invertebrates tolerate even higher temperatures. Certain spe-

cies of beetles (*Melolontha melolontha*), butterflies (the caterpillar of *Platyedra gossipiella*), and Orthoptera (*Calliptamus italicus*) can withstand temperatures of 45–50° C from 20 minutes to over an hour; they can live for a long time at 40–45° C (butterflies) and even at 46° C. The speed with which amphibians lose their muscular irritability when the environmental temperature is raised is greater in inverse proportion to the natural temperature of their habitats. In the grass frog muscular irritability begins to diminish at 30° C, in *Rana macrocnemis* and *R. ridibunda* at 32–33° C, and in *Hyla arborea* at 35° C (Ushakov, 1955) (Fig. 72). Reptiles can withstand heating up to 45° C without harm, but they seldom tolerate greater temperatures. The southern species generally withstand higher temperatures than do the northern ones. An animal's resistance to high temperatures increases in inverse proportion to the amount of water in the body. Dry cysts of the ciliate protozoon *Colpoda cucculus* can withstand 3 days of 100° C temperatures in dry air (G. L. Clarke, 1954).

The lower temperature limits in poikilothermic animals are more changeable. Death caused by low temperatures may occur as a result of the formation of ice crystals, dehydration of the tissues, destruction of the inter- and intracellular structures, or the previously mentioned disharmony of metabolism, which is just as common at low temperatures as at high ones. The cold resistance of many animals is associated with their

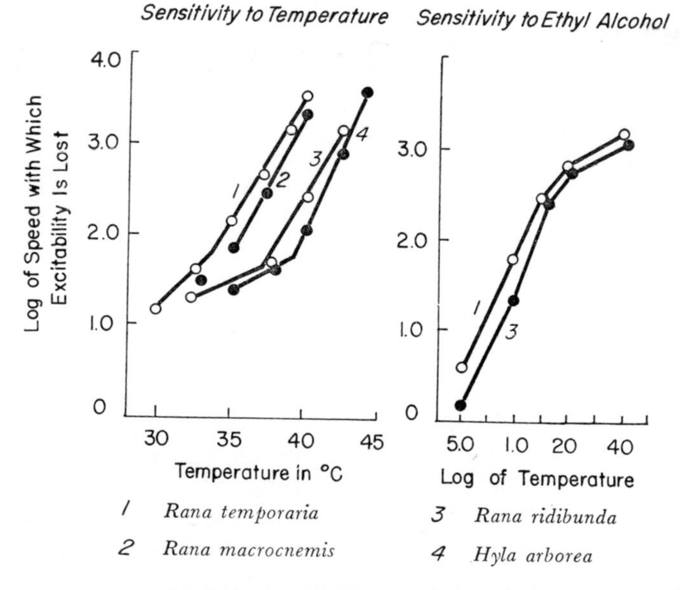

Fig. 72. Temperature sensitivity and sensitivity to ethyl alcohol of the musculature of a frog (after Ushakov, 1955).

conditions of habitation and way of life; it varies according to the phase of development. Protozoa withstand cooling down to —15° C; *Paramecium caudatum* has been observed with its cilia still moving at —12° C (Wolfson, 1935). When dehydrated, the rotifers are capable of being cooled down to —60° C. Terrestrial mollusks cannot be cooled below —8° C. Their ability to withstand low temperatures changes with the seasons and attains its maximum in the middle of winter (Table 22).

TABLE 22. *Cold resistance in the snail* Helix caesarina depressa *in different seasons (after Bodenheimer, 1934).*

	Dec.	*Jan.*	*Feb.*	*Mar.*
Number of individuals in experiment	10	5	6	2
Average temperature of supercooling in °C	−3.1	−6.4	−0.7	−0.4

The cooling temperatures of adult butterflies of the genus *Pyrameis,* the cabbage moth, the nettle moth, and others fluctuate from —8 to —20° C. But the wintering caterpillars *Loxostege sticticalis* and *Pyrausta nubilalis* withstand —20° C, the larvae of the wood borer *Anobium* —22.4° C, the eggs of the Asiatic *Locusta migratoria* —30° C, and the larvae of *Scolytus multistriatus* —53° C (Bachmetjev, 1901; Kalabukhov, 1946; Yermolayev, 1950; Lozina-Lozinsky, 1952). In *S. multistriatus* cold resistance diminishes upon repeated lowering of temperature. When the first cooling took place and the temperature was dropped to —30° C, 20 percent died after 48 hours; after 2 hours at 13–14° C and a repeat freeze down to —30° C, 75 percent died; after a third freeze 83.3 percent died (Pantyukhov, 1958). Small fish have been successfully cooled down to 3° C in water, while amphibians have been taken down to —1.4° C (Schmidt, Person, and Platonov, 1936; Kalabukhov, 1946). Ruff and perch fry have been cooled down to —14.8 and —16.8° C (Lozina-Lozinsky, 1952).

The significance of lowered temperatures for animals depends on the duration of their actions. Eggs of *Ceratitis capitata* at 7° C die only toward the end of 7 weeks, at 4° C they die in 4 weeks, and at 1° C they can live as long as 2 weeks.

Lowering the "fluid coefficient" (the amount of water in the body) or raising the amount of fixed water and the salt content of the body fluids raises the level of cold resistance (Fig. 73). It increases also as glycogen and fat are accumulated. In the owl butterfly and the capricorn beetle no ice formation took place at —5.75° C when fat supplies make up 14.4–18.2 percent of the body weight, but at —11.1° C 3.53–28.90 percent of the water turned to ice. In bees and May beetles with a fat content of 2.76–6.10 percent of the body weight, 37–74 percent of the water contained in the body froze at —5.75° C. Reducible matter and glycogen play an important

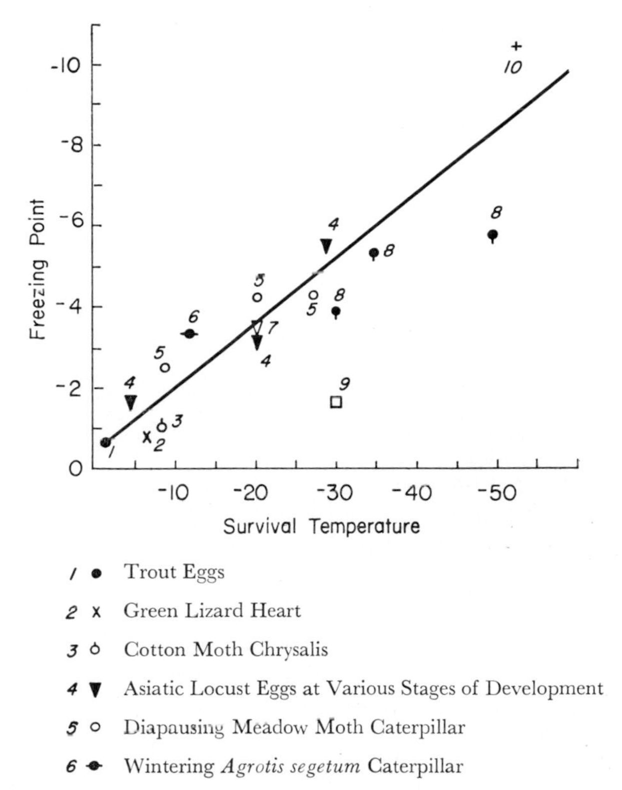

1 ● Trout Eggs

2 x Green Lizard Heart

3 ♁ Cotton Moth Chrysalis

4 ▼ Asiatic Locust Eggs at Various Stages of Development

5 ○ Diapausing Meadow Moth Caterpillar

6 ◆ Wintering *Agrotis segetum* Caterpillar

7 ▽ *Laspeiresia pomonella* Caterpillar

8 ♦ Diapausing Corn Moth Caterpillar

9 □ Wintering Garden Moth Caterpillar

10 + Wintering *Scolytus multistriatus* Larvae

Fig. 73. Relationship between freezing point of body fluids and survival of animals during cooling (after Lozina-Lozinsky, 1952).

role in providing cold resistance in species which winter in the open (*Pieris brassicae*). When they winter under unusual circumstances in temperatures above freezing, water accumulates in their bodies and they lose their resistance to cold (Ushatinskaya, 1950). Supercooling without freezing of the body fluids and subsequent death is more likely when the speed of freezing is kept low. This depends on the animal's size and the state of its protective covering; the formation of ice crystals on the surface accelerates freezing in the body fluids.

The thermal limits of life and activity of a species (the so-called "temperature constants") change when the thermal regimen of the environment

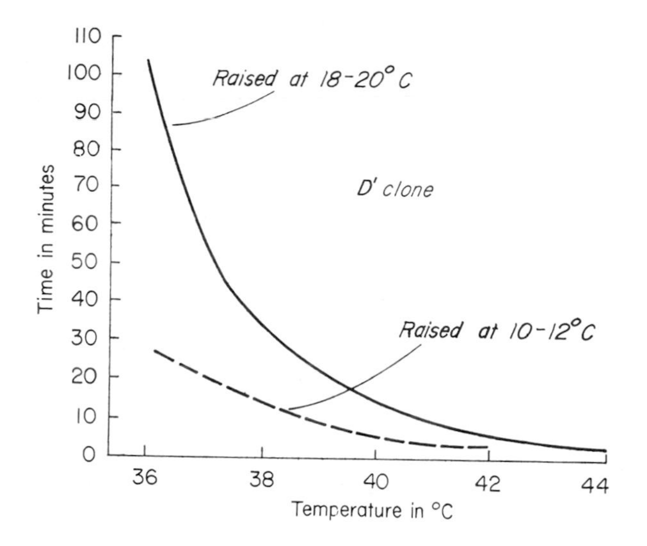

Fig. 74. Frequency of death among two lines of *Paramecium caudatum* at various temperatures (after Polyansky, 1957).

is altered. The infusorian *Paramecium caudatum* acquires great heat resistance in the first hours after its temperature is raised; later this resistance rises more slowly. At a temperature of 40° C paramecia which have acquired heat resistance survive three to five times better than those individuals which were not "primed" with earlier applications of heat (Fig. 74). The resistance which they build up to high temperatures is transmitted to 80 succeeding generations, even if they are returned to the usual medium, but later it diminishes and finally disappears after the 80th generation (Polyansky, 1957). Keeping houseflies at different temperatures for several days shifts the zone of their activity (Derbeneva-Ukhova, 1940), while keeping several generations at a high or low temperature correspondingly raises or lowers the level of the temperature constant (Fig. 75).

The level of matter exchange of poikilothermic animals within the zone of the active state changes along with changes in the temperature of the environment (Fig. 76); the minimum magnitudes are reached during supercooling. The caterpillars of *Mamestra* require 10–16 cu. mm of oxygen per g of live weight per hour in the active state (temperature 12.0–13.6° C); during torpor this amount is reduced to 7–9 cu. mm, and at 10° C it is only 0.36 cu. mm per g per hour (Lozina-Lozinsky, 1952). In *Eremias* lizards which have been supercooled to —7.8° C, respiration does not stop, although the animals required only 4.2 cu. cm of oxygen per kg of body weight (Rodionov, 1938). Changing the level of gas exchange is

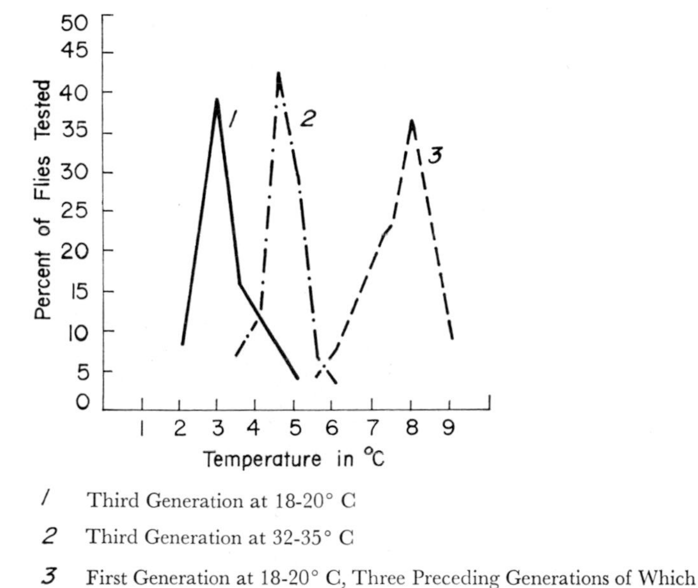

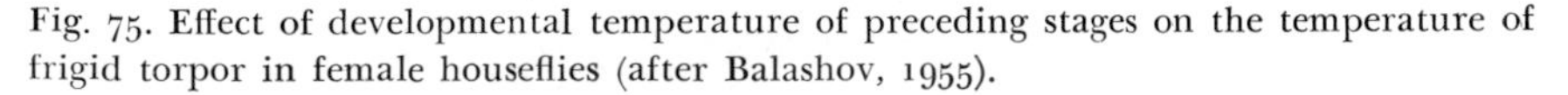

Fig. 75. Effect of developmental temperature of preceding stages on the temperature of frigid torpor in female houseflies (after Balashov, 1955).

accompanied by qualitative shifts which are reflected in the respiratory coefficient. In the active state almost all carbohydrate nutritives are oxidized (the respiratory coefficient approaching unity). When torpor sets in, however, fat reserves are expended, and this process can go on with smaller amounts of oxygen.

With regard to environmental temperature, poikilothermic animals are divided into eurythermic types, with a low thermal minimum, high maximum, and a broad zone of activity states, and stenothermic types. Both groups have a series of transitional forms. There are few eurythermic types among the aquatic animals, since temperature fluctuations in the aquatic environment are not very great. The more eurythermic types are the plankton animals, which make large-scale daily and seasonal migrations, and the migratory fishes. Resistance to temperature fluctuations is greater among the terrestrial animals.

The stenothermic animals are divided into cold- and heat-loving forms—the cryophiles and the thermophiles. The cryophiles are primarily inhabitants of the temperate and high latitudes, the thermophiles of the tropics. Temperature tolerability (valence) of the paddle-footed crayfish *Copilia mirabilis*—a true thermophile—is limited to a total of 6 degrees (from 23

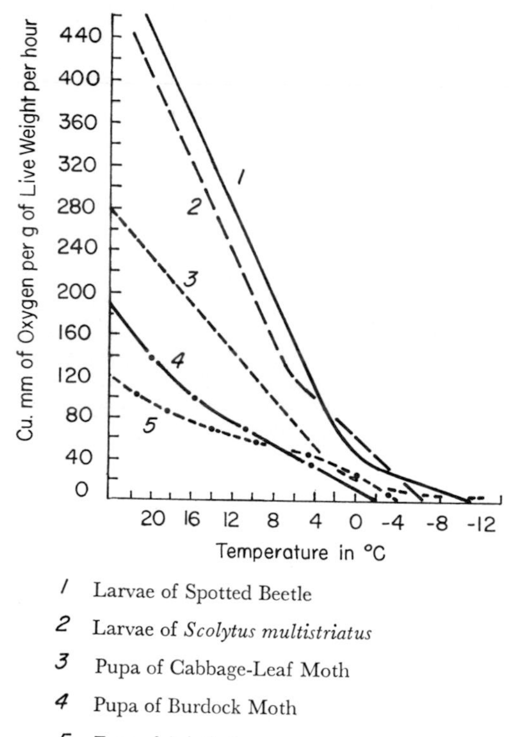

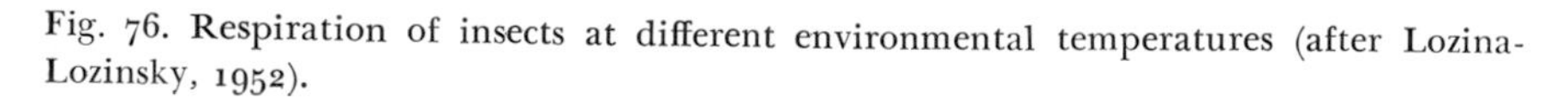
Fig. 76. Respiration of insects at different environmental temperatures (after Lozina-Lozinsky, 1952).

to 29° C), while the planaria *Planaria alpina,* which lives in cold streams and is an undoubted cryophile, tolerates a range of 10 degrees. Thermo- and cryophilia are associated with the properties of gas exchange. Thus an identical magnitude of oxygen consumption occurs in polar species at 0° C, in the boreal seas at 8° C, in the Mediterranean Sea species at 12° C, and in the tropical seas at 27° C (Thorson, 1950).

When temperature changes occur in a habitat and exceed the limits of tolerability of the poikilothermic species, mass death occurs. On the border between cold and warm currents (Spitzbergen, Japan, certain regions of the east coast of South America), where a great mixing of hot and cold water is likely to occur, the stenothermic animals are not rare. In 1882 a great storm caused an influx of cold waters along the shores of North America, killing masses of heat-loving fish; for hundreds of kilometers the surface of the sea was covered with their carcasses. In the Black Sea the anchovy, which usually flees the Crimean shore when the water temper-

ature drops, dies en masse if the temperature falls suddenly (Zernov, 1949).

The development of poikilothermic animals depends directly on the temperature of the environment. It begins with a definite "threshold of development" called "biological zero" (Table 23, Fig. 77). In the frog *Rana*

TABLE 23. *Lower temperature thresholds of development for various phases of certain insects (from Kozhanchikov, 1937, with changes).*

	TEMPERATURE THRESHOLD IN °C		
Species	*Egg*	*Caterpillar*	*Pupa*
Habrobracon juglandis	13.0–14.0	–	–
Lymantria monacha	4.5– 5.9	3.2–7.8	8.4
Panolis flammea	6.0	7.0	–
Loxostege sticticalis	11.2	9.6	12.0–13.0
Agrotis segetum	10.0	9.1	10.0

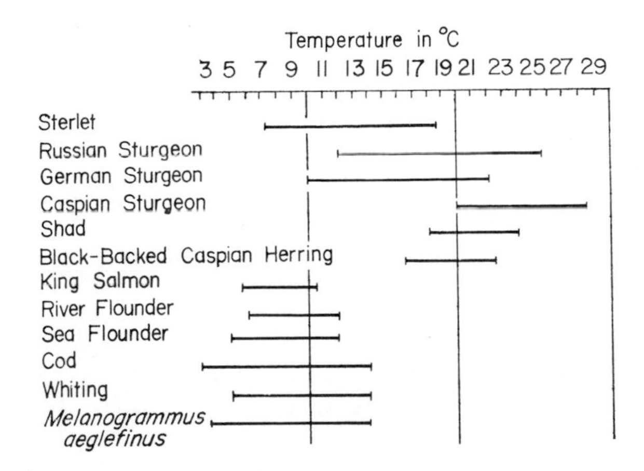

Fig. 77. Upper and lower temperature limits for development of fish (after Derzhavin, 1922).

fusca biological zero is close to 0° C, while in the toad *Bufo lentiginosus* it is 6° C. Between the lower threshold of development and the lower limit of life the space may be of significant magnitude. During the transition to the gastrula stage the fertile egg of the hen tolerates cooling down to 4° C; later its ability to withstand a low temperature is decreased.[2] The resistance of the embryo to high temperatures usually increases as development proceeds. Sensitivity to high temperatures in developing fish eggs goes through several stages of magnitude corresponding to the passing of the separate phases.

2 Hen embryos behave like poikilothermic organisms.

Increase in the temperature of the environment accelerates development and growth only in the zone of optimum temperature; above this it restrains development (Fig. 78). The correlation between speed of development and temperature in the optimum zone may be expressed by the so-called "thermal constant," or the sum of the daily average temperatures during the time of development. The temperature at which development (or growth) takes place with the greatest rapidity does not always correspond to the optimum temperature. Mackerel eggs develop fastest at 21° C, while the greatest number of larvae hatch at 15° C. The water flea *Daphnia magna* lives longest (108 days) at 8° C (under experimental conditions), but its offspring develop fastest at 23° C, and the maximum number of offspring come from those living at 18° C (MacArthur and Bailey, 1929). By the same token, the significance of individuals in the life of the species is not identical. At comparatively low temperatures they often attain a greater size and are more long-lived, and the population therefore renews itself slowly, its abundance being relatively regular. At temperatures three times higher the individuals are short-lived and the population is subject to fluctuation, but its abundance may quickly re-establish itself after curtailment.

The data in Table 24 prove the need for a definite amount of external heat to bring about a complete cycle of development or any stage thereof. The amount of heat needed is measured during observations in the wild by the sum of the daily average temperatures (according to data from the

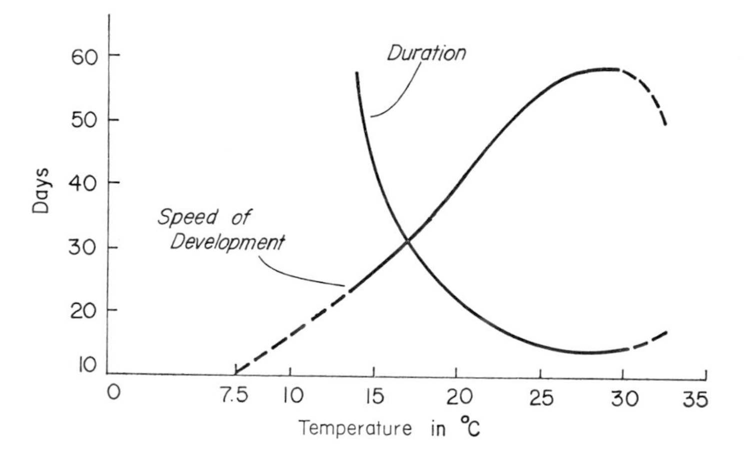

Fig. 78. Correlation between temperature of the environment and average duration and speed of development of the malarial mosquito *Anopheles maculipennis* from egg-laying until development of the wings (after Beklemishev, 1944). Dashed lines represent approximate data.

TABLE 24. *Correlation between speed of processes and external temperature conditions (from Allee* et al., *1949).*

Species	*Process*	TEMPERATURE IN °C *Low a*	*b*	*c*	*High d*	*Limits of Temperature Change in °C*	*Author*
Proteus anguineus	number of days external gills are retained	30	22	20	8	8–21	Rubner, 1908
	coefficient of nitrogen use in food	0.75	0.73	0.75	0.75		
Beetle (*Tenebrio molitor*)	total production of carbon dioxide in l per kg of live weight during time of development	59.6	59.1	58.0	59.3	20.9–32.3	Krogh, 1914
Daphnia magna	duration of life in days	108.2	41.7	25.6	–	8–28	MacArthur and Bailey, 1929
	production of heart-beat rhythm per life span	182.5	177.5	175.0	–	–	
Locust (*Melanoplus atlantis*)	duration of nymphal stage in days	94	54	27	25	22–37	Parker, 1930
	consumption of dry food for entire period in mg	4,079	4,311	4,098	3,988	–	
Meadow moth (*Loxostege sticticalis*)	time of pupation of 50 percent of the individuals in days	58.0	22.0	12.5	11.0	15–30	Filinger, 1931
	consumption of food in mg	868	870	815	735	–	

Note: The first line shows the speed of a given process for the species in relation to the temperature of the environment. The second line shows the index which characterizes the completed process as a whole (the amount of food consumed, carbon dioxide eliminated, etc.).

meteorological network), or the sum of "effective temperatures," which represents the same average daily temperatures but above the threshold of development (biological zero). According to the data in Table 25, it is quite apparent that the changeability of the first indicator, characterizing the speed of development, resists the constancy of the second, which expresses the demands of the process as a whole. Table 26 clearly shows the varied amounts of heat necessary for completion of the individual phases or stages of development in various species.

This index is a convenient and sufficiently precise form of evaluation of the need for heat of the various species of animals and the possibilities for their existence, development, and reproduction in different regions (according to the sum of daily average temperatures collected over many years) or in separate years (according to the same sums of daily temper-

TABLE 25. *Total heat needed for completion of the individual stages of development in the meadow moth* (Loxostege sticticalis) *and* Agrotis segetum *(from Kozhanchikov, 1937).*

	TOTAL HEAT NEEDED[a]	
Stage	*L. sticticalis*	*A. segetum*
Egg	39.0	77.4
Caterpillar	167.5	582.2
Pronymph	38.0	79.6
Cocoon	140.2	26.2

[a] Influence of effective temperature on number of days of development.

TABLE 26. *Total heat needed for development of different insects (from Kozhanchikov, 1937).*

Species	*Total Heat*[a]
Hippotion celerio	316.2
Polychrosis botrana	378.2
Tineola biseliella	893.0
Plutella maculipennis	377.0
Loxostege sticticalis	450.0
Agrotis segetum	1,000.0
Lymantria monacha	1,240.0
Cryptolaemus montruzieri	612.0
Calandra oryzae	358.0
C. granaria	523.0
Ceratitis capitata	250.0
Blattella germanica	1,895.4

[a] Influence of effective temperature on number of days of development.

atures but for a given year). The following simple formula is used to evaluate the temperature conditions for development:

$$\frac{C}{C_1} = I,$$

where C is the sum of heat for a given year or the average of many years, C_1 is the sum of heat demanded by a given species for a complete developmental cycle of one generation, and I is the index showing the probability of development. If I is less than 1, successful development of the species is highly improbable or impossible; an index greater than 1 indicates a sufficient degree of heat for the development of one generation; an index of more than 2 indicates sufficient heat for two generations, etc.

In actuality, many harmful insects in the northern part of the nonchernozem belt of the European USSR yield one generation, while those in the south yield two, those in the steppe zone three, etc. Their harmful

effects can be expressed in the north only under the most favorable conditions. In the northern Ukraine, where the annual sum of effective temperatures (above 10° C) is 930 degrees, *Laspeiresia pomonella* develops one generation; in the forest-steppe zone of the Ukraine it develops two; in the southern Ukraine, where the total effective degrees come to 1,870, it develops three (Vasil'yev, 1950). The correlation between reproduction and external temperatures is used to predict the abundance level of destructive insects.[3] The same kinds of examples are known for fish; the roach (*Rutilus rutilus*) takes 5 or 6 years to mature in Finland, 4–5 years in Central Europe, and 3 years in southern Europe.

Animal development in the wild does not and never has occurred at constant temperature. In the process of natural selection development has become adapted to seasonal, daily, and other fluctuations of temperature, and these fluctuations have become one of the conditions of life and development for organisms. In many species temperature fluctuations are known to have a positive effect on development. *Porasigrotis* develops faster under changing temperatures, gives birth to fewer malformed offspring, and survives better than under constant temperature, even though the latter is optimal. Daily fluctuations in temperature accelerate the development of the eggs of *Laspeiresia pomonella* by 7 percent (Shelford, 1929). Development of the eggs and larvae of the locust *Melanoplus atlantis* is more than doubled when the temperature fluctuates. At the same time egg development is impeded if they are not first subjected to a false spring (20–30 days of low (0° C) temperatures). The same fact was established for the Australian *Austroicetes cruciata* locust (Fig. 79). Similar phenomena have been observed in the incubation of wild bird eggs. Constant temperature increases the number of "stillbirths," while fluctuations in the incubator temperature raise the number of healthy offspring (Rol'nik, 1939).

Despite the undoubtedly passive correlation between body temperature and the thermal state of the environment in poikilothermic animals, they still have certain forms of thermoregulation. When solar radiation is absorbed by their bodies, causing the body temperature to rise, many cold-blooded animals become temporarily "warm-blooded." The snow flea *Colembola*, which appears on the snow in winter and more often in spring, uses solar radiation and the favorable microclimate of cracks and fissures in tree bark for habitation during transition to the active state. These

[3] However, the temperature of the environment, even though it is a very important condition of development, always acts in conjunction with other factors which may accelerate or weaken its effects. Moreover, the abundance level of insects changes under the influence of not only physico-chemical but also biotic factors, such as those created by predators, parasites, concurrents, etc.

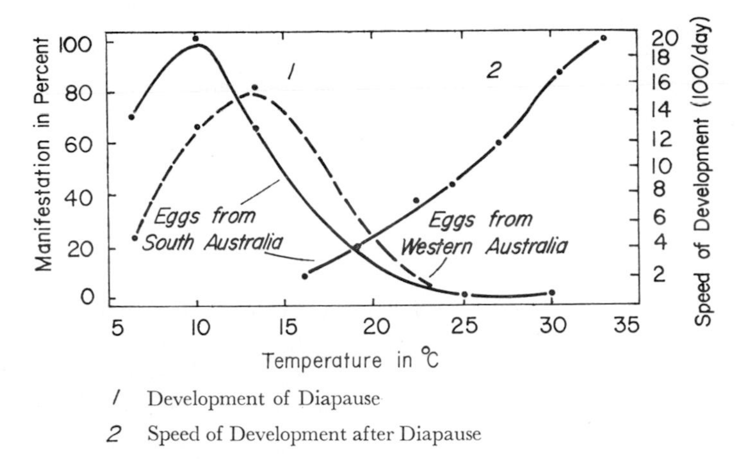

Fig. 79. Effect of temperature on development of diapause and speed of development after diapause in eggs of the locust *Austroicetes cruciata* (after Andrewartha, 1952).

so-called "sun-using" species warm themselves in the sun before beginning their daily activities. When this happens, their body temperatures rise 10 or more degrees, and the animals become active (Fig. 80). Ultimately, the relatively high body temperature is maintained by muscular work. In the flying insects (moths, dragonflies, bumblebees) the body temperature rises 15–20° C in flight. The Asiatic locust has a body temperature of 30–37° C when flying in air heated to 17–20° C, but the same species at rest has a temperature only slightly different from that of the environment (Yurgen-

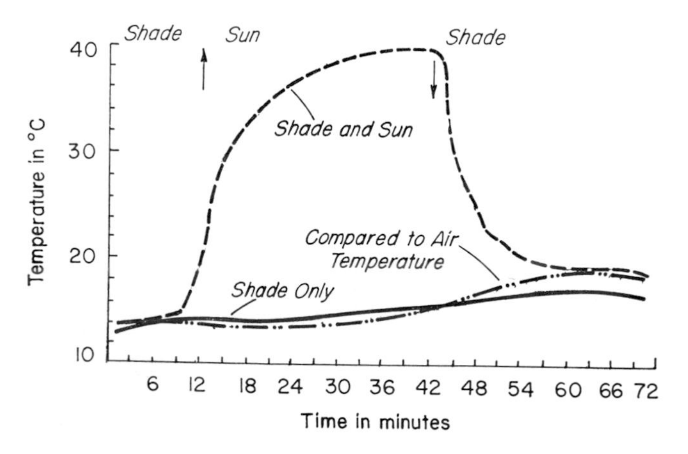

Fig. 80. Body temperature of caterpillars of *Vanessa urticae* found in shade and in sun and only in shade, compared to air temperature (after Strel'nikov, 1948).

son, 1950). P. I. Bakhmet'yev (1899) observed how *Dielephila elpenor* raised its body temperature from 19.6 to 30.3° C in 11 minutes by flapping its wings. In the mountains of the Caucasus at 2,300 m above sea level at an air temperature of 4–8° C, bumblebees reached a temperature of 38–44° C and maintained this level even in a snow storm with an ambient temperature of 2° C. In poor fliers (the meadow moth) the body temperature rises only 2–5° C in flight (Strel'nikov, 1935, 1940, 1948). In a good swimmer like the tuna the temperature is maintained at 37° C during movement (temperature of the water, 27.2° C).

Absorption of solar heat is aided by the coloring and formation of the body surface (Fig. 81); in the Arctic and in the high mountains the majority of insects are dark. A glistening "metallic" (reflective) surface occurs very often in species which stay in the hot sunlight. The air pocket under the elytra is regarded not only as a hydro- but also as a thermo-isolating layer hindering both supercooling and superheating of the body. In the sun the temperature of such insects is lower than that of others with a different construction and skin color. The dark pigmentation of the eggs of several species aids in raising the temperature.

The seasonal differences in the character of metabolism observed in poikilothermic animals are elements of heat-exchange regulation. The optimum temperature for the mosquito *Anopheles hyrcanus* in summer (15–20° C) is higher than in autumn (3.0–17.5° C), which expresses the seasonal reconstruction of the organism's energetics (Chagin, 1948). The "winter" frog *Rana terrestris* consumes less oxygen than the "summer" one (Fig. 82). As is apparent, the curve of oxygen consumption by "sum-

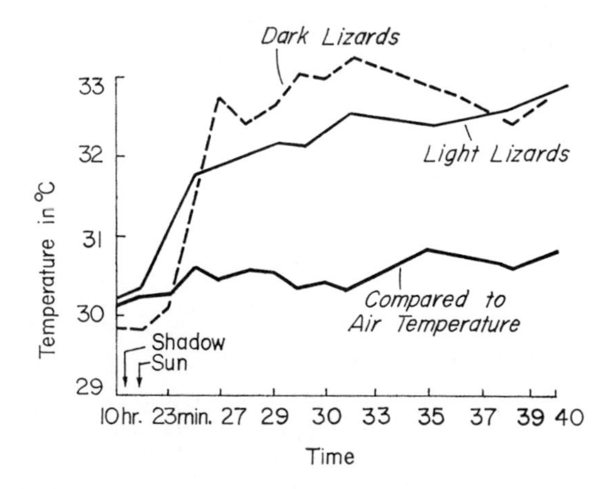

Fig. 81. Body temperatures of dark- and light-colored lizards in the sun and in the shade, compared to air temperature (after Strel'nikov, 1954).

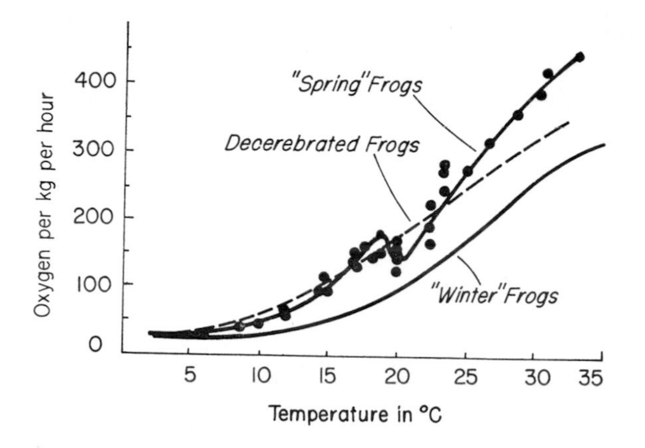

Fig. 82. Seasonal properties of gas exchange in *Rana terrestris* at different environmental temperatures (after Kozhanchikov, 1948).

mer" frogs has a dip in the 20° C zone, recalling the optimum temperature of homoiothermic animals; it is absent in the "winter" torpid frogs (Kozhanchikov, 1948).

Heat losses are also regulated. The body temperature of the caterpillar of the cotton moth *Chloridea obsoleta* does not rise more than 5–8° C above the air temperature in sunlight because of evaporation from the surface of the body and cooling from contact with plants (Strel'nikov, 1948). The black-bodied *Adesmia* beetle actually has a lower body temperature in sunlight than when it is dead. The difference reaches 2–9° C (Beckston, 1932).

HOMOIOTHERMIC ANIMALS

The development and perfection of heat-regulating mechanisms and the formation of a heat center in the brain in birds and mammals have led to the appearance of a second type of heat exchange, homoiothermy, which furnishes the body with a relatively stable temperature. It is still not very resistant in the primitive cloacal and marsupial mammals, but in carnivores, ruminants, and primates it attains great stability. Even among these animals there are seasonal and daily fluctuations.

In birds and mammals homoiothermy is the most complex form of heat relation with the environment. In homoiothermy a constantly high level of matter exchange is furnished by the maintenance of a high, stable body temperature; the body is made less dependent on fluctuations of the ex-

ternal temperature. This has been achieved by the development and perfection of the organs of locomotion and the blood-building system, the evolution of a complex central nervous system, increase in heat, insulation in the form of a fat layer and a furry or feathered coat, and a general raising of the energy of metabolism. Chemical and physical heat regulation are distinguished. The former is associated with the production, the second with the distribution and dissipation of heat. This division, convenient for practical purposes, is a conditional one, since both aspects of heat exchange are closely interconnected.

Chemical thermoregulation consists in changes in the level of oxidizing processes corresponding to the temperature of the environment. The average magnitude of thermoregulation, which can be evaluated by the amount of oxygen consumed, is associated with the magnitude of heat elimination; the latter becomes greater as the size of the animal decreases. Because the mass (volume) of the animal grows in cubic proportions while its surface area is merely squared, a unit of mass in a small animal is covered by a relatively greater amount of skin than it would be in a large animal. Therefore, the relative dissipation of heat in small animals is greater than in large ones and is compensated for by greater heat production (Table 27).

TABLE 27. *Consumption of oxygen by animals of different sizes (after Buddenbrock, 1928).*

Animal	*Weight in g*	*Experiment Temperature in °C*	*Consumption of Oxygen per kg of Weight in ml per hour*
Horse	400,000.0	–	220
Sheep (ram)	50,000.0	–	284
Rabbit	3,000.0	29	478
Rat	115.5	29	1,800
Mouse	12.9	29	4,130
Hen	–	19	746

Richet (1889) and Rubner (1902) attempted to make this purely physical correlation (the law of surface) the basis of the concept of thermoregulation. But heat production and dissipation are not simple functions of an animal's mass and body surface; they also depend on properties of matter exchange, neurohumoral regulation, body form, and character of covering. For example, certain parts of the surface are supplied in various ways with blood vessels and thus give off more heat. In bats heat production from the surface of the body is comparatively low because the enormous membranes of the wings have a low temperature and give off little heat (Table 28). Properties of thermoregulation also depend on the animals'

TABLE 28. *Gas exchange and body surface in bats and small rodents (after Slonim, 1952).*

Species	*Body Weight in g*	*Body Surface in sq. cm*	*Carbon Dioxide Eliminated per minute*	CALORIES	
				hour/kg	*hour/cu. m*
Rhinolophus ferroequinum	22	451.4	0.95	12.90	27.5
Myotis myotis	26	504.6	1.02	11.99	23.8
Apodemus flavicollis	20	66.3	2.30	345.00	520.0

biological types and differences in metabolism (Table 29). Thus *Vanellus vanellus*, *Turdus*, and *Erithacus rubecula* are distinguished by their poor heat regulation, high body temperature and heat dissipation, demand for large amounts of food, energetic digestion, high level of gas exchange, and little resistance to hunger. Pelicans, geese, ducks, chickens, and crows have a well-developed form of thermoregulation, with low body temperature, little heat dissipation, small food demands, slow digestion, and high resistance to hunger (Groebbels, 1922, 1928). Similar differences also exist among the mammals.

TABLE 29. *Elimination of heat by various birds of different sizes (after Groebbels, 1928).*

Species	*Body Weight in g*	*Experiment Temperature in °C*	HEAT DISSIPATED IN K-CAL.	
			Per kg of Weight	*Per cu. m of Skin*
Pelican (*Pelecanus conspicillatus*)	5,090	17.3	70.00	1,267.0
Domestic goose (fed)	3,097–3,578	27.0	–	1,043.0–1,135.0
Domestic duck (fed)	924	22.0	–	1,182.0–1,246.0
Domestic hen (fed)	1,150–1,920	22.8–24.0	–	1,533.4
Strix aluco	198	26.0	76.45	874.8
Vanellus vanellus	196	25.2	200.00	1,428.0
Corvus corone (6 months old)	340	22.5	129.00	1,719.0
Turdus merula	106	25.2	382.00	2,129.7
T. ericetorium	88	25.0	511.00	2,812.0
Erithacus rubecula	17	25.0	1,223.00	3,276.0

All this reveals how complexly the simple physical principle is actually manifested in the body and why we should not accept the law of surface in the simple interpretation of Richet and Rubner. Ecologists and zoogeographers associate this law with the so-called Bergmann's rule, according to which northern forms (subspecies or closely related species) of

warm-blooded animals are larger in size than their southern counterparts (Table 30). A study of the geographic races of European birds has shown that among the Sturnidae and Oriolidae there are no exceptions to this rule. Among the Fringillidae only 7 percent and among the Corvidae only 17 percent of the species do not follow the rule (Rensch, 1939). This rule is justified also for many invertebrates (see Table 19).

TABLE 30. *Geographic changes in the length of the body in two species of mammals (after Betticher, 1915).*

Species and Subspecies	*Geographic Distribution*	*Body Length in mm*	*Jan. Isotherm, Area of Feeding, in °C*
Mustela erminea erminea L.	northern and central Scandinavia	app. 300	0 to −10
M. e. aestiva Kerr.	from southern Sweden to Alps and Pyrenees	251–292	6 to −4
M. e. stabilis Bar.	England, Scotland	254–280	6 to −2
M. e. ricinae Mill.	southwestern Scotland	220–270	6 to 4
M. e. hibernica Thos.	Ireland, Isle of Man	228–283	8 to 4
Felis concolor hippolestes Merr.	Colorado	2,025	5 to −5
F. c. olympus Merr.	Washington, British Columbia, Oregon to California	1,320	15 to 0
F. c. coryi Bangs	Florida	1,248	20 to 10
F. c. costaricensis Merr.	Costa Rica, Panama	1,000	25 to 30

An explanation of the observable increase in animal dimensions in areas with a colder climate has been sought in the influence of environmental temperature on the speed of growth and development (Kalabukhov, 1938; Vinberg, 1950, 1959). It has been found experimentally that under high temperatures genetically uniform populations of poikilothermic and homoiothermic animals consist of smaller individuals. This is to be explained by the fact that raising the temperature accelerates development and maturation to a greater degree than it does feeding; therefore, individuals born under high environmental temperature conditions suffer from hunger (Bodenheimer, 1951). Thus, when a genetically homogeneous population of mice was bred, the results shown in Table 31 were obtained

TABLE 31.

Experiment Temperature in °C	*Weight of Mice in g*	LENGTH IN MM			*Weight of Hair in mg*
		Tail	*Ears*	*Hind Legs*	
26.3	17.0	93.2	16.0	21.4	264.6
6.2	17.9	75.9	15.9	20.9	294.8

(after Sumner, 1909). In the high latitudes the increase in blood formation usually causes a growth in the relative size of the heart. In the polar owl *Nyctea scandiaca* the heart makes up 0.91 percent of the body weight, while it accounts for only 0.47 percent in the eagle owl *Bubo bubo*. In the northern *Buteo lagopus* it is greater (0.84 percent) than in the common *B. buteo* (0.71 percent).

The mechanisms of physical thermoregulation include the isolating coverings (fur, feathers, fat layers), vascular regulation of blood formation (deep and superficial blood flow), activities of the sweat glands, and superficial increase in respiration (polypnoea), which increases heat dissipation by evaporation from the surface of the respiratory passages. When the external temperature changes, metabolism is not altered in the homoiothermic animals as it is in the poikilothermic ones (Fig. 83; compare with Figs. 8 and 71). The oxidation processes and the heat production associated with them are minimal in the homoiothermic animals when external temperature is at its optimum (the critical point). This is the zone of physical regulation. When the external temperature drops, heat production rises to a definite limit (the zone of chemical thermoregulation), all the while compensating for the growing losses of heat. Then supercooling of the animal's body sets in, accompanied by a general disturbance of metabolism and a sharp drop in body temperature. Raising the external temperature beyond the zone of physical regulation leads to overheating of the body and death of the animal.

In certain species the mechanisms for regulating heat exchange are different. The works of the famed naturalist and anthropologist N. N. Miklukho-Maklay (1883, 1884) indicated long ago the poor temperature resistance of the cloacal and marsupial animals. Heat regulation is also

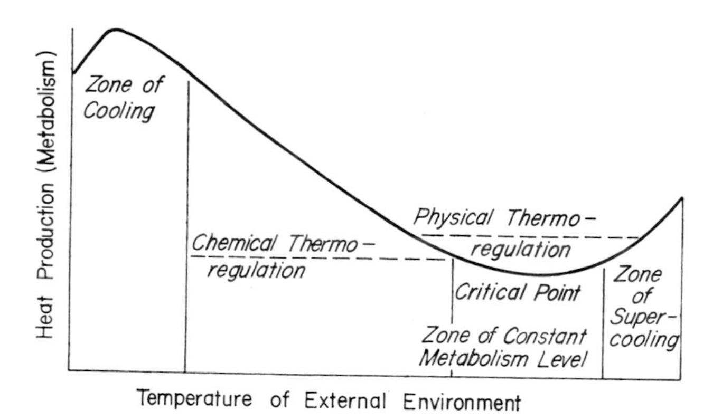

Fig. 83. Changes in gas exchange (chemical heat regulation) under the influence of temperature (after Slonim, 1952).

poorly developed in the Cheiroptera and Insectivora, but their temperature resistance is higher. Even in these animals chemical regulation plays a basic role. The large and widespread order of rodents of differing biological types still retains significant irregularity of body temperature and a large amount of chemical regulation, but vascular regulation is well expressed, and the furry covering is dense and warm. When external temperatures are high, the body temperature rises steadily, which increases heat dissipation. "Panting" (polypnoea) develops. Species of southern origin leading a burrowing or aquatic way of life have a higher "critical point" and poorer chemical heat regulation (Slonim, 1952) (Fig. 84).

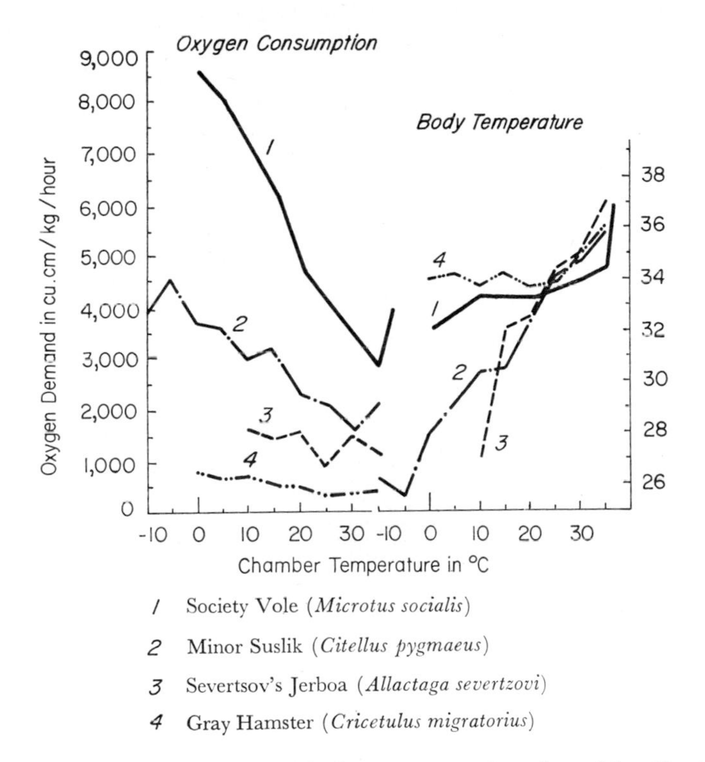

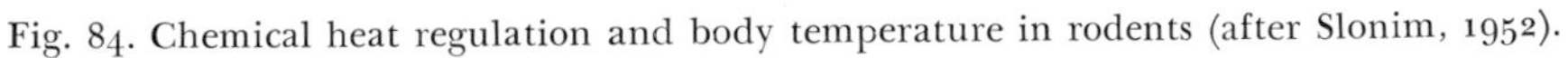
Fig. 84. Chemical heat regulation and body temperature in rodents (after Slonim, 1952).

Heat regulation is best expressed in the carnivores, primates, and ruminants. Carnivores have well-developed vascular regulation. They are resistant to cold because of their dense, warm fur, but high external temperatures are tolerated only with difficulty, since the perspiration system is poorly developed in these animals, and panting does not fully eliminate the dangers of overheating. Because of their ability to sweat freely,

the primates are better able to deal with high temperatures, but they are less resistant to low temperatures (Fig. 85).

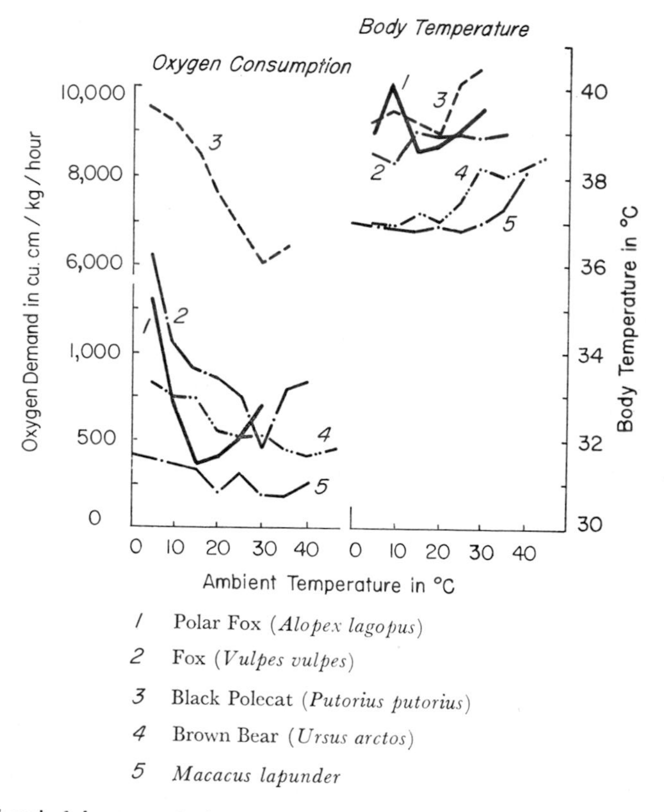

Fig. 85. Chemical heat regulation and body temperature in carnivores and primates (after Slonim, 1952).

Chemical thermoregulation apparently arose prior to physical regulation during evolution (Table 32). The parallel existence of both systems and the subsequent development of physical regulation brought a progressive rise in the body temperature and a growth in the body's resistance. The

TABLE 32. *Change in consumption of oxygen by mammals (per °C) when temperature of environment is changed (after Slonim, 1952).*

When External Temperature Is Changed from	*Insectivores*	*Bats*	*Rodents*	*Carnivores*	*Primates*
20 to 10° C	16.7%	7.9%	5.9%	3.8%	2.9%
25 to 15° C	10.1	–	5.8	3.1	2.5

widely distributed species have a geographic changeability in thermoregulation. The northern populations of the gray rat *Rattus norvegicus* have a lower "critical point," poorer chemical thermoregulation, and greater fluctuations in body temperature than the southern ones (Ruttenburg, 1953).

The influence of environmental temperature on the development and growth of homoiothermic animals is more complex than it is in poikilothermic animals. It has been shown that external temperatures are associated not only with the amount of weight gain but also with the character of the differentiations which affect the proportions of the separate organs and of the body as a whole (Sumner, 1913; Prizhibran, 1923; Sakharov, 1933, 1949). Low environmental temperatures impede growth, but because of belated sexual maturity growth takes longer, and the animals reach larger sizes. Mice and rats living in refrigerators are larger than those raised in houses or fields; their fecundity is also higher.[4] Mammalian growth at low temperatures is accompanied by a relative decrease in tail and extremity length and ear size (Allen's rule) and by better development of the hairy covering. In animals with an unperfected thermoregulation system the effect of external temperature on growth and development is even greater. Thus the duration of embryo development in bats depends on the fluctuating body temperature and the environmental temperature and may vary between 50 and 70 days (Eisentraut, 1937). Young and small rodents grow almost ten times slower in the autumn and winter months than in summer (N. P. Naumov, 1948). Animal breeders and experimenters have used the effect of temperature on growth and development to increase productivity, tolerability, and resistance to infection. A combination of full-valued feeds and priming with cold (keeping the growing animals under conditions of relatively low temperature) is used to attain greater size in domestic animals and to raise their resistance to infection.

ADAPTIVE BEHAVIOR

Normal heat exchange with the environment is brought about not only by regulating the temperature of the body but also by selecting or artificially creating favorable thermal conditions in the environment. The

[4] In a refrigerator in which eggs had been stored at an air temperature of −5° C, gray rats were larger (average body length, 219 mm), heavier (333 g), and more fecund (average number of embryos per female, 8.5) than rats raised in an oat-filled grain elevator. In the elevator the average body length was 214 mm, weight was 262 g, and average number of embryos per female was 8.1. Air temperature in the elevator was maintained between 10 and 15° C (Shepeleva, 1950).

adaptive behavior of animals includes migration to areas of favorable temperature, alternation of periods of rest and activity (daily activity rhythm), special habits (behavior properties), and, finally, the use of existing and the creation of special refuges with a favorable microclimate.

The bases of active searching for favorable environmental temperatures are the thermokineses and thermotaxes.[5] Thermotaxes have been studied in aquatic and terrestrial animals with the aid of the thermogradient device suggested by Herter (1929) and perfected by Kalabukhov (1951), in which a temperature gradient or substrate is created by which the animal, the air, or the water is heated or cooled (usually it is a metal plate cooled at one end and heated at the other).

The animal's preferred substrate temperature (thermotactic optimum) or environmental temperature (thermal optimum) proves to be associated with the animal species' heat-exchange properties. Homoiothermic small animals with a high level of exchange select higher temperatures than do large animals; young individuals prefer higher temperatures than do adults. The same occurs with species known in both southern and northern regions. The tiny forest mouse has a temperature optimum of 26–29° C, while the larger yellow-throated mouse prefers 20.3° C. Forest mice from around Moscow choose a temperature of 26.4° C (average), while individuals of the same species from the Pre-Caucasus choose 29.2° C. Polar foxes from the tundra prefer a temperature of 18.8 °C, while foxes of other species prefer an artificial zone with an average temperature of 23.6° C (Kalabukhov, 1950). The black polecat *Putorius putorius* has a higher preferred temperature than its steppe-zone relative *P. eversmanni*, whose range is associated with the more severe continental climates. In the active state animals select a somewhat lower temperature than they do at rest. Homoiothermic animals prefer an air temperature somewhat lower than their critical point, while the preferred substrate temperature is equal to it (Fig. 86). In all cases animals select those thermal conditions under which they can most easily balance their heat exchange, i.e., the zone of "temperature comfort," in which energy losses to basic exchange are minimal (Kalabukhov, 1953).

Among the poikilothermic animals "heat balance" with the environment may be obtained at any temperature within the limits of the active state if the animal's metabolism has adapted to the latter during the period preceding the experiment (Slonim, 1952). This explains the difference in daily and seasonal preferred temperatures in insects; the selection is associated with daily and seasonal changes in their metabolism (Likventov,

[5] Thermokinesis is a nonoriented augmentation of the movement of an animal inside a zone of unfavorable temperature; thermotaxis is the directing of an animal's movements from an unfavorable to a favorable temperature.

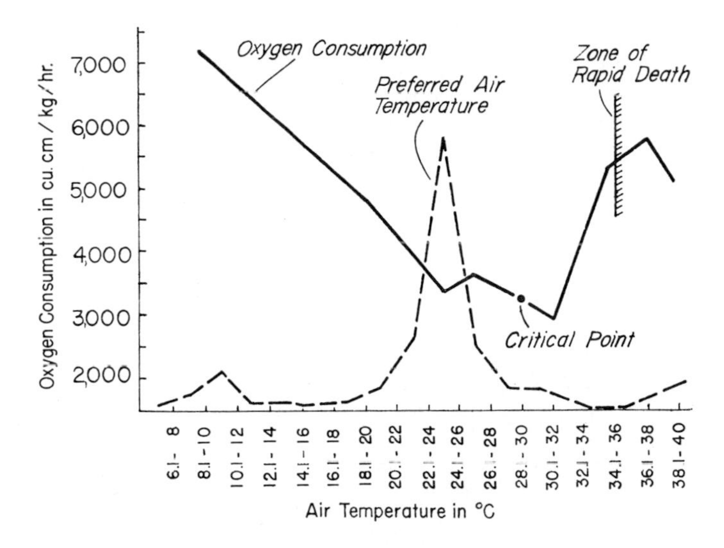

Fig. 86. Relationship between preferred air temperature and oxygen consumption at various environmental temperatures (after Bashenina, 1958).

1950). But even here the northern species usually select a lower temperature in the gradient device. The jumping lizard, which inhabits the southern portion of the forest and the northern part of the steppe zone, and *Lacerta muralis* in southern Europe, Asia Minor, Iran, the Crimea, and the Caucasus, prefer a temperature of 38.5° C. *Lacerta serpa,* which lives in warmer and dryer areas, prefers a temperature of 40° C and approaches the preferred temperature of the steppe *Agama* (Melakhat Lutfi, 1938). The same correlation between thermotactic optimum, character of specialization, and geographic distribution occurs among the insects. The larvae of malarial mosquitoes have two maxima of mobility in the active state (zones of "thermal and frigid excitation"). Between them is a zone of minimal movement, corresponding to the optimum temperature, in which the larva apparently does not obtain enough temperature disturbance and is less motile (Fig. 87). The same thing has been found for fish (Fig. 88) and has also been associated with their distribution (Fig. 89).

Reactions to substrate or air temperature also affect the distributional patterns of poikilothermic animals in the wild. The red pine sawfly, whose average preferred temperature in June is 21.3° C, moves to the 15–26° C zone in the gradient device and in the wild inhabits the entire corona of the forest. At the same time the birch sawfly *Croesus septentrionalis,* which prefers a temperature of 18.6° C (on the average) and inhabits the gradient device only in the 16.4–20.8° C zone, prefers only the lower portion of the tree corona in the wild (Likventov, 1960). These examples show that

poikilothermic animals have no real thermotaxis. Increasing their motility under the influence of unfavorable temperatures, i.e., as a result of thermokinesis, causes them to move automatically to places where their motility is decreased and the temperature is most favorable (Sullivan, 1954; Ivlev, 1960).

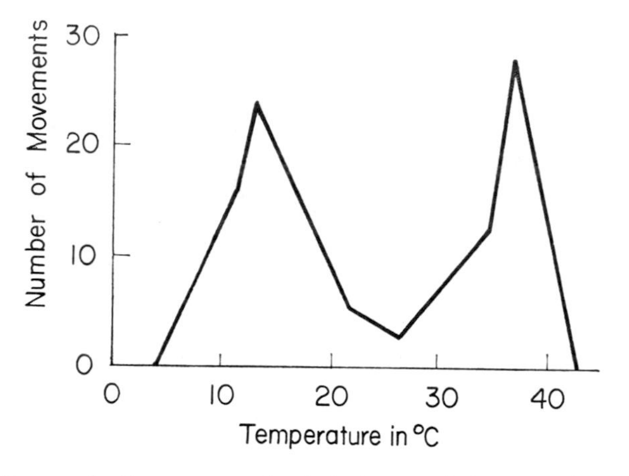

Fig. 87. Spontaneous darting movements of the larvae of the mosquito *Anopheles maculipennis* at various environmental temperatures (after Ivanova, from Beklemishev, 1944).

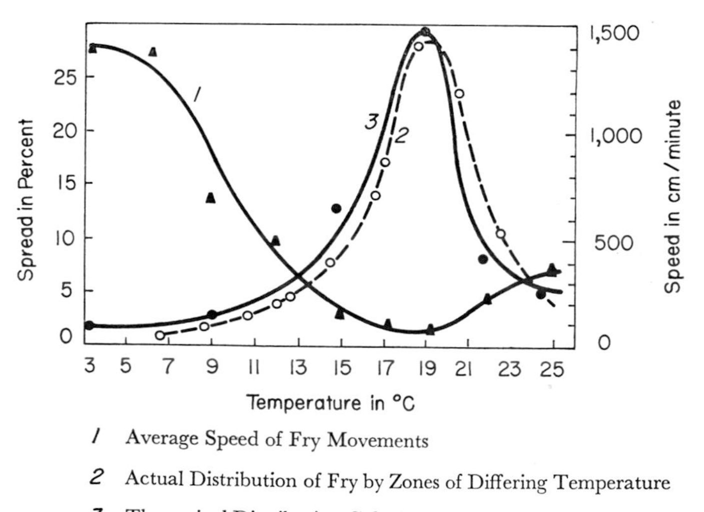

Fig. 88. Motility and distribution of *Salmo salar* fry in relation to temperature of the environment (after Ivlev, 1960).

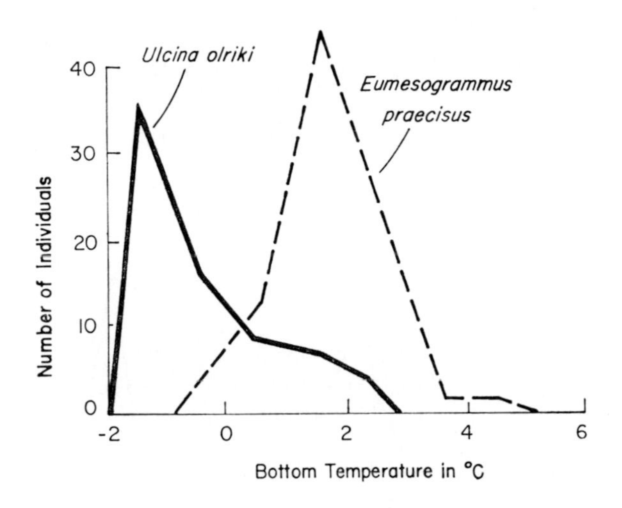

Fig. 89. Distribution of *Ulcina olriki* (Cottoidei) and *Eumesogrammus praecisus* (Blennioidei) in relation to natural temperature (after Andriyashev, 1939).

The first presentations of a thermotactic optimum as a reliable species index apparent in a hybrid ancestor (Herter, 1940) were not subsequently confirmed. In both homoiothermic and poikilothermic animals it changes with the seasons, with changes in food, and in conjunction with other physiological reorganizations of the body. The thermotactic and thermal optima also change with weather fluctuations (Sun Ju-yung, 1958).

When the threat of overheating occurs, most animals will hole up in their refuges, get under cover, or bury themselves in the ground. The danger of overheating eggs or fledglings among the open-nesting birds is met by covering them from direct solar rays, by setting, or by the parent's standing over the nest with wings spread like an umbrella. On hot days pelicans often leave their nests and enter the water to moisten their plumage and cool their feathers by evaporation. Predatory birds do not watch their prey from the ground on hot days; rather, they cruise at high altitudes (Formozov, 1934). The long-eared lizard of the desert crawls onto the crest of a sand hill on hot days and cools itself in the wind. Many insects make periodic flights for the purpose of cooling off.

The temperature of the environment often serves as a signal for animal migrations. Movements of the migratory locust usually begin after the soil temperature has reached 31–35° C. The spawning migrations of fish are governed in many cases by temperature differences in the water. The herring of Kiel Bay go to lay their eggs in the Schley River in March and April, when the water temperature there is 1.5° C higher. Their return to the bay occurs when the water there becomes warmer than the water in the

river. In both cases the fish move in the direction of the higher temperature. The same is known for the herring off the coast of Murmansk. The migration of the Black Sea mackerel away from the shore occurs when the temperature of the inshore waters falls to 10–12° C. The summer flounder *Paralichtys dentatus* in Great South Bay off New England swims into the rivers in June and leaves in October; the winter flounder *Pseudopleuronectes americanus* is abundant in the streams during the cold part of the year and leaves in summer. Pacific sardines approach the banks at a temperature of 8° C and stay there en masse as long as the temperature of the inshore waters remains between 10 and 20° C. When it falls, the fish leave the shore, and this phenomenon is expressed in the size of the catch. Temperature works in the same way in the distribution of herring in Kandalaksha Bay (Fig. 90) and Sorok Gulf. The sensitivity of fish to temperature change is great, but it is not the same in all species. Herring react acutely to a temperature differential of 0.2° C, the cod *Gadus morrhua* to 0.05° C, the flounder *Pleuronectes platessa* to 0.06° C, and *Pholis gummelus* and *Loarces viviparus* to 0.03° C. Salmon react to temperature gradients and will keep to the stream with the same temperature as that which they are in.

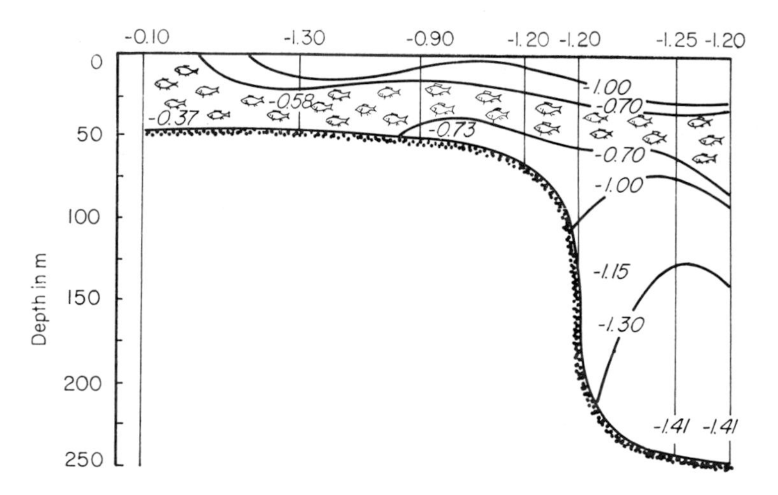

Fig. 90. Winter distribution of herring in Kandalaksha Roads in relation to temperature (after Nadezhdin, 1959).

Migrations of terrestrial animals are also associated with temperature signals. Such signals are very important in the migration of birds. The autumnal drop in temperature in both the soil and the air causes mass movements of mice and voles into built-up regions (Fig. 91).

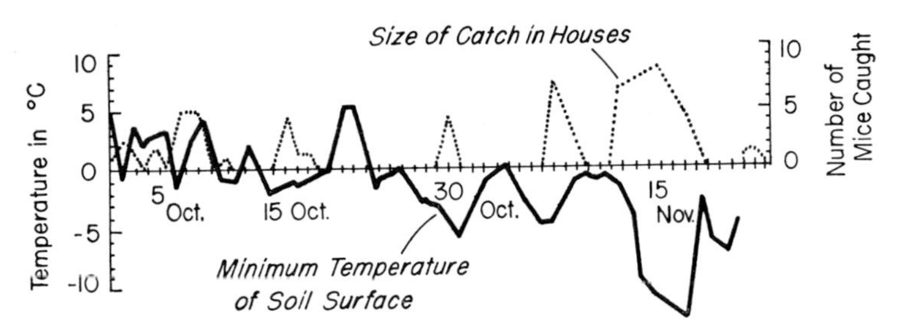

Fig. 91. Migration of house mice from fields into built-up areas in autumn (after Tupikova, 1947).

For protection against predators and inclement weather many animals employ existing refuges or build special ones, creating a favorable temperature environment. In ant- and termite hills, beehives, insect burrows, worm holes, reptile pits, and animal lairs, in closed and even open bird nests, the temperature and humidity differ from those of the surrounding environment in their lesser amplitude of changes and their relatively high level, while movements of air are absent or weak. When building refuges, the animals react strongly to local features. In the north the anthills of *Formica rufa*, the red ant, are situated only on the south sides of trees, which serve as a heat screen. The black ant *Lasius niger* builds supraterrestrial hills in shady and relatively cold spots, and in the heat of the sun it limits itself to underground nests. In shady forests this same species lives in rotting logs.

EFFECTS OF TEMPERATURE ON REPRODUCTION AND MORTALITY

Copulation, development of eggs, and egg-laying of various species of insects all take place at different temperatures. The meadow moth normally copulates when the temperature of the grass (its normal place of habitation during the butterfly stage) is 10–15° C. Egg-laying takes place at 14–30° C and is greatest at 25° C. The tsetse fly *Glossina palpalis* gives birth to its larvae, which develop in the mother's body, only at 25–30° C. In birds and mammals the time for reproduction is usually confined to a period of optimal temperatures, deviations from which are usually accompanied by a loss in intensity or a cessation of reproduction. The correlation is of a distinctly signal character (Fig. 92). White laboratory mice multiply intensively at 18° C, but at a temperature higher than 30–31° C their fecundity falls or reproduction stops.

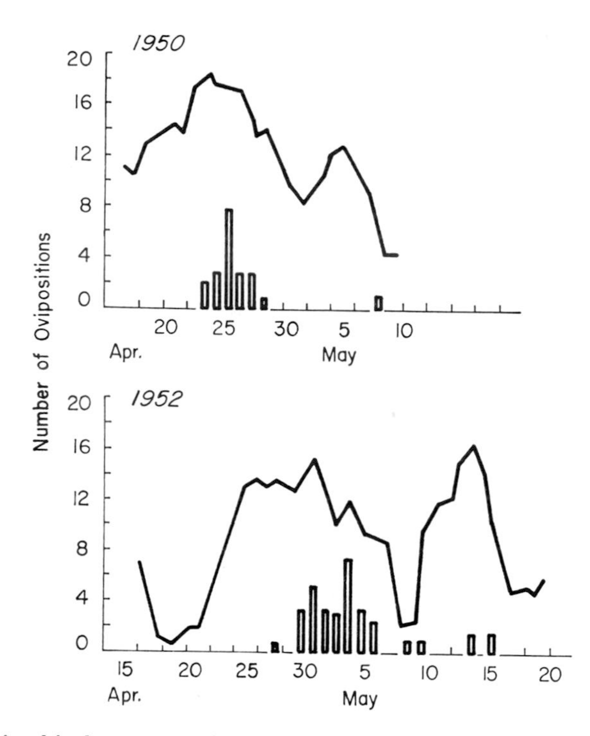

Fig. 92. Relationship between egg-laying in the large tomtit (columns) and changes in air temperature (lines) (after Likhachev, 1953).

Among the desert rodents (gerbils, voles) and many of the steppe species reproduction is interrupted by the onset of the hot, dry season and starts up again when the temperature drops in autumn. Deviations of temperature (rise or fall) from the optimum are accompanied by curtailment of fecundity or by cessation of multiplication among the common and society voles (Strel'nikov, 1940). The same is known to be true of birds (Kabak and Tereza, 1939). Curtailment of spermatogenesis and ovulation in males and females respectively during a cool spell or a hot period may depend on the effects of environmental temperatures on the gonads or on the heat-exchange regimen of the body as a whole (Bukhman and Andriyevsky, 1940; Vasil'yev, 1947).

The limiting effect of temperature on animal abundance is most often observed in the north. The success of reproduction and the survival of black and wood grouse near the northern border of their range in the Lapland Preserve depends on the temperature conditions during the period of egg-laying and setting (June). Heavy frosts during this month, when the females often leave the nest, kill the eggs. In such years the per-

centage of young birds in the population is not large, and the total size of the flock falls (Fig. 93). The white and tundra ptarmigans, which nest later, do not usually suffer from frost.

In the north changes in the abundance of small rodents (with their less-perfected heat regulation) are strongly correlated with the temperature

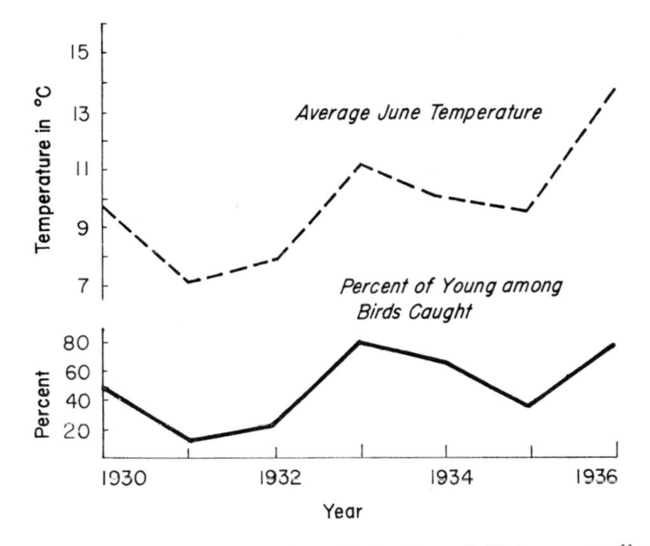

Fig. 93. Correlation between success of multiplication of *Tetrao urogallus* and June temperatures in the Lapland Preserve (after O. I. Semënov-Tyan-Shansky, 1938).

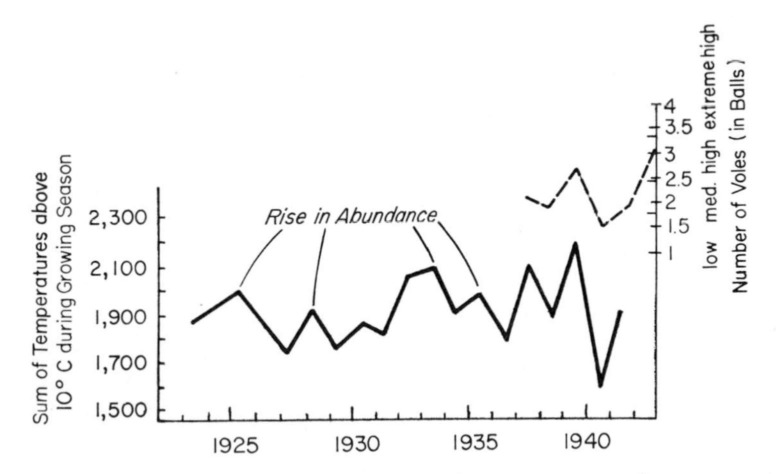

Fig. 94. Changes in abundance of forest voles (*Clethrionomys* and *Microtus oeconomus*) in relation to total heat during the growing season in the Kirov oblast' (after Bashenina, 1949).

regimen (Fig. 94). The resolving point occurs in May, when the temperature can make or break reproduction (Naumov and Folitarek, 1945; Formozov, 1948; N. P. Naumov, 1951). A strengthening of the flow of the Gulf Stream and the consequent warming of northern and eastern Europe are accompanied by the mass appearance of the Norwegian lemming *Lemmus lemmus* and the Ob' lemming *L. obensis* (C. S. Elton, 1942; Nasimovich, Novikov, and Semënov-Tyan-Shansky, 1948).

Destruction of thermal exchange is often the cause of animal deaths. Mass kills in severe winters usually occur after poor harvests or when food supplies are unattainable and the animals must wander in search of them. A return of cold weather in spring, especially in May, usually causes death among insectivorous birds which eat summer insects (flycatchers, chiffchaff, redstart, etc.)

CHAPTER 7

Motion and Pressure in the Environment

The immediate environment which surrounds an animal may be water, air, soil, or, for parasites, another organism. Particles of the aquatic and aerial environments are usually in motion. There may be movement of molecules in space (flowing, inflow and outflow, wind, falling or precipitation) or, if we consider the character of mechanical fluctuations, sound and ultrasound. Both play an important role in the life of animals.

WATER FLOW

Currents in the aquatic environment bring animals and plants their necessary oxygen and food and remove the products of metabolism. They aid the union of male and female sexual cells, serve as a mixing agent for poorly motile animals, and disperse the young, but they may cause harm when they act with great force. In fast-flowing, usually chilly streams and rivers the rheophilic species inhabiting them are distinguished by several adaptations. These include a cigar-shaped, somewhat laterally flattened body in the nektonic species. Benthic animals, on the other hand, are flattened dorsoventrally, and their extremities and bodies are armed with hooks and anchors for holding onto the ground or submerged objects. The form of the refuge used by the stream-dwelling *Hydropsyche* depends on the speed of the current (Fig. 95). Similar protective adaptations are obligatory for all inhabitants of the surf and even the littoral zones.

Protection is also furnished by rheotaxis, a movement usually directed against the current and permitting the animal to hold itself in place in a

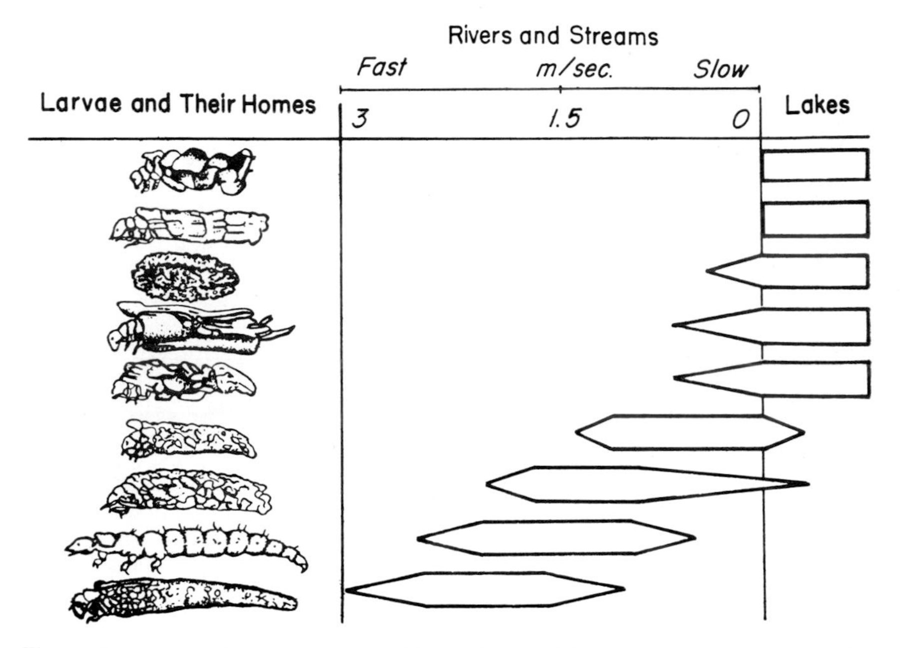

Fig. 95. Properties of case structure and larval forms of *Hydropsyche* in relation to their distribution by bodies of water with different current speeds (after Dodds and Hizo, from G. L. Clarke, 1954).

stream of water. In many of the fishes this reaction is governed by vision (glimpses of the bottom) or is subordinate to water pressure. Because fast-flowing cold-water rivers are oversupplied with oxygen, their inhabitants (rheophiles) have a somewhat reduced set of gills, which lessens the danger of oxygen narcosis (see Chapter 4).

The limnophilous species, which inhabit lakes, ponds, slow streams and rivers, and their backwaters, are distinguished by another form of body structure—often flattened laterally in the fishes but not drawn out (crucian, bream) or flattened dorsoventrally (sheat). As waters become poorer in oxygen, the fish living in them tend to have larger gills, their blood has greater ability to buffer high concentrations of carbon dioxide, and their nerve centers have high tolerance and resistance to the same gas.

Even in standing or slowly flowing waters many animals are sessile or nearly sessile. They are often distinguished by radial symmetry, which is profitable in that it allows food to be taken from all directions, and such animals usually live colonially. Sessile animals as a rule grow more quickly in places with even the slightest current, but some of them can live only in a rapid current. In such places the larvae of *Hydropsyche* construct nests

oriented against the flow, with a funnel-shaped entrance into which food particles flow. The larvae of *Chirotenetes* attach themselves to the bottom with their heads against the current, stick out their anterior legs, which are covered with filaments, and use them to catch their food as if with a net.

The plankton animals are a special form of aquatic life in that they spend their entire lives in the water but (unlike the free-swimming nekton organisms such as the fishes and the majority of the cephalopods) are incapable of overcoming the current. They tolerate it passively as it carries them from place to place and make only the most limited movements of their own. This kind of movement (flotation) is associated with the form of many of the plankton animals, the appendages and equipment of which act as a unique type of sail (Fig. 96). The commonly observed irregular vertical distribution of plankton may depend on the differing rate at which plankton animals settle and on the density of the water. As a result, plankton organisms cluster over the impenetrable layers; this phenomenon accounts for the so-called "food layers" or zones.

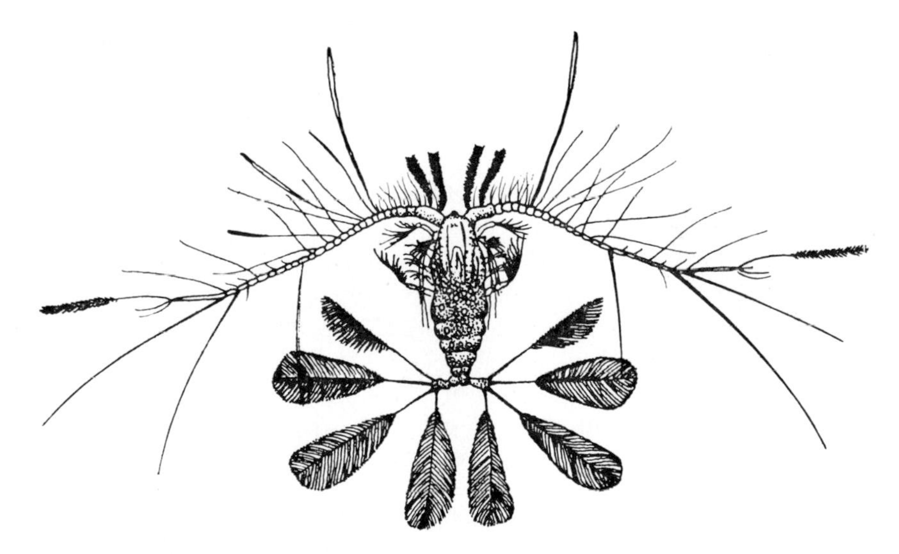

Fig. 96. Tropical plankton *Calocalanus pavo* (body length 2.5 mm) (from G. L. Clarke, 1954).

Both marine and river currents serve as important conditions of existence and as means of dispersing animals. Marine plankton can disperse only through these means. Some plankton organisms are so characteristic of a given current that their presence is taken as an indicator of the current's direction; in the English Channel there is a population of the chaetognath

Sagitta sagitta. Another species, *S. elegans,* is characteristic of the mixed waters of the ocean and the channel, while *S. serratodentata* is encountered only in purely oceanic waters (Russel, 1939). The nekton forms are also transported enormous distances by currents. The dispersion of the conger eel is of this type. The adults move to their spawning grounds (in the Sargasso Sea) with the aid of deep-water currents, while the young are carried back to the shores of Europe and America by relatively shallow currents, in particular by the Gulf Stream (Lebedev, 1959) (Fig. 97). Transport by water is the most important means of dispersing young aquatic animals. In flowing waters it occurs steadily in one direction, which may threaten the survival of populations living upstream. Oysters and other inhabitants of estuaries, whose planktonic larvae are carried out to sea,

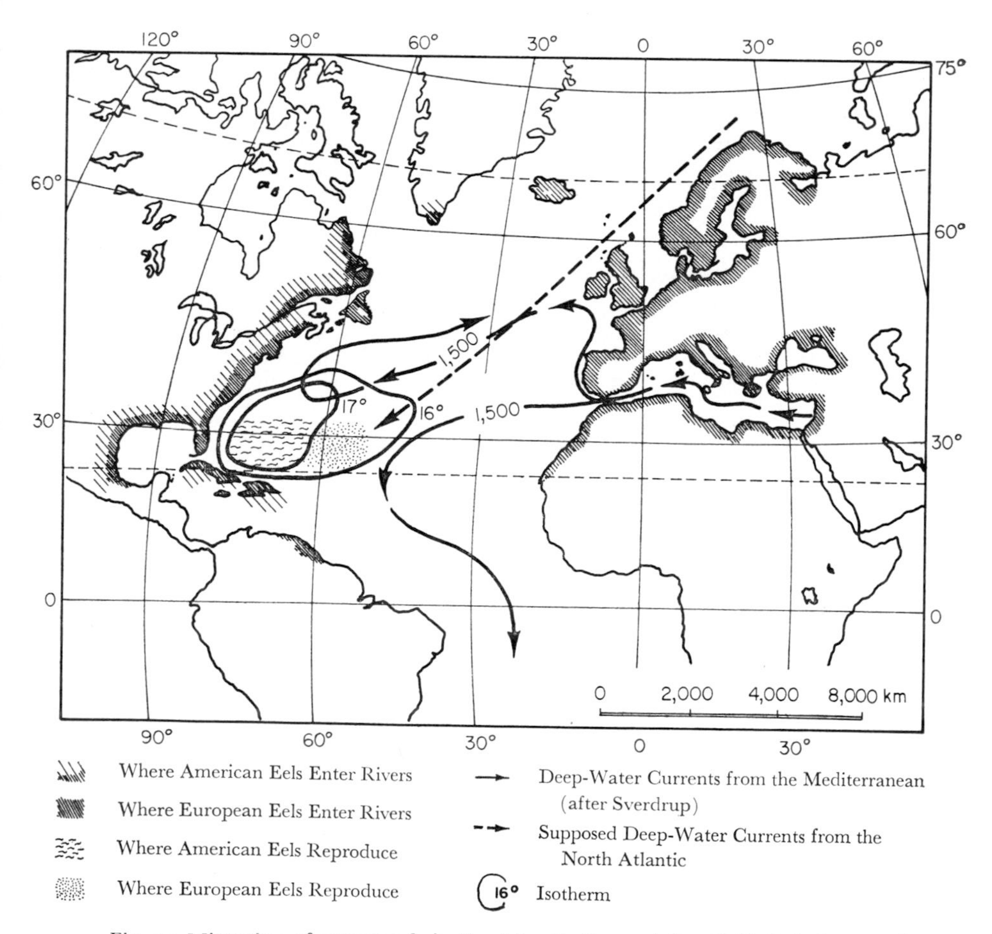

Fig. 97. Migration of conger eels in the Atlantic Ocean (after V. D. Lebedev, 1959).

use the tides to return. Before the transition to the sessile stage their larvae sit on the bottom during the outgoing tide and catch the incoming one to return, thus resettling the upper part of the estuary step by step (Carriker, 1951).

Incoming and outgoing tides, which occur in a significant part of the littoral along the shores of the oceans and seas, determine the regimens of these habitats. Their inhabitants are adapted to periodic changes in the water level and to the existence of a surf. They are capable of spending long periods of time outside the water (*Balanus* and other sessile littoral animals), they are protected against mechanical destruction by the surf, and their periods of reproduction are timed for the periods of maximum ebb tides. The ebb-tide zone, despite the severity of its conditions (regular flooding and exposure twice a day), is thickly settled by a variety of animals and plants. The reasons for the abundance of life here are the variety of food and the abundant light and oxygen in the shallows. For this reason many terrestrial animals (birds and mammals) come here to feed. Inhabitants of the deep (fish, certain crustaceans, etc.) migrate here at high water and return. The permanent inhabitants of the tidal zone are only aquatic. Some of them are capable of withstanding substantial water loss. Sea anemones can last 18 days outside the water, after which they dry out completely, but when the water rises, they revive. Animals with shells settle thickly in these areas and in such a state may exist at relatively low temperatures for more than 3 months. Slugs and starfish crawl into crevices when the tide goes out, while worms and similar forms dig into the ground.

Many forms living in the tidal zone find that alternative residence in the air and in the water is more useful than life in the water alone. This is well known to oyster farmers; they prefer to raise them into the air at least once a day. The danger of the gills drying out is great among inhabitants of the tidal zone. Thus the crab's gills are enclosed in a special cavity, while the number of gills and their relative surface is very small. In species living below the minimum line, such as *Menippe mercenaria* and *Callinectes sapidus*, the number of gills is 18 and 16 respectively, and the ratio between body size and gill surface is 36:1 and 23:1. In *Uca pugilator* and *U. minax*, which live in the tidal zone, there are only 12 gills, and the ratio of body size to gill surface is 49:1 and 40:1. The dry-land crab *Ocypode albicans*, which lives above the high-tide zone, has the same number of gills; the ratio between body size and gill surface is 67:1 (Pearse, 1950). The daily activities of tidal inhabitants are associated with the rise and fall of the tides. The eider duck *Somateria mollissima* feeds between the tides, but it rests during ebb tide (about 4 hours) and the outgoing tide (about 1 hour) (Flint, 1955).

WIND

Wind is not very important to aquatic animals, but it may be a significant factor in the lives of amphibiotic forms and those inhabiting a littoral. During strong and especially long-lasting storms the inhabitants of the tidal zone and the littoral die in large numbers. At the same time the strong winds tend to aerate shallow water, thus preventing death and halting the unfavorable progress of the habitat.

The significance of wind for terrestrial animals is more varied. In raising the amount of evaporation and increasing heat dissipation, wind influences the water and heat exchange of dry-land animals.[1] Thus birds, mammals, and other inhabitants of open areas with strong winds usually have a covering material of great density. Their feathers or hair is relatively tough and solidly pressed against the body, which makes it difficult for wind to reach the surface of the body and dissipate heat. Such is the plumage of the desert grouse, pigeon, white grouse, and tundra grouse. The feathers of forest birds (hazel hens, robins [*Erithacus rubecula*—Ed.], redstarts, warblers, and others) are ruffled and relatively soft. These birds pass their entire lives in the branches of trees, in bushes, or on the ground, where the effects of wind are insignificant. It is interesting that the Tibetan butterfly *Parnassius*, which lives in very windy places, has relatively short wings, and its body is covered with hairs. The same is typical also of certain insects in such places. But some movement of the air is necessary for terrestrial animals, even for spiders, the larvae of the ant lion, and certain beetles. The wind removes carbon dioxide eliminated by such animals.

The geographic and stage distribution of several species depends on the character of the winds. In regions with constantly strong winds the species assortment of small flying insects is impoverished, since the wind simply carries them away. Nonflying species or those which fly very well are the only types which can survive. On ocean coasts and islands and in the steppes, deserts, and tundras, the fewer the number of winged insects, the stronger the winds. On the islands of Kerguelen, Auckland, and South Georgia, where there is great danger of insects being blown into the sea, flying insects are an exception. In the Crozet Islands there are only three genera of flying insects. There are almost no butterflies on the islands in the Aral Sea. There are no butterflies such as *Pieris, Hippoarchias,* and *Vanessa* or flies of the genera *Emphis, Anthrax,* and *Eristalis* in the Fries-

[1] Studies of the combined effects of temperature, humidity, and air movement (wind) on the temperature of an object are conducted with a catathermometer, which computes the speed at which an object becomes cold (Kalabukhov, 1951).

land Islands in the North Sea, even though the neighboring continent has all of them.

Along the lower Amazon, where the winds are very strong, 19 genera containing 100 species of butterflies are absent, even though they occur in nearby regions where the wind is weaker. In deserts with poor grass cover and a low air humidity there are few butterflies and flies; the prevailing insects are the beetles, Orthoptera, chinch bugs, and good fliers like the Hymenoptera (bumblebees, certain species of wasps, etc.). These types are much less endangered by excessive loss of water and transport to unfavorable areas. Bats are fewer in number and species in areas of strong wind, since flying insects are in short supply. The species which dwell here are the long-winged type (Miniopterini and others) that make relatively quick flights, enabling them to hunt in the wind and cover large distances. Broad-winged bats (*Rhinolophus, Myotis, Murina,* and others) hunt in calm weather, avoid areas of strong wind, and are most widespread in protected locations.

Other animals, on the other hand, make use of the wind for movement and occur in areas where the wind is strongest. The frigate bird, albatross, and stormy petrel, which are found in the open ocean, can glide over the water for hours by using rising air currents, even the insignificant ones which rise over wave crests. They fly little during calm weather, and their range of distribution conforms to the regions of constant winds. Many predatory birds glide for long periods of time.

Wind is an important means of distribution; for the tiny terrestrial species it is almost the only one. Passive transport of freshwater animals through the air usually takes place when bodies of water dry up and the wind blows away the bottom or shoreline materials. This phenomenon is so widespread that it is the only way we can explain the amazing uniformity of freshwater invertebrates and the usual absence of geographic races among these species. Transport occurs at all stages of development, thus raising its effectiveness. Many cases of wind transport of even larger animals have been recorded, even of vertebrates. Powerful waterspouts carry mollusks (including the large *Anodonta anatina*) into the air. Frogs and fishes have been dropped at a distance of 20 km from the point where they were seized by a waterspout. A similar "rain of fish" was observed during the hurricane of 23 October 1947 in Marxville; some of the 100 fish which fell 100 km away were still alive. The similarity of the amphibian, reptilian, and other small-animal fauna of the Greater Antilles and the fauna of Central America is best explained as being due to transport by hurricanes, even across the ocean (Darlington, 1938).

Of the terrestrial animals the most often transported are the tiny ones, the gliding insects and their larvae. Cases are known in which small an-

imals have been carried from Scandinavia to Spitzbergen, a distance of approximately 1,300–1,800 km (C. S. Elton, 1939). Many spiders regularly make use of the wind for dispersion via their webs; they usually "fly" in a relatively light wind, but strong winds can transport them for great distances, usually to places unfit for them. They have been observed, for example, in the open sea more than 400 km from land. In some individual cases this has aided in the settlement of new regions. The spider fauna of the Hawaiian Islands, which lie 3,700 km from the American continent, consists of these "flying" species. Mosquitoes are dispersed great distances by the wind; they have been observed in Egypt 47 km from their nearest known habitats after a hurricane (Garret-Jones, 1950).

Actively flying insects are also carried by the wind. This is aided by the fact that many species (wasps, bumblebees, flies) climb to altitudes of 1,000 and even 1,800 m, where they are picked up by strong and constant wind currents. In Heligoland and England *Plusia gamma* and *Psilura monacha* are constantly being blown over from the continent. Cases are known in which the butterfly *Acherontia atropus* has been blown to the island of St. Helena, and *Erebus odora* has been transported to Tristan da Cunha. Flocks of the dragonfly *Pantula flavescens* often fly to the Cocoa Islands, 900 km from Java, where this species lives. The so-called flights made by birds are often merely a case of wind transport and are usually observed after heavy storms. Flamingos from the steppes of southern Kazakhstan have been carried as far as Tomsk. Some 37 species of birds have "flown" from Europe to America, and 50 species of American birds have been blown to Europe on the winds. An Australian crane, *Grus rubicundus*, which leads a sessile life, was once caught near Yakutsk.

Besides accidental transport of animals, the wind plays an important part in the dispersion of many insects. Mass flight of newly emerged butterflies of the meadow moth takes place, as a rule, in the warm sector of a cyclone, while the direction of the flight corresponds to that of the winds. The distance of the butterflies' flight depends on the strength of the cyclone; it may reach 400–600 km or more. The butterflies settle in that part of the cyclonic system where the warm front presses against cold air, creating increased cloudiness, rainfall, low temperature, and a drop in wind strength. This creates favorable conditions for multiplication of the settled insects. In this way large masses of insects may be able to reproduce in an area far from their original range and create a new nidus (Mel'nichenko, 1936).

The winds play a most important role in the migration of the desert locust *Schistocerca gregaria*. Its movements are directed from a zone of high pressure into one of barometric minimum, where rainfall is greater, and in general correspond to the direction of the winds. At a wind speed

of as low as 7 km per hour the desert locust always moves with the wind, even if the insects attempt to fly against it (Rainey, 1951). Migration with the aid of wind is necessary for dispersion and maintenance of the huge and varied distributional range of *Schistocerca*, whose mobility is not comparable to that of other locusts (Bodenheimer, 1958).

Zones in which different air currents are encountered are very important for the desert locust. Such an intertropical zone of concentration lies between the monsoons and the trades. Here, in a comparatively narrow belt, the locusts concentrate from many thousands of square kilometers around to form their staging phase. The constancy of rainfall in these areas explains their preference for the region and the steadiness of locust outbreaks here, while concentration with the aid of winds is looked on as the basic mechanism of locust clustering. Late-winter flights of the locust to the Middle East correspond to the mass influx of warm air from the south which occurs in certain years and which at the same time brings in flocks of the butterfly *Pyrameis cardui*.

The daily breeze directions are associated with the dispersion of malarial mosquitoes. In the lower course of Baksan Gorge (Kabardino-Balkaria) they fly not horizontally after hatching but in the direction of the nearest cattle shed containing domestic animals (their food sources), while they settle in houses and barns situated higher, climbing almost 30–50 m vertically over the floodplain to reach their objective. This is explained by the fact that the evening breeze blows down from above, bringing to the hatching area (a body of water) the odors of the more highly situated settlements and the smells of the barns beside them (Fig. 98).

Among mammals the winds which bring odors are an important means of orientation. Tagged polar foxes on the Yamal Peninsula disperse in various directions in a fan-shaped pattern which corresponds to the wind rose.[2] This can be explained by the animals' movement against the wind, as predators usually do when they search for prey (Fig. 99). Northern reindeer also move against the wind, since in our tundras the winds in spring and summer usually blow from the seas onto the continent and in the reverse direction during the fall and winter.

During the warm part of the year, when swarms of small, bloodsucking flies appear (horseflies, blackflies, mosquitoes, midges, etc.), ruminant animals move to open spaces where the swarm is blown away by the wind; if this is ineffective, they run toward the wind. In the tundra and in the mountains reindeer, sheep, and goats stand on patches of snow during this period or else stay on the glaciers, where the combination of strong winds and low temperatures protects them from parasite attacks. Taiga in-

[2] The "wind rose" is a graphic representation of the frequency and strength of the winds in various directions in a given space of time (month, season, year).

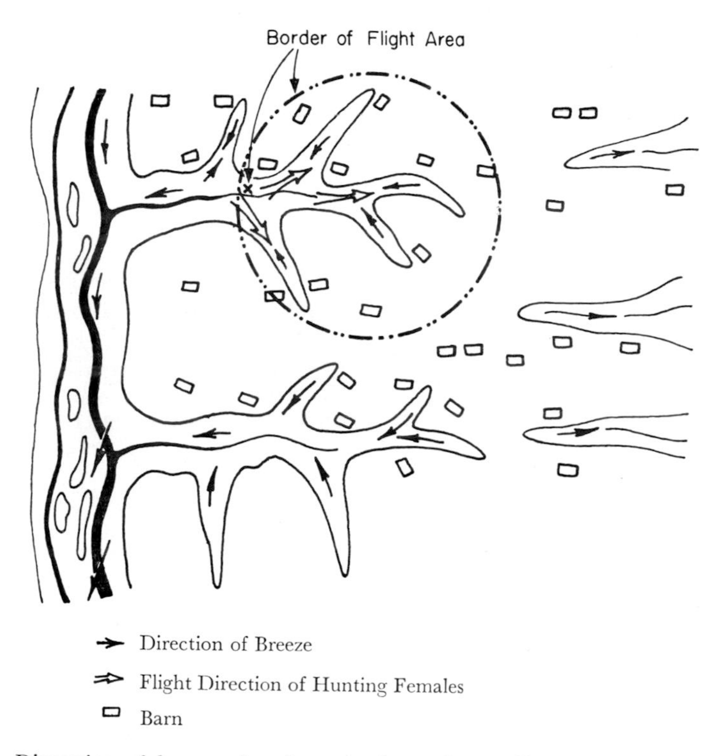

Fig. 98. Dispersion of hungry female mosquitoes along valleys and strongly inter-cut areas as they head into the evening breeze (after Beklemishev, 1949).

habitants (bear, elk, and others) go down to the shores of rivers, where they also make use of the winds or go into the water. Birds in cold and windy areas build their nests away from cover (soil irregularities, bushes and grass, etc.) and sometimes even equip them with protection in the form of cane or reed walls.

PRESSURE

In the aquatic environment pressure, like the distribution of light, is a most important factor in the distribution of marine inhabitants. All the forms living in deep waters have special adaptations permitting them to exist under pressure, which at a depth of 10 km reaches 1 metric ton per cu. cm.

The direct effect of atmospheric pressure on animals is associated with

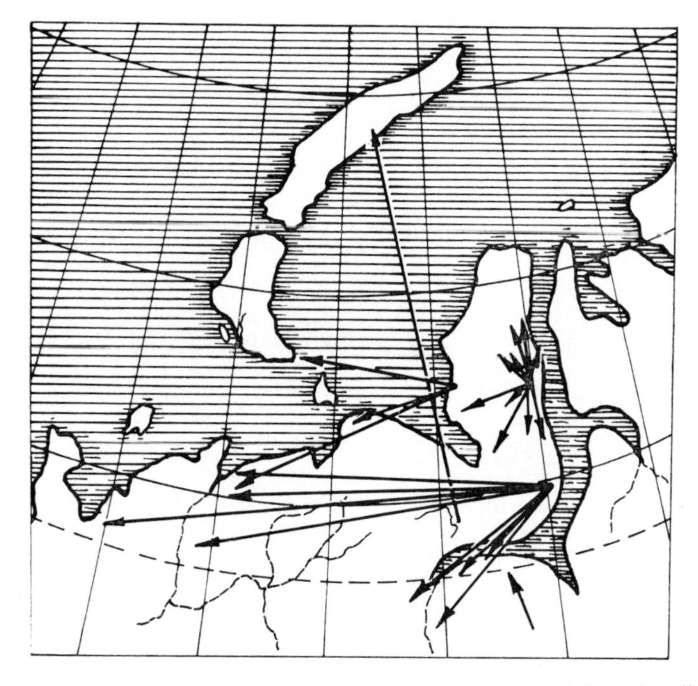

Fig. 99. Migrations of polar foxes tagged on the Yamal Peninsula (after Sdobnikov, 1941).

a decrease in the partial pressure of oxygen at high altitudes, which prevents metabolism. But atmospheric pressure and changes in it are also important as signals. Even at low altitudes a decrease in pressure is accompanied by a change in animal behavior. Intensity of singing in birds, their level of activity, the motility of amphibians, fish, insects, and other terrestrial and aquatic animals—all change with fluctuations in atmospheric pressure, which in this case are signals denoting a change in weather. Periodic phenomena in the lives of animals are also associated with pressure changes. Thus the time which migratory birds spend in our latitudes corresponds to the prevalence of the barometric minimum, while their absence (winter) is typified by high pressure. The times for immigration and emigration of birds correspond well with the baric state (Shpakovsky, 1937). Such also is the connection between pressure changes in the atmosphere and fluctuations in blood pressure in rabbits, milk yield in cattle, and certain other phenomena in domestic animals.

CHAPTER 8

The Substrate

SIGNIFICANCE OF THE SUBSTRATE

The substrate is unimportant only for those animals which spend their whole lives in a state of suspension (plankton and, to some extent, nekton). All other organisms are associated with it as a place of attachment (sessile forms) or support, a food source, a surface upon which movement can take place, or a background against which existence occurs. The character of the substrate is associated with the structure and function of the organs of locomotion and attachment, body form, body coloring, etc.

The substrate of terrestrial animals may be the soil or the continental rocks, snow and ice cover, dead or living plants; for parasites it is the surface of the body or the internal tissues of the host organism. The substrate of aquatic animals is the bottom (ground) of a body of water or objects which have fallen into the water. For "neuston" (the group of organisms which breath atmospheric air but which are sometimes nourished through a film of water) it is a surface water film. The larvae of mosquitoes and some mollusks attach themselves to the surface film from underneath and perforate it with special adaptations, while water bugs, water beetles, and others walk on it but remain in the air. The soil (on dry land), the bottom (in a body of water), and the snow cover may simultaneously serve as a permanent or temporary habitat for certain animals.

It is characteristic of most animals that they search actively for the necessary substrate. These searches are usually directed by the so-called "thigmotaxes," positive or negative relationships to various substrates. Many insects and worms keep moving until their bodies are in contact with the accustomed type of substrate. Caterpillars may be found in any position in which their legs can touch the leaves of a plant, but when they are

placed on their sides or backs and this sensation disappears, they begin to turn around until they again find the accustomed substrate.

SOIL

The soil is the habitat of the soil-dwelling animals and acts as the substrate and background for the terrestrial species. Both indirectly and through vegetation the chemical and physical properties of the soils have an immense influence on the animal population, the content and richness of which greatly depend on properties of soil cover and the amount of macro- and microelements contained in it. The mollusk fauna on the limestone soils near Basel is rich and contains 81 species; these are all species which need lime to form their shells. They include *Buliminus detritus, Cillostima, Pomatria,* and *Xerophila,* which need warmth and normally live only near the Mediterranean Sea. Only 35 species of mollusks have been recorded from all the rest of Switzerland. Limestone soils are also responsible for the northward penetration of such southern insects as the praying mantis and the cicada *Tibicen.* Antelope horns develop better in areas with limestone soils, and the local animals appear stronger and more resistant to infection.

Salt flats and salt marshes have their own unique fauna of invertebrate halophilic animals. In Central Europe alone some 20 genera of beetles have been recorded which are encountered primarily or exclusively on salt marshes or salty soil. Pastures harboring halophilic plants and grasses are highly valued for use in fattening animals. The distribution of terrestrial mollusks has a manifest dependence on the pH of the soil; in Ireland only 4 species live on acid soils, 19 on neutral soils, and 13 on slightly alkaline ones. Finally, the so-called petrobiont animals occur in areas with rocky ground (boulders, screes). They include about 3,000–4,000 species of European beetles, *Parnassius* butterflies, and the mollusks *Patula* and *Clausilia.* Among the lizards several geckos are more or less associated with rocks *(Alsophylax pipiens, Gymnodactylus kotschyi*). And similar examples among the birds and mammals attest to the breadth and variety of the connections between the soil and the animal population. For many groups of animals the soil has been the historic intermediate environment by means of which the transition from an aquatic to a terrestrial way of life was established (Gilyarov, 1949).

The structure and mechanical makeup of the soil determine its basic physical characteristics (aeration, moisture capacity, moisture, heat-holding

capacity, and thermal regime).[1] The possibilities of animals penetrating the soil, moving through it, and digging in it are determined by its structure and density. Tiny animals make use of existing holes and chinks in the soil, and this is aided by their worm-shaped body structure. Protection from invasion is achieved by thickening of the cuticular cover of the body, while flexibility is furnished by segmentation (Fig. 100) and by preservation of the anterior pair of extremities, especially important in such methods of locomotion as "pulling oneself along."

When the soil is "mined" in this way, particles of dirt are displaced by the body of the animal so that behind it either a tunnel is formed, with slightly flattened walls (in moist soils), or else the path falls in and disappears. Many insects and their larvae, earthworms, centipedes, certain snakes (the sand boa), and, among mammals, the moles, shrews, and certain rodents use this method of locomotion. The comb-fingered jerboa *Paradipus ctenodactylus* digs itself into the sand with such speed that it cannot be caught with a spade (B. S. Vinogradov, 1937). Tunneling is often associated with a periodic change in the transverse diameter and length of the body. When this happens, the position of the posterior end of the body is first fixed; later lengthening of the body occurs, with the anterior end being carried forward and fixed in its new position. Finally, the posterior end becomes stretched. This type of locomotion is characteristic of animals with a wormlike shape and an elastic skin (earthworms, soil larvae of the Tipulidae). Changing of the form and size of individual segments is accomplished by redistribution of the cavity fluid (Gilyarov, 1949).

Mining of the soil by animals incapable of changing the shape of their bodies is accomplished by pushing the body along, usually by fixing it with the aid of the fore and, less often, the hind legs or by means of special structures on the sides or the end of the body. One form of adaptation to such locomotion is the roller-shaped body, wedge-shaped in the front, such as is very well expressed in the burrowing rodents, insectivores, etc., and the specially armed extremities shaped like spades and used as thrusting

[1] The mechanical content depends on the amount of continental rock fragments (stones, boulders, etc.), pebbles and gravel (diameter 2–3 cm to 3 mm), sand (3.00 to 0.25 mm), dust (0.05 to 0.01 mm), and silt (particles smaller than 0.011 mm) in the soil. Sandy soils contain more than 90 percent sand particles, sandy loams 80–90 percent, light loams 70–80 percent, medium loams 55–70 percent, heavy loams 40–55 percent, light clays 30–40 percent, medium clays 20–30 percent, and heavy clays less than 20 percent. Between the hard particles there are cavities filled with air or water; in addition, water forms a molecular film on the surfaces of the particles. Colloidal substances bind separate particles together, forming aggregates which change the structure of the soil and the relationships among its solid, liquid, and gaseous phases. The sizes of the soil clods formed permit soils to be distinguished as lumpy, nutlike, or large- and small-grained.

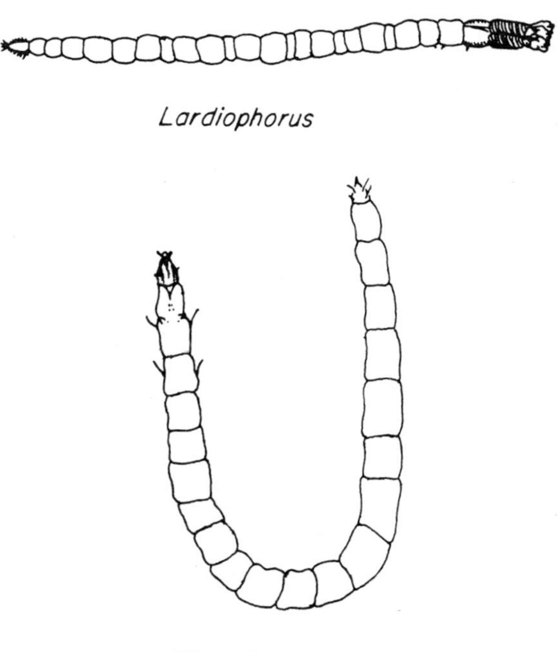

Fig. 100. Larvae of *Lardiophorus* and Therevidae (after Gilyarov, 1949). Supplementary segmentation of the body is easily seen.

levers, which the animal shoves against the side walls and uses to propel itself forward. Such are the amazingly similar structures on the anterior extremities of the mole *Gryllotalpa* and many soil-dwelling insect larvae. Hard (chitinous or otherwise) body coverings used to protect the body may also be adapted to such use.

Among the inhabitants of the upper soil layers and plant litter the roller-shaped body is often changed into a flat one (many myriapods, ticks, etc.). This form exists also among the sand lizards, which dig into the sand with alternating lateral movements of the body. Hairs, horny plates, and scales, such as the extremities of sand-dwelling animals (comb-fingered and shaggy-legged jerboas, long-eared and sand lizards, *Stybaropus*, and others) are armed with, are important as means of support when running along the sand, and the comblike apparatus is helpful in digging the sandy soil.

The "grubbing" form of digging, using powerful claws (black-beetle larvae, the turtle, the zokor [*Myospalax*—Ed.], and, to a lesser degree, shrews, marmots, susliks, mice, and voles) or chisel-shaped spiny outgrowths on the head (click-beetle larvae), is widespread in the harder soils. When

digging, the body of the animal is fixed by the hind legs (vertebrates) or by special formations on the hind end of the body (urogomphi, spines, the "tuggers" of insect larvae), while the grubbing apparatus breaks up the soil to be combed away by the extremities and shoved back by movements of the body. Then the body shoves itself forward with its forward extremities and fixes itself again. A tunnel remains behind the animal which can be traversed many times, while insects put the dirt back and cover their paths. M. S. Gilyarov (1949) feels this to be an important advantage, since predators cannot follow the larvae down the tunnel. Burrowing by birds is a special kind of digging; they use their claws and beaks to prepare their sometimes long and complex burrow nests in earthen outcroppings (bee eaters, cliff swallows, etc.).

Even tougher soils are overcome with the aid of the so-called "boring" method of digging, in which the basic tool is the maxillary apparatus, which "gnaws" into the soil. Removal of the individual soil particles is accomplished by the extremities (most frequent) or by the same maxillary organs and the head as a whole. Bees and wasps use this method in preparing their nesting holes; the larvae of many beetles (*Carabus, Calosoma, Agriotes, Pleonomus,* etc.) and certain other insects dig in this manner. All of them are characterized by a sclerotized covering, especially on the head, a powerful mandible, and supporting formations on the posterior end of the body (Fig. 101).

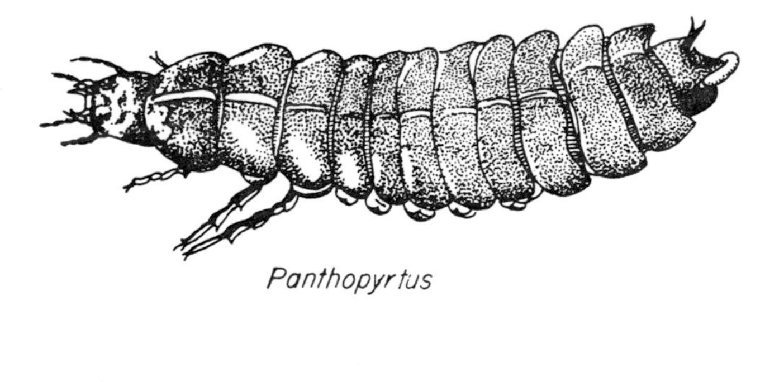

Fig. 101. Larvae of the *Carabus* beetles (*Panthopyrtus*) and of *Calosoma* (*Callisthenes*) (after Gilyarov, 1949).

The best diggers among the mammals are the moles, mole rats, jerboas, and steppe and desert voles, which usually employ the same method. The part of the jaw which does the digging is especially strong; in the burrowing rodents this role is played by the incisors. They protrude from the mouth, and their angle of emergence with regard to the axis of the skull is magnified. In order to protect the oral cavity from the penetration of soil particles, a fur-covered piece of skin is bent onto the diastema, forming a unique hairy covering which separates the oral cavity from the incisors (Vinogradov, 1926). In the most specialized species the incisors are almost exclusively used for breaking the soil and do not even participate in catching or processing food. In the drilling type of digging the animal changes its position almost constantly, turning from side to side and even on its back, almost as if it were screwing itself into the soil.

Comparatively few digging animals are narrowly specialized; more often they alter their method of digging depending on the character of the ground. The preference of an animal for a definite type of soil structure—a well-known phenomenon—is associated with the animal's digging technique. Solid soils are required by many Hymenoptera, Orthoptera, and click beetles, while darting beetles and the Hymenoptera *Cerceris, Bembex*, and *Ammophila* usually prefer friable soils. Quicksand is no hazard to psammophilic animals, many of which have extremities armed with so-called "sand skis"; these not only ease the difficulty of moving over shifty ground but are also used in digging and mining.

All soil inhabitants are endangered by flooding of burrows and water seepage in the ground. In this connection the penetrability of the soil by moisture and the structure of the surface (micro-relief) are important. Burrow entrances are usually found on slopes, seldom on horizontal surfaces. The tendency of foxes and badgers to build their burrow entrances on the sunny slopes of gullies, ravines, and valleys is well known. Here there are usually fewer layers of clayey soil, and sandy layers are closer to the surface. The burrows of marmots and susliks are also built on the sides of hills (Fig. 102). Among the jerboas burrows are often associated with road ruts and other irregularities in the soil. The most permanently inhabited burrows of voles and mice are situated along the walls of canals, boundaries, embankments, and other well-drained areas. The complex burrows of the large gerbil are found in the lower third of the slope of a sand hill, where the sandy soil is denser, the moisture horizon closer, and the vegetation richer.

The tendency of refuges to be located among definite elements of relief explains the irregularity of animal mixing. In regions with a varied relief the location of many species is stretched along valleys, boundary lines, ravines, or other depressions or else occurs in separate islands. The latter

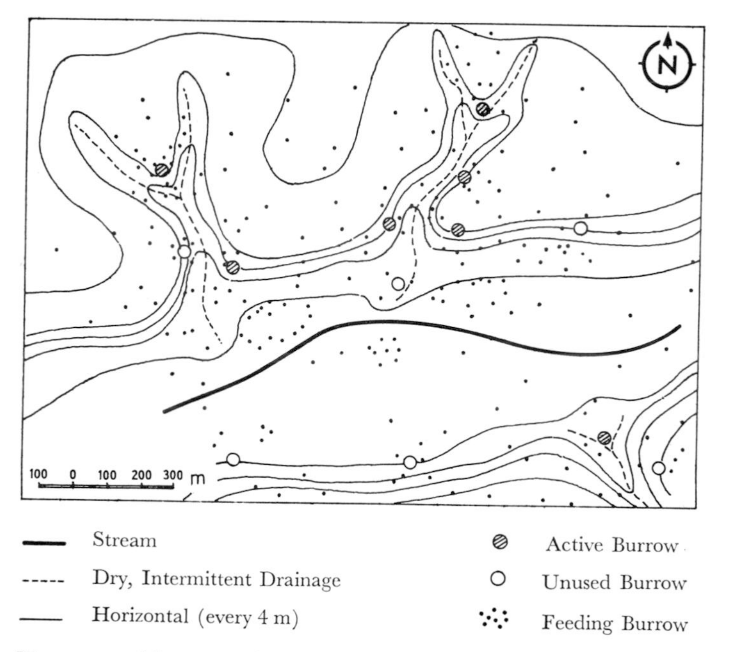

Fig. 102. Placement of burrows of gray marmots (*Marmota baibacina*) in hollows in the Tyan'-Shan' (original).

case is referred to as ribbon-type or island-type, unlike the solid or diffuse location pattern encountered on even and uniform surfaces or in a mosaic pattern (N. P. Naumov, 1954).

The insular colonial location pattern is used by ground-nesting birds: *Merops apiaster* and *M. superciliosus, Riparia riparia, Coracias garrula,* and certain others. The absence of *Merops* in many areas of the flat Pre-Caucasus is explained by the rarity of gullies, which are necessary for building its nesting burrows. In the mountain regions the quality of the fine-grained soil levels plays an important role. It explains the tendency of marmots and susliks to build their burrows at the foot of a slope or in the lower third of it. For animals which are not directly associated with the soil the structure of the surface is still very important, since it determines the local climatic properties, or microclimate. The variety of climatic conditions arising out of this is widely used by animals and permits immediate and close habitation by species with varying relationships to climatic factors.

Soil moisture influences the abundance and displacement of the soil's inhabitants; when it fluctuates, the latter move vertically within the soil layer (Fig. 103). Such periodic migrations are well known for the larvae

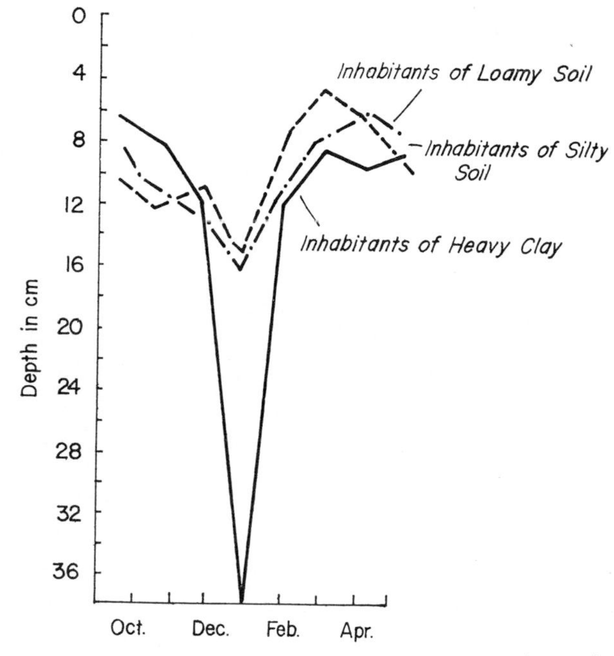

Fig. 103. Vertical migrations of soil invertebrates into deeper layers in autumn and winter (after Daudi, from Allee *et al.*, 1949).

of click beetles. When high levels of moisture are present, they climb to the upper levels of the soil; when moisture drops, they descend to a depth of up to 1 m. Thus their destructiveness is high in spring and less in summer; it grows as moisture increases. Some species select a different moisture level for laying their eggs. When all or most of the spaces between soil particles are filled up with water, soil-dwelling animals are destroyed. Therefore, long and abundant rains are usually accompanied by mass destruction of soil inhabitants.

The thermal regimen of the soil is also important to its inhabitants. Light, dry, sandy, and loamy soils are more easily overheated. Therefore, many insects (the winter moth, larvae of the European marble beetle *Polyphylla fullo,* the May beetle, and the click beetle *Agriotes obscurus*) live only on light soils in the northern parts of their ranges (Gilyarov, 1949). The most northern nidi in which the Asiatic locust reproduces also consist of sands, while the southern nesting grounds are associated with various types of soils. The thick-tailed, three-toed jerboa *Scirtopoda telum,* which lives on hard soils in Kazakhstan, is found only on the Aleshkin Sands at the mouth of the Dnieper. The yellow suslik (*Citellus fulvus*),

which is improperly called a sand dweller, actually lives on the sands but only in the north, while in the deserts of Kazakhstan and Central Asia it prefers hard soils.

The character of the vegetation is of great importance in determining the thermal regimen of the soil. In the high Tyan'-Shan' the temperature of the soil in summer is significantly higher in the sparse-grass steppes than it is under the solid cover of cobresia on the meadows. Better heating of steppe soils favors the existence of fleas and increases their abundance on their marmot hosts, which in turn stimulates the transmission of the plague microbe among the marmots and causes another natural nidus of plague in these areas (Ioff, 1949; Ioff *et al.*, 1951).

The general decrease in temperature as one goes north or as one rises in altitude occurs more slowly in the soil than in the air. Therefore, the soil is usually warmer than the air at high latitudes and altitudes. At an altitude of 1,600 m in the Alps the soil temperature exceeds that of the air by 2.4° C, at an altitude of 1,900 m by 3.0° C, and at an altitude of 2,200 m by 3.6° C (Berg, 1936). Life in such regions is pressed into the soil, one might say, in order to use its heat. The activities of the overwhelming majority of insects and other invertebrates are confined to the surface of the soil and the layer of air immediately adjacent to it. More than 50 percent of the bird species nest on the ground in these areas, while in the central belt of the European USSR they are much fewer (Novikov, 1944).

The soil is the substrate for animal locomotion. Movement over different substrates calls for different body extremities. Three basic types of locomotion are possible along the ground; striding (trotting, ambling), galloping, and hopping are possible on grounds of varying structure and hardness. The quickest form of movement, the gallop of hoofed animals, carnivores, and certain rodents (hares) and the hopping of jerboas on their two hind legs, is provided by an increase in the length of the extremities. Since acceleration of locomotion is associated with strengthening of the pushing motion, the protective surface of the legs and the number of digits are decreased (in the vertebrates). However, this decrease is possible only on a hard support. So it is not surprising that the best runners are associated with hard soils. The known data on speed of movement among the mammals are presented in Table 33. The figures therein have only relative value, but they still show that the best runners live in the steppe and desert regions. Speeds in the water are significantly lower than on land, primarily because of the viscosity of water, which is 60 times greater than that of air. The highest speeds are attained by flying fish, which reach 56–80 km per hour on leaving the water. The speed of the blue shark and the tuna is limited to 35–45 km per hour.

TABLE 33. *Maximum speed of various mammals (after Howell, from Burle, 1951).*

Species	*Speed in km per hour*
Banteng (*Bibos sondaicus*)	25
Rabbit (*Oryctolagus cuniculus*)	38
African elephant (*Loxodonta africana*)	25–40
Rhinoceros (*Diceros bicornis*)	45
Wolf (*Canis lupus*)	45
Giraffe (*Giraffa camelopardus*)	45–50
Jackal (*Thos aureus*)	55
Wild ass (*Equus hemonius*)	50–55
African buffalo (*Sinceros cafer*)	55
European hare (*Lepus europaeus*)	55–70
Zebra (*Equus zebra*)	65
Coyote (*Canis latrans*)	65
Grant's gazelle (*Gazella granti*)	65–80
Thompson's gazelle (G. *thompsoni*)	80
Gnu (*Connochaetes gnu*)	80
Lion (*Felis leo*)	80
Cheetah (*Acinonyx jubatus*)	104–112

As was mentioned, movement over friable dry soils is facilitated by special adaptations of the "sand ski" type. Adaptations for locomotion over stony or rocky substrates are also unique. Hooks for snagging irregularities in the soil and suckers are found on many insects and reptiles living on rocks. Birds which clamber along rocks have very sharp claws and an irregular, rough surface on the bottom of the claw. Similar adaptations are seen in the climbing wood-dwelling forms (woodpeckers, nuthatches, dormice, squirrels, etc.). Among the mountain mammals (mountain goats, chamois, sheep) the strong striking of the hooves on hard soil is compensated for by the abundant growth of a very hard, horny cover. The central part of the hoof is filled with an elastic material, similar to rubber, which reduces leg vibration when the animal is moving over a smooth surface.

BOTTOMS OF BODIES OF WATER

The ground has the same basic role in the life of aquatic animals—a place of attachment, a medium of habitation, a place to find food, and a substrate for locomotion. Depending on the means of using the ground, the form of the animal's body, the structural properties of its organs of locomotion, the character of its covering (in particular, its sclerotization), and its modes of behavior may vary. As on dry land, the abundance and variety of the animal population, which is especially rich on silty bottoms and poor

on stony bottoms, depend on the character of the ground (stone, sand, clay, silt), the degree of its mobility, the richness of the food stored within it, and its suitability for various species of animals. Every type of substrate is inhabited by its own characteristic flora and fauna. The inhabitants often destroy their own substrate. Certain mollusks on sandy shores dig burrows for themselves, and in some places the bottom looks like a beehive. The reverse process is common too, i.e., the creation of a substrate, such as the growth of coral reefs and the deposition of organogenic precipitates.

SNOW COVER

The significance of snow cover in the life and evolution of terrestrial animals is similar to the role of the soil and has received fine treatment in the work of Professor A. N. Formozov (1946). It has proven to be more than a common meteorological factor, since snow cover serves both as a background and substrate on which some animals live and move and as a cover under which others spend a considerable part of their time. In concealing supplies of food, the snow cover diminishes their availability, while for the species living in the soil or on its surface the snow is a good thermoisolator, protecting the animals from sharp freezes.

With regard to snow cover, Formozov has suggested that a distinction be made between animals which are hindered by snow—the chionophobes or "snow haters"—and the chinophiles or "snow lovers." The former live above the snow and the latter below it. An intermediate position is occupied by the so-called chioneuphora, which are well adapted to moving over the snow and feeding on it (the varying hare, white and tundra grouse, and similar "snow-tolerating species"). The same depth of snow may affect these species in various ways. In the cold but nearly snowless winter of 1944/45 moles, voles, and other subsnow animals or subterranean species died en masse, but this same winter had an unusually favorable effect on hoofed animals. On the other hand, a winter of heavy snow, such as occurred in 1940/41, is accompanied by mass extermination of the roe deer but is very favorable for rodents and moles (Kirikov, 1950).

The white coloration of suprachionous animals evolved against a background of snow. In winter many of the inhabitants of open, snowy landscapes (Arctic foxes, hoofed animals, lemmings, white grouse, etc.) turn white or become lighter in color; in the forest, however, light coloring seldom occurs. The nonwhitening winter animals live wherever the snow cover is relatively short-lived and irregular. *Mustela nivalis,* which turns white in most areas, retains the brown coloring on its back during the

winter in the south, where there is no snow or where it lies on the ground for only a short time.

Snow cover hinders movement and catching of food on the ground. The long extremities of the large and heavy beasts permit them to overcome this difficulty. The elk runs freely along friable snow no greater than 40–50 cm deep, its powerful legs punching through the snow down to the ground, but even its movements are hindered by greater depths. The shorter-legged roe deer, wild ass, and antelopes find 40–50 cm of snow too deep to overcome. Among small and medium suprachionous animals the supporting surface of the extremities is enlarged, and these animals move almost exclusively at a gallop, setting down two or all four legs at once (hares, squirrels, martens, mice). In the species which run well on snow the amount of weight resting on 1 sq. cm of supporting surface usually does not exceed 10–15 g. A load of 30–40 g per sq. cm impedes locomotion over friable snow, while large size makes it impossible for short-legged animals to move over such snow (Tables 34, 35).

An increase in the supporting surface of the legs in winter is achieved by the growing of thick hairs, spines, feathers, or horny shields. These

TABLE 34. *Weight load on sole of foot of various mammals (after Formozov, 1946; Teplov, 1947).*

	LOAD IN G PER SQ. CM OF FOOT SURFACE		
Species	*Minimum*	*Maximum*	*Average*
Boar	–	–	903.0
Capricorn sheep	551.0	996.0	848.0
Elk	440.0	570.0	app. 500
Roe deer	–	–	320.0
Wolf	110.0	240.0	190.0
Dog (Husky)	140.0	143.0	141.0
Wild northern reindeer	–	–	140.0
Lynx	–	–	42.0
Fox	–	–	23.0
Northern pika	–	–	26.0
Rusak hare	–	–	22.0–24.0
Large-eared pika	11.0	19.1	15.8
Narrow-skulled vole	–	–	13.5
Forest marten and sable	8.0	18.0	12.0
Ermine	5.8	11.0	8.0–10.0
Red vole	–	–	9.0
Common squirrel	–	–	8.0
Rust-colored vole	7.1	8.2	7.7
Forest mouse	–	–	7.5
Common shrew	–	–	7.3
Economist vole	–	–	6.5
House mouse	–	–	5.5
Wolverine	20.0	29.0	22.0
Varying hare	–	–	19.0

TABLE 35. *Weight load on sole of foot of various birds (after Formozov, 1946; Teplov, 1947; Zimina, 1952).*

Species	LOAD IN G PER SQ. CM OF FOOT SURFACE		
	Minimum	*Maximum*	*Average*
Snow cock	65.5	100.8	84.3
Capercailzie (male)	48.0	67.0	59.0
Capercailzie (female)	38.0	58.0	49.0
Pheasant	57.1	59.5	58.3
Black grouse	40.0	46.0	43.0
Gray partridge	–	–	40.0–41.0
Hazel grouse	30.0	36.0	34.0
Willow ptarmigan	11.0	15.0	12.0

wintertime "skis" are found on many inhabitants of areas with a deep, friable snow cover (Fig. 104). Among the reindeer only the northern *Rangifer tarandus* has a comparatively large hoof area. In winter it increases in size owing to the growth of long, resilient, spiny hairs; toward winter the edges of the foot puff out, while the soft spot grows smaller. These parts become glass-shaped and acquire a high degree of durability. The long,

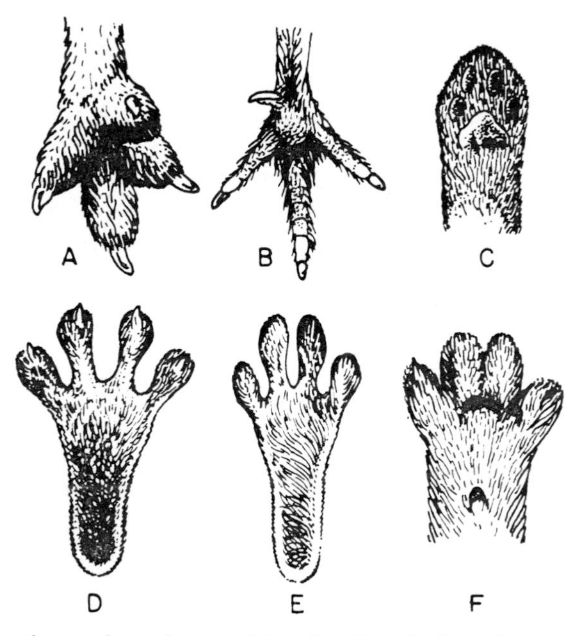

Fig. 104. Supporting surface of extremities of various birds and mammals (after Formozov, 1946). A: "snow skis" of the willow ptarmigan; B: the same foot in summer; C: fore foot of Caucasus cat in summer; D: hind foot of varying hare (weight 3.4 kg); E: hind foot of European hare (weight 4.2 kg); F: fore foot of lynx.

sharp-edged, and spiny hooves of the northern reindeer are above all a tool for scraping away snow to expose moss; they work like hoes (P. P. Tarasov, 1956). The equids, antelopes, goats, and sheep have little supporting surface and therefore live in regions with a shallow, short-lived, or nonexistent snow cover.

Among the different geographic races of the northern reindeer the size of the animal, the relative length of the extremities, and the strength of the horns all increase with the depth of the snow cover. The small, hornless reindeer with narrow, sharp hooves—the musk deer (*Moschus moschiferus*)—lives inside our borders only on the relatively snowless ranges of eastern Siberia (beyond the Yenisei). Snow limits the geographic distribution of another species of small reindeer, the roe deer (*Capreolus capreolus*) (Fig. 105).

The distribution of the narrow-pawed, toe-walking cats tends toward regions of little snow, and only the American lynx lives in the taiga. The weight load on the supporting surfaces of the extremities in this species is five times less than it is for the tiger. Among the Canidae the most narrow-pawed species are the korsac, which lives in the desert and semidesert, and the polar fox, which inhabits the tundra. Both zones are distinguished for their shallow, solid snow cover. The red fox and the wolf are only slightly better adapted to locomotion on deep, friable snow, but they are not abundant in the forest zone, with its deep, friable snow cover. On the other hand, the Mustelidae, which have a relatively large supporting surface on their paws, are rather numerous in the area of powdery forest snows. The ability of the smaller species (weasel, ermine) to mine the snow and move through it aids them in living within the forest zone.

The catching of food by suprachionous animals is complicated both by deep snow cover and by shallow but solid cover. Domestic cattle successfully scrape away snow which is 20–30 cm deep, but sheep and goats may be pastured only where the snow is 10–15 cm deep. Northern reindeer may be pastured in early winter if the snow is no deeper than 40 cm and is not too dense. In the later stages of winter, when the snow cover is harder, they cannot be sent to pasture unless the snow is less than 20 cm deep (Aleksandrova, 1937). Therefore, year-round confinement of cattle to a diet available on the ground is practiced only on the almost snowless steppes, deserts, forest tundra, and tundra. It also occurs in the taiga of Yakutia, where the snow cover is not deep.

Many animals migrate when snow falls. Most of the ground-feeding birds abandon areas which have even the slightest snow cover in winter. The sessile species of mammals and birds change from summer food to some different variety when the snow flies: the capercailzie, the heath cock, and the hazel hen change to pine needles and tree buds, the reindeer and

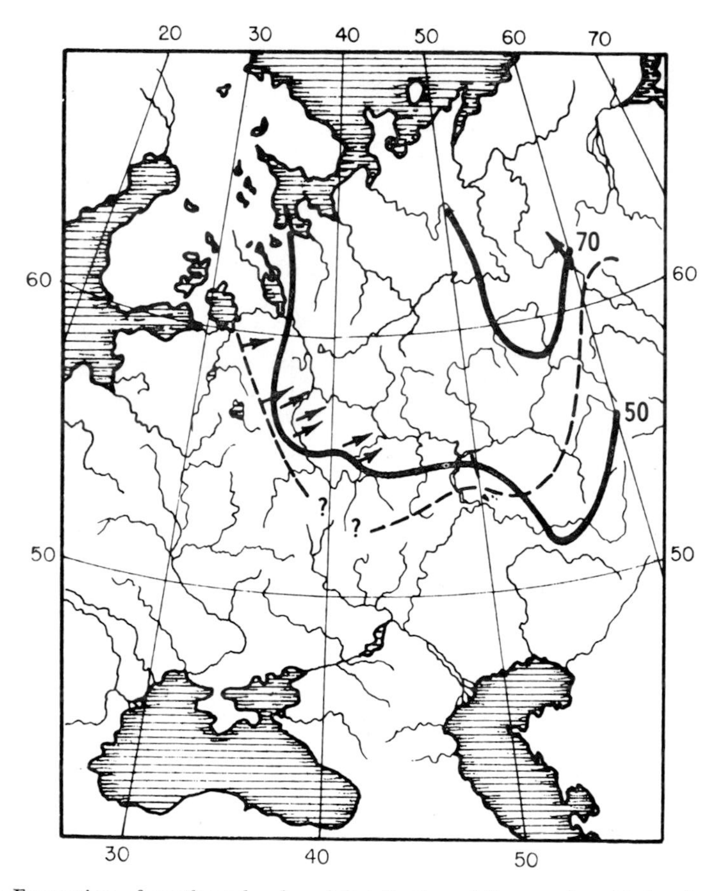

Fig. 105. Formation of northern border of distribution of the roe deer (*Capreolus capreolus*) in the European part of the USSR (dashed line) and isolines of average maximum depth of snow cover at 50 and 70 cm (solid line) (after Formozov, 1946). Arrows indicate areas of most distant solitary migrations of roe deer into snowy regions.

hares to sprigs and shoots.[2] The diet of predators also changes; in winter they catch primarily chionophobes.

The abundance of the inhabitants of the snowy zones changes frequently depending on the amounts of snow in different years; these changes are usually opposite for chionophobes and chionophiles (Fig. 106). When the snow is deep and the weather includes blizzards and sometimes ice storms, many ruminants and birds die en masse. In such years goitered gazelles

[2] In years with a great deal of snow about 85 percent of the varying hare's diet consists of woody and brushy vegetation; when there is little snow, it is only 40 percent (S. P. Naumov, 1947).

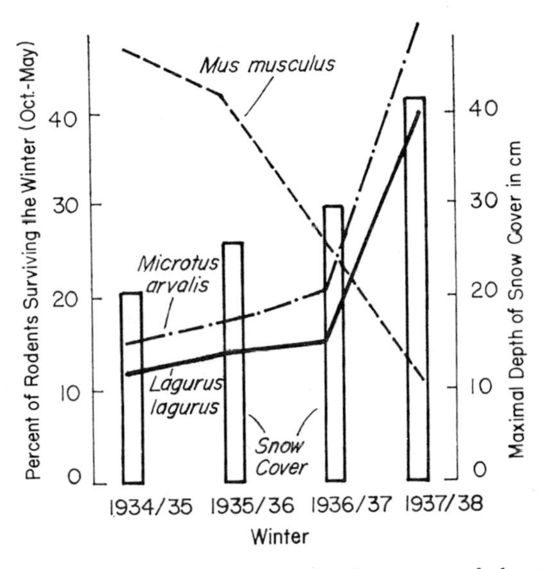

Fig. 106. Survival of winter by voles and mice in the steppes of the southern Ukraine depending on depth of snow (after N. P. Naumov, 1948).

and boars have migrated to built-up areas and even to the cities of the Trans-Caucasus (1942), while European hares have turned up on the ice of the Azov and Caspian seas dozens of kilometers from shore (Kolosov and Bakeyev, 1946). A storm which deposits ice on top of deep snow is especially horrifying. In earlier times the nomadic animal-breeding industry suffered cruelly from this curse. In such years there are mass kills of hoofed animals, bustards, sterepet *(Otis tetrax)*, francolins, skylarks, and other wintering birds which feed on the ground. Among wild ruminants mortality is especially high among the adult males, which are emaciated from the autumn drives.

Catching of food is complicated and the danger of death from predators is heightened during the yearly formation of the spring ice crust and during ice storms. During severe winters with a deep and continuous snow crust the small passerine birds die off even in England. During ordinary years they usually winter successfully (C. S. Elton, 1927). Deep snow is also fatal to predators which hunt animals that live under the snow. In such years polar owls and rough-legged buzzards do not stay in the central belt of the USSR or even the Ukraine but fly farther south and west, while mass death occurs among the wintering "mouse eaters" (owls, foxes, etc.).

Chioneuphorous animals adapted to existence in areas with a deep, friable, and continuous snow cover suffer less from deep snow. Varying hares feed on bark and the twigs of the numerous bushes and young trees,

while white and tundra grouse feed on the twigs and buds of bilberry and blueberry bushes and on the shoots of bushes and woody plants. The polar foxes have a more difficult time; part of them go to the seashore and out onto the ice, and part of them migrate into the forest tundra and the taiga, where they find the friable snow very unfavorable. Only an extreme catholicity of tastes and the eating of substitute foods permit even a part of the population to escape death during heavy snow. In years of very heavy snow even the varying hare and the partridge suffer from famine, since the snows cover the bushes (M. V. Popov, 1960).

Chionophilous animals use the snow cover as a protection from predators and unfavorable weather conditions. This is done by the small mammals (rodents and insectivores), which in winter carry on their activities under the snow. Because of its poor ability to conduct heat,[3] snow is an excellent thermo-isolator; strongly reflecting heat away from its surface, it protects the deeper layers of snow and soil from cooling (Voyeykov, 1889). During a freeze 20 cm of snow keeps the soil even warmer than it would be in the open air, and temperature fluctuations are weaker. At a depth of 60 cm they are very slight (Fig. 107).

By using these properties of the snow cover, the small animals (voles and shrews, including the shrew *Sorex tscherskii,* which weighs nearly 2 g) remain active in winter under conditions of great heat dissipation. They

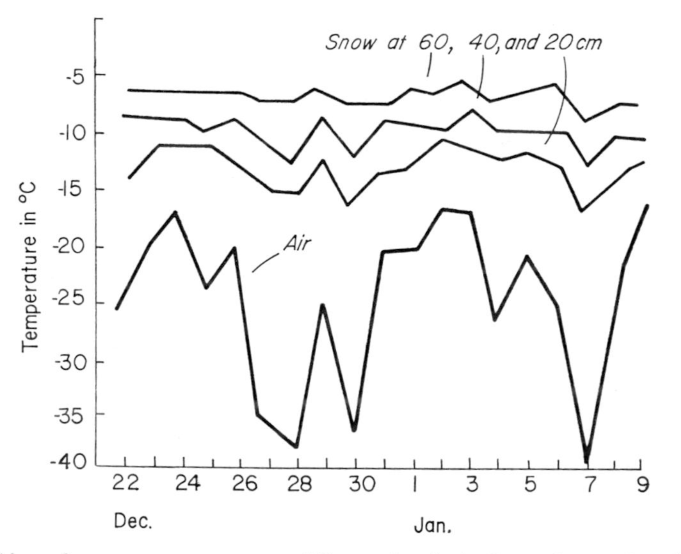

Fig. 107. Air and snow temperatures at different depths in December 1938 and January 1939 in the Pechoro-Ilych Preserve (after Teplova and Teplov, 1947).

[3] Approximately 20 times less than the heat conductivity of ice.

build many paths under the snow and use them to seek food (larvae and adult insects, worms, seeds, etc.). In these subsnow galleries the common voles construct large, thick-walled, warm nests, where they actually multiply in warm years when there is abundant food. Changes in the abundance of such species, as Fig. 108 shows, are closely connected with the depth of the snow cover, which determines the level of the rivers in the forest zone. Chionophilous animals occur primarily in regions with a regular and comparatively deep snow cover.

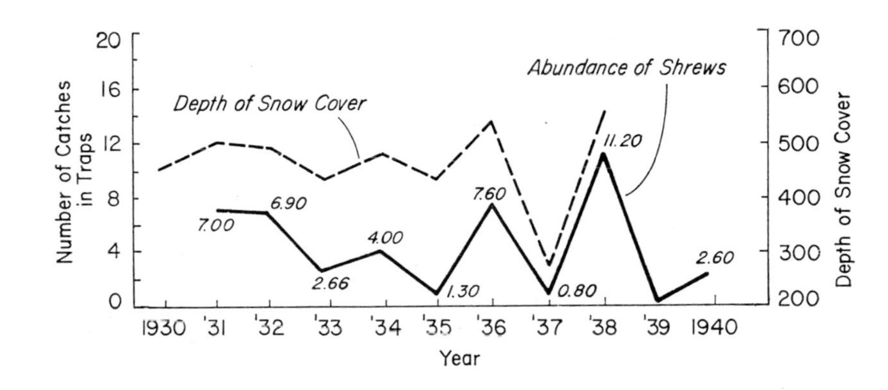

Fig. 108. Relationship between abundance of shrews in the forests of Privetluzh'ye and height of spring floods along the Vetluga River, depending on depth of snow cover (after Formozov, 1948).

Snow is also used as a temporary refuge during frosts by the grouse-type birds, which live in snow burrows during the cold weather (Fig. 109). Even in heavy freezes (—40° C) the temperature gets no lower than —10° C here. The larger the size of the bird and the higher the relative heat dissipation, the more often they dig into the snow for warmth. The only exception is the willow ptarmigan, which lives in open spaces where it is difficult to build snow burrows.

Small snowfalls are unfavorable for chionophiles. After a winter of little snow in which the soil has frozen to a significant depth, the number of soil inhabitants is markedly reduced. Many of them die of hunger (moles, water rats, shrews); others are frozen to death. Under these circumstances most of the locust eggs disappear, and in the following summer the number of locusts decreases markedly. This in turn affects the lives of many insectivorous birds (bustard, rosy starling, merlin, and others). The appearance of Asian locusts north of their usual areas of mass reproduction corresponds to winters with a great deal of snow and favorable conditions for the wintering stages of these insects. Among the Siberian species only

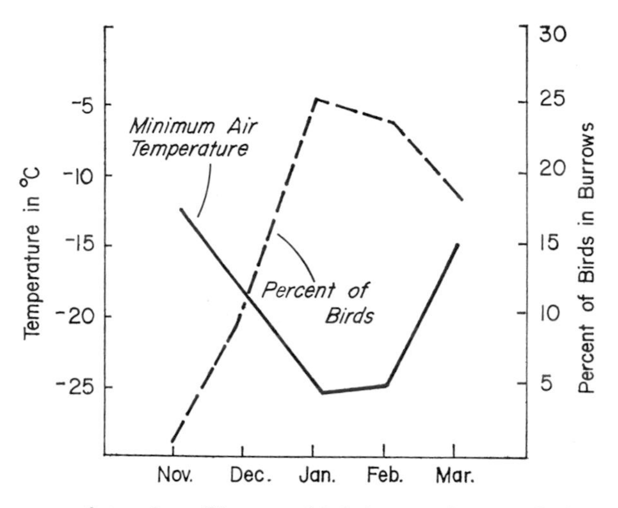

Fig. 109. Frequency of stays by gallinaceous birds in snow burrows during the day in relation to air temperature (after Teplova and Teplov, 1947).

0.9 percent of the eggs are viable and 86.3 percent are completely dead in the spring after a snowless winter, while the majority survive in snowy areas (Rubtsov, 1938). The distribution of the turnip moth along the lower Volga tends toward areas of deep snow (Sakharov, 1928, 1930).

In years of little snow it is observed that the small rodents, shrews, ptarmigans, hedgehogs, and many other species which live all winter under the snow or use it as a temporary refuge die off. The absence of amphibians in Iceland is explained by their having died out in the cold, snowless winter of 1829/30 (Geptner, 1936). The depth of the snow cover influences the freezing of bodies of water and the survival of fish. After cold, snowless winters in the small streams and rivers of the Urals there is a sharp decrease in the number of *Thymallus thymallus* (Kirikov, 1946).

PERMANENTLY FROZEN GROUND AND ICE COVER

Permanently frozen ground is also associated with snow cover. This phenomenon exists in regions of little snow and a severely continental climate. The permanently frozen soil layer occurs because heat losses (light radiation) exceed heat absorption. The depth of the frost layer is directly proportional to the magnitude of the heat deficit and reaches approximately 18 m near Mezen', 50 m in Olekminsk, and more than 114 m near

Yakutsk. The presence of permanently frozen layers is also responsible for wintertime accumulations of unfrozen water under the snow on many of the Siberian rivers, the swelling of ice mounds in swamps, and a series of other phenomena. The water accumulations are sometimes a dangerous trap for small animals, which may fall through the delicate layer of dry snow (roe and musk deer, fox, wolf, and several others).

Permanently frozen ground limits the possibilities for existence among soil inhabitants. The soil fauna in such regions is severely impoverished. Earthworms and the larvae of lamellicorn beetles almost completely disappear. The larvae of the click beetles (Elateridae) and certain other animals appear to be the most tolerant. The northern limit of distribution for the mole in Siberia correlates rather well with the distribution of solid permafrost. This creature is not numerous even in regions where permafrost is insular in character; here the animal keeps to the so-called taliks, or areas of thawed soil amid the permanently frozen ground. The voles and shrews living in the permafrost zone use forest litter and moss cover, spending little time trying to dig into the shallow layer of thawed earth in the summer. The lemmings do the same in the tundra. The suslik *Citellus undulatus* and the marmot *Marmota camtschatica,* which penetrate the permafrost zone to dig burrows in which they hibernate, choose either islands of thawed soil or deep-melting, sunlit slopes and sandy hills (Ognev, 1926). S. A. Buturlin (1913) observed the tendency of susliks to settle beneath fields bare of construction in the region of the lower Lena and Kolyma rivers. Polar foxes dig their rather complex burrows in the sunlit, deep-melting slopes of hills made of well-heated sandy and loamy soils.

Temporary (seasonal) and permanent ice covers are distinguished. The former consists of the ice on rivers, lakes, and seas in the temperate and high latitudes. In the Arctic seas it remains year-round, which may be spoken of as the icing of the polar basin. Permanent ice cover on dry land is found in the form of glaciers. Approximately 16 million sq. km (or 11 percent of the dry land) of the earth's surface are covered by them. The small-sized continental and insular glaciers are not barren of life, since various species of animals inhabit them, more or less regularly, in summer, using them for protection from heat and biting insects.

The ice on the polar seas is permanent. Historically, it played a part in the development of whales and pinnipeds, many of whose structural and behavioral properties are associated with their existence in the icy marine landscape. They are adapted in different degrees to habitation on ice; for at least some of them the ice has aided in the transition from a terrestrial to an aquatic way of life. The seals, though long considered aquatic animals, have not lost their connection with hard substrates, since mating, calving (childbirth), and molting in these species may take place either

on the seashore (the geophiles—*Callorhinus ursinus,* the sea lion *Zalophus californicus, Eumetopias jubatus,* and certain others) or on the ice (the pagophiles—Greenland and striped seals *Histriophoca groenlandia, H. fasciata, Erignathus barbatus,* etc.). The whales have lost their connection with dry land, and even their young are born in the water.

Some terrestrial carnivores (foxes, especially the polar foxes) go quite far out onto the sea ice and live there for long periods of time. F. V. Wrangel (1841) encountered a fox on the ice floes more than 100 km from shore. Foxes have been encountered by different polar explorers at even farther distances from land. Distant passages of banded foxes from continents to islands in the polar basin (from the Yamal Peninsula to Novaya Zemlya, a distance of more than 1,000 km; Sdobnikov, 1940) and encounters with foxes by drifting polar stations show that an icy landscape does not frighten these creatures away. They must find the conditions of existence on the ice profitable. The foxes are usually accompanied on the ice floes by polar bears, who leave remains of their catches for the foxes to eat.

Polar bears spend their entire lives on the ice and hunt on it, but their young are born on dry land. The mothers bear their young in lairs on the islands of the polar basin, while the males continue to lead an active life even in winter. We must recall here that brown bears, both male and female, hibernate in the winter, since these heavy, short-legged beasts cannot move through light, deep snow in the forests.

The ice which covers a large part of the continental waters in the temperate and high latitudes in winter is of great significance in the lives of the hydro- and amphibionts and such terrestrial but water-oriented species as the waterfowl, snipe, certain feathered and four-legged predators, water voles and shrews, the desman, etc. For the hydrobionts the effects of ice cover are associated with a diminution in the water's oxygen supply, which brings on an acute form of mass death.

For dry-land animals which get their food in the water, ice cover interferes with the availability of food and forces the animals into long migrations or short wanderings. For this reason many snipe and waterfowl leave our latitudes, but Cracow ducks which for one reason or another have not flown south will sometimes spend the winter in the central belt on unfrozen springs. In several cities, including Moscow, sessile populations of Cracow and other ducks have sprung up on nonfreezing ponds.

When swamps, rivers, and lakes freeze over, the water voles take to a subterranean way of life and feed on plant roots, tubers, and bulbs. They abandon the shores of bodies of water, where they live in the warmer periods, and wander to the fields and meadows. In winter another type of small creature—*Neomys fodiens*—leaves the water area. *Lutreola lutreola*

and *Lutra lutra,* which live only along rivers in winter (where they can obtain feather grass), have burrows with a permanent connection with the water. The amount and distribution of feather grass determine the number of wintering animals (Novikov, 1939; V. A. Popov, 1949). This explains the rarity of the otter and the absence of burrows on the Siberian rivers, with their long-term, heavy ice coatings and ice storms. The desman *Myogale moschata* and the muskrat *Ondatra zibethica,* which are to a great extent adapted to an aquatic way of life, do not need feather grass, since they live in burrows or houses with an exit opening directly into the water. However, if the thickness of the ice cover is great and the zone of shoreline vegetation freezes right to the bottom, muskrats cannot live. Their existence in the zone of permanently frozen ground is made difficult by ice crusts, which increase the strength of the ice cover and flood the dwellings in which they live during the winter (Lavrov, 1946, 1957). The thicker and more resistant the ice cover of a body of water, the poorer will be its fauna.

CHAPTER 9

Biological Cycles

INFLUENCE OF CLIMATE AS A WHOLE

The environment in which animals exist is changing constantly; these changes are of a dual type. Under the influence of the organic world the biosphere in all its parts changed irreversibly, though slowly. With the appearance of man and the development of agriculture and industry, the tempo of these changes was speeded up to an extraordinary degree. Against this background and in close association with the irreversible changes, periodic or strictly cyclic changes occur in the state of the biosphere which are determined by cosmic causes—the movements of the heavenly bodies in the solar system. Such are the daily, seasonal, and epochal fluctuations in climate and the even longer, but still periodically repeated, changes on the face of the earth which are governed not only by cosmic but also by tectonic processes (e.g., the cycles of mountain-building).

These fluctuations of the environment existed before the rise of life and accompanied life from the first day of its existence. Naturally, all the organisms developed adaptations to them which were expressed in the alternation of biological phenomena occurring periodically and timed to the circumstances most favorable to them. In this way the bioclimatic cycles were formed, so called because they are based upon adaptations to periodic changes of the sum of climatic factors. Such are the daily-activity rhythm and the seasonal and lifetime biological cycles. They are the reactions of animals (and plants) to changes in the climate as a whole.

We now refer to climate as the regime of weather characteristic of a given place and determined by solar radiation, the character of the underlying surface, and the associated circulations of the atmosphere (Alisov,

Drozov, and Rubinshteyn, 1952).[1] Climate is used to distinguish regions situated at various latitudes (climatic zonality) and at various distances from the oceans. The mountains have altitudinal climatic belts analogous to climatic zones.

The climate of the aquatic environment, in which fluctuations in temperature, pressure, and movement of the water are less strong than the corresponding atmospheric fluctuations, is more regular and favorable for life. But even here latitudinal zonality is distinct, though it may be masked by currents. An intermediate position between the steady climate of bodies of water and the very changeable climate of the atmosphere is occupied by the climate of the soils. The hot and, in places, moist climate of the tropical zone of the earth is also favorable for life; it is distinguished by the stability of both its daily and its seasonal fluctuations, even though this stability is less than that found in the seas and oceans. The climates of the temperate and high latitudes, however, not only have a distinct seasonality but also have been less stable historically, changing much more than the tropical climates over large intervals of time.

Although the alternation of seasons is constant, like the alternation of day and night, the character of the weather in each season in the same place is not the same from year to year. Dry years alternate with moist ones, warm ones with cold ones, etc. Thus the daily and seasonal climatic cyclicity never appears as a complete and precise repetition of cycles anywhere, since the cosmic phenomena governed by the movement of the earth around the sun are complicated by changeable meteorological phenomena. It is for this reason that no day or season is ever a perfect repetition of a preceding one. This applies to an even greater degree to the great climatic epochs; the repetition is even more indefinite.

When we analyze a phenomenon as complex as the influence of climate on animals, the need arises for a comparison of certain biological phenomena with the weather phenomena so as to explain the reasons for their relationship. When this happens, we obtain more precise results if we contrast the biological phenomena not against individual meteorological elements but against an aggregate of them. Technically, it is easier to contrast a phenomenon with a pair of interacting factors capable of being illustrated by a plane diagram with a system of two coordinates. Trimetric and even more complex graphs (stereograms) have not obtained wide acceptance in animal ecology.

Valuable results are obtained by contrasting a biological phenomenon with the combined effect of temperature and humidity (thermohygrograms). Several methods of drawing bioclimagrams have been suggested.

[1] A regime of weather is a totality of various types of weather phenomena which repeat themselves with unidentical frequency in given areas.

Bremer's climagrams have been most successful, and Formozov's climagrams are convenient, simple, and expressive. The annual temperature curve serves as the basis of these graphs. Precipitation, however, is marked below the zero line during the warm period (primarily in the form of rain) and above it during the cold period (in the form of snow). Ball-Cook's climagrams have achieved the widest distribution (Fig. 110). Temperature on these graphs can be associated with the amount of precipitation, air humidity, moisture deficit, or any other index. Sometimes temperature is compared with illumination. Fig. 111 shows the extent to which the common bunting (*Emberiza citrinella*) and the garden bunting (*E. hortulana*) differ in their ability to withstand fluctuations in the photoregime and the temperature. The near-sessile common bunting lives under varying light and temperature conditions in both summer and winter, while the migratory garden bunting lives year-round at an essentially identical temperature and under a little-fluctuating photoregime (Walgren, 1954). It is obvious that the migrating species is more sensitive to external influences than is the sessile one.

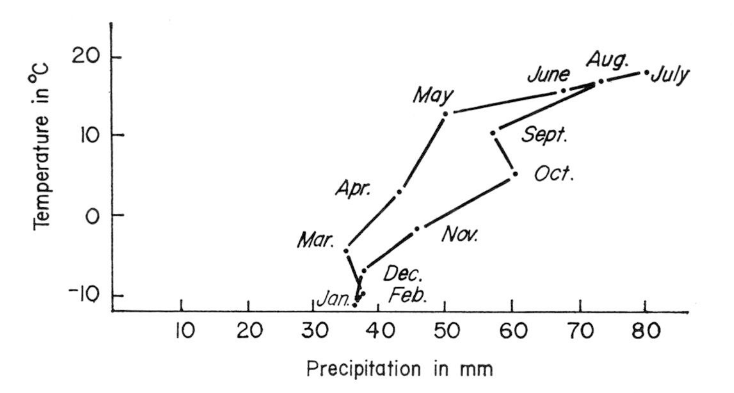

Fig. 110. Climagram of Moscow (after Ball and Cook, original).

V. E. Shelford (1952, 1955) expanded the number of indices capable of being used in bioclimagrams by comparing ultraviolet radiation, duration of sunshine, and many other factors with temperature, precipitation, and humidity. He has also been successful in using bioclimagrams for homoiothermic animals, explaining their most sensitive periods (months) when the state of the weather has a salient effect on their reproduction, growth, and survival. The prairie chicken (*Tympanuchus cupido*) in Illinois is most responsive during the period of gonad growth, which takes place primarily in April. At this time the combination of 50–65 percent of the

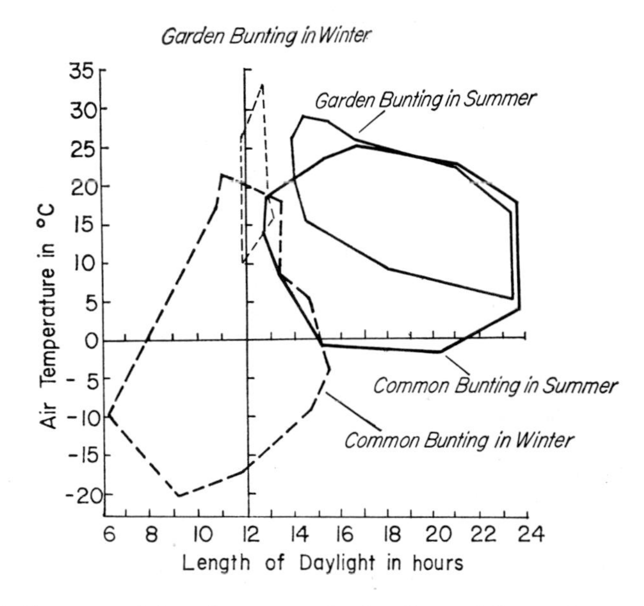

Fig. 111. Bioclimagram of the migratory garden bunting and the sessile common bunting (after Walgren, 1954).

hours of possible sunshine and 50–310 mm of precipitation provides the greatest number of mating males. Temperature plays an important role during the period when the young hatch and begin to feed (optimum is 20.5–23.5° C). So does precipitation (64–117 mm in June).

In the antelope *Antelope cervicapra* and the cottontail *Sylvilagus* the period in which the gonads ripen has also proven to be critical. At this time favorable humidity and an abundance of ultraviolet rays (clear weather) make for large litters and hardy offspring (Shelford, 1954). When the limits of the conditions providing for the development and existence of a species are transferred to a system of climagram coordinates, the suitability of a given climate for a given species can be evaluated (Fig. 112).

DAILY CYCLES

The daily cycles are a rhythmic alternation of periods of rest and periods of activity. Nocturnal, diurnal, twilight, and round-the-clock activity periods are distinguished (Fig. 113). All are characterized by an alternation between time of rest and time of activity, but in some animals both of these periods occupy significant intervals of time, sometimes the entire

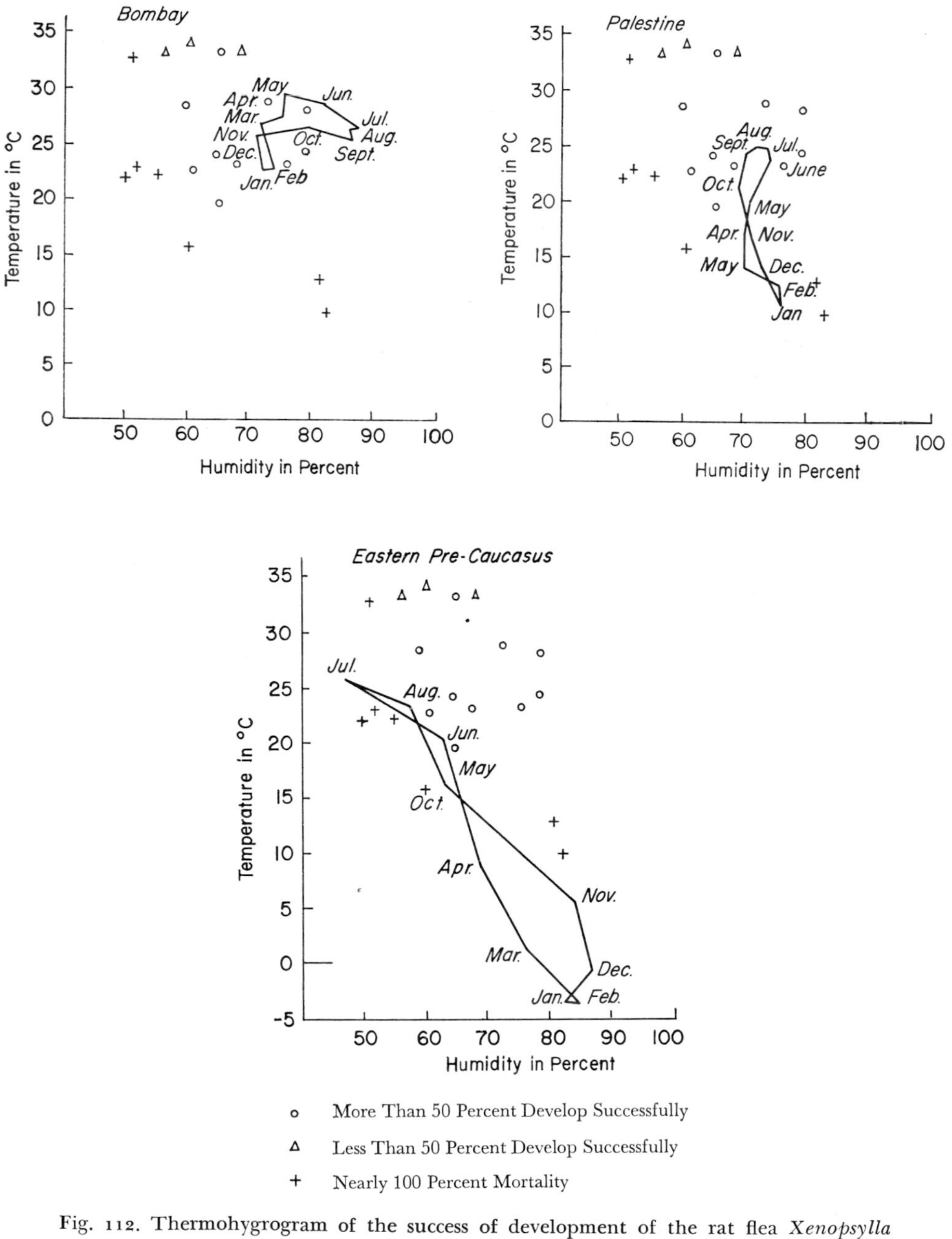

Fig. 112. Thermohygrogram of the success of development of the rat flea *Xenopsylla cheopis* compared with the climagrams of Bombay, Palestine, and the eastern Pre-Caucasus (cities of Stavropol' and Budennovsk and the village of Divnoye) (after Ioff, 1941).

night or day (nocturnal and diurnal animals). Round-the-clock animals have short periods of activity alternating with equally short rest periods

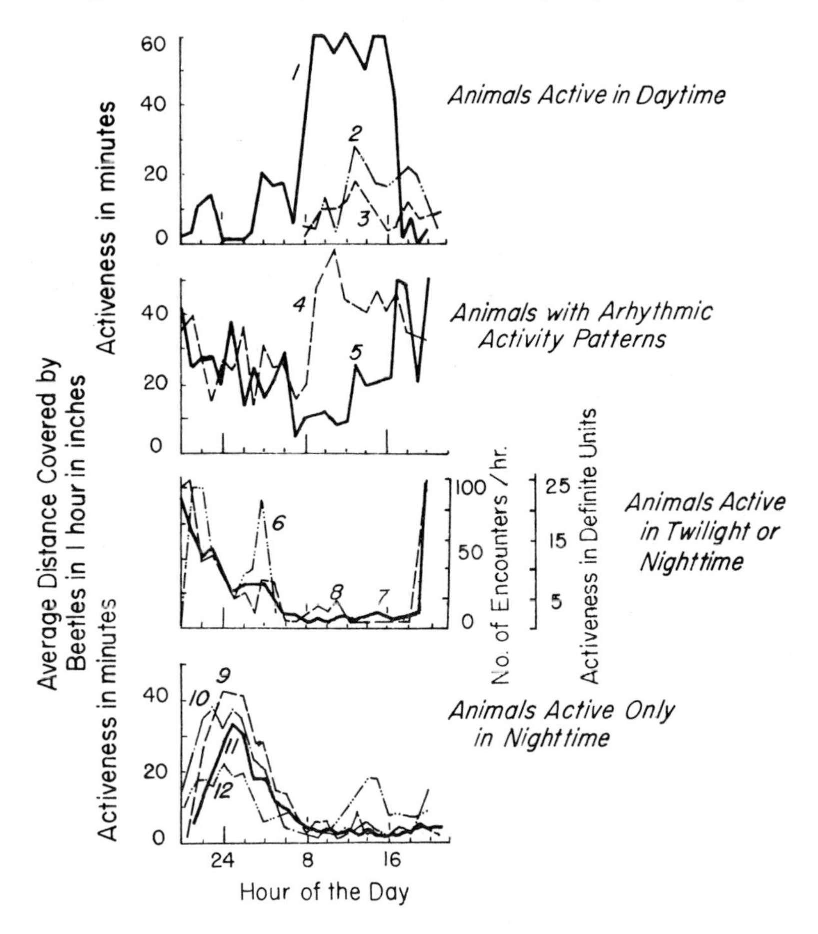

1 *Vulpes vulpes*
2 Male *Citellus pygmaeus*
3 Female *Citellus pygmaeus*
4 *Vulpes fulva*
5 *Microtus agrestis*
6 *Rana temporaria*
7 Beetle (*Megalodacne heros*)
8 Lizard (*Thecadactylus rapicaudus*)
9 Male *Apodemus flavicollis*
10 Female *Apodemus flavicollis*
11 *Putorius putorius*
12 *Putorius eversmanni*

Fig. 113. Daily activity cycles of various animals (after Kalabukhov, 1940, 1951, with changes).

at all hours of the day and night. Alternation of active and quiescent states is accompanied by changes in metabolism, the level of which is higher in an active animal. For this reason the oxygen consumption and body temperature of diurnal species increase during the daytime (Fig. 114), while in nocturnal species these phenomena occur at night. In the gray owl *Strix aluco* a high consumption of oxygen is observed for 2–3 hours in the predawn period (2.0–2.5 cu. cm per g per hour), in the evening (1.9–2.6 cu. cm per g per hour), and at midnight (1.4–1.9 cu. cm per g per hour). During the intervals gas exchange stays at a low and approximately identical level (Segal', 1946, 1958).

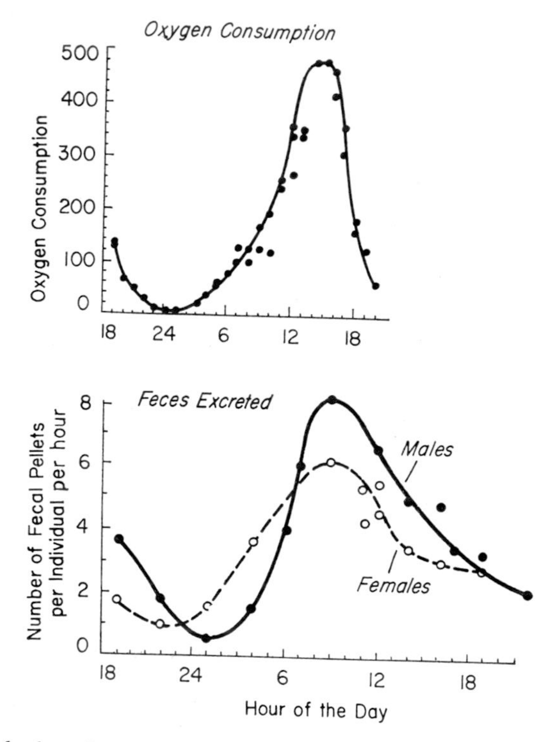

Fig. 114. Daily rhythm of oxygen consumption and feces excretion of the shrimp *Idothea baltica* (after Gayevskaya, 1958).

The relationship to the temperature of the environment changes at various hours of the day. Grain beetles and the carabid beetle *Pseudophonus pubescens* lead a nocturnal life and prefer a daytime temperature (between 10:00 A.M. and 7:00 P.M.) of 15–28°C, while at night they distribute themselves between limits of 2 and 30°C when placed in a thermogradient

device, avoiding the hot portion of the machine (over 30° C) (Fig. 115). At the same time the average preferred temperature changes from 17.5–19.5° C at night to 20.4–24.4° C in the daytime. The beetle *Opatrum sabulosum,* which leads a diurnal life, clusters in a definite temperature range at night, while broad dispersal occurs in the daytime (Likventov, 1960).

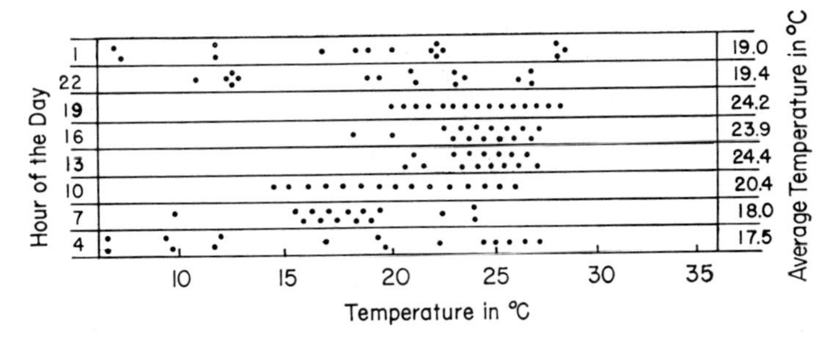

Fig. 115. Distribution of the beetle *Pseudophonus pubescens* in a thermogradient device at various hours during the day (after Likventov, 1960).

The daily rhythm of activity in animals of the same species generally differs in different areas and changes when the feeding conditions are altered. In the gastropods of the Black Sea a distinct alternation of rest and activity periods occurs when food is abundant, but it disappears completely during famines, when the individuals are highly active both day and night. Species which live on low-calorie food and which must eat it in large quantities (the voles, for example), or small warm-blooded animals with a high specific heat dissipation and a high consumption of food (the small insectivores), are forced by the necessity of frequent eating to operate all day long.

Daily activity is not identical in different environmental temperature regimes. The duration of the active period changes differently in different species. In winter large gerbils come out of their burrows only on clear, relatively warm days and even then not for long. In spring and autumn they are active for the entire period in which the sun shines, especially in the middle of the day (10:00 A.M to 4:00 P.M.). In summer the active maxima come in the morning and evening hours; the creatures sit out the hottest part of the day in their burrows (Fig. 116). Red-tailed and southern gerbils are strictly nocturnal in summer, but in autumn, winter, and spring they often leave their burrows by day (Rall', 1938; Shilova, 1953). Among the jerboas and susliks daily activity in the desert is also distinctly associated with temperature, and it changes with the seasons (Fig. 117). In winter the ants in the Palestinian deserts are active at midday and for a total of about 6

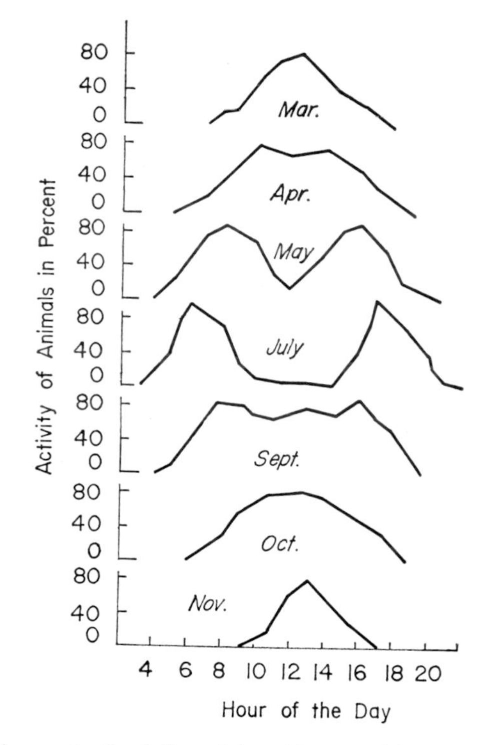

Fig. 116. Seasonal changes in the daily activity of large gerbils (after Shekhanov, 1952).

hours out of each 24; in spring and early summer they are active up to 12 hours a day, primarily in the evening and early part of the night; in summer the level of activity rises to 16 hours, but the maximum occurs at night. In high latitudes the daily periodicity is retained even during the uninterrupted summer day (Novikov, 1949). The activity of the mosquito *Anopheles hyrcanus* corresponds with the night and with relatively low temperatures (Fig. 118). A temperature of 2–16° C favors the activity of the Karelian population of *Ixodes persulcatus* ticks (Kheysin, 1953). Their daily activity is associated with fluctuations in temperature and humidity (Fig. 119).

Many physiological phenomena are associated with the daily changes in matter exchange. The common *Bombyx* moth generally emerges from its cocoon between 5:00 and 8:00 P.M., but the lime-leaf hawkmoth comes out at noon and the death's head moth somewhere between 4:00 and 7:00 P.M., while the one-day butterfly comes out between 8:00 and 10:00 P.M. Daily changes in susceptibility to poisons are known.

The rhythm of daily activity is a complex biological adaptation to many

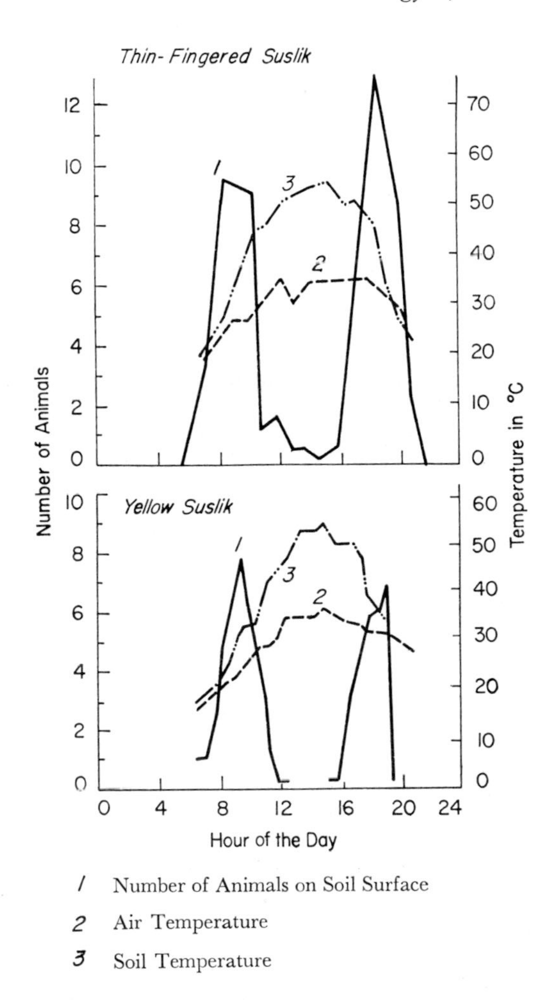

Fig. 117. Distribution of daily activity of thin-fingered and yellow susliks in May in Turkmenia (after Kalabukhov, Nurgel'dyyev, and Skvortsov, 1958).

environmental factors which change rhythmically during the course of 24 hours (temperature, precipitation, humidity, wind, etc.). Many factors in the environment may become regulators of the daily cycle (food and the conditions under which it is caught, temperature or humidity for poikilothermic animals). But because of its regularity, the photoregime is the basic signal-factor regulating daily rhythms in the majority of animals. The alternation of rest and activity periods during the course of the day may be changed by artificial replacement of the light regime (shown experimentally for monkeys, forest mice, canaries, oak *Bombyx*, cater-

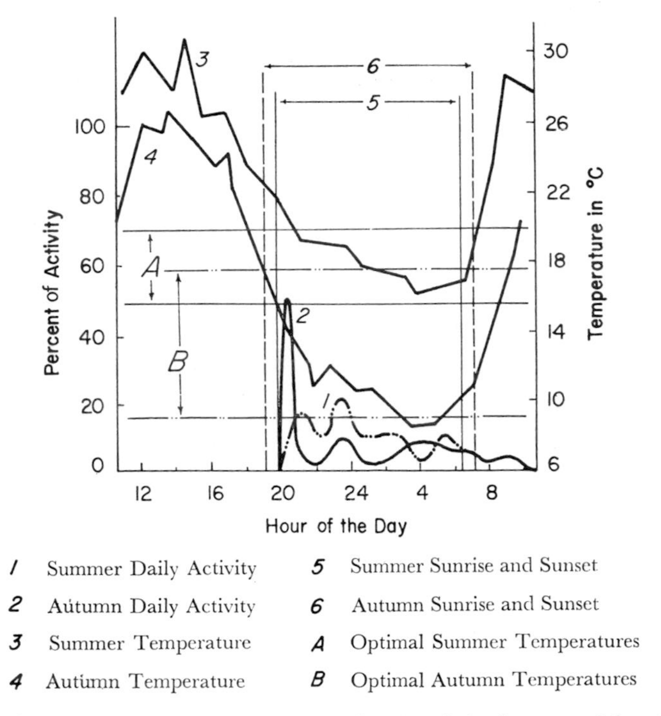

Fig. 118. Daily rhythm of activity of the mosquito *Anopheles hyrcanus* (after Pavlovsky, Pervomaysky, and Chagin, 1951).

pillars, etc.) (Shcherbakova, 1938; Kalabukhov, 1940; Shakhbazov and Sirotenko, 1949), changes in feeding (Grodzinsky, 1958), temperature, and other factors (Byunning, 1961; Lobashev and Savvateyev, 1959).

Movements of animals are closely associated with the rhythm of daily activity. The daily migrations of many animals—aquatic, terrestrial, and soil—are performed under the influence of temperature, which changes during the course of the day, humidity, and chemistry of the surrounding environment, but they are regulated especially often by the change from day to night and back again (Figs. 120, 121). The mysid shrimp in the Caspian stay at a depth of 250–350 m in the daytime, never rising above 150 m, but at night they can be found in the surface layers of the water. *Limnocalanus* migrates along with the mysids, as do the larvae of the Caspian sprat and other animals. In these migrations the mysids and their fellow travelers move from layers with a relatively low temperature, low oxygen content, and high pressure into warmer and oxygen-rich layers with less pressure (Knipovich, 1938). The shrimp *Calanus finmarchicus* migrates as much as 500 m along the vertical plane in the English Channel.

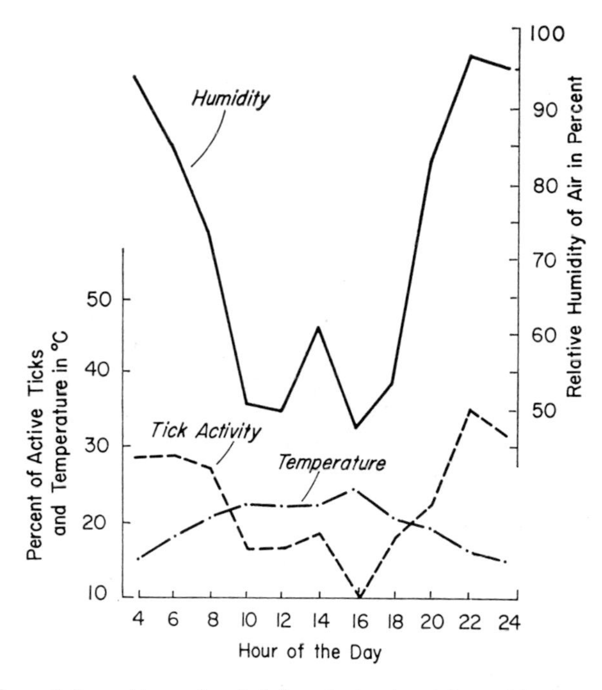

Fig. 119. Daily activity pattern of unfed female *Ixodes ricinus* ticks in a meadow (after Lutta and Shul'man, 1958).

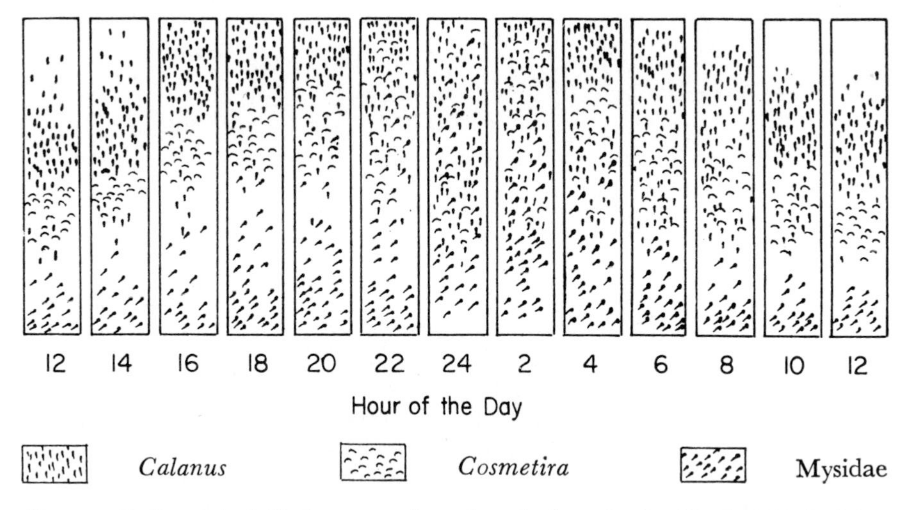

Fig. 120. Daily vertical displacement of marine plankton in the North Atlantic (after Russel and Jonge, 1928).

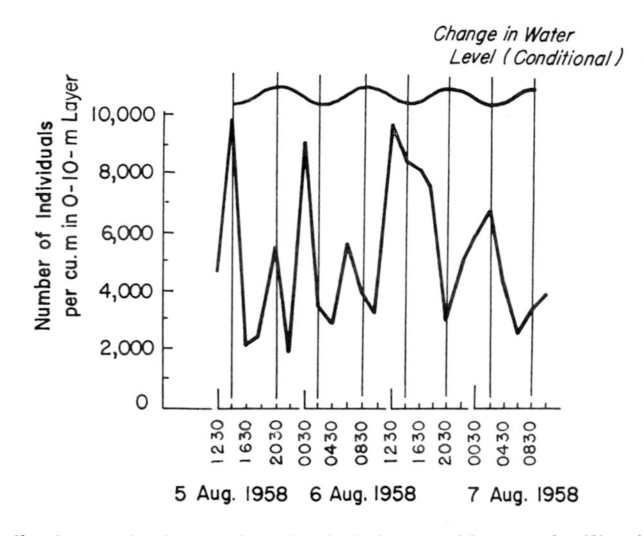

Fig. 121. Daily changes in the number of pelagic larvae of bottom-dwelling invertebrates in the upper layers of the water (0–10 m) in Great Salma Gulf on the White Sea (after Mileykovsky, 1960).

On the shores of Novaya Zemlya this same species makes no daily migration at all during the uninterrupted daylight of summer (Bogorov, 1938). Daily migrations of aquatic invertebrates change at different seasons (Fig. 122). Many fishes making a daily trip of 400 m twice every 24 hours subject themselves to a pressure change of 40 atmospheres. Plankton shrimp migrate to a depth of 600 m.

The daily migrations of terrestrial animals take place along both the vertical and the horizontal planes. They are associated with alternations in the areas where food is caught and rest takes place and are regulated primarily by light, temperature, humidity, and other factors. Among the terrestrial invertebrates "there are none which is completely sessile, i.e., of the type whose center of settlement could not experience daily displacement" (Beklemishev, 1934). All of them undertake movements of one scale or another, moving from one soil layer to another at various times during the day or from one tier of vegetation associations to another as they change their place of feeding and rest. The migrations sometimes prove to be quite small in size.

In the midlands the oribatid mites, vectors of helminths of farm ruminants (*Moniezia expansa, M. benedeni, Thysaniezia ovilla,* and others), migrate from the ground litter into the grass to a height of 2–3 cm each evening and return to the surface of the soil at the beginning of day, when

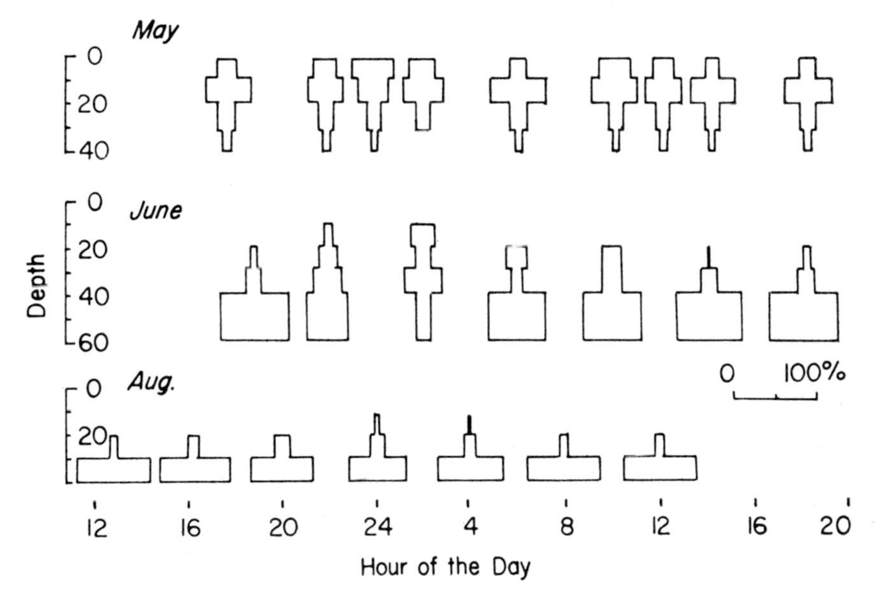

Fig. 122. Daily changes in the vertical distribution of the shrimp *Limnocalanus grimaldii* in the Gulf of Finland in May, June, and August (after Bityukov, 1960).

the temperature rises and the air humidity drops. On dry and unusually hot days the mites flee into the soil to a depth of 2–3 cm. The daily vertical migrations of the oribatids are not expressed identically in each biotope. In damp meadows the mites stay in the grassy tier permanently, even in hot and dry weather, nor will they often leave friable soil which is rich in decaying remains (Shaldybina, 1956). In spring and fall the mountain goat *Capra sibirica* makes daily horizontal trips, in the summer vertical ones, because the high daily temperatures of summer force the beasts to search for high, cool spots (Yegorov, 1952). As was pointed out, temperature plays an especially important role in the daily displacement of soil inhabitants.

SEASONAL CYCLES

Animals adapt to seasonal changes in the basic conditions of existence (feeding, heat, water, and gas exchange, etc.). At the basis of these changes lies climatic seasonality. Climatic seasonality is weakly expressed in the lower latitudes (the tropics), where the fluctuations in average monthly temperature for the year (1–6° C) are less than even the usual daily ampli-

tude. In the subtropics seasonal changes in atmospheric pressure, temperature, humidity, wind, and amount of precipitation are slightly greater, but with the exception of the tropics and subtropical savanna and desert, the fluctuations are expressed relatively weakly even here. In the temperate and high latitudes, especially in the Arctic, light, heat, and other conditions are markedly altered from season to season. The yearly amplitude of temperatures (65.6° C at Verkhoyansk) here exceeds the daily fluctuation by no less than five times.

The seasonal dynamics of supplies of animal and vegetable foods are likewise distinguished. In the nontropical regions the summertime abundance of foods is replaced by the poverty or unavailability of food in winter. The grassy plants end their growing season and die off; the leafy trees become bare of foliage. Most of the fruits and seeds have fallen, and only a few late-maturing ones remain in winter, while reserve food has been stored in the bark and subterranean parts of plants. When mating is finished, many animals die, migrate, or go into hibernation. A small number of newcomers from even higher latitudes cannot be expected to make up the difference completely. In the snowy regions even the remaining food, often significant amounts of it, remains unavailable under the snow cover.

Fluctuations in the supplies of vegetable and animal food in the tropics are distinguished by their lesser scope, since seasonality of fruit-bearing in plants and of reproduction in animals is less expressed here, and hibernations and migrations are rare or absent altogether. In conjunction with favorable hydrothermic regimes, the stability of tropical food supplies, unlike the situation in the temperate and high latitudes, creates conditions which are almost constantly favorable for feeding and matter exchange. Only the tropical deserts are an exception.

The seasonal biological cycles are an alternation from feeding on certain types of foods to feeding on others, from a period of activity to one of hibernation, from a period of reproduction to a period of sexual quiet, from a sessile way of life to a migratory one, etc. At the basis of these cycles lie complex physiological mechanisms which aid the animal in meeting the oncoming seasonal changes in weather and other factors of existence in a prepared state. This preliminary adaptation rests on a system of complex reflex dependencies. Preparation of the necessary changes in metabolism is furnished by signal excitations; these may consist of any external phenomena (changes in the photoregime, temperature, humidity, amount of precipitation, appearance or disappearance of certain types of foods), any one of which or any combination of which has the power to prepare the animal for large and important seasonal changes in the external environment. The signal-factor perceived by the receptors as in-

formation capable of being processed in the nerve centers[2] affects the secretory glands associated with the latter, and the secretory glands produce the necessary hormones in accordance with the information received. The glands act on other links in the endocrine system; this action ultimately leads to the reconstruction of metabolism and the necessary change in the state of the animal. This scheme is responsible for such complex periodic phenomena as reproduction, fat accumulation, hibernation, migration, etc.

With this kind of activity going on, it becomes clear that the signal-factor need not be connected directly to that phenomenon in nature which is to be handled by the adaptation it causes; it is necessary only that it reliably "inform" the animal beforehand. It is only natural that the most complex forms of animal-environment relationships, which do not partake of the usual physico-chemical principles, must have arisen as a result of a long period of natural selection. They do not yield to explanation if we accept only the direct and simple influence of the environment on the organism. In the regularities of the seasonal biological cycles, especially in the phenomena of preparatory adaptation to approaching changes (preadaptation), we see distinctly the specifics of the biological forms of material movement which are not covered merely by physico-chemical laws.

Alternations in feeding associated with changes in the amount, quality, and availability of food are the most important component of the seasonal biological cycles. They are less expressed at the low latitudes, but they are very distinct at high latitudes, especially in zones where a snow cover appears in winter. There the majority of terrestrial invertebrates and poikilothermic vertebrates go into hibernation during cold weather and refuse food; certain fish restrict their activities to a minimum (Fig. 123). During the warm period the food content changes depending on the abundance and availability of various foods (Fig. 124) or on changes in habitat. Thus in the Aral Sea the bream stays close to the shore in spring and autumn and hence consumes more mollusks, while in summer it feeds mainly on crustaceans. The roach living in this same body of water does not rise to the upper layers of water in summer because of the high temperature and therefore catches no insects in June and July (0.8–1.6 percent of its food); in May and September insects make up 33.8 and 53.0 percent of its ration (Nikol'sky, 1944). The intensity with which fish feed is associated with maturing of the sex organs; at this time intensity falls somewhat, but it rises again sharply after they have matured.

The feeding habits of warm-blooded animals which hibernate (bats, mice, marmots, susliks, jerboas, hedgehogs) change little, since their active

2 In vertebrates, in the hypothalamus of the midbrain.

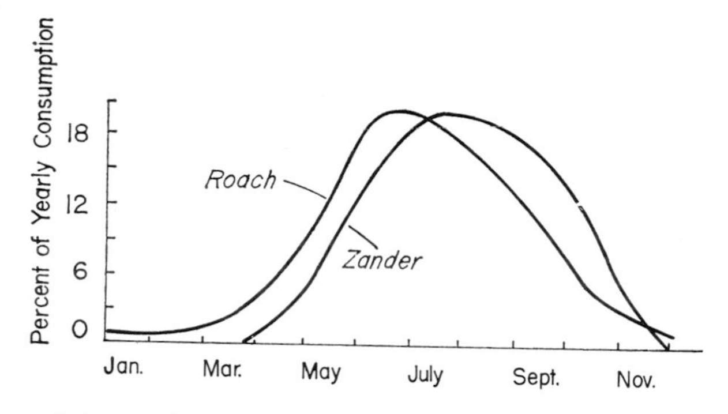

Fig. 123. Seasonal changes in the monthly consumption of food (in percentage of yearly consumption) by the Caspian roach and the zander in the northern Caspian Sea (after Shorygin, 1952).

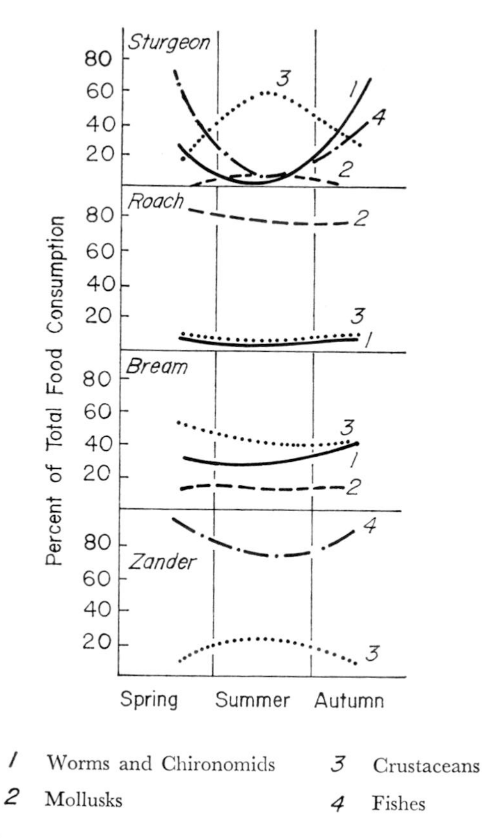

Fig. 124. Seasonal changes in the food content of the sturgeon, roach, bream, and zander in the northern Caspian Sea (after Shorygin, 1952).

state corresponds in time with the season of abundant food, and the disappearance of this food sends them to sleep (Kalabukhov, 1956). In other species the variety of food during the different seasons depends on the abundance and availability of various kinds of food.

As a rule high-calorie foods predominate in winter, in particular, foods rich in fats. Many small birds which in summer feed primarily on tiny animals switch to a diet of seeds in winter (Figs. 125, 126). During the snowless period *Nucifraga caryocatactes macrorhynchus*, which lives on the Pechora River, eats insects (43 percent), cedar "nuts" (41 percent), berries (13 percent), and vertebrate animals (3 percent). In winter 96 percent of its food consists of cedar "nuts" and only 4 percent of other types of food (Ye. and V. Teplov(a)). After the snow falls, the feeding habits of many predators change, since small animals are almost entirely unavailable to them during this period. Because of this, owls find fewer animals which live under the snow (voles) and switch to animals living above it (mice and *Peromyscus*) (Hendrikson and Swan, 1938). The same course is taken by the predatory mammals (Fig. 127; see also Fig. 15). This almost always corresponds in time to the general deterioration in feeding conditions

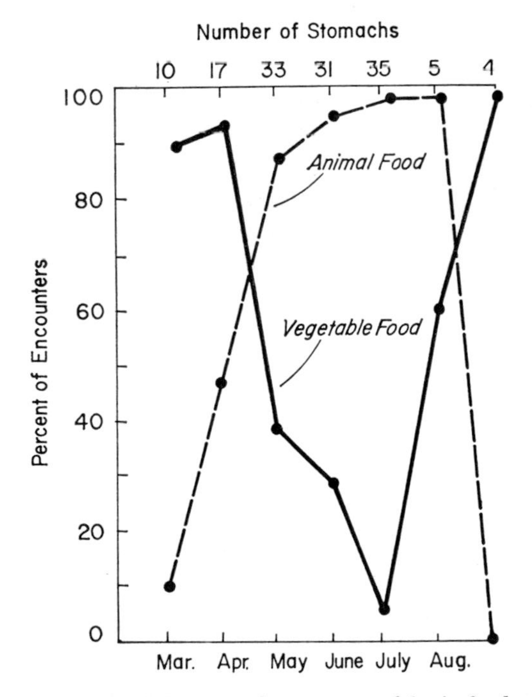

Fig. 125. Seasonal changes in frequency of encounter of basic food groups among the finches (*Fringilla coelebs*) (after Osmolovskaya and Formozov, 1950).

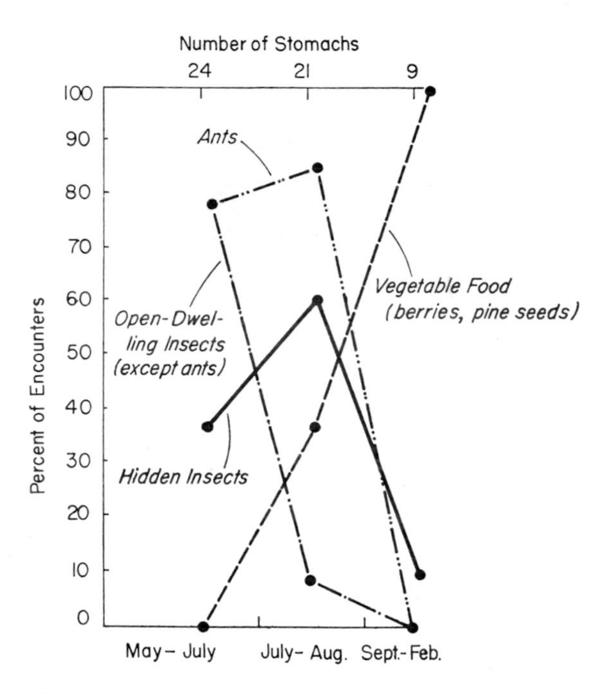

Fig. 126. Seasonal changes in encounterability of basic food groups for the great striped woodpecker (*Dryobates major*) in the forests of the Kostroma oblast' (after Osmolovskaya and Formozov, 1950).

and to the growth in the use of undervalued foodstuffs (such as carrion and garbage). In Arctic foxes this is associated with a drop in the fatness index from 1.7 in the near-snowless tundra to 0.3 in the deep-snow taiga (Pereleshin, 1943). The transfer from summer to winter feeding patterns among the omnivores is accompanied by a change in the animals' "tastes." Usually the food most preferred is the one most abundant and most easily available at a given season; in other words, a change in preference corresponds with the natural dynamics of the food supply in nature.

Certain foods have specific tonic or therapeutic significance; these include plants containing tannin, turpentine, ethers, or alkaloids. They are eaten in significant amounts before hibernation and in spring after awakening, i.e., when changes in food regimes are taking place. Among the insectivorous birds the warblers willingly eat honeysuckle and elderberries in autumn; thrushes eat mountain-ash berries, guelder-rose berries, and juniper. The grouse also transfer to berries in autumn and spring after a snowfall, and many carnivores (ermines, foxes) include berries in their autumn-winter ration. Grass-eating mammals (ruminants and rodents) eat

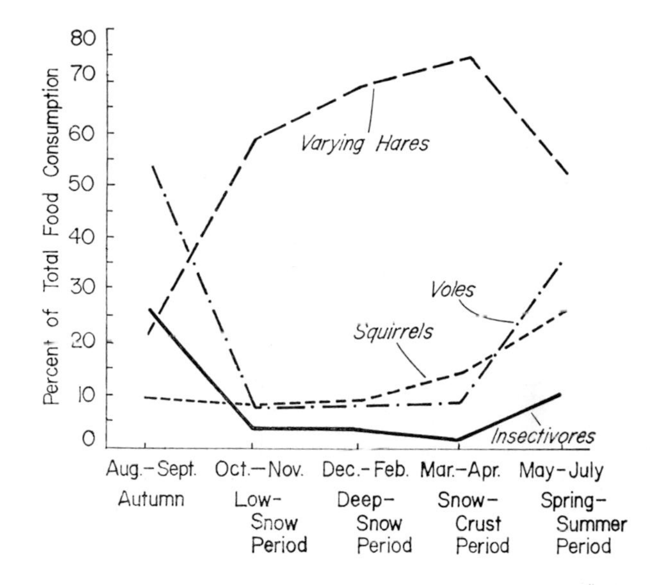

Fig. 127. Relationship between the eating of small mammals by foxes and the snow cover in the Pechoro-Ilych Preserve (after Teplova, 1947).

the ether-rich feather grasses, which sometimes contain toxic alkaloids. Elk consume a great deal of *Menyanthes trifoliata* and in spring and autumn eat the bark of woody plants. In winter bark is also eaten by rodents (voles, less often mice), since at this time the bark is rich in nutritive matter (sugar, etc.).

Part of the seasonal foods is apparently eaten as a vermifuge. In the white partridge and *Tetrao urogallus* ascarid infection changes with the food regime. The percentage of infection in the birds reaches its maximum in June and September. After the transition to berry-type foods, and then to buds and shoots of bushes, the infection rate falls and reaches its minimum in the December-March period. Cestodes in these birds have been known to appear from March through October and disappear from November through February.

On the Kola Peninsula and on the Pechora River the drop in cestode infections and reduction in trematodes in *Tetrao urogallus* and the willow ptarmigan in September and October corresponds in time to the transition to a berry diet, then to pine needles and buds. In winter the percentage of bird infections rises again, apparently associated with incomplete dehelminthization caused by the heads of the worms being left in the walls of the intestine (Huus, 1928; O. I. Semënov-Tyan-Shansky, 1938, 1960).

The character of seasonal alternations of food changes depending on

the "harvest" or "poor harvest" of basic foods (Fig. 128). Seasonal changes in the abundance or availability of food in the wild are associated with the accumulation of reserve substances (fat storage), collection of food supplies, hibernation, migration, periodicity of reproduction, and other seasonal phenomena. All of them are governed by instincts aided by the already-mentioned signal-factors of the environment, but the food itself (the changing of its supplies in the wild) rarely acts as such a signal.

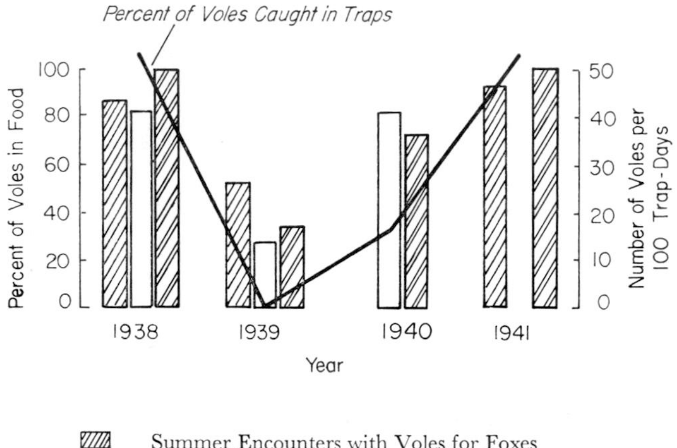

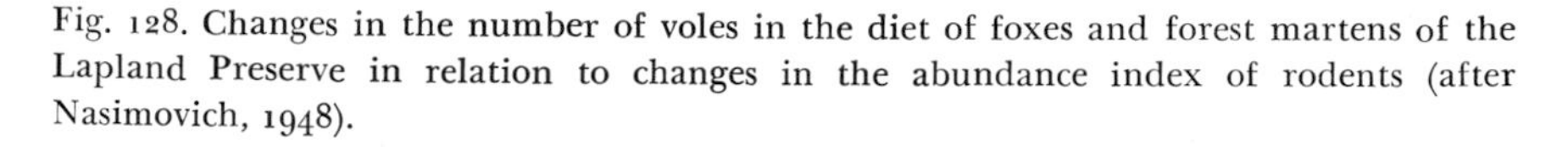

Fig. 128. Changes in the number of voles in the diet of foxes and forest martens of the Lapland Preserve in relation to changes in the abundance index of rodents (after Nasimovich, 1948).

Fattening of animals or the accumulation of nutritional supplies in the body precedes the season of unfavorable feeding conditions; it is a phenomenon widespread among animals (Fig. 129). Fat is stored up as a reserve in the majority of animals (Fig. 130). In *Eurigaster integriceps* starch grains are stored in the midgut in autumn as a supplement to fat supplies, and at this particular time the midgut occupies a large part of the body cavity (Fedotov, 1944, 1949).

In fish the maximum degree of fattening usually appears just before spawning time. In terrestrial animals it occurs right before hibernation (insects, vertebrates), migration (birds and mammals), or the wintering state (see Fig. 133). *Marmota sibirica* passes the winter successfully after having accumulated only 2.0–2.5 kg (adult) or 1.0–0.9 kg (young) of fat before hibernation. If the animal has not fattened itself sufficiently before

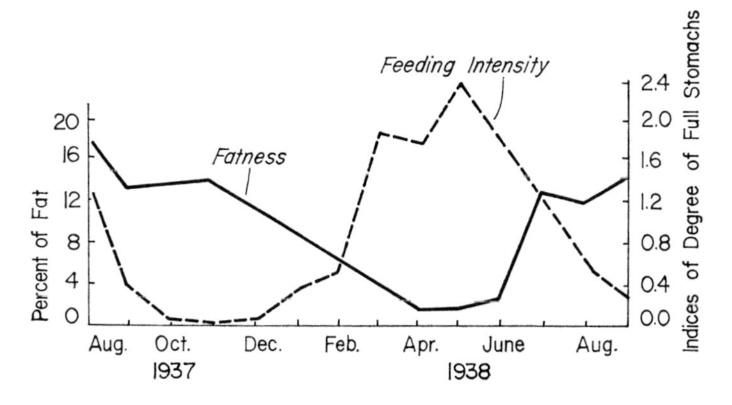

Fig. 129. Dynamics of feeding intensity and fatness in the Murmansk herring (after Tikhonov, 1939).

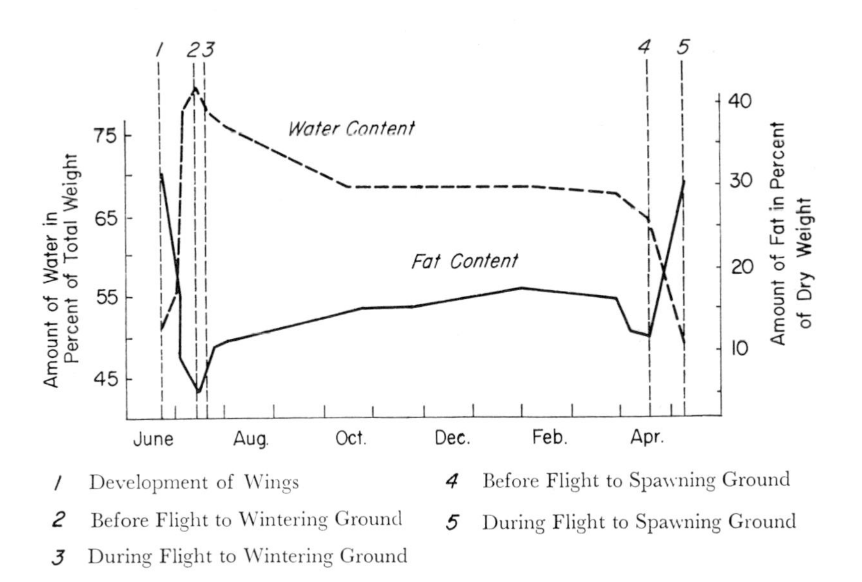

Fig. 130. Changes in the fat and water content of the injurious bug over a yearly cycle (Krasnodar kray, 1951–1952) (after Strogaya, 1954).

going to sleep, its fat supply will run out before the winter is over, if the spring is long and cold, and the animal will die. Fat storage is made difficult by a dry summer, and the following hibernation is accompanied by a higher mortality, especially if the growing season begins late in spring.

Storage of food is an adaptation which permits existence even when the basic food supply has disappeared. Collection of food supplies is char-

acteristic primarily of the higher animal groups, with their complex nervous activity. Among the social animals (bees and ants) it is associated with polymorphism and "division of labor." Among the insects larvae often serve as food. Wasps and bees lay their eggs on larvae of insects which have been collected in a special refuge. The paralyzed but still living insect serves as food for the larva after its emergence from the egg (Fig. 131). The beetle *Deporans tristis* twists the leaves of food plants into little tubes and then lays its eggs inside. *Onthophagus, Ceratophyes, Geotrupes, Lethrus,* and other dung beetles lay their eggs in spheres of excrement and then place the latter in specially built and sometimes complex burrows.

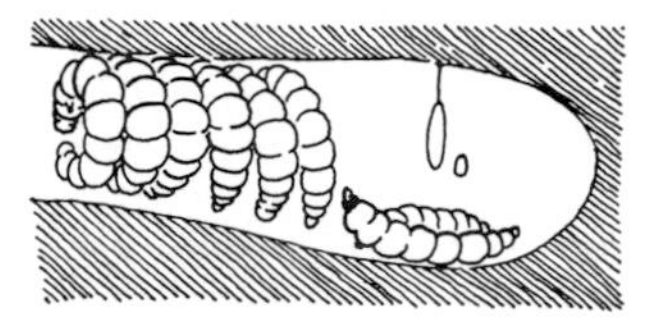

Earthen Nest of *Odynerus reniformis*

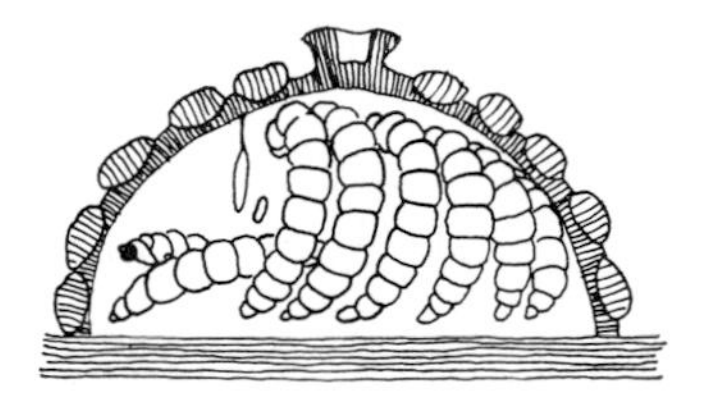

Free Nest of *Eumenes arbustorum*

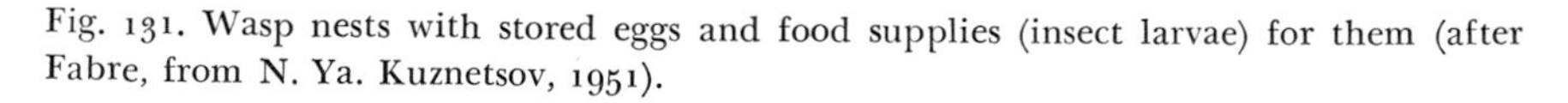

Fig. 131. Wasp nests with stored eggs and food supplies (insect larvae) for them (after Fabre, from N. Ya. Kuznetsov, 1951).

Food storing is either absent or accidental among the other invertebrates and the lower vertebrates. The instinct is weakly developed among the birds; their supplies are usually not very large, while their technique of collecting and preserving is primitive. Nuthatches pack tree seeds into cracks in the bark of trees. Large shrikes store insects and small vertebrates on thorns or sharp ends of branches. Nutcrackers and jays stuff the seeds of woody plants (cedar "nuts," acorns) into moss, soil cracks, or under tree roots, thereby aiding the distribution of these species, since a significant part of the stored seeds remain unfound. Certain owls (the sparrow owl) put small rodents away in the hollows of trees during the autumn and winter. In the Tula Preserve, where this owl lives, 81 red voles, 3 common voles, and 2 small shrews were found in a hollow of the aspen

tree used by one bird (Likhachev, 1952). The owl collects supplies only at the start of winter during the first snowfalls (Shilov).

In mammals the instinct for storage is highly developed. The nomadic and wandering animals do not store food; any supplies among carnivores and insectivores are accidental. But without collecting for winter, the mice, gerbils, pikas, squirrels, chipmunks, hamsters, and many others would not survive. Storage of the collected food takes place in special warehouse chambers in burrows, in the hollows of trees, or in natural refuges. The pika *Ochotona daurica* stores dried-out plants under overhanging rocks or in piles standing in the open not far from its burrow entrance. *Mus musculus hortulanus,* which lives in the Ukrainian and Hungarian steppes, collects cereal grains and weed seeds and stores them in perfect conical piles covered with earth (Fig. 132). Brandt's vole (*Microtus brandti*) lives only on summer-collected foods during the winter on the snowless steppes of Mongolia. The food supplies of other voles are usually uneaten remains of food stored in burrows or at feeding spots. Among the water rats, for example, the supplies actually grow by spring, when the burrow is abandoned (Vishnyakov, 1957).

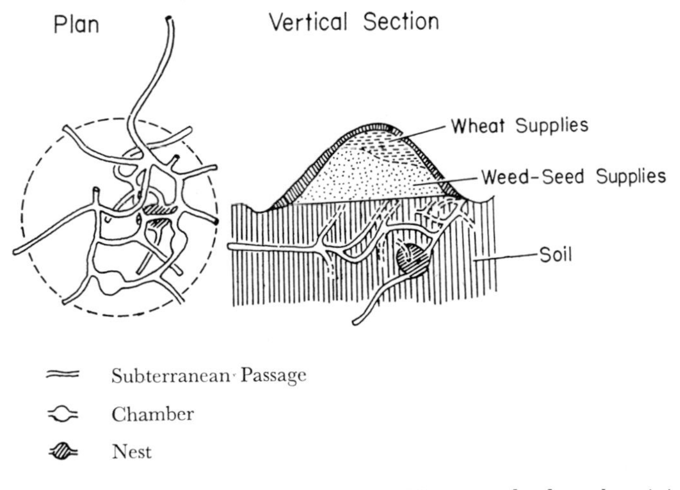

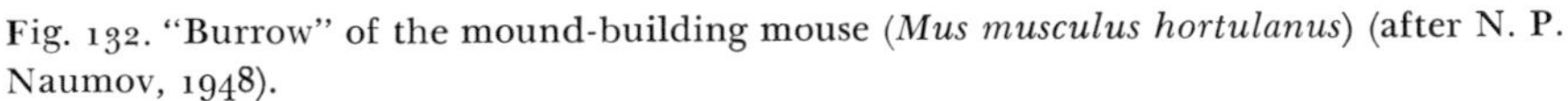

Fig. 132. "Burrow" of the mound-building mouse (*Mus musculus hortulanus*) (after N. P. Naumov, 1948).

The biological significance of stored food supplies is still undoubted, since they serve as a reserve in case fresh food cannot be obtained because of freezing of the ground, a long, drawn-out snow storm, etc. Such piling up of food remains and their use in unfavorable situations were probably the historical origin of the food-accumulating instinct (Formozov, 1939).

The instinctive character of storage is undoubted in most species. Thus squirrels, for example, hang mushrooms out to dry even while they are on the move. If they are of no use to the given individual, they are still important for the survival of the population as a whole.

The beginning of the collecting period corresponds in time with the end of reproduction and the gathering of the creatures into their wintertime groups. Observations and experiments have shown that an abundance of food, a drop in temperature, and a shortening of the day-length aid in the appearance of the storage instinct (Sviridenko, 1957). At this time the behavior of the creatures undergoes a change, and the food which used to be eaten on the spot is stuffed away in storage areas. The speed at which rodents store the seeds of woody plants, especially when the rodents are abundant, is so great in the autumn that in the Tula oak forest 99 percent of the acorns specially placed under leaves were taken and stored away within 4 days (N. P. Naumov, 1948). The size of the food supply laid away by large gerbil families (usually five or six individuals) sometimes exceeds 50 kg. The mound-building mice have taken as much as 20 kg of ears and seeds of cereal grains, while the storing voles (Brandt's) have carried off and stored up to 30 kg of sub- and supraterranean plant parts. Piles accumulated by the Daursk pika range from 0.5 to 1.0 m in diameter and 30–40 cm in height. The amount of food storage can even be computed by region for a given species. Large gerbils collect good-sized piles in Mongolia and Kazakhstan, where the winters are severe and windy, while in the deserts, which have relatively light winters, supplies are small or are not accumulated at all (Shekhanov, 1952).

The instinct to store is especially strong among species with specialized feeding habits. In P. A. Sviridenko's (1944) experiments a forest mouse hid 1,483 acorns in 6 days, a male yellow-throated mouse accumulated 1,492 acorns in 15 days, and a female gathered 589 acorns in 10 days. Both species are specialized seed eaters. But field mice (male and female) hid only 39 acorns in their nest over a period of 13 days; their feeding habits are more varied.

Seasonal changes in body weight and growth are associated with the properties of feeding and fat storage. Among species living on a stable food base, such changes are absent or weak. Among the others fattening usually reaches its maximum at the end of the period when food is most abundant; the lightest weight is attained at the end of the winter or at the end of the reproductive period. Among ducks of the subfamily Anatinae maximum weight is achieved in September and October. During migration and on the wintering grounds they grow thin from the transition to dry or comparatively hard-to-digest seeds and vegetative parts of dry-land plants. In spring, when full-valued foods and tender green shoots appear, weight

increases rapidly, and by the time reproduction occurs, the birds have attained full fatness. The weight of the female at this time exceeds that of the male (Fig. 133). Among the marine Nyrocae (subfamily Fuligulinae), which eat the same food year-round, seasonal changes in weight are substantially absent. But during the wintering period changeability arises, since only birds wintering on the sea eat live food, and those forced to winter on waters in the interior are less well fed (Isakov, 1940).

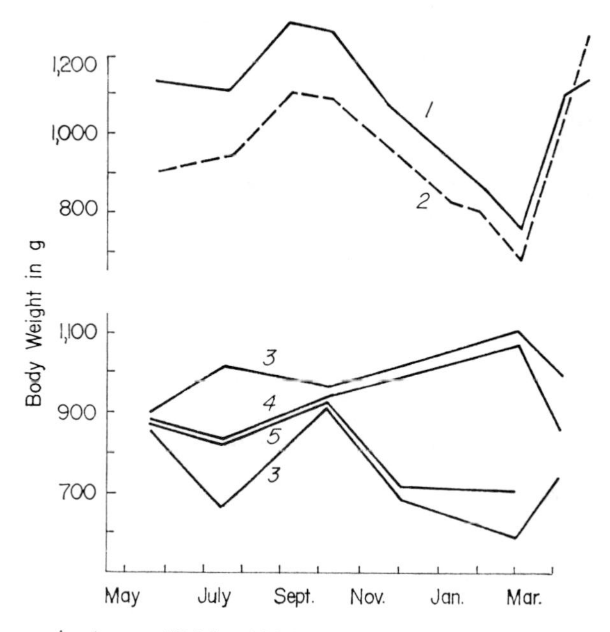

1 Average Weight of Male Cracow Duck (*Anas platyrhynchos*)

2 Average Weight of Female Cracow Duck

3 Extreme Weights of Red-Crested Pochard (*Netta rufina*)

4 Average Weight of Group of Normally Fed Individuals

5 Average Weight of Group of Biologically Weak Individuals

Fig. 133. Seasonal changes in the body weight of ducks (after Isakov, 1940).

Seasonal reconstruction of metabolism is associated with changes in the temperature of the environment. Poikilothermic animals in the temperate and high latitudes fall into torpor when the temperature drops. At the same time species wintering in the open, subject to the effects of low temperatures, undergo a rise in cold resistance which is accompanied by accumulation of reserve matter (fat) and a drop in the amount of water in the body (Table 36).

TABLE 36. *Changes in the ability of the larvae of* Synchroa punctata *to resist cold as related to the amount of water in the body (after Pane, from Kalabukhov, 1956).*

Month	*Average Temperature of Supercooling in °C*	*Amount of Water as Percent of Body Weight*
July	−2.0	–
Aug.	−3.5	54
Sept.	−3.8	42
Oct.	−7.8	40
Nov.	−8.8	38
Dec.	−12.0	32
Jan.	−17.0	–
Feb.	−22.0	31
Mar.	−14.2	–
Apr.	−9.8	–
May	−8.2	–
June	−2.8	–

Seasonal changes in cold resistance are not identical in the different ecological groups of insects. Those which winter in the trunks of trees and are subject to the influence of low temperatures, such as the wood-borer beetle, develop a high degree of cold resistance toward winter. Cold resistance fluctuates very little in species which winter in closed areas. But the majority of aquatic insects (swimming beetles, water bugs of the genus *Notonecta*, the larvae of *Hydropsyche*, etc.) are distinguished by their constantly low resistance to cold. Before hibernation the heterothermic animals (such as the small susliks) reduce their consumption of oxygen, and under the influence of external temperatures oxygen consumption changes less noticeably than in summer (Fig. 134). The sand crab *Emerita talpoida*, an inhabitant of the tidal zone, has an oxygen consumption at 3° C in winter four times greater than at the same temperature in summer. As a result the animal continues to grow actively even in winter, when the other inhabitants of the tidal zone are in a state of torpor (Edwards and Irving, 1943).

Among the freshwater fishes both the number of erythrocytes and the hemoglobin level change during the wintering period (Fig. 135). The wintertime changes in the zander are hardly noticeable, and in *Aspius* sp. the number of erythrocytes begins to increase only toward spring. Both species maintain their active way of life in winter, but the zander's activity is less, and some of the fish of this species actually form winter clusters. The bream, which always clusters in pits during the winter and which curtails its feeding, suffers a sharp drop in the number of erythrocytes. The number rises again as spring returns, but the fish do not go into a deep torpor in winter. The carp has the greatest fluctuation in the number of erythrocytes. In winter a genuine torpor sets in, with suppression of all

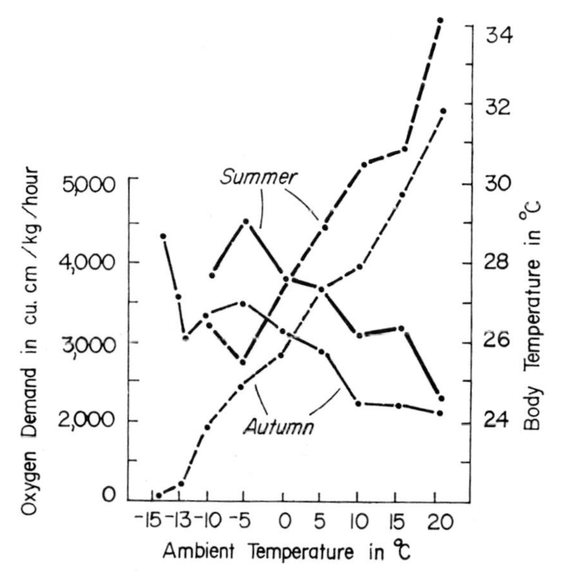

Fig. 134. Oxygen consumption (solid lines) and body temperature (dashed lines) of small susliks at low environmental temperatures (after Shcheglova, 1953).

the basic vital functions. At first the lack of oxygen is compensated for by an increase in the number of erythrocytes, but the profound deficit of this gas leads to a serious disturbance of the blood's contents. Finally, in the sheatfish, which falls into a torpor even deeper than that of the carp and which lives under conditions of even greater oxygen deprivation, the wintertime rise in the number of erythrocytes serves as an adaptation to prevent extreme asphyxiation and death. As is apparent, the seasonal changes in oxygen capacity of the blood correspond to the ecological properties of the species and depend on the state of the surrounding environment (Ivlev, 1955).

Among homoiothermic animals the levels of matter exchange and heat production change. In species with imperfect physical thermoregulation the levels of oxidation processes and heat production rise noticeably during the cold part of the year. Such is the metabolic picture in the mouse-like hamster *Calomyscus* and the house mouse. In these species the autumn-winter level of exchange is higher than the spring-summer level (Fig. 136). Among gray rats it is especially high at the end of winter and in spring, falls in summer (toward autumn), and increases again in winter (Fig. 137). The same thing is observed in the European hare, the common vole (Volchanetskaya, 1954), and the jerboas. On the other hand, in the yellow-throated mouse, the rust vole, the gray hamster, the northern carnivores

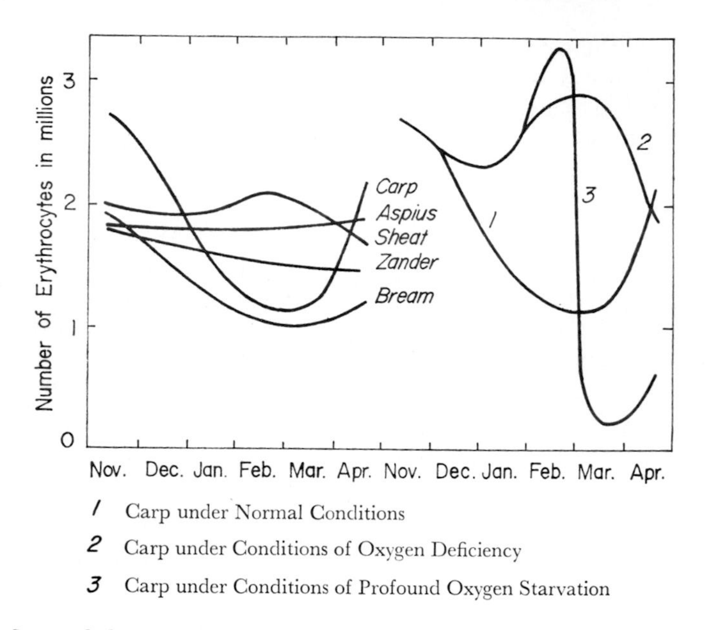

Fig. 135. Seasonal changes in the number of erythrocytes in fishes (after Ivlev, 1955).

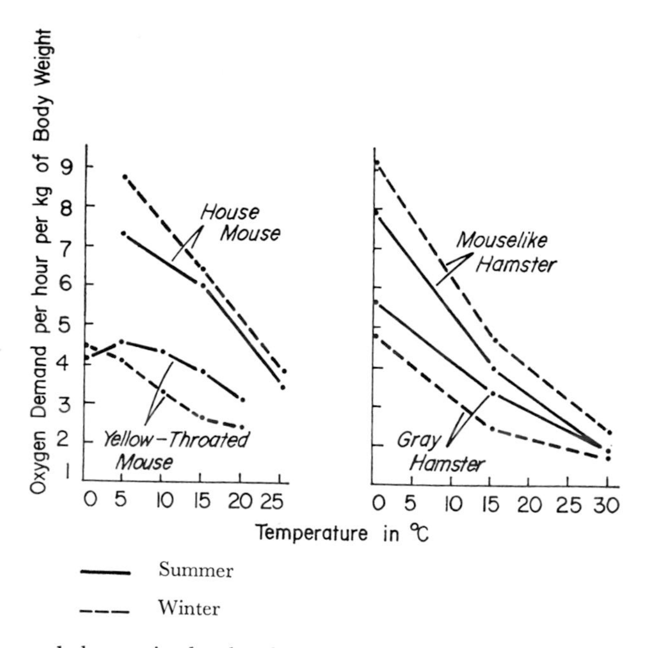

Fig. 136. Seasonal changes in the chemical thermoregulation of mice and hamsters (after Kalabukhov, 1951).

(Fig. 138), and the varying hare the relationship is reversed. A drop in the consumption of oxygen in winter is possible because of the magnificent fur which protects these animals from heat loss and permits them to curtail their food consumption, which is very important in the severe snowy climates.

In zones where snow cover is absent or not significant the amount of food available in winter differs little from that in summer. Often, however, winter is more favorable than summer, since it is at the start of winter that fruits and seeds mature. Under such conditions a rise in heat production is possible and biologically desirable, both among the subchionous voles and among the storing mice, which winter in piles containing large amounts of accumulated seeds (N. P. Naumov, 1939, 1940). The wintertime feeding habits of the gray hamster, rust vole, and yellow-throated mouse are not always provided for. This is even more true for the red and polar foxes, for which a rise in the level of exchange is simply impossible in winter. The inhabitants of high latitudes are distinguished not only by a low level of exchange in winter but also by little seasonal fluctuation in exchange.

In all cases consumption of oxygen by birds and mammals rises when the ground cover changes (spring and autumn). In birds thermo-isolation in winter is increased not only by thicker plumage but also by fat deposits and by feathering of portions of the body which were naked in summer. All this, together with increased metabolism, strongly raises resistance to low temperatures (Table 37).

TABLE 37. *Seasonal changes in the duration of survival (in hours) of the house sparrow* Passer domesticus *without food and at different environmental temperatures (after Baldwin and Kendeigh, 1932).*

	AIR TEMPERATURE IN °C						
Time of Observation	*13.9–14.9*	*9.6–10.2*	*0.3–1.5*	*19.8–22.2*	*33.4–33.5*	*35.9–36.5*	*38.4–39.1*
Summer (June, July, Aug.)	11.4	10.5	17.9	36.6	47.9	32.6	13.6
Winter (Dec., Jan., Feb.)	21.4	20.5	23.7	63.3	76.1	45.5	9.9

Because of the low level of exchange of matter in winter, birds may survive without food for a long time at any temperature except a high one; when high temperatures occur, summer birds survive longer than winter ones. This means that seasonal changes in heat exchange have not only a quantitative but a qualitative character. Similar phenomena are observed even among frogs; the maximum temperature capable of being tolerated by a "winter" frog is 26–27° C, while "summer" frogs have borne

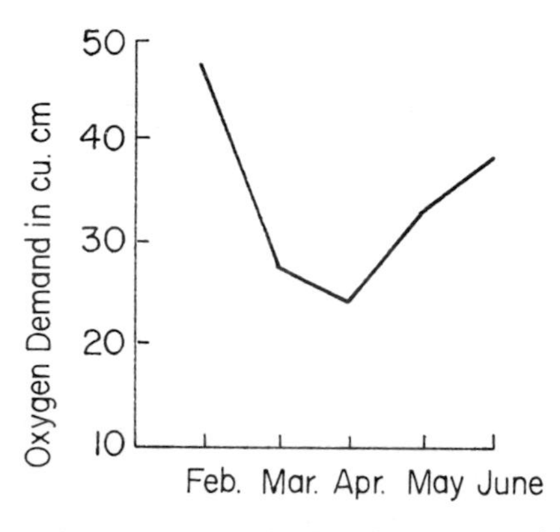

Fig. 137. Seasonal changes in gas exchange in the laboratory rat (after Slonim, 1952).

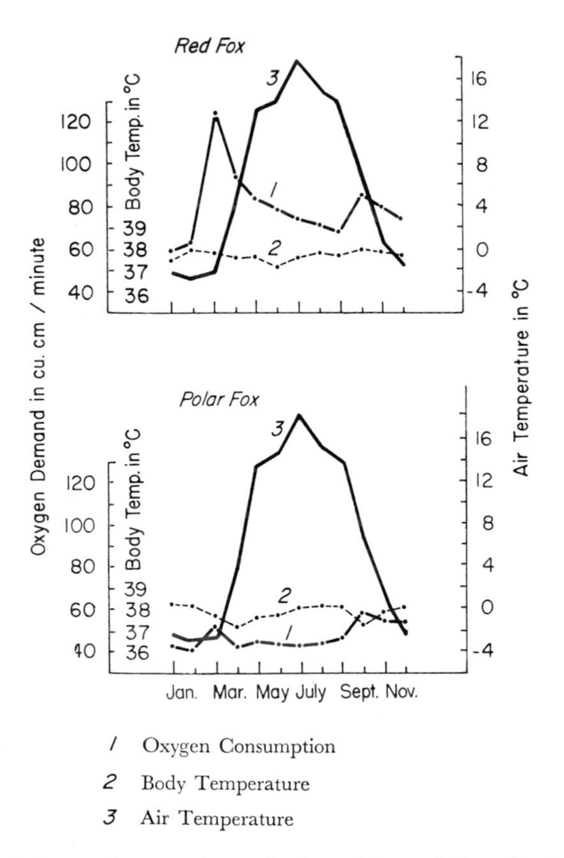

Fig. 138. Seasonal changes in gas exchange in the red fox and the polar fox (after Slonim, 1952; Isaakyan and Akchurin, 1953).

as much as 35° C. However, injection of thyroxin or a pituitary extract into "winter" frogs raises their resistance to high temperatures, and in the females it causes ovulation (Stier and Taylor, 1939). The changes in the temperatures preferred by the animal, usually lower in winter, change in close connection with seasonal alterations in metabolism (Fig. 139).

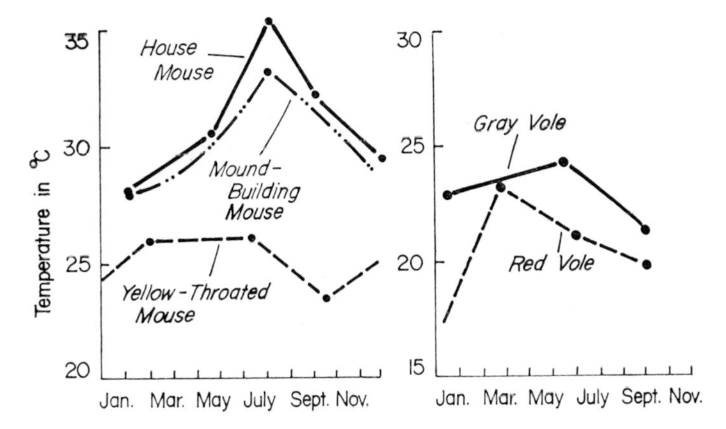

Fig. 139. Seasonal changes in temperature preference among mice and voles (after Kalabukhov, 1955).

Animal hibernation is a special form of adaptation to seasonal changes in the conditions of existence. The metabolism of the animal is changed and the activity and temperature of the body are lowered, usually even before the onset of the unfavorable season. Preparatory changes in the body take place under the influence of external exciters (changes in the photoregime, the content and amount of food, supplies of water, etc.).

Summer and winter hibernation are different. In the latter the temperature of the animal's body drops, and the creature falls into a torpor during the cold part of the year. In heterothermic animals this is accompanied by a loss of homoiothermy, as a result of which the temperature of the body stays close to that of the environment. Animals go into summer hibernation during periods of high temperatures; usually the phenomenon corresponds in time with a drought. Summer hibernation is also characterized by a drop in body temperature, loss of homoiothermy (in warm-blooded animals), and torpor. In many species the summer hibernation turns directly into a winter one, and the line between the two is difficult to draw. In both cases hibernation is characterized by a drop in the level of matter exchange and is used as an adaptation for living through a period of unfavorable conditions. It is accompanied by loss of activity (torpor) and curtailment of feeding, and it may be governed by

conditions unfavorable for heat exchange as well as by a dearth of food, poor water exchange, or other causes.

Hibernation is widespread among animals of the high and temperate latitudes. In winter, and to some extent in summer, most of the terrestrial invertebrates and soil inhabitants, almost all of the aquatic invertebrates, and some of the fish fall into torpor. In winter the reptiles, amphibians, certain birds, and many mammals fall into a torpor. The number of animals going into torpor in winter in the temperate and high latitudes exceeds the number which stay awake (Kalabukhov, 1956). Especially large is the percentage of species which hibernate among the animals of the desert, steppe, and forest zones. The proportion of these animals diminishes both to the south (the subtropics and tropics), as the contrast between the seasonal climates decreases, and to the north, where in the tundra and the Arctic the unfavorable season is so long and the warm period so short that possibilities for accumulating reserve fat for surviving a long period of torpor are insufficient. This is one of the reasons for the relative scarcity of terrestrial invertebrates, reptiles, and amphibians in these areas.

The forms of hibernation differ in the extent to which the energy processes of the organism are suppressed; the smallest magnitude is attained in animals which do not go to sleep but merely curtail their activities in winter, i.e., the mammals. In the more severe periods, during heavy snow or hoarfrost (rime) on the branches of trees, squirrels do not come out of their nests for weeks at a time. Among the tiny rodents (voles), with their imperfect thermoregulation, activity is curtailed as a result of the general drop in the level of matter exchange. Body temperature drops, and the speed of growth is cut ten times. In winter reproduction stops in fields, meadows, and forests but continues in haylofts and grain piles, where the creatures feed on high-calorie food (plant seeds). There the body temperature is somewhat higher (3–4° C) than that of animals living in fields and meadows. We must point out that the average temperatures of haystacks and other places of habitation do not differ substantially (Strel'nikov, 1940).

In the so-called winter sleep of predatory mammals (badgers, bears, raccoon dogs, and certain others) the drop in the level of matter exchange is not great. Their bodies have a temperature only slightly below normal, while exchange is cut by 30–35 percent. Shifts in matter exchange consist of a drop in the nervous regulation of feeding patterns, the inclusion of specific food excitement effects (rise of heat production and body temperature after ingestion of food), and an acute limitation of muscular activity. In such a state the body "sleeps" during the winter, after which the effect of food ingestion on matter exchange becomes stronger, the level of mus-

cular activity increases, and the degree of matter exchange itself begins to rise. During winter sleep brown bears even give birth to their young, but the size of the young is very small.[3] This permits them to be fed by the mother even under conditions of her reduced matter exchange and while she herself is not feeding. Interrupted winter sleep is known among the rodents in the case of hamsters and chipmunks. They store food in the chambers of their burrows, which they consume during warm winter days and especially in spring.

Many mammals, reptiles, amphibians, certain fishes, and the majority of invertebrates in the temperate and high latitudes go into a long and deep torpor in winter, a true sleep in which the character of heat exchange between the animal and the environment is profoundly altered.

Reserve matter is stored up before the drop into torpor. This matter supplies the organism's demands not only during the hibernation period but also during the first days after awakening, when supplies of food in nature are insufficient or unavailable. Therefore, the success of hibernation is determined by the necessary amount of reserve matter. Accumulations of fat sometimes double the initial weight. The speed with which reserve matter is expended is determined by the temperature of the refuge. Weight losses in mammals over a period of 120–170 days (the average duration of the winter hibernation) reach 30—50 percent of body weight and occur at the expense not only of adipose tissue but also of the liver, muscles, and even lungs. Among insects the fat content of the body prior to torpor may reach 5–28 percent of total body weight and may be 16–60 percent of the dry weight (in mosquitoes). After emergence from torpor it quickly drops (see Fig. 130). At the same time that reserve matter is being stored, the amount of water in the body diminishes. In the suslik *Citellus pygmaeus* it has been established that ascorbic acid is stored in the adrenals, the thyroid gland, the sex glands, and the blood (Kratinov *et al.*, 1947; Slonim, 1952; Kayser, 1961).

Changes in matter exchange are usually accompanied by a decrease in the respiratory coefficient, since in the state of torpor it is primarily fat which is being oxidized. The level of exchange also changes; it drops especially sharply among the heterothermic animals. Consumption of oxygen by insects, frogs, and lizards is curtailed by as much as 5–20 times (Kozhanchikov, 1937; Lozina-Lozinsky, 1952, 1956).

Thus even before the animal falls into torpor, a preparatory change in the organism's energy scheme takes place. A hypofunctional state of the thyroid gland and especially of the midbrain centers is of great importance in preparing vertebrate animals. An external stimulus may serve as a direct form of jolting the animal into hibernation (Faleyeva, 1959). As we know,

[3] A mother weighing about 150 kg gives birth to a cub weighing only 400 g.

the thyroid gland is controlled by the anterior lobe of the pituitary, the thyrotropic hormone of which is in turn associated with the stimulating (signal) effect of changes in the photoregime and certain other external conditions. The photoregime of the environment actually proves to be an excitor, stimulating the animal to total preparation for hibernation. Among others, the important factors are changes in temperature, chemistry, and moisture content in the food, carbon dioxide in the air (in the burrows), and general changes in the landscape. These determine the actual times when hibernation will set in, which explains the well-known correlation between the time of hibernation and the seasonal phenomena of nature (Kalabukhov, 1956).

Under unusual circumstances of existence (for example, when mountain animals are transported into the lowlands) the absence of the usual external signals does not always disturb preparation for hibernation, which is to a significant degree an automatic process regulated not only by external but by internal stimuli. Hibernation proceeds normally when the temperature in the refuge is a definite one, usually close to 0° C. A noticeable drop in the temperature leads to the animal's falling into a deep sleep.

The periodicity of reproduction is a direct consequence of the seasonal changes in metabolism. Attempts to breed fur-bearing animals and others not yet completely domesticated have shown that preparation for reproduction demands a definite degree of fatness. If this degree has not yet been attained, the possibility of reproduction is lessened or eliminated entirely. In the temperate and high latitudes most animals multiply according to a strict seasonal pattern. The exceptions are certain fecund inhabitants of fields and human dwellings, human symbionts, and parasites, since they are provided with food and a favorable microclimate or reliable refuges all year round.

The time when reproduction begins depends on the length of time needed for the development of offspring, since the appearance of the young is usually timed to occur when food is abundant and favorable conditions for development are present. Therefore, species with a short cycle begin to multiply in the spring, so that laying and development of the eggs and hatching or birth of the young will occur in the spring-summer period. Among the migratory birds the first to return home and begin nesting are the starlings, thrushes, and wagtails, which feed on ground-dwelling larvae and adult insects, slugs, worms, and other invertebrates abundant in soil litter and the upper layers of the soil. Those birds which live on flying insects return home and begin multiplying later. The latest of all are the warblers and the martlets. Among the predatory birds the osprey (a bird specialized for eating fish) reproduces at the end of April and the beginning of May. The fledglings of the steppe eagle in Ka-

zakhstan hatch by the time the young susliks emerge from their burrows (the end of April and the beginning of May). In the Trans-Baykal, where the fledglings feed on young marmots, hatching occurs at the end of May and the beginning of June, when the young tarbagans begin to appear outside their burrows. Later (at the end of May and in June) the kestrel, buzzard, and polar owl, which feed their young primarily small rodents, begin to hatch offspring. Later still, when the Orthoptera, wasps, and bees are abundant, the so-called entomophagous birds hatch their young.

Finally, the ornithophagous falcon *Falco eleonore,* which lives in the Cyclades Islands, nests only in August, while *F. concolor,* which lives in the Sahara, raises its young about the time when migratory birds are making mass flights to their wintering grounds. The emperor penguin (*Aptenodytes forsteri*) lays its eggs and sits on them during the polar autumn and winter, when the temperature fluctuates between —25 and —55° C. The actual growth of the young takes place in spring, when availability of food and temperature conditions are especially favorable. In years when pine and fir seeds are abundant, the crossbill lays its eggs and sits on them even in winter when food is abundant. In autumn many fish species living in the high latitudes of the northern hemisphere (salmon and various whitefish) go upstream to spawn. A winter spawn is known to be characteristic of *Hippoglossus hippoglossus,* the cod, and certain others. Finally, autumn is mating time for ruminants which give birth to their young in spring and summer.

The different mating seasons of the various mammalian species depend not only on the speed with which the embryos develop (8–12 days among the opossums, 12 months among the great cats, and 24 months among the elephants) but also on the presence of diapause, or the so-called latent period of pregnancy (an interruption in the development of the embryo), which drags out the birth for 6–8 months (among the Altay moles, elk, badger, ermine, sable, and weasel). Because of this, pregnancy lasts nearly a year in the American weasel *Martes pennanti.* Certain of our bats mate in autumn; among the mated females the sperm remains inactive in the vagina all winter, and true conception occurs only in spring, after which the embryo develops quickly.

Among desert and semidesert mammals (gerbils, voles, jerboas, mice, ruminants, and others) there are one or two cycles of reproduction, corresponding in time with one or two periods of moisture. Among the forest mice in the broad-leafed forests of the Caucasus, multiplication proceeds all year round and reaches its maximum in the autumn-winter period, when beech nuts are abundant and the supply of acorns is high. In summer, when food is less abundant, reproduction is at its minimum (Zharkov,

1938). In the oak forests of the Tula oblast' forest mice reproduce only in spring, summer, and autumn, since the deep snow cover in winter interferes with searches for food and diminishes the general fatness of the animals (N. P. Naumov, 1948).

Reproduction of individual species is periodic not only in the high and middle latitudes but also in the tropics. However, even there it is timed to occur in various months, creating the impression that there is no seasonality associated with it. This phenomenon can easily be seen in Fig. 140, which shows the distribution of egg-laying by various birds according to month and latitude. In the north (60–70° N latitude) the majority of species prefer June, while below 30–60° N they prefer May. South of 30° N cases of bird reproduction are distributed almost equally by months. The reproduction periods of diurnal predators and waterfowl correspond to each other in the high and temperate latitudes; among both groups the multiplication season becomes later and later as one goes north. In the tropics (20° N to 20° S) birds of northern origin reproduce in December and January, i.e., the northern winter, while birds nesting there which originated in the southern hemisphere reproduce in June and July, the southern winter (true for predatory birds). Therefore, in the tropics the predatory diurnal birds have two seasons of reproduction, corresponding to the relatively cold periods (the time of the winter solstice). A similar picture obtains in the case of waterfowl, but the reproduction season for species of northern origin distributed primarily north of the equator comes during the northern summer, while among the southerners it occurs during the southern summer (December). Because of this, the earliest repro-

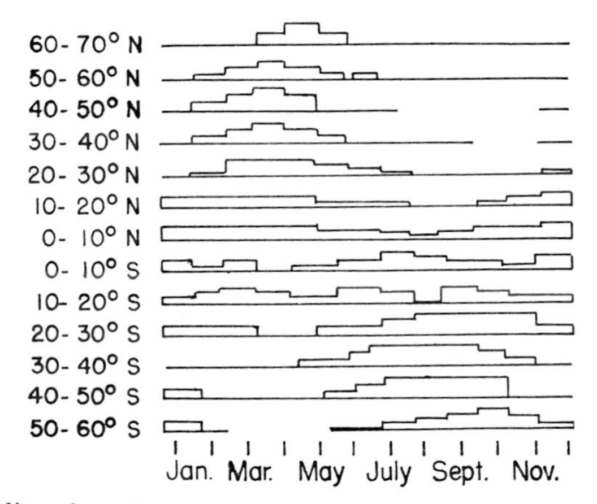

Fig. 140. Seasonality of egg-laying among birds at various geographic latitudes (after Baker, from Bullough, 1951). Size of block indicates number of species.

duction periods occur among waterfowl living mostly south of 40° N (Fig. 141).

The mechanism of seasonal periodicity of reproduction among vertebrates is associated with the activity of the pituitary, the anterior lobe of which secretes gonadotropic hormones. These stimulate the development of both the germinal epithelium and the interstitial tissues of the gonads, which give rise to the sexual cycle. An indispensable part of the secretion of these hormones by the pituitary is the internal state of the organism—its degree of fatness and the level of its matter exchange. However, this condition alone is not sufficient; external stimulation of the pituitary is necessary as well, and this stimulation occurs through the organs of sensation and the central nervous system. By this method correspondence is achieved between the internal state of the organism and the external state of the environment, so that optimum conditions occur in both at the same time.

Such signal connections are an evolutionary adaptation of enormous biological importance. The conditions of existence serve as signals for the different species. Most often one species makes use of several different external factors. Thus in the house sparrow not only the day-length but also the spring rise in temperature influences the speed with which the testes develop. It was these factors which enabled us to determine that the testes

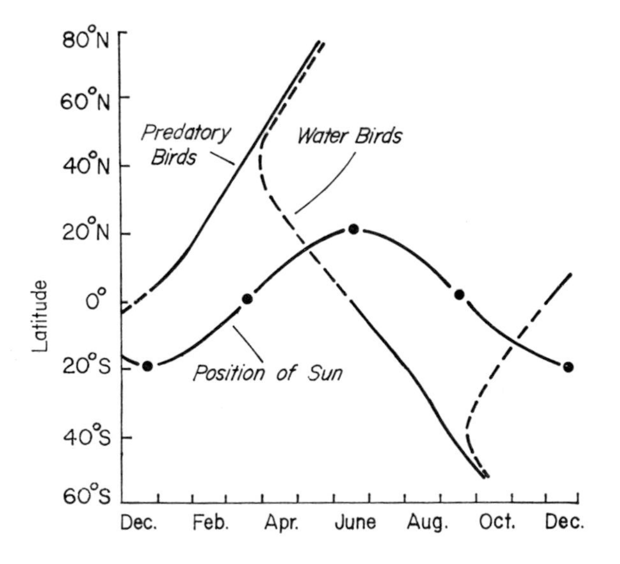

Fig. 141. Times for beginning of egg-laying among diurnal predatory and water birds at various latitudes, compared with the position of the sun (after Baker, from Bullough, 1951).

did not develop in 1956 and 1957 (Fig. 142). In addition to physical factors, psychic factors (the type of nesting place, the presence of building material, the nest structure itself, courting of females by males, etc.) play a special role in the completion of the sexual cycle, but even these factors act in conjunction with the temperature of the environment (Fig. 143) and the light regime (Il'yenko, 1958).

For many species the basic stimulator of reproduction is the light regime (length of daylight and its seasonal alternations). This has been known to students of the problem for quite some time. In Holland and many other countries singing birds have long been kept in darkness and then brought into the light just before attempts to catch others. In Japan birds were forced to sing in winter by submitting them to extra illumination in the autumn (3–4 hours a day). The same technique has long been applied in Europe to make domestic fowl increase their egg-laying (Bullough, 1951). The significance of the length of illumination in stimulating

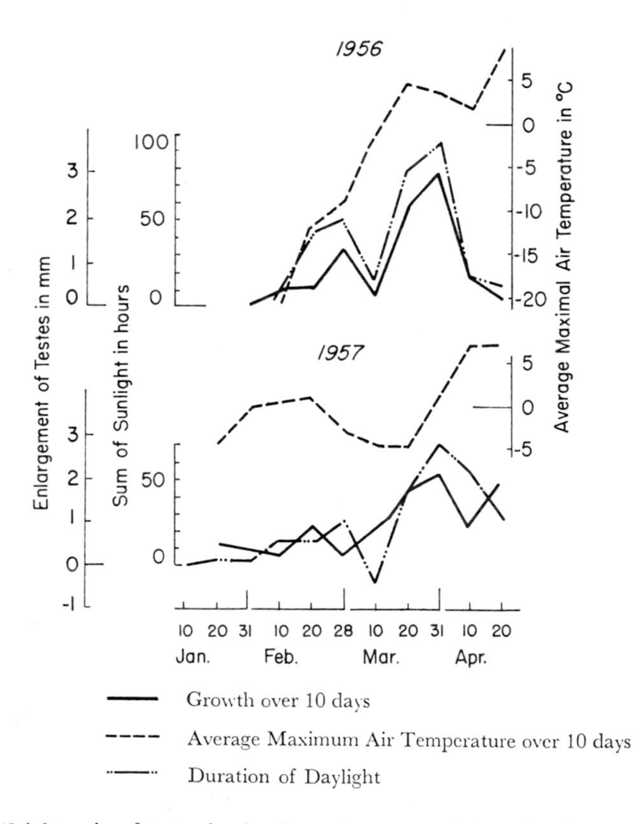

Fig. 142. Weight gain of testes in the domestic sparrow (after Il'yenko, 1958).

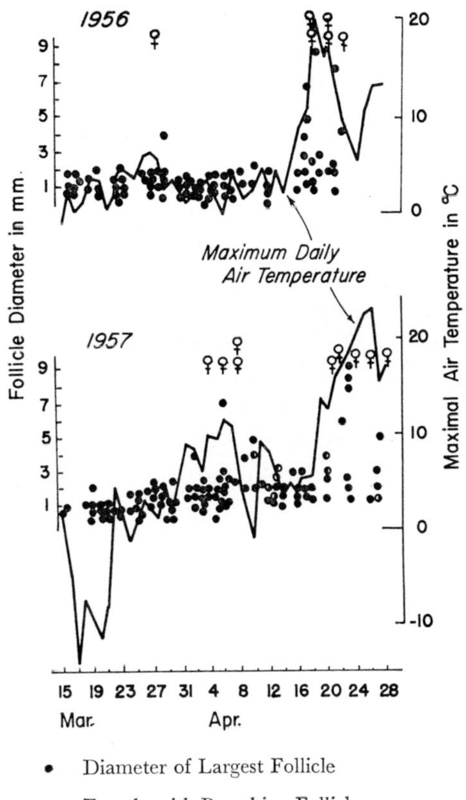

Fig. 143. Development of follicles in the ovaries of the house sparrow (after Il'yenko, 1958).

the sex glands, first found experimentally by W. Rowan in 1926, was later confirmed for several types of fishes, amphibians, reptiles, birds, and mammals. The critical factor was found to be the direct influence of the light regime itself, not the increase in motility, activity, and feeding of the animals subjected to supplementary illumination, as was earlier supposed. In wintertime experiments with starlings T. H. Bissonette (1936; Bissonette and Wadlund, 1932) overfed the birds and artificially increased their motility, but he failed to increase the size of the gonads. When he increased their exposure to light, however, the gonads increased in size.

The significance of individual portions of the light spectrum has proved to be varied. Among domestic drakes the testes increased unusually well

when the birds were subjected to supplementary illumination with red and yellow rays. With blue light they remained practically the same as before. In experiments with starlings the best results were also obtained when illumination was done with red light.

Stimulation of gonad development in birds and mammals which copulate takes place when the length of day increases, but the autumn curtailment in illumination exercises the same effect on ruminants and pinnipeds. Moonlight has the same effect on nocturnal animals. The connection between reproduction and the lunar months is well known in the case of the polychaete *Eunice viridis.* Egg-laying by the nightjar (goatsucker) usually occurs during the fourth quarter of the lunar cycle, so that the hatching and feeding of the young occurs during the succeeding half-moon phase, when hunting may be carried on all night long.

The preference for light signals over all other forms is due to their constancy, since it tips off the birds to the coming season in which the organism must be totally prepared for reproduction. It is produced by the sensing of illumination phenomena (the light regime), which are astronomically correct in their recurrence. If the signal-factor were like temperature change, pressure, air humidity, precipitation, or something similar, its inconsistency would cause "mistakes" to be made in the sexual cycle: either unnecessary acceleration or deceleration in sexual development. As has been shown above, the importance of these phenomena lies in their functions as "triggers" which lead immediately to the necessary action on the part of the organism, just as a cocked pistol is ready to begin operation as soon as the trigger is pulled.

But changes in the length of the daylight period are not the basic factor for all species of animals. They are of little use in the tropics, where day-length undergoes little change. The absence of a stimulating photoeffect has been proved in the case of hedgehogs, the suslik *Citellus tridecemlineatus,* reptiles, amphibians, etc. The light regime is of little use to species whose abundance and reproductive practices are limited by changes in the food supply; in such cases the size of the food supply usually serves as the chief signal. A good example of this is the crossbill, which reproduces even on short winter days if the basic food supply (pine seeds) is abundant. When food is in short supply, these birds do not bother to reproduce, or else only a few individuals in the population take part in multiplication. Intensity of reproduction among pink starlings depends on the abundance of locust-type insects. Cliff swallows, which eat flying insects, multiply poorly in cold weather or in a cold, moist summer, when mosquitoes and Ephemeroptera are few.

In the tropics the reproduction of many species is timed to occur in the rainy season, while the Australian *Kakatoe roseicapilla* does not multiply

at all during years of drought. In Western Australia the forest marten multiplies only in February–March or in June–July, or in both these seasons, depending on the duration and abundance of rainfall (Backer, 1938). The rains determine the multiplication period of certain amphibians (the toads *Scaphiopus bombifrons* and *Bufo cognatus*). The rainy season governs the reproductive habits of the gazelles: their young are born when rain is falling and when the plants are in their growing season. The rhythms formed in the native land have proved to be fixed by heredity and are preserved even in captivity (Fig. 144).

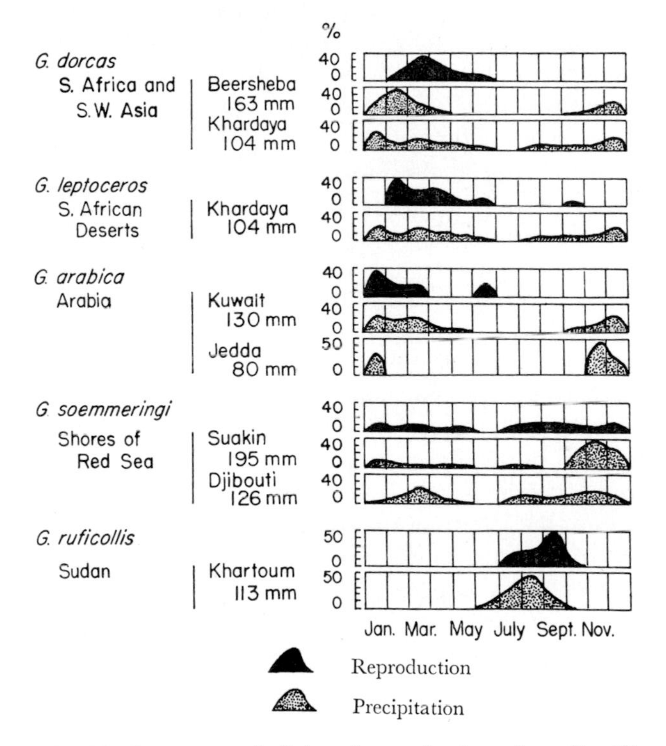

Fig. 144. Relationship between periodicity of reproduction of gazelles (*Gazella*) in the Cairo Zoo and seasonality of precipitation in their native regions (after Genn, from Kashkarov, 1945).

For the majority of amphibians and reptiles, as well as for many insects and other invertebrates, the role of signal-factor is usually played by temperature. Temperature strictly controls reproduction and other periodic phenomena in the life of the malarial mosquito. Eggs are first laid in spring at an average daily temperature of 7–9° C; at the same time the water in which the eggs are laid should have a temperature of 11–13° C.

In this connection the times for periodic phenomena in certain populations of malarial mosquitoes, especially those inhabiting different zones, differ substantially (Fig. 145). In fishes the spawning period corresponds in time with the onset of a definite water temperature at the spawning grounds, a temperature which is different for each species. The period during which the grass frog lays its eggs changes depending on the climate (Table 38). Fluctuations in these periods during various years in England have correlated well with the average February temperature and the total precipitation for that month.

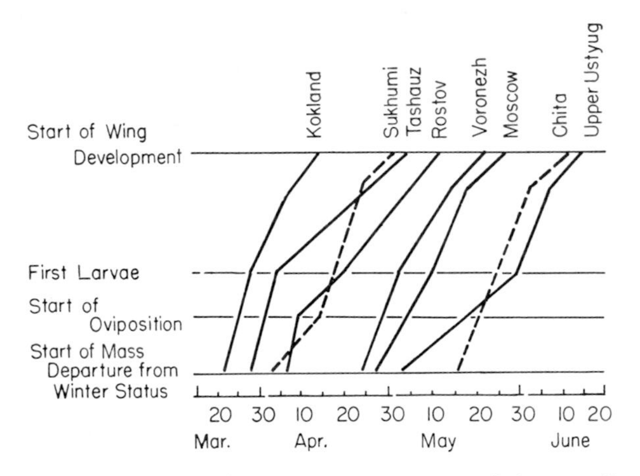

Fig. 145. Times of emergence of the first generation of *Anopheles maculipennis* in 1946 (after Beklemishev, 1949).

Temperature plays an important but usually secondary role in the reproduction of almost all animals; the actual times for reproduction are associated with it (see Figs. 92, 142, 143). Among both blue titmice and sparrows the actual egg-laying periods are determined by the intensity of solar radiation and by the temperature. Many aspects of intraspecies relationships (concentrations of individuals among colonial species, mating play, pairing in the migratory flock) may serve as factors determining the actual time when reproduction will occur. The customary native landscape acts similarly, playing the role of a signal, especially for migrating animals as they return to their reproduction area. For many species the basic factor may be such elements of the landscape as nesting refuges, places suitable for their construction, or many others (Promptov, 1940, 1941, 1946, 1956). A strange landscape and the state of the endocrine system of migratory birds make them insensitive or little sensitive to changes in the light regime during the wintertime, so that northern species

of birds which fly far into the southern hemisphere sometimes fail to reproduce. Finally, periodicity of reproduction is also governed by internal cyclic phenomena in the organism associated with the restructuring of the endocrine system; but this last, as has been shown, is associated with external signals.

TABLE 38. *Time of egg-laying for the grass frog* Rana temporaria *in various places (after Terent'yev, 1950).*

Place	*Latitude*	*Time of Egg-Laying*
Brittany	48°	Jan.
London	53°30′	Feb.–Apr.
Kiev	50°30′	7 Mar.
Kurst oblast'	52°	20 May
Moscow	55°40′	24 Apr.
Novgorod	58°30′	26 Apr.
Leningrad	60°	29 Apr.
Finland	60°30′	26 Apr.
Finland	62°	14 May
Finland	64°20′	25 May
Finland	65°10′	24 May
Finland	66°20′	29 May
Lapland Preserve	67°40′	27 May

Thus both the periodicity of multiplication and the mechanisms which aid in its establishment are complex adaptations to seasonal changes in climate, at the basis of which lie instincts or successive nonconditioned reactions of the organisms to definite (signal) stimuli (environmental factors). They are specific for each species of animal and reflect those conditions of the environment under which the animal developed evolutionarily and to which it has become adapted. The other periodic phenomena in the lives of animals (cyclomorphosis, molting, seasonal migration, etc.) are also associated with seasonal periodicity.

The developmental cycles of many insects and other invertebrates are distributed over the calendar year and are regulated by seasonal shifts in the light regime or other factors and may be changed artificially by changes in illumination, temperature regime, etc. In the aphid *Macrosiphum* the winged generation appears in nature in autumn, when food is in short supply, temperatures are low, and day and night are of approximately equal length. Using the winged generation, the aphids disperse and change food plants. Under laboratory conditions of confinement at low temperatures and of food shortage, but with continuous illumination or darkness, winged generations are not produced. On the other hand, they are born en masse when 12–14 hours of darkness are alternated with

10–12 hours of illumination. A rise in temperature, an improvement in the food situation, or a lengthening of the light period is accompanied by the disappearance of winged individuals.

Among insects a change in the length of day is one of the basic regulators of the seasonal developmental cycle, especially of the appearance of diapause.[4] In many species development proceeds only during long light periods (summer days), while diapause sets in when the days get short. Length of day during the period when diapause sets in is different for each species and even for different populations of a single species. It has been found experimentally that in this case the factor acting on the insects is not light alone (absorption of solar energy) but the rhythmic alternation of light and dark, especially their changes as sensed by the organism. By changing the illumination, we can cause diapause in insects at any time of the year (Danilevsky, 1950, 1961). The light regime proves to be a more powerful factor than temperature and even food, although the latter do exert their own influences. The spider mite *Tetranychus urtice,* which lives in protected ground, has a diapause among adult fertilized females in August. The diapausing females differ from the summer ones by their coloration; they also do not feed, are negatively phototactic, and are more resistant to low temperatures and to toxins. Females fall into the state of diapause under the influence of a shorter day-length, which may be accelerated by keeping the mites in darkness and lowering the temperature (10—14° C). On the other hand, diapause can be slowed down by the action of high temperatures; at a temperature of 34° C the mites do not undergo diapause at all (Bondarenko, 1958).

Among the inhabitants of temperate and high latitudes a drop in temperature may be a factor in diapause, while in the desert a drought may play the same role. Diapause in mammals (a cessation in the development of the embryo) is known in roe deer, Altay moles, badgers, weasels, and sables. Among the Altay moles the females are impregnated in June or July, but the embryo does not begin to develop until spring. As a result, the litters are born in April and May of the year following conception (Borodulina, 1951). Female sables and weasels which mate in June and July give birth to their young in March and April of the following year. Birth is preceded by the so-called false pregnancy (in February), corresponding in time to the implantation of the offspring, which prior to this lie free in the uterine cavity. Diapause in mammals also occurs in connection with the light regime; supplementary illumination makes it possible for pregnancy to be cut to 4 or 5 months in weasels.

The shedding of hair or feathers is associated with changes in the length

[4] Diapause is an apparent halting in development, usually occurring during an unfavorable period.

of daylight, since these act on the pituitary. The thyrotropic hormone which it secretes acts on the thyroid gland; under the influence of the latter's secretions the change in covering takes place. Artificial changes in the light regime have been used to change the speed and time of molting in the Dzhungaria hamster *Phodopus zongarus,* the varying hare *Lepus timidus,* the fox, the willow ptarmigan, the domestic hen (Fig. 146), the weasel, the mink, and several others. Anticipatory whitening of the varying hare in winters when the snowless period lasts unusually long is generally accompanied by a rise in mortality among these creatures and might be regarded as a unique "mistake" in the light signal system.

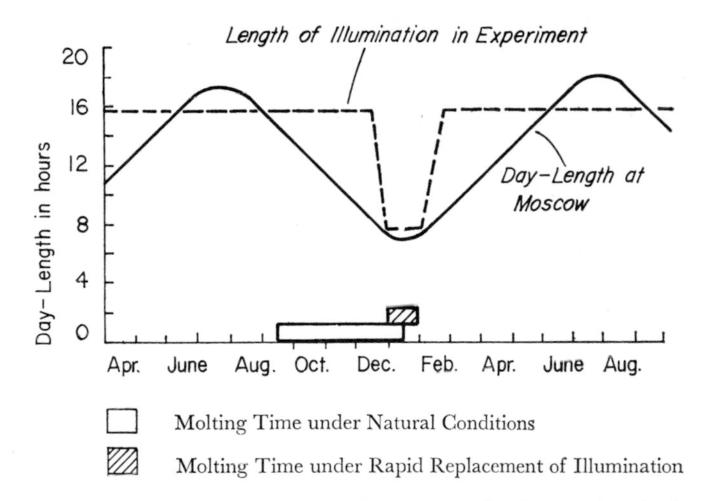

Fig. 146. Changes in molting times of the chicken when the light regime is altered (after Larionov, 1945).

The relationship between multiplication and the light regime permits us to increase fecundity and regulate the times for reproduction and molting in domestic birds, fur-bearing animals, and farm animals just by changing the photoregime (D. K. Belyayev, 1950).

Seasonal migrations are of different character in different animals. They are just as complex a biological phenomenon, created with the aid of instinct, as are feeding and reproduction. The chainlike reflex associations with the environment which lie at the basis of migration may be governed by various external signals. Seasonal migrations are especially often associated with a change in feeding and a shift in foraging areas. The cause of these alterations may be the disappearance of one food and the appearance of another. Such "minor migrations," usually covering only short distances, also take place after a deterioration in the quality or a decrease

in the general quantity of a food. After reproduction many birds which are sessile in the summertime take to wandering because their food has either diminished or disappeared, forcing them to search elsewhere. Voles, susliks, and other rodents migrate to lowlands for food after the vegetation in the highlands has dried out. Long periods of rain or drought, mowing, plowing, crop harvesting, and many other changes in the food supply are also accompanied by movements of a similar type. As a result, animal dispersal is usually found to be closely associated with the distribution of food supplies (Figs. 147, 148).

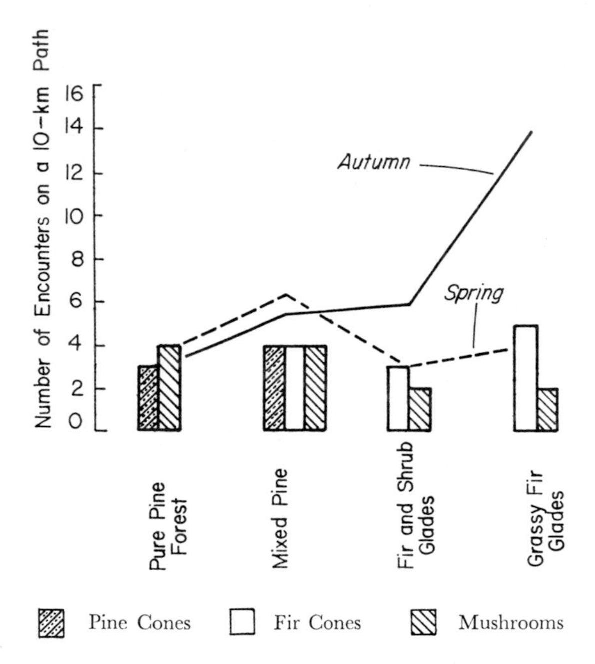

Fig. 147. Autumn and spring distribution of squirrels (*Sciurus vulgaris*) in different biotopes depending on abundance of food (original). Number of squirrels encountered on a 10-km path.

The importance of the light regime has been found experimentally for the American crow *Corvus corone brachyrynus* and the junco *Junco hyemalis*. Birds which were submitted to a preliminary supplementary illumination and were then let loose in the winter flew north (200 miles in 10 days). At the same time control birds, which had not been subjected to supplementary illumination, flew south or stayed in the same place (Rowan, 1930, 1932; Wolfson, 1942). The influence of temperature is well known (Fig. 149). Seasonal migrations of birds are produced by general

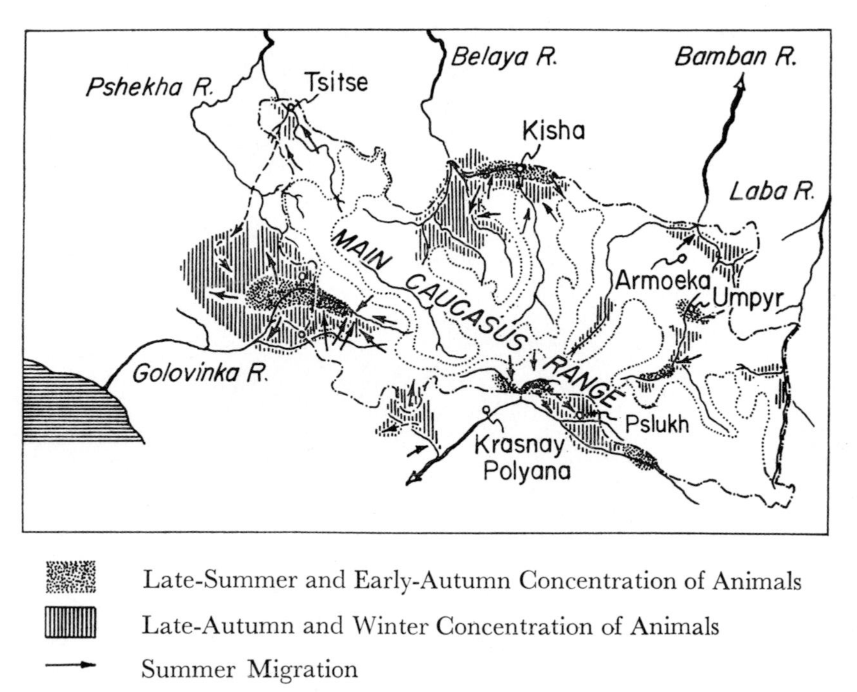

Fig. 148. Displacement of bears and boars in the Caucasus Preserve in connection with distribution of the fruit and nut harvest in the autumn of 1951 and the winter of 1951/52 (after Chernyavskaya, 1956).

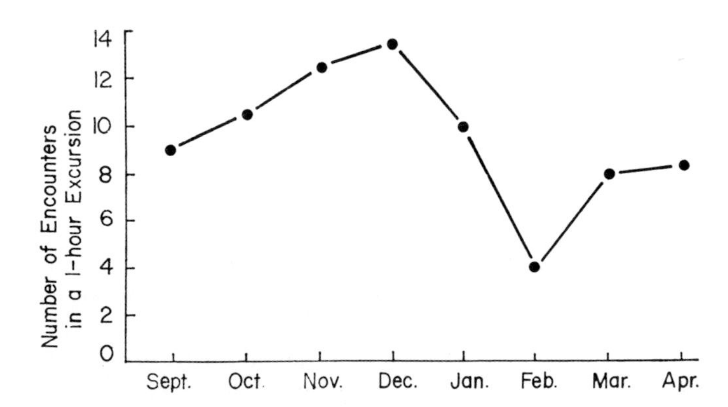

Fig. 149. Seasonal changes in the number of tits (*Parus atricapillus*) near Moscow in 1919–1922 (original).

changes in nature in autumn and spring. The light regime plays a special role in this phenomenon, but the actual course of migration is determined by external temperature changes (Fig. 150).

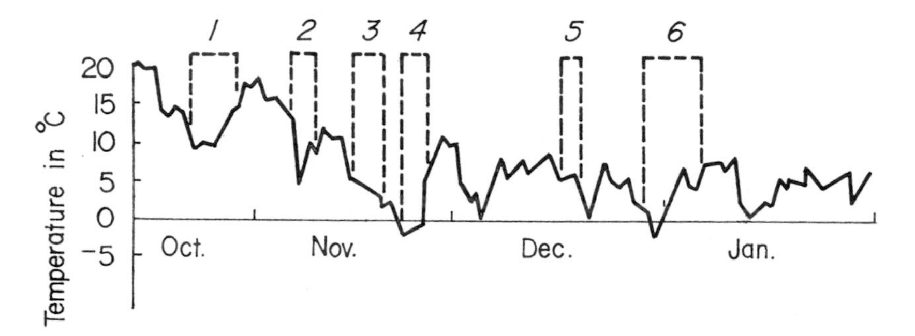

Fig. 150. Coincidence of basic stages of arrival of birds at the wintering ground in the southern Caspian (Gassan-Kuli) with minimum temperatures in Krasnovodsk in the winter of 1937/38 (after Isakov, 1940). 1: termination of flights of *Vanellus vanellus, Limosa limosa,* and *Hirundo rustica*; 2: arrival of pintails (first wave) and first geese, increase in number of coots; 3: mass flight of red-crested pochards (*Netta rufina*), flight of geese, shorebirds (*Crocethia*), and sheldrakes, mass arrival of coots; 4: arrival of first smews, departure of whistling teal, decrease in number of *Netta rufina*; 5: arrival of wild ducks, sea scaup, smews, and pintails (second wave); 6: arrival of trumpeter swans, mass arrival of male tufted scaup.

The permanently existing vertical temperature gradient and its daily and seasonal reverses have a great importance in the lives of soil-dwelling animals. Temperature fluctuations gradually diminish as one goes deeper into the soil. Daily fluctuations penetrate to a depth as great as 1 m but are sensed very little at 30–40 cm; yearly fluctuations penetrate to 25 m, but heating and cooling occur very slowly. So the soil "summer" comes in October and November, while the soil "winter" comes in March and April. The vertical migrations of soil inhabitants are associated with these seasonal reverses in soil temperature (see Fig. 103). The stenothermic animals show this by the relative constancy of their thermal exchange. Larvae of the Trans-Caucasian marble beetle *Polyphylla olivieri,* an important destroyer of vine crops, are displaced regularly from one layer to another as the layers become warmer or cooler (Fig. 151).

The migrations of many animals are associated with snowfall. In the forest regions the hoofed animals move into areas where the pine-tree cover is so thick that snow is kept in the tree crowns, keeping the ground relatively free. Jays, nutcrackers, tits, and squirrels are also attracted to such spots. In the mountains the ruminants move to the less snowy areas as the snow cover gets deeper; they settle on sunlit slopes and wind-swept

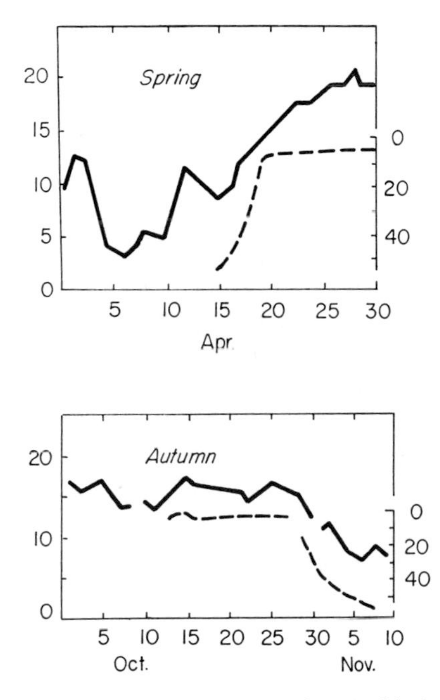

Fig. 151. Movements of larvae of the Trans-Caucasian marble beetle *Polyphylla olivieri* (dashed lines) in connection with changes in soil temperature (solid lines) (after Prints, 1937).

areas. The larger and more powerful males may remain in the areas of deeper snow. The male reindeer lives on snowy patches with a depth of up to a meter, while the female finds a half meter difficult to traverse (Nasimovich, 1957). When deep snow falls, the roe deer wander in the Urals, West Siberia, and the Far East; in the almost snowless regions of Central Asia (the Tyan'-Shan') these beasts are sessile. In the taiga of West Siberia and the Urals, an area of much snow, even the elk wanders. In years of deep snow the motility of hoofed animals increases and their mortality rate rises, leading to a decrease in population levels (Chernyavskaya, 1956).

Birds are also affected by snowfall. White grouse feed on the buds of the dwarf birch in snowless years, but after shoots have been covered with snow, they switch to scrub willow (Teplova and Teplov, 1947). The tundra grouse on the Kola Peninsula leaves the hills during ice storms (when there is an ice crust on the top of the snow) and moves into the birch forests. Its numbers are sharply curtailed during snowy winters (O. I. Semënov-Tyan-Shansky, 1938, 1960). The bustard *Otis tarda* makes seasonal migrations from the steppes of the Ukraine, with their relatively

deep snow cover. It spends the winter in the almost snowless regions of the Pre-Caucasus and lives a sessile life in the snowless Trans-Baykal steppes, although the winter there is distinguished by its great severity (low temperatures and strong winds). The autumn out-migration of the goose *Anser fabalis* is associated with the time of the first snowfall (Galakhov, 1937). When the birds return in spring, many of them fly over the still-snowbound Ural massif or go around it (Kirikov, 1952).

Many migrations are associated with the drinking-water supply in the wild and the state of water holes. Regular daily migrations from nesting grounds and feeding areas to water holes are common among the inhabitants of the steppe and the desert (grouse, gerfalcon, pratincole). They are absent wherever there are long distances between springs, streams, and other fresh or at least brackish sources of water. Grouse will cover as much as 80 km per day while making regular migrations to seek water. Thousands of sand grouse and grouse, their different flocks alternating throughout the whole 24 hours, collect on the few and far-flung bodies of water found in the deserts of Kazakhstan and Mongolia.

The drying up of water holes causes extensive wanderings among birds and mammals; migrations of steppe and desert ruminants during years of great drought are accompanied by mass death among these animals. For many of them the demand for water holes is a seasonal phenomenon. The gazelle *Gazella subgutturosa* and the saiga (*Saiga tartarica*) are satisfied with snow or the moisture in plants during the winter, spring, and beginning of summer and wander widely over the desert during this period. In summer they approach the freshwater sources or move to the banks of the Caspian or Aral seas, the brackish water of which they drink willingly during the hot part of the year.

The aggregate of seasonal changes in the state and activity of the organism also forms the annual cycle characteristic of each given species. Annual cycles may not be similar in the different populations of a single species. Let us recall that the character of an animal's attachment to its surroundings (migratory or sessile life) may differ in different parts of its distributional range; animals which hibernate in the winter in the north may be active year round in the south. The more acute seasonal changes in conditions of existence in the middle and high latitudes are matched by the greater seasonal changeability of the animal's way of life and population size. In the tropics these changes are relatively small in most species. In some species they depend on the character of reproduction (polycyclic or monocyclic) and are different at various latitudes.

The greater seasonal changeability of the weather in the temperate zone, especially in the high latitudes, is reflected in the broad adaptability of the terrestrial inhabitants. Such adaptability is expressed in changes in

their metabolism and energetics, seasonal dimorphism of the hair or feathers, changes in food, fat storage, hibernation, food accumulation, migration, and seasonality of reproduction, and seasonal changes in the intraspecies organizational scheme. Similar but less strongly expressed phenomena take place in the aquatic environment too.

In the high and temperate latitudes the warm period of the year is accompanied by an increase in numbers, while the winter sees a decrease in the total population. The minimum occurs just before the beginning, and the maximum at the end, of the reproduction period. These seasonal increases and decreases in the population level are augmented by in- and out-migrations among the migratory species. However, certain species which multiply throughout the year have no distinct maxima and minima. In the tropics changes in the abundance of individual species occur according to the same seasonal scheme, but since the time for multiplication of the various species does not coincide, the total abundance of animals has no distinct seasonal rise and fall.

CYCLES GREATER THAN ONE YEAR

In different years the weather conditions, food supplies, and hence intensity of reproduction and mortality are not identical. As a result, the time and even the character of the periodic phenomena change, as does the seasonal progress of the abundance index. The result is an increase or decrease in the abundance of individuals in one year as compared with the previous year. The success or failure of reproduction is usually determined by the conditions during the critical period. They are not the same for all species. For the butterflies *Choristoneura fumiferana* and *Malacosoma disstria* (Canadian forest pests) the most important factor is the weather in April (the time when eggs are beginning to ripen) and June (the time for development and basic death among caterpillars; Wellington *et al.*, 1950; Wellington, 1952), but *Choristoneura* appears en masse during dry, hot weather (domination of southerly and southeasterly air masses). Mass appearance of the butterflies usually occurs after several years of weather favoring one species or another. The abundance of squirrels is governed by their winter food supplies (Formozóv, Naumov, and Kiris, 1934), that of voles by autumn temperatures (October and November), depth of snow in winter, and temperatures in April and May (Formozov, 1946; N. P. Naumov, 1948).

Alternation of years of abundance and years of paucity of species having a close ecological character, i.e., adapted to similar conditions, is deter-

mined by changes in the weather, regular changes in types of weather, and the influence of fluctuations in solar activity (see Figs. 3, 4, 5).

The primary influence of climatic changes is explained by the deep and many-faceted action of this factor on the mass and quality of vegetation, the size of the harvest, and the animals' metabolism, state, level of activity, motility, and mutual relationships. However, perennial periodicity of biological phenomena, in particular of fluctuations in abundance, is not a simple reflection of climatic periodicity. The more or less regular repetition every 10 years of "large waves" in the abundance of mouselike rodents (Vinogradov, 1934), the 10-year cycles in the abundance of certain birds and mammals in northern Europe and America (Lek, 1957), the 14-15-year repetitions in the abundance of Brandt's voles, and examples of many other animals with different types of periodicity attest to the complex relationships between climate and abundance. These relationships are interpreted through inter- and intraspecies associations and through the connections between animals and plants. On the basis of such associations we can set up a biological periodicity independent of climatic periodicity (see Fig. 6). Finally, climatic periodicity itself, as has been shown above, is of a complex character which permits distinctions to be made between short- and long-term fluctuations.

As a result, the perennial periodicity of biological phenomena and, above all, the changes in animal abundance are of such a complex character that we can distinguish within them increases and decreases in the number of animals stretching over many years, accompanied by expansion and contraction of the distributional ranges of many species, and even over shorter periods of fluctuation (20 years, 10 years, or less). If it were actually possible to observe a close correlation between these fluctuations and the various climatic periods, we might have grounds for establishing a direct connection, but so far this type of observation has not been made successfully, and it is impossible to speak of such a connection.

CHAPTER 10

Spatial Orientation

GENERAL CONCEPT OF ORIENTATION

The active behavior of animals and their selection of favorable conditions are established with the aid of the mechanisms of spatial orientation. With its aid, animals find their necessary chemical and hydrothermic conditions, places for nests or other refuges, and food, seek out and recognize their partners, recognize individuals of their own or other species (concurrents, predators), orient themselves at migration time, and attain the most profitable spatial placement.

The functions of the sense organs—vision, hearing, smell, touch, taste—lie at the basis of orientation. The ability of animals to sense and, in some cases, to produce electric and electromagnetic fluctuations also belongs here. Sensations are the windows through which the animal "sees" the world which surrounds him. The more perfected the organs of sensation, the greater is the ability of the animal to orient itself in space.

The primary role in orientation is played by nervous activity—instincts and "memory"—on which depends the possibility of accumulating and synthesizing the representations of the external world and the various phenomena taking place within it. The more perfected the instincts and the conditioned-reflex activities, the greater are the abilities and the more complex are the mechanisms of orientation. The most complex forms of orientation are characteristic of the more highly organized animals: among the invertebrates, the insects; and among the terrestrial vertebrates, the birds and the mammals. In the process of natural selection, evolution, and the rise of species, their ways of life and the properties of their behavior became established. The more perfected forms of spatial orienta-

tion characterize the motile species, which make long migrations, and the active predators.

The animal's ideas about the phenomena of the external world are formed on the basis of an analysis of the stimulus signals received by the sense organs. The circle of biologically significant signals is limited. It was formed in the course of evolution, but it is to a limited degree specific for each species as well as for each of the species' populations and even for each individual. Thus the appearance of a predator is usually accompanied by a defense reaction on the part of the victim, while the appearance of a harmless species causes no reaction. The rustling of reeds in the wind does not disturb ducks, but the type of rustling caused by the movement of a predator or a man causes them to move and fly away.

The physico-chemical properties of the environment are of primary importance in the area of orientation. In this sense the conditions of existence of aquatic, terrestrial, and soil inhabitants are not identical. In water the activities of the organs of vision are sharply curtailed because of water's optical properties. But the possibility of sensing sounds, not only with the organs of hearing but with the entire surface of the body, is significantly increased. At the same time losses in the intensity of a sound are not great, owing to the greater uniformity of the contacting environment. Because of the excellent electrical conductivity of the water environment, a new active means for orientation, protection, and attack appears here—electric currents. The physical properties of water explain the widespread use of electric-sounding and echo-ranging for orientation among aquatic organisms.[1] When animals live in the air, their organs of vision become much more important because of air's greater optical properties and because of the great range of vision allowed in the sky (especially true in the case of birds). Among the flying mammals the leading method of orientation is echo-ranging. Among the subterranean forms the importance of vision diminishes almost to nil, and smell, touch, and hearing become most important. Among the terrestrial animals the greater changeability and variety of living conditions are reflected in the broad gamut of orientation practices.

The factors impinging on the activities of the sense organs are divided into visual, aural, chemical, and cutaneous. We also recognize electric and electromagnetic factors, but most animals have no specialized organs for sensing them. A special position is devoted to the "sense of time," at the basis of which lie the rhythms of the physiological processes and muscular

[1] Echo-ranging is the sending of waves reflected from various objects so as to enable the animal to determine its position with regard to these objects.

sensitivity, an important characteristic of which is the animal's kinesthetic (motion) "memory."

LIGHT SENSITIVITY AND VISION

Practically all living organisms react to light. Even protozoa, which have no special structures for this, react to changes in the intensity of illumination. But even these organisms sometimes have certain areas of special sensitivity to light, such as the "eyespots" of the protozoon *Euglena viridis*. Starting with the hydroid polyps and the scyphoid medusae, special light-sensitive organs begin to appear among the animals. The photosensitive organs and structures are extraordinarily varied—simple eyespots, complex faceted compound eyes, chambered eyes, syncytial organs, light-sensitive spots on the skin, and special pigmented patches.

Photovisual orientation is a complex phenomenon. It is formed on the basis of distinguishing between intensities of illumination, sensing and "recording" of viewed forms, light and the positions of objects in space, their relationship to surrounding objects and phenomena—all against a background of sensitivity to polarized light and orientation of the heavenly bodies in many animals.

Reaction to light lies at the basis of the phototropisms characteristic of many animals. It is a means of orientation in many coelenterates, flatworms, roundworms, annelids, and nemertines. Among those animals capable of distinguishing form, light sensitivity of other organs and structures is often preserved as a supplement to the eyes (insect, lamprey, and frog larvae). Sometimes simple photosensitivity is acquired anew as a result of secondary reduction of the eyes in species which have lost the ability to recognize forms visually (*Proteus*, moles). Animals capable of distinguishing only light and dark or of differentiating various degrees of illumination use light signals in order to approach or leave an area of illumination. Inhabitants of the soil which get into the sunlight return to the ground, and plankton organisms, depending on their ecology, either rise toward the light or descend to the depths to escape from it. As a general rule, simple photosensitivity prevails among the soil animals, plankton organisms, and inhabitants of the marine deeps, caves, etc.

The appearance of the ability to distinguish forms in the surrounding environment is a new stage in the evolution of the animal world. The enormous variety of visual signals and the multiplicity of their combinations expand immensely the possibilities for orientation. Color sense is an

even greater addition to the visual recognition of form. It brings about a further expansion in the means of orientation and markedly increases the number of possible visual orientations.

Sensitivity to light is far from equal among the animals. Some aquatic invertebrates react to intensities of light which are not capable of being distinguished by the human eye. In their daily migrations certain species of marine plankton orient themselves by a change in illumination at a depth of 800–1,000 m. Many fishes sense light rays having 1/10,000,000,000th the intensity of full sunlight. This threshold is close to the threshold of sensitivity of the human eye. The eye of many birds are highly sensitive to light, especially the nocturnal species (owls). They observe their prey at a distance of 2 m with an illumination of 7.3×10^{-7} to 5.0×10^{-7} and even 1.5×10^{-7} footcandles (Dice, 1945).

One of the most important orientation reactions using the developed eye is the distinguishing of form and the spatial positioning of objects. This ability appears already in scorpions and spiders. The former can see a female at a distance of 2 cm, the latter at about 30 cm. The complex faceted eyes of crustaceans and insects do not deliver a distinct outline. They show the form of an object in the shape of a mosaic, much like a rough newspaper photograph. Among the invertebrates only squids and octopi have a chambered eye capable of accommodating and creating distinct outlines.

Most insects react best to moving objects: bees willingly settle on flowers which are in motion, dragonfly larvae attack only moving prey, and the males of many butterflies follow only moving females. Predators which catch their prey in flight see even better. Certain insects (flies, wasps, beetles) are capable of distinguishing the number and form of simple geometric shapes and can orient themselves by them in order to take their food or a bait more quickly and precisely. The outlines of figures and combinations of light and dark strips serve to orient many insects and their larvae. Caterpillars, ladybirds, and aphids move along the edges of light and dark strips. Diving beetles and notonectids orient the direction of their movement along dark strips. Bees and other Hymenoptera learn all the pertinent properties of the flowers on which they are about to settle—size, number of parts, sometimes the petals—and orient themselves accordingly. Digger wasps do not seem to be able to find the entrance to their nests if the usual objects surrounding it, such as sticks, reeds, etc., are removed. Most insects are near-sighted. The darting beetle reacts to movements made no more than 1.5 m away, the dragonfly to movements within 2.0 m, the butterfly to movements within 1.5 m.

Cephalopods recognize the form of objects better than insects. They

easily distinguish simple geometric figures (squares, rectangles), while octopi even recognize combinations of various objects. They recognize even more easily figures which have been extended higher or strips which have been positioned vertically. The development of form recognition goes further among the fishes. Many of them not only distinguish a circle from an oval, a triangle from a square, and a cube from a pyramid, but also letters, numbers, and even combinations of them, if the latter are placed in a row. Like human beings, fish are subject to optical illusions. A dark circle surrounded by small white circles appears larger to them than the same figure surrounded by large white circles. Fish, like insects and cephalopods, can be trained to signals made of vertical strokes faster than of horizontal ones. Fish are normally near-sighted. Their eyes see well at distances of approximately 1 m, while accommodation, which is not identical for all species, can usually tune the eye to see about 10–15 m.

Amphibians distinguish the forms of objects poorly, and their reactions to visible objects are primitive. The frog often seizes scraps of moving paper before its conditioned reflexes establish the fact that the target is inedible. The frog can distinguish the movement of large objects at distances of more than 2 m. Distinguishing of forms is not identical among the different species of reptiles. Geckos may distinguish the form of motionless prey at a distance of 8–10 cm. Snakes see human beings from a distance of 5 m or more. Experiments with turtles have not yielded any positive results.

Most birds distinguish objects well. However, their conditioned reflexes establish the motion and coloration of the object better than its form, even in such simple cases as circles, squares, and triangles; in the last case the reflexes are less solid. Birds are able to see various distances. Eagles, falcons, condors, and vultures inspect their prey from great heights, displaying not only an amazing acuity of vision but also an ability to adjust their vision to a change of distance with surprising rapidity. The vision of some birds is four or five times sharper than that of man. The ability to accommodate is enormous in some species. In the cormorant it has been evaluated at 40–50 diopters, while in man it is only 14–15. In chickens it is only 8–12 diopters, while in owls it is 2–4. In connection with this, owls seldom see anything in their immediate proximity.

The development of vision in mammals is also varied. In several animals the ability to distinguish form exceeds that of man. Dogs can distinguish a circle from an ellipse even if the ratio of the axes is 9:8. The distance which a mammal can see, as in the case of birds, depends on its way of life. Distant vision is shared by several carnivores and ruminants, primarily inhabitants of the mountains and the open expanses.

COLOR VISION

Studies of the ability of animals to see colors has shown that they often react not so much to the color of an object as to its brightness. Failure to take this property into consideration sheds doubt on the results of many investigations.

A great many experiments on color vision have been carried out with insects. Their sensitivity to the visible portions of the spectrum is not identical and depends on their way of life. Most of them sense rays in the range of 6,500–350 Å, but many of them react to red light just as they would to darkness. However, ants are sensitive to infrared rays and deposit their pupae at the margin of these rays. The Diptera (suborder Brachycera), Lepidoptera, Heteroptera, and Trichoptera react especially well to ultraviolet light; *Culex* mosquitoes and simuliid flies are weakly attracted (Pogodin and Saf'yanova, 1957). Ultraviolet rays are sensed by insects as a definite color which we, unfortunately, cannot evaluate. Apparently the world is not colored the same for insects as it appears to us, especially if we succeed in providing ultraviolet illumination for certain plants and animals (Lutz, 1933; Ash, 1961). Insects are especially sensitive to ultraviolet rays (350–390 mμ). Bees distinguish four basic colors: yellow-green, blue-green, blue-violet, and ultraviolet. They cannot distinguish yellow from green or blue from violet, and are blind to red. *Bombylius* and *Macroglossa* are blind to red. This is associated with the absence of pure reds in the Central European flora. Red sensitivity has been found for bumblebees and certain butterflies—for example, the cabbage butterfly, family Pieridae—as well as for beetles, mosquitoes, and the wasp *Pseudovespa rufa.* Many moths can distinguish blue from yellow or orange but do not distinguish green. Dung beetles distinguish blue from yellow, blue from orange, and violet from blue but cannot distinguish dark green and light green and simply cannot distinguish color saturation. The pine *Bombyx* caterpillar, the white cabbage worm, and the willow worm can distinctly see various parts of the spectrum, reacting to violet light as if to white. They accept red light as darkness, while their reaction to intermediate parts of the spectrum is transitional (Fig. 152). The bug *Troilus luridus* has no color vision. Certain *Deilephila* which feed during the twilight hours distinguish yellow-greens from blue-violets even in darkness too deep for human beings to distinguish colors (Knoll, 1926).

The ability to sense ultraviolet rays has been found in fish. A study of fish vision has refuted suspicions that they are color-blind (Froloff, 1922; Frisch, 1933). The abilities of fish to distinguish colors are not identical among inhabitants of different depths; this is a consequence of the

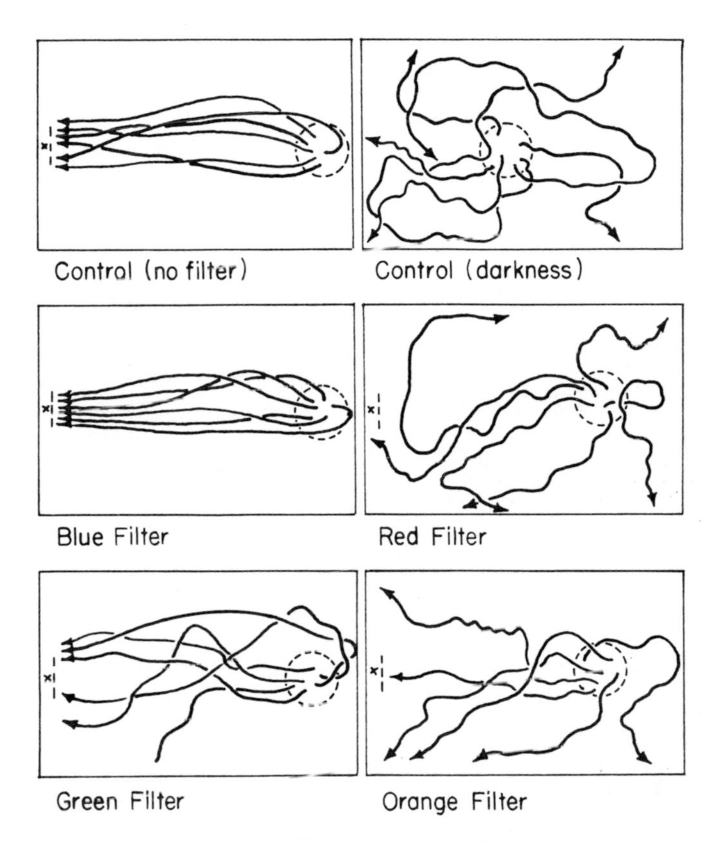

Fig. 152. The pine *Bombyx*'s (caterpillar) choice of various rays of the spectrum (after Geyspits, 1957).

optical properties of the water and the character of water's ability to absorb light rays. Water absorbs red rays (25 percent at 1 m) best of all. Violet rays penetrate more deeply than any others, but at a depth of 100 m even they cannot be distinguished. Therefore, fish at great depths do not distinguish light colors. Species living in the coastal zone and in the upper layers of the water have a broader scope of light sensitivity (Protasov and Golubtsov, 1960).

The problem of the color vision of amphibians has only recently begun to be explained. The works of E. Thomas (1955) and certain other investigators have established that color vision is absent in the nocturnal forms (toads, toad frogs [Pelobatidae—Ed.]). Diurnal forms such as *Rana esculenta, R. pipiens, R. clamitans,* and *Bufo bombina* distinguish all the basic colors. *Discoglossus pictus* misses only yellow, and *R. latastei* does not react to yellow or yellow-green. Properties in distinguishing various parts

of the spectrum are also known for other species. A 1960 study by G. Wade [?—TR.] has shown that color vision is better developed in the salamander than in the tritons.

Color vision is well developed in many reptiles (turtles, lizards). *Emys europaea, Lacerta agilis, L. viridis,* and *Anolis caroli* distinguish five to eight colors. For other species the optimum is a mixture in the red-yellow part of the spectrum. Color vision in birds is similar in its capabilities to that of man; the sensation of color is in many cases very close. Numerous experiments with pigeons have shown that these birds distinguish significantly fewer shadings (about 20) than man (up to 160).

The data on mammalian color vision are contradictory. Many investigators deny its existence in all species except the primates and man. Attempts to work out conditioned reflexes to color in cats did not give positive results until recently (Gunter, 1954; Meyer, Miles, and Ratoosh, 1954). Experiments with cats have been negative in almost all cases. V. K. Shepeleva (1953, 1954) finally succeeded in producing a conditioned reflex to green in a dog, but only after 650–700 trials. In her opinion forest skunks are able to distinguish between red and green. In his most recent summary E. Wade noted the presence of color vision in the horse, zebu, squirrel, guinea pig, and certain other mammals.

In the process of evolution many animals have developed the ability to orient themselves by polarized light and the heavenly bodies. Orientation by polarized light was discovered in 1948 (Frisch, 1948, 1949). The light is polarized by the blue sky, and the polarized light, especially if the sun is not visible, may act as a unique compass for animals. The faceted eyes of crustaceans and insects, the single eyes of arachnids and insect larvae, and the chambered eyes of certain vertebrates react to polarized light. The crustaceans, crab scorpions, arachnids, and insects use this polarized light for orientation. Many vertebrates whose way of life is associated with the necessity of looking from the air into the water or vice versa are capable of blocking out the polarized light reflected from the surface of a body of water. Fishing birds, frogs, and toads apparently sense and partially block out polarized rays (Minner, from Buddenbrock, 1952).

Scouting bees taking nectar back to the hive show other bees the direction in which to fly by doing their so-called dance, which is also oriented by polarized light. In sensing polarized light, animals select a definite angle of movement in relation to it. If we artificially alter the direction of the plane of polarization above the animal, the latter will turn around with it, trying to preserve the original azimuth. During cloudy weather this method of orientation breaks down.

Orientation by the heavenly bodies, especially the sun, is widespread

among animals. It has been established for the arachnids, insects, reptiles, birds, and mammals. Animals not only select a definite angle of movement with relation to the sun but also constantly correct it during the day, "computing" the daily movement of the heavens. They make their correction with the aid of an "internal clock" which operates on the basis of the rhythm of the physiological processes. Thus, if a lightproof cover is placed over an ant or other insect, the ant corrects its movement as soon as the cover is removed in a manner corresponding to the movement of the sun during this period. During the time of seasonal migrations solar orientation becomes especially important. Here it becomes one of the most important means of determining the direction of the route. In recent years works have appeared concerning the ability of nocturnal animals (birds) to orient themselves by the stars. These data, however, require further confirmation.

SOUNDS AND HEARING

It is necessary for an animal to distinguish from among the sounds of its environment those sounds actually made by animals themselves. These sounds are, in turn, both specific and nonspecific. The former consist of the "voices" of animals in the broadest sense—sounds used to signal or communicate with individuals of the same or different species, and many other types of sounds as well. The latter have no special meaning as a rule and may even be unfavorable sounds which unmask the animal. Examples of the latter are chattering of teeth, chomping, fluttering of wings, etc.

Animals sense sound waves as both biogenic and nonbiogenic phenomena. In addition to the usual range of sounds, many of them are capable of sensing and producing ultrafrequency waves, while some arachnids, many insects and fish, and possibly the urodeles can sense sounds of less than 20 cycles per second. Penguins can apparently predict the approach of far-distant storms with the aid of their infrasonar sensitivity. The organs which detect sounds in these animals are usually also the organs of hearing. Fish pick up ultrafrequency sounds with their lateral lines. The swim bladder, which serves as a resonator, also plays a substantial role in sound-sensing. Insects pick up sounds through their auditory bristles and chordotonal and tympanic organs.

Sound is produced by various means. Spiny lobsters emit a chirring sound by using special adaptations on their antennae. Alpheid crabs rub their claws. The fiddler crab, in addition, beats its claws on the hard ground

to call its companions from their burrows. Soldier termites and hornets make sounds by striking the substrate. Locusts chirr by rubbing the vein of the posterior wing against the edge of the anterior one or the surface of the femur against the edge of the same wing. Ants emit ultrasounds by friction. The bottom of the abdomen of the male cicada has a protruding membrane which emits a strong sound when a special muscle contracts quickly. The death's-head moth produces sounds by forcing air from the esophagus through the proboscis. The buzzing sounds of female bees are made by jetting air through the spiracles. The ticking of beetles in their burrows, the chirring of the Orthoptera, and the clicking noises of the hay-eating Psocoptera are significant as signals of the opposite sex.

Many fish emit sounds with the aid of the swim bladder, which has special musculature (the umbrine and the wrasse). In some of the sheatfish the organ of sound is the pectoral fins in combination with the bones of the humeral girdle. In the Tetradontidae sounds are emitted by the glottal and maxillary teeth. The voices of fish resemble gnashing, whistlings, chirps, snorts, chompings, drum rattles, etc. These sounds serve as signals of alarm, threat, pleasure, etc. Their biological significance is apparently not very great. In the sheatfish, umbrine, and certain others the sound is intensified during the period of reproduction or when the school is feeding.

Many amphibians are mute. Others produce sounds by using resonators or special forms of tissues. Their varied voices are distinguished by a rather complex assortment of sounds and by significant modulation of tones, frequencies, and rhythms. Not much is known of their biological significance. They are primarily a form of attraction and excitement during mating and possibly an expression of emotion. As a rule, reptiles are also mute. The hissing and wheezing of turtles, snakes, and crocodiles and the hissing of certain snakes and lizards are all achieved by contraction of the glottal passages during expiration. Rattlesnakes make their sound by shaking their rattles. All these sounds are primarily threats.

The sounds emitted by birds are more varied. They consist of singing, whistling, or crowing, all produced by various adaptations in the throat and oral cavity; snapping the beak (*Tetrao urogallus,* owl, stork); flapping and fluttering the feathers (pigeons, nightjar); rustling and whistling the feathers in flight (snipe, great snipe); and drumming the beak against wood (woodpecker). The modulation abilities of bird voices exceed those of all other animals. In addition to the complex signals used to attract the opposite sex (singing, cooing, mating calls), the bird voice expresses varied and often emotion-charged sounds, and a complex system of signaling and social intercourse is put into operation. At the present time more than 30 types of sounds have been described, each of which has its own significance. The most important sounds are those used to express danger,

threat, fear, satisfaction, hunger, etc. Voices are important in the commerce carried on between different species. In this connection the widespread ability of birds to imitate the sounds of other species is very interesting.

The voices of mammals are immensely varied: different types of screaming, hissing, growling, barking, whistling, etc. emitted by the throat and oral cavity, as well as gnashing of teeth, rustling of quills, and striking of extremities on the substrate (hoofed animals and certain rodents). The biological significance of the sounds emitted by mammals is also very complex. This is especially true of the gregarious animals. More than 30 group signals have been recorded for the hamadryad baboons. These signals accompany different moments in the life of the group as a whole and in the lives of individuals. Especially distinct are the sounds associated with eating, defense reactions, sexual reactions, aggression, relations between offspring and mother, and many others. The modulational capacity of the mammalian voice is less than that of birds, however.

The mechanisms producing ultrasounds have not been studied sufficiently. In many cases they are similar to the production of sounds in the ordinary range (bats, small rodents).

The conditions under which animals sense sound waves are not always the same. In the air sound waves travel at a speed of 331 m per second. In water, depending on temperature, depth, and salinity, their speed may exceed 1,500 m per second. The pressure of sound waves is also greater in water. The transition of sound from one environment to another is accompanied by changes. Sounds penetrate poorly from air into a more dense medium (water, soil, mountain rocks), since a significant part is reflected off the surface of the denser medium. When this happens, their intensity also decreases. Aquatic animals easily detect underwater sounds and fluctuations in the substrate, but sounds made in the air are more difficult for them to hear. Subterranean animals hear only a small part of the sounds made in the air. Because parts of their bodies are in contact with the soil, terrestrial animals can also pick up sounds being spread through the substrate. This is the basis of the "seismic hearing" of certain animals—the ability to hear sounds made in the soil; it is a complex interaction of cutaneous, muscular, and aural analyzers. It is most difficult to draw a line separating these sensations in the insects, many of which sense tactile disturbances as aural ones. Seismic hearing is especially important among the "deaf" forms (insects, amphibians, and reptiles), whose sound-conducting apparatus is either undeveloped or is not as perfected as it is in the animals which hear well. Caterpillars, which have no aural organs, react beautifully to many sounds. Snakes know of the approach of their enemies long before they actually appear.

The limits and sensitivity of hearing are not the same among representatives of the various groups (Table 39). They depend on the ecology of the species, individual properties, age of the animal, its physiological state, etc. As a rule, small animals are more sensitive to high sounds, large animals to low sounds. However, there is a series of exceptions, especially in the prevalence of seismic sound and in the use of echo-ranging (certain whales, dolphins). Hearing is better developed among those nocturnal forms which have poor vision.

TABLE 39. *Frequency of sound waves detected by various animals.*

Species	*Lower Threshold in cps*	*Upper Threshold in cps*	*Author*
Insects:			
Cockroach (*Blatta*)	8	6,000	Pamfrey and Rowdon-Smith, 1939
Caterpillar of *Vanessa antiopa*	32	1,024	Minnich, 1921
Gryllidae	600	16,300	Pamfrey and Rowdon-Smith, 1939
Conocephalia	2,000	20,000	Autrum, 1940
Pholidoptera	870	50,000	Regen, 1912
Anacridium	2,480	40,000	Oje and Fessard, 1925 (?)
Katydidae	?	90,000	Weaver, 1933–1935
Night moth (*Prodenia eridania*)	3,000	240,000	Roder and Trit, 1957
Fishes:			
Anabas scandens	?	over 659	Diesselhorst, 1938
Anguilla vulgaris	36	488–650	Diesselhorst, 1938
Corvina nigra	?	1,000	Dijkgraaf, 1952
Sargus annularis	?	1,250	Dijkgraaf, 1952
Sebistes reticulatus	44	1,200–2,068	Farkas, 1936
Semotilis atromaculatus	20	5,750	Kleerecoper and Chagnon, 1954
Phoxinus laevis	16	5,000–6,000	Stetter, 1932
Sheat (*Ameiurus nebulosus*)	?	13,139	Stetter, 1932
Amphibians:			
Rana catesbiana	50	1,000–24,000	Stroter, 1959
Ambystoma (larvae)	?	244	Ferhat-Akat, 1939
Salamandra maculosa (larvae)	?	1,034	Ferhat-Akat, 1939
Anura	?	15,000	Bykov *et al.*, 1954
Reptiles:			
Emys orbicularis	?	1,200	K. L. Polyakov, 1930
Terrapene	?	1,000	Weaver and Vernon, 1956
Tachydromus	?	15,000	Kuroda, 1933
Birds:			
Sturnus vulgaris	100	15,000	Granit, 1941
Loxia curvirostra	?	20,000	Knecht, 1940
Fringilla coelebs	200	29,000	Schwartzkopf, 1955

Species	*Lower Threshold in cps*	*Upper Threshold in cps*	*Author*
Pica pica	100	21,000	Schwartzkopf, 1955
Erithacus rubecula	?	21,000	Granit, 1955
Asio otus	100	21,000	Schwartzkopf, 1955
Strix aluco	100	21,000	Schwartzkopf, 1955
Falco sparverius	300	10,000	Traner, 1946
Mammals:			
Erinaceus europaeus	64	18,000	Lindemann, 1951
Dipodomys	100	20,000	Katasukü and Davis, 1954
Equus caballus	?	21,000	Safonov, 1957
Equus caballus	?	40,000	Ikoyev, 1957
Canis familiaris	?	38,000	Andreyev, 1936
Rattus norvegicus alb.	?	40,000	Schleidt, 1952
Chiroptera	?	98,000	Galambos and Griffin, 1942
Tursiops truncata	?	80,000	Kellog, Kohla, and Morris, 1953
Tursiops truncata	?	150,000	Chevell and Lawrence, 1955

Many insects are deaf. Among the others it is difficult to tell whether they pick up sound through hearing or through tactile sensations. Sometimes species which produce sounds, such as the *Vanessa* butterfly, lack hearing organs, while species which have such organs cannot produce sounds audible to man. Moths are capable of picking up and reacting to ultrafrequency sounds emitted by bats (some of them fly faster, others fall, those on the ground take off, etc.). This shows that the role of hearing in the orientation of insects is not directed entirely at attraction of the sexes and the expression of elementary emotions.

The lack of sound-conducting apparatus (outer and inner ears) in fish is not a mark of primitiveness but is associated with the specific requirements of life in an aquatic environment. Insignificant emissions of sound passing through water and into the body, combined with the fine sonar penetrability of the tissues (owing to similarity in density of the contacting media), provide the fish with the ability to sense sound through its entire body. Sensing of sound and determination of its direction under water and in the air are substantially different. Using the organs of hearing ("ears"), fish cannot determine the direction of a sound. Orientation over short distances is provided by skin receptors, while orientation to weak and distant sounds is made by trial and error. At present it is felt that fish sense sounds no higher than 15,000 cps.

Sensitivity to the strength of a sound is relatively high in many species. The threshold of detection in the minnow, for example, is 21 decibels; in the electric eel it is 35–40. In the interval between 350 and 850 cps the minnow can distinguish differences in sound down to one-third of a tone. Stetter (1929) found that the minnow *Phoxinus* is able to recall a given tone and react to it after 1–9 months. Certain individuals may simultaneous-

ly remember up to five tones and discriminate between one "food" tone and two or three "no-food" tones. Anura easily sense air tones in the range between 50 and 15,000 cps. The larvae of the axolotl and the salamander react only to the lower tones. This latter group apparently includes the majority of the urodeles which live in water.

Hearing is well developed among the lizards. Some of them detect sounds higher than 8,000 cps and easily catch an insect or other animal just from the rustling it makes in the grass. Snakes are deaf to air sounds; seismic hearing is their basic means of orientation. Signals of up to 3,000 cps have been registered as having biopotential for the box turtle (Weaver and Vernon, 1956), but these experiments demand further confirmation.

Birds yield nothing to the mammals in their sensitivity to sound. Many species have been found to detect ultrafrequency sounds above 20,000 cps. The owls have very delicate hearing. According to some information, they can detect ultrafrequency sounds up to and beyond 60,000 cps. Hearing plays an important role in the lives of mammals. Recent studies have shown that many of them, especially the nocturnal species and the carnivores, detect sounds far higher than those which can be heard by man. It has been established that ultrahigh-frequency sounds are emitted even by such species as laboratory rats, guinea pigs, marsupials, certain South American monkeys, and many others.

Sonar and ultrasonar echo-ranging is important to many animals. Ultrahigh-frequency echo-ranging is the more common. At present echo-ranging is known for many of the passerine birds, the curlew, the owls, the guajaro, and the edible-nest swift. Among the mammals it has been established for insectivores, fruit- and fish-eating bats, dolphins, whales, and certain others. The possibility that fish may use echo-ranging for orientation has also been mentioned (Griffin, 1955).

It is difficult to determine the character of echo-ranging in fish and insects because of the difficulty in distinguishing sound reception from tactile sensation. It is known, for example, that gyrinid beetles use reflections of waves of water for this purpose. The same has been observed in the case of fish which detect reflected waves of water with their lateral lines, even detecting frequencies of 6 cps. However, it is unknown whether these waves are sensed as sound or as pressure. Nevertheless, the location of a reflected wave permits fish to stay with the school and to determine the position of the bottom (in benthic fishes the lateral line often merges with the lower surface of the body).

Echo-ranging has been best studied in the bats. They use echo-ranging not only for orientation in space but also for catching insects. *Eptesicus fuscus* in straight-line flight and under natural conditions emits ultrahigh-frequency impulses lasting 10–13 milliseconds and repeats them

8–10 times per second. If the straight-line flight is changed (while catching insects), the character of the ultrahigh-frequency sounds changes. The repetitions are sharply increased (to tiny fractions of a second and as many as 200 per second), while the length of each impulse drops to 1 millisecond. The sounds emitted by *Asellia tridens* are midway in length and frequency between those emitted by the smooth-nosed and the horseshoe-nosed bats. They combine two distinct parts: an initial one, characterized by even amplitudes and a very high frequency (120,000 cps), and a final part, a short "crack" of lower frequency (about 50–70 kcps). The first part is similar to the ultrahigh-frequency sounds of the horseshoe-nosed bats, the second to the sound of the smooth-nosed ones. Depending on whether the bat is trying to detect distant or close objects, one type of high-frequency sound or the other will predominate. The sensing analyzers of the bat are extraordinarily acute in their ability to detect reflected signals. Fish-eating bats are apparently capable of detecting even the insignificant sounds reflected from the body of a fish in the water, which lose 90 percent of their intensity in passing from the air into the water and back again.

Roussettus emits a unique metallic sound, very much like that of the insectivorous bats, which is part of the audible range of the ultrahigh-frequency impulse. Unlike the bats, *Roussettus* uses ultrahigh-frequency echo-ranging only in darkness. When light is present, it orients itself by vision. It produces its ultrahigh-frequency sounds not from its wide-open mouth, as the insectivorous bats do, but almost exclusively through the nose. Dolphins, which emit ultrahigh-frequency sounds above 80,000 cps, orient themselves perfectly even in muddy water and darkness, never failing to determine correctly the location of food or obstacles. They can distinguish a fish from a stick of wood, etc.

CHEMICAL SENSITIVITY

This category of sensitivity, in addition to the direct sensing of environmental chemistry, also includes taste and smell. In this connection we sometimes recognize two types of sensitivity: chemical sensation from contact and chemical sensation at a distance. The latter is associated with the fact that in some animals it is difficult to distinguish taste from smell (crustaceans, insects, etc.).

Chemical sensitivity appears first in the protozoa. Infusorians have positive chemotaxis toward weak acids and negative chemotaxis toward bases and salts of the alkaline-earth metals. Resting and nonglowing *Noctiluca* begin to glow when boiled salt or sugar is introduced into

their medium, but alcohol cuts short their illumination. Only chemotaxis can explain at this stage such phenomena in the life of protozoa as observation of food, searching for similar organisms for reproduction, and many others.

Separate organs of chemical sensitivity are already found in the coelenterates, then in the worms, echinoderms, mollusks, insects, and all chordate animals. Among moths, bees, and flies they are situated on the palpae, antennae, and tarsi; among the fishes they are concentrated in the nasal cavity, the oral cavity, or along the surface of the body. In the terrestrial vertebrates they are found in the oral and nasal cavities. Odor and taste receptivity are not only not identical among the different species, but changes occur within each of them depending on external circumstances, physiological state, age, etc. Sensitivity usually decreases in satiated individuals but becomes very acute in hungry ones. It becomes blunted with age. It is more highly developed in the males of some species than in the females.

Smell is one of the most important sensations associated with chemical analysis of the environment. For many animals it is the leading factor in orientation. It is already well developed in the Turbellaria, which can sense their food from a distance. Starfish crawl along the bottom of an aquarium after a moving chunk of meat. Blinded octopi can observe fish 1.5 m away in an aquarium. Distant olfactory receptors play an important role in the biology of the insects. Among the Hymenoptera, Coleoptera, Orthoptera, and many others they are situated on the antennae. When the antennae are removed, the insects usually fail to respond to odors. In some species the olfactory organs which sense nearby smells are located on the palpae.

The olfactory sense is the leading factor in orientation among the bees and ants. It is characteristic of them that the olfactory and tactile senses merge into one sensation. The combined tactilo-olfactory organ has been termed topochemical. The bee's sensitivity to smell is approximately equal to that of man, but bees distinguish fragrances better. Small traces of such repulsive substances as indol and scatol are highly attractive to bees. Bees have a good "memory" for smells. Ants use the sense of smell to recognize individuals from their own colony. When an ant's antennae are removed, it cannot find its way home or recognize adults or larvae of its own species; ants of different species join together, sometimes eat one another, or attack individuals of their own species. Fragrant traces play a very important role in orienting the blind termites. Female ichneumon flies apparently recognize by smell eggs already infected by other females of the same species and do not lay their own eggs in them.

Only an especially acute sense of smell can explain the ability of male

Saturnia butterflies to find females and catch them over distances of several kilometers (according to some data, from as far away as 8–11 km). Many butterflies and their caterpillars recognize their food primarily by smell. The caterpillars of the butterfly *Danaus plexippus,* which feed exclusively on *Asclepias* leaves, recognize them even when the leaves are smeared with strong-smelling camouflage preparations and are scattered among leaves of other species. They distinguish their food better by smell than by taste (Dethier, 1937). The gypsy moth and oak *Bombyx* are governed by the organ of smell and taste in selecting food (Getsova and Lozina-Lozinsky, 1955). Malarial mosquitoes seek their prey by smell and usually approach it from the downwind side.

Olfactory sensitivity is characteristic of fish, although this was once doubted. Sharks, sheatfish, and fish of other species with artificially covered nostrils cannot find food hidden in grass or tied up in sacks, but one need only uncover their nostrils for them to swim directly to the catch. Blinded minnows (*Phoxinus*) sense eugenol and phenylethyl alcohol 150–200 times better than man; their sense of smell easily tells them the difference between their own species and their enemies (perch, pike, etc). The olfactory sense of the eel is no worse than that of the dog, and it is 1,000 times more sensitive than that of the minnow. It has been found that the eel can sense a smell when even one molecule of the fragrant substance has touched its nasal cavity (Teichmann, 1959). It has been assumed that smell serves the eel as its primary tool in orienting itself to currents. Smell is the basic sense which orients salmon during their migration. *Oncorhynchus kisutch* cannot find its way back to its native stream if its olfactory organs are removed (R. B. Miller, 1954; Wisby and Hasler, 1954). Species seeking their food under weak illumination orient themselves with the aid either of the taste organ (carp, tench) or of the olfactory organ (eel, sheatfish).

The olfactory sense plays a substantial role in food-finding among the urodeles. Blinded tritons easily sense their food from a distance; they can smell both in water and in the air. Turtles distinguish their food by smell. Snakes even find their partners, as well as their prey, by smelling the traces. They crawl to their wintering places by smelling traces left by other individuals of the same species. Despite previous opinion, the olfactory sense of many birds is satisfactorily developed. It is weak in the pigeons, jays, and starlings. The singing birds easily distinguish the odors of essential oils, benzaldehyde, and scatol. Ducks can smell their food at a distance of 1.5 m. Night birds with poor vision (the stormy petrels *Fulmarus* and *Pogodroma*) usually have well-developed organs of smell.

Olfaction plays a very important role in the life and orientation of many of the mammals. Species with a well-developed olfactory sense are called macrosmatic (carnivores, ruminants), and those with weak development

are called microsmatic (monkeys, many insectivores, etc.). We distinguish animals with upper, long-distance sensing powers and those with lower, primarily short-range discrimination. Many species orient themselves by both methods. The upper powers are especially important in orienting the animal over long distances and are thus characteristic of carnivores and herd animals. Among mammals the olfactory sense serves an important function in helping the animals to recognize their own kind. This is the basis for the widespread occurrence of odoriferous and musk glands among them. The secretions of these glands often serve as markings of individual areas, and their odors are used as signals to wandering individuals. The role of olfaction is especially great in the food-hunting activities of mammals. In still air dogs can find their food by sniffing from a distance of 12–14 cm, cats from 3–4 cm, rats from 1–2 cm (Shepeleva, 1959). The mole will move toward an earthworm from a distance of no greater than 2–3 cm (Geptner, 1957). When the air is in motion, the distance over which food can be smelled rises markedly.

Taste has no significance for distant orientation. Its role is in analyzing stimuli with which the animal is in immediate contact. The Turbellaria can be used as a starting point for a discussion of taste differentiation. Among the arthropods the physiology of taste has been best studied among the insects (bees, butterflies). Bees are capable of distinguishing sweet, bitter, sour, and salty, and they react to very small concentrations. Their antenna receptors are highly sensitive to sugar (they react to a ½ normal solution); their tarsal receptors and oral organs can detect a 1 normal solution of sucrose. Bees, like many other animals, react differently to different sugars. Thus sucrose and maltose appear the same to bees, while fructose and inosite seem less sweet than they do to man. Saccharin does not appear sweet to them. They cannot distinguish saccharin from pure water, nor can birds (hens, pigeons), lymnaeid mollusks, many insects, and crustaceans. Besides man, certain fishes, such as the gudgeon, are sensitive to saccharin. The tarsi of butterflies are very acutely sensitive to various sugars. Hungry tortoise-shell butterflies can detect 200 types, and the Danaidae are 2,000 times more sensitive to sugar than man (Anderson, 1932).

The tastes of fish are highly developed. They can distinguish sweet, sour, bitter, and salty. Their sensitivity to bitter substances is somewhat less than that of man, but their sensitivity to sweet tastes is greater. In some fishes the taste buds are present only in the oral cavity (pike, stickleback); in sheatfish, codfish, and carps they are found about the jaws and upon the extended spines of the fins (burbot, whiting). The gurnard has taste buds on the extended spines of its pectoral fins, on which it "walks" along the bottom seeking its food. In comparison with other types, the

blind fish living in caves have taste buds hundreds of times better developed (Humbach, 1960).

The fact that amphibians eat ants, ladybirds, and other sour and bitter insects attests to their lowered sensitivity to taste stimuli. The taste organs of reptiles are apparently developed more highly. Taste plays a small part in the lives of birds. Canaries will drink water made bitter with aloe juice just as readily as pure water, but they will not drink salty or acid solutions and detect these substances about the same way man does. Other birds have great sensitivity to sour and salty tastes, but they often fail to react to sweet tastes and have a reduced sensitivity to bitter ones. Cats, rabbits, and rats are highly sensitive to bitter tastes. The decrease in sensitivity to other substances occurs in the following order: sour, salty, sweet. Reduced sensitivity to bitter tastes is found in the goat and certain rodents (gerbils). Many ruminants are known to be attracted by salt.

An animal's cutaneous analyzers occupy an enormous amount of surface area; any disturbance of them arouses a series of sensations—tactile, temperature, and pain. Reactions to tactile stimuli are observable as far down the scale as the protozoa; mechanical stimuli cause them to contract their bodies or pull in their cilia, pseudopods, etc. Touch is the basic means of orientation among spiders with poor vision.

The integument of insects is rich in platelets and filaments, one of whose functions is the sensing of mechanical fluctuations, permitting the insect to sense even the tiniest changes in the soil, air, water, or plants. *Tettigonia* is able to react to soil tremors so small that they cannot be measured (such as those transmitted by growing plants), tremors with an amplitude less than half the diameter of an atom of hydrogen (Autrum, 1940). Ichneumon flies can feel tremors in the bark of a tree produced by the larvae of those insects in which they lay their eggs. Gyrinid water beetles use their special antennae to sense not only ripples spreading on the water surface from various animals but also the echo waves which they themselves have made. Removal of the antennae leads to a loss of orientation in this species; here we are dealing with true echo-ranging, the basis of which is tactile sensitivity (Eggers, 1922).

Touch is especially important in the orientation of deep-water and benthic fish and of those species which are not very motile. Most of them have special organs and tactile papillae scattered along the surface of the body (the sheats, cods, and sturgeons). The body surface of the sheatfish is acutely sensitive. This fish always tries to keep as many points as possible on the surface of its body in contact with surrounding objects. The European triggerfish is most sensitive to touch in the area of the fins and the anterior upper spiny surfaces. The dogfish is most sensitive in the area

of the outer gill openings and spiracles, near the eyes and the eyelids. The dorsal fin of the marine eel is very sensitive to touch.

Among fishes the lateral line is of great importance in detecting fluctuations in the water. It enables the fish to sense even very small currents of water and use them for orientation (rheotaxis). Similar seismosensory organs are found in most of the urodeles living continuously in water. The clawed frog has a true lateral line. Sensory elements are often scattered along the entire exposed skin of an amphibian. Toads have special sensory papillae, while the Apoda have a special palp. Cutaneous organs of touch are found in all reptiles. Snakes and lizards have an especially well-developed tactile sensitivity. Snakes can determine the presence of solid bodies at a distance. It is supposed that snakes use their tongues to detect reflections of air waves set in motion by the wagging motion of the tongue itself. In reptiles with well-developed legs the pectoral and groin areas are very sensitive to contact.

Until recent times the tactile sensitivity of birds was considered to be very poorly developed. This was associated with the presence of the downy and feathery covering. At present it is supposed that this sensitivity is delicate, and in many cases birds can even detect sound waves if their feathers are used toward this end. The tactile sensors of the bird which have already been studied are the tongue, cere, and soft membranes of the bill. The birds with the most highly developed tactile sense are the woodcock, snipe, duck, flamingo, and certain others.

Among mammals touch plays a large role among species with poorly developed vision (shrews, moles, desman, etc.). Among rodents and carnivores the tactile receptors are concentrated at the base of the "feelers," the hairs of which are very sensitive to contact (these are the "moustaches" on the upper lips of such animals). Among ruminants the tactile elements are strongly developed on the upper lip, while in the elephants the end of the proboscis carries the responsibility. This permits the animals to distinguish and sort their food.

HYGROTHERMIC SENSITIVITY

In many animals detection of fluctuations in the temperature of the environment is closely associated with detection of fluctuations in humidity. This is associated with the fact that changes in both factors alter the conditions of water and heat exchange in animals in the same way. Hygrothermic sensitivity permits the animal to choose actively the most favor-

able living conditions and is therefore an important means of orientation in space. Its role is especially important in pessimal zones (cold, hot, dry, or overmoist areas). An animal's choice of one set of hygrothermic conditions over another is not constant and changes not only in different species but also among various populations and single individuals, and even depends on their state.

Even protozoa actually react to changes in temperature, though they have no special thermoreceptors. When the temperature changes, they speed up or slow down their movements, concentrate in places of more favorable temperature, etc. This depends on the special properties of their thermokinetic reactions (see Chapter 6). Special thermoreceptors have not been found among the worms or arthropods. But insects are extraordinarily sensitive to temperature fluctuations. They react most distinctly to diffuse heat. Ticks, mosquitoes, and other bloodsuckers find their hosts successfully by means of heat.

The thermoreceptors of fish are concentrated in the head area. It has been established that some of them can distinguish temperature fluctuations of 0.4° C. They are governed by changes in temperature in finding their feeding places, spawning grounds, etc. Among terrestrial vertebrates thermoreceptors are scattered along the entire surface of the body but are more numerous in the facial area, spine, abdomen, and lower extremities. The latter is very necessary for orientation of the animals in burrows and refuges (Kalabukhov, 1950). Certain species of reptiles (pythons, pit vipers) can detect temperature changes as small as 0.2° C. Temperature orientation is of special significance for snakes; it permits them to determine the location of warm-blooded prey and to attack it successfully even in the dark. Bullock and Diecke (1956) have found that rattlesnakes are able to react to temperature changes in several thousandths of a degree.

Terrestrial insects detect the humidity of the air with the aid of special hygroreceptors, usually situated on the antennae. Experiments have revealed that in some species (*Tribolium confusum, Tenebrio molitor,* the cockroach *Blatta germanica,* the mosquito *Aëdes aegypti,* and others) the humidity receptors are of the hygrometer type (Pielou and Hunn, 1940), while in others (the larvae of *Agriotes*) they are similar to an evaporimeter. In the first case the larvae react to a relative humidity gradient and in the second to a deficit in saturation (see Fig. 50). Insects are sensitive both to the degree of humidity in the atmosphere and to the presence of water. Water insects (swimming beetles and others) orient themselves by air humidity in seeking out bodies of water.

Hygroreception serves frogs as an important tool for finding water in

which to lay their eggs. A very subtle hygroreceptivity explains the ability of desert ruminants to find bodies of water.

ELECTRIC AND ELECTROMAGNETIC SENSITIVITY

The ability of an animal to detect and, in certain cases, to produce electric and electromagnetic fluctuations has been established for a large number of species. Even the protozoa are sensitive to electric stimuli and have a definite galvanotaxis. Passing a weak electric current through a culture of these animals serves to orient them. Paramecia and amoebae usually turn toward the cathode, while *Cryptomonas* and *Polytoma* turn to the anode. Under the same conditions *Spirostomum* will turn with its long axis at right angles to the direction of the current.

Aquatic vertebrates, especially the electric fishes, make wide use of electric currents. The strong currents which they produce serve for attack and defense (electric eels, skates, sheats, etc.). Weak currents, as we have recently learned, are used for orientation, in particular in echo-ranging. The Mormyridae orient themselves in this manner. When the anterior end of the body is charged during a search for food in the silt, the electric generator in the posterior end puts out a weak constant current, fluctuating at up to 100 impulses per minute, which is then reflected off the surrounding objects and detected by a special organ situated at the base of the fin. Electric echo-ranging helps the Mormyridae to orient themselves even in muddy water and enables them to sense danger no matter which direction it may approach from. Weak electric organs are found in the stargazers, certain skates, and the American Gymnotidae; their electrical impulses do not exceed 0.3 volt. These species produce discharges regularly and with great frequency, regardless of whether they are in motion or are lying quiescent on the bottom (Coats, 1954).

The electric eel gives off small discharges during quiet movements, enabling it to determine the location of objects in the surrounding environment (the organs which detect reflected waves are located in the head). Electric echo-ranging, along with the sense of smell, is apparently the most important orientation device of this animal. A weak electromagnetic field (about 300 millivolts) is present about the head end of the marine lamprey and is often regarded as an adaptation for location of approaching objects.

The majority of the fishes react keenly to electric and electromagnetic fluctuations. Fish can be successfully trained to react to an electromagnetic

field, which attests to their ability to detect electromagnetic waves (Kholodov, 1958). Fish with well-developed skin receptors are the most sensitive to electric currents (Bodrova and Krayukhin, 1958). Three phases of a fish's reaction to electric currents are distinguished, based on the strength of the current. In the field of a weak current the fish tries to flee or to take up a position in which the axis of its body is parallel to the current. In the second phase the fish turns toward the anode and swims in its direction. If the current is increased further, the fish falls into a state of galvanonarcosis or dies. These properties of fish behavior are being more and more widely applied in the fishing industry, in the protection of irrigated fields from fish, and in many other areas.

Suggestions have been made about the possibility of electromagnetic orientation among butterflies. Orthoptera, Lepidoptera, and other insects react to electromagnetic waves with movements of the head and abdominal appendages (Busnel, Dumortier, and Busnel, 1960). More than once different variants of the magnetic theory have been advanced concerning magnetic orientation of birds (Middendorf, 1855; Moren, 1923; Jaegli, 1951). Further research, however, has not upheld these suppositions (Dorst, 1935; Danier, 1936; Cramer, 1948; Orgel and Smith, 1954; and others). Radar stations influence the orientation of birds. When flocks of birds (daws, crows, seagulls, geese, and others) enter the zone of a radar unit, the cohesiveness of the flock is broken up; the birds fly off in different directions and reunite only after they have flown beyond the range of the sending unit (Dorst, 1959). Sessile birds are not affected by radar even at short distances. In areas where radar stations are working, the loss of orientation among birds often leads to mass death. Despite numerous observations and hypotheses, electric and electromagnetic orientation has been found only in aquatic animals. But the use of electric and electromagnetic phenomena for orientation is apparently more widespread than we imagine.

ANALYSIS OF TIME AND SPACE

The coordinating significance of animal orientation belongs to the time-space analyzers. The basis of time analysis is the general rhythm of the physiological processes; the external analyzers—visual, aural, etc.—also play a significant role. Among animals having a variable body temperature, the leading mechanism of time computation is the rhythm of matter exchange. By changing the intensity of exchange in such animals (arachnids, insects, bats), we can cause a shift in the progress of the internal "clock." In animals with a constant temperature, analysis of time is made by the

central nervous system, which integrates the many internal and external signals. The establishment of reflex connections between the "internal clock" and the various phenomena of the external environment (appearance of food or enemies, motions of the sun, etc.) does much to perfect the mechanisms of animal orientation.

In the analysis of space a significant role, in addition to the already-discussed analyzers (visual, aural, etc.), is played by muscular (motile) analysis in conjunction with the "sense of balance." Destruction of their functions leads to a loss of orientation. In individual cases the destruction of one analyzer may be compensated for by restructuring of the work of the others. When the organs of balance are destroyed, the animal's ability to orient itself and to position its body correctly is often handled by the coordinated work of the motile and visual analyzers.

ORIENTATION AS A TOTAL PROCESS

An animal's spatial orientation is a complex and involved phenomenon. As the animal world evolved, the sense organs and mechanisms of space orientation also developed. A definite connection exists between the development of vision on the one hand and of the other sense organs on the other. Along with the evolution of the sense organs and psychonervous activity—from simple to complex—a parallel process of adaptation to concrete conditions of existence and way of life occurs for each separate group and species of animals. Orientation is more perfected among insects than among the lower orders of invertebrates. Among insects the best-developed organs are those of vision, smell, and touch, but there are not a few insects which are deaf and blind. Simplification in this instance came as the result of specialization, or adaptation to special conditions of existence.

The basic form of orientation in the protozoa is their general chemical sense. Among the arthropods the mechanisms of orientation become more complicated. In some insects (bees, ants, butterflies) the sense of smell plays the leading role in the seeking of food and of other individuals. In others it is touch, which does not develop such a high level of sensitivity in any other group (spiders, Hymenoptera, etc.). In a third group the leading role is played by vision, the perfection of which is associated with the appearance of compound eyes, the possibility of orientation according to the plane of polarized light, and the sense of color.

Among vertebrates further perfection of psychonervous activity produces other mechanisms of orientation, still mostly unstudied, such as the "sense of geographic position," "sense of direction," "kinetic memory,"

and others. Among fishes (eels, sheats, salmon) the sense of smell plays the leading role in orientation; touch and hearing are also important. Electric and possibly electromagnetic orientation and sonar echo-ranging also appear.

The way of life of contemporary amphibians and reptiles does not demand such perfected means of orientation. The prevailing method in certain amphibians is hygroreception, while in some lizards and snakes it is the sense of smell. Certain amphibians are obviously able to orient themselves by polarized light, and lizards can orient themselves by the sun. Vision predominates among the birds, and only a few species with a nocturnal way of life orient themselves by ear. Birds may orient themselves by various objects, detect and remember the general features of a landscape, etc. The speed with which conditioned reflexes arise, and their regularity and stability, increase the flexibility of their adaptation to different conditions of orientation.

The orientation mechanisms of birds are of special interest during the long seasonal flights. According to modern findings, they include (a) a congenital "sense of direction" of flight, which in some species is supplemented by teaching of younger birds by older ones; (b) visual orientation by separate objects and their relationships to the total outlines of the landscape; (c) "kinetic memory," i.e., an ability to remember and repeat automatically a movement completed earlier; (d) photocompass orientation by the sun and, for nocturnal forms, apparently by the stars (astronavigation); (e) an apparent electric and electromagnetic orientation which has still not obtained convincing confirmation. At definite points on the route orientation occurs by wind direction, by direction of wave crests on flights over the sea, etc. The possibility of species orienting themselves by polarized light has not been eliminated.

As a result of markedly differing ways of life, the orientation mechanisms of mammals are extraordinarily varied. Among carnivores it is established chiefly by the senses of smell and hearing; in the primates and man it is mainly by vision. Among aquatic and flying mammals the basic method is sonar and ultrahigh-frequency echo-ranging, while in subterranean forms it is smell, touch, and sometimes hearing. Orientation of mammals during periodic wanderings over long distances is similar in many ways to the orientation of birds. Nonperiodic migrations observed during mass reproduction of certain species (rodents) are directed to all sides; i.e., they are radial. Directed migrations of this type arise under the influence of the distribution in space of vitally important conditions: food, water, the protective properties of a territory, its penetrability, etc.

PART II

THE ECOLOGY OF POPULATIONS

CHAPTER 11

The Species as an Ecological System

INTRASPECIES STRUCTURE

Separate individuals exist only as parts of a whole species. They are united in a natural aggregation (the population) which changes the character of their interrelations with the environment and increases the possibilities for exploiting it. This is a qualitatively different way of uniting individuals and differs from other aggregations of blood-related individuals which form a single system and aid its existence through reproduction. These individuals inhabit a territory (range) suitable for the existence of the species; in the associations (biocenoses) located thereon they occupy a position determined by its previous evolution. The role and significance of the species in the biosphere are determined by its place and specific weight in the biological turnover of matter. The disappearance (dying out, extinction) of one species or the appearance of a new one changes the turnover of matter to a greater or lesser degree but still, by necessity, in the areas where the species lives. The existence of a species is furnished by the connections between its individuals, which form the basis of the intraspecies structure. These connections have become positioned during the course of evolution according to the formation of the species and the establishment of its morpho-physiological and ecological properties. The intraspecies organization which arose on this basis has been the crucible of natural selection and maintains the integrity of the species.

Since this conception was first determined scientifically by J. Ray and C. Linnaeus, it has undergone substantial changes. To the biologists of the seventeenth and eighteenth centuries it presented itself as a collection of independent, only temporarily paired families, flocks, or colonies of individuals, of a single type throughout the entire range. Only age and sex

differences were recognized as being regular. Individual mutability was considered accidental and attracted no attention, and no one knew anything of geographic differences. However, P. S. Pallas as far back as the 1760s, A. A. Kaverznev in 1775, and L. von Buch in 1825 had mentioned the possibility of a new species being formed by geographic removal of a race (subspecies) from the species type. But it was only with the publication of Darwin's *The Origin of Species* (1859) that interest in group (i.e., geographic and ecological) and individual mutability began to grow. This interest was strengthened by the formulation of the basic positions of the theory of inheritance.

Accumulation of data led to a replacement of ideas on the monotypic (unique within itself) species by ideas of the species as a system (aggregate) of individuals associated with one another. Such populations were to be distinguished by their morpho-physiological, ecological, and genetic features and occupied either different geographic regions (subspecies) or definite ecological niches (ecotypes, biological races) (A. A. Semënov-Tyan-Shansky, 1910; N. I. Vavilov, 1931; Huxley, 1940; Komarov, 1940; Mayr, 1947). Most biologists consider the lowest systematic unit to be the subspecies, or group of individuals inhabiting a definite geographic part of the range and differing structurally to the extent that no less than 75 percent of the individuals studied can be assigned to the given subspecies (the "75 percent rule"). In addition to subspecies, even smaller taxonomic units have been demarcated, such as A. A. Semënov-Tyan-Shansky's tribe, or natio, which can be distinguished objectively in many cases only with difficulty or not at all.

Accumulation of ecological reports and information has shown that the real structure of a species is more complex than a simple system of geographic forms, since each subspecies in turn breaks down into a series of even smaller hierarchically subordinated groups of individuals or populations. Among them are distinguished territorial populations and biological groupings (races, ecotypes, seasonal phases, etc.). Certain of them are inherently determined and ontogenetically irreversible; others are reversible in the first generation (Kholodkovsky, 1908, 1910; Mordvilko, 1939; Kozhanchikov, 1946, 1951, 1956; Beklemishev, 1959, 1960; Rubtsov, 1960). Unfortunately, the term "population" is often used in different senses and has been used to designate every group of individuals united by common properties or connections. Most people understand a population to be a group of individuals which inhabits a definite territory, entering into an association (biocenose), existing there, and representing the form of the species existing under a concrete set of landscape-geographic conditions (N. P. Naumov, 1936, 1948, 1956; Arnol'di, 1957; Beklemishev, 1959, 1960).

According to M. S. Gilyarov (1954), "In each concrete habitat (biotope)

the population of a given species enters into the concrete biocenose, and usually the boundaries of the population of the species are the boundaries of a given biocenose into which the given species enters." The interpretation of a population as a natural spatial grouping of individuals is more correct. It is actually made the title of this chapter. Allee *et al.* (1949) regarded population as a generation which is born, develops, and subsequently dies out. Geneticists understand population as an aggregation of hereditarily like individuals.

In actuality, the species consists of a complex aggregation of morpho-physiologically, genetically, and ecologically differing groups of territorial, ecological, and "hierarchical" character. All of them were developed by adapting to local conditions of existence (territorial forms), on the basis of division of functions (sex groups), by splitting up into different ecological niches (biological races), or as a means of utilizing territories ("structure and hierarchical organization"). Among the territorial groupings the major ones—the subspecies—occupy a geographically unique part of the range. They break down into geographic populations, each of which settles a definite landscape-geographic region which differs first in its landscape-climatic properties. The geographic population breaks down into ecological populations, forming settlements of animals in definite biotopes. Finally, the ecological population, if it is not unique, may be dismembered into even smaller groups inhabiting elements of a mosaic landscape—the elementary population. Population organization may be shown as in the diagram below.

Each of the territorial groupings, or populations, usually consists of different generations, in many cases differing markedly from one another in their morpho-physiological properties, and sex groups (males and

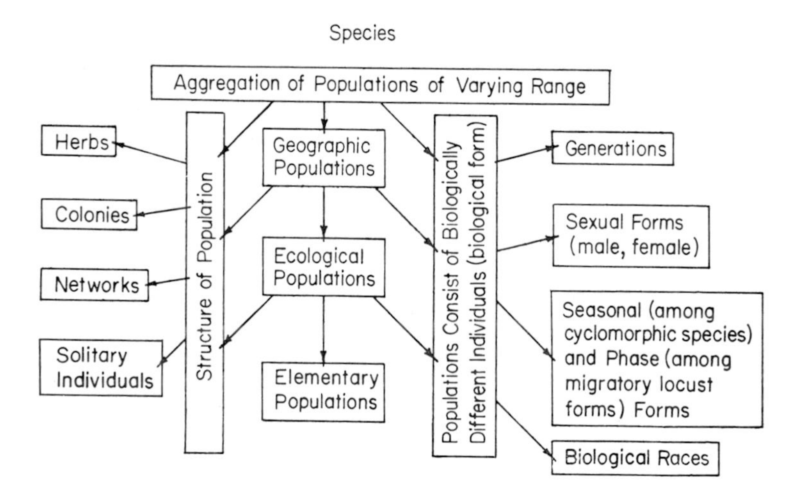

females). Among the social insects and other forms they differ also in division of labor. Through breakdowns into various ecological niches, so-called biological races are formed (the food races of insects, races according to host species among parasites, spring and winter races among fish, etc.). All these ecological forms increase biological variety and polymorphism within the species, complicate its structure, and enrich the forms of its relationships with the environment. Finally, within the species population there exist groupings of individuals associated with ways in which the species exploits the territory (families, herds, colonies) and which determine the structure of its populations and its way of life (solitary, herd, colonial).

SUBSPECIES

The largest territorial groupings of individuals are the subspecies. Their territorial dimensions depend on the landscape variety of the territory (its geographic differentiation), the mobility of the animals, and the properties of their intrapopulational relationships.

In hilly regions and archipelagoes subspecies usually occupy smaller areas than they do on the plains of continents. In the New Guinea bird *Myiolestes megarhymchos* 25 well-distinguished subspecies have been described from the main island of New Guinea and the surrounding smaller islands. The flycatcher *Monarcha castaneoventris* from the Solomon Islands has six subspecies (Fig. 153). The kingfisher *Tanysiptera galatea* has eight subspecies on the main and neighboring islands of New Guinea, and some of them occupy very minor territories.

The mobility of animals and the degree of their intercourse with one another have an especially great influence on the dimensions of the area occupied by the subspecies. The nearly immobile land snails occupy very small territories. On the mountainous island of Oahu (Hawaii) 15 structurally distinguishable populations of the snail *Achatinella mustelina* were observed in an area of 6 sq. km. Among the European garden snails *Cepaea nemoralis* and *C. hortensis* stable anatomic differences occur even in neighboring colonies. In the more motile marine species *Littorina obtusata* we can distinguish one anatomic population near the Norwegian coast, two populations on the British coast, and eight North American ones.

Populations of the cod in the North Atlantic differ in number of vertebrae. One of them occupies a region from the west coast of Greenland to the banks of the Kola Peninsula and the Baltic Sea, another lives west

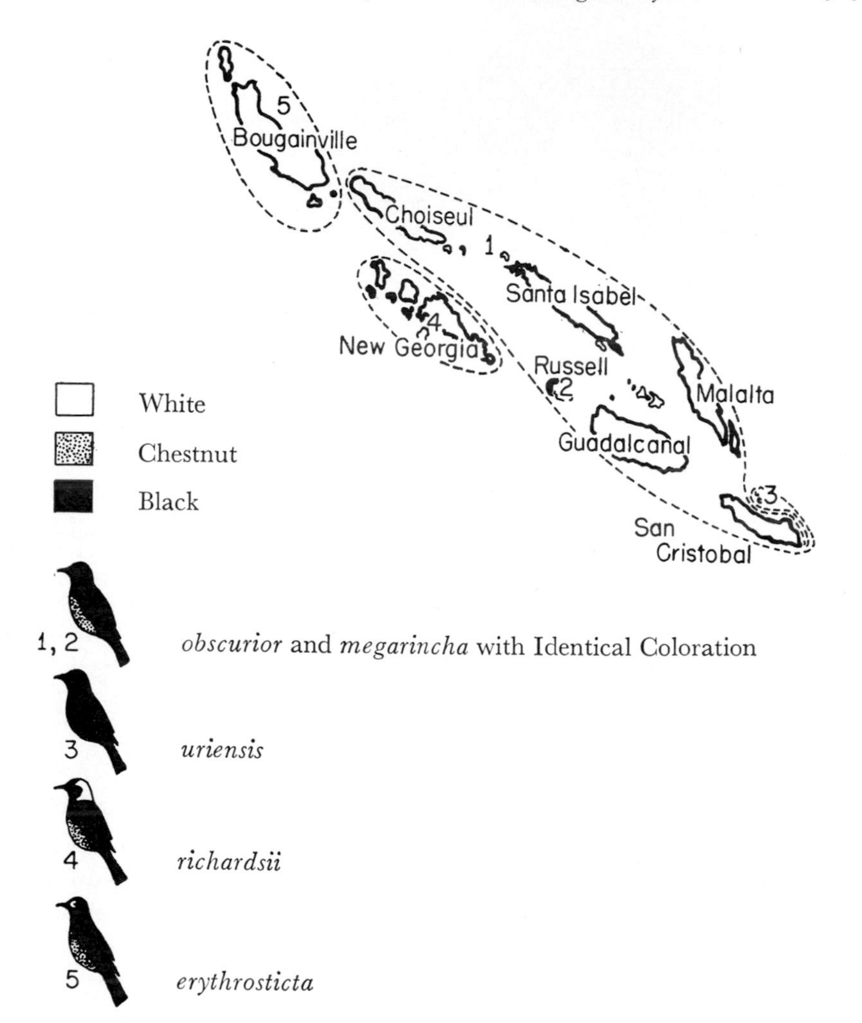

Fig. 153. Insular geographic races of the flycatcher *Monarcha castaneoventris* in the Solomon Islands (after Mayr, 1947).

of the British Isles, and the North Sea race occupies a smaller area. The regions occupied by the Murmansk and Norway herrings are just as large (Fig. 154). The subspecies of the continental birds are scattered even more widely (Fig. 155).

The areas occupied by the common squirrel are minimal in southern Siberia and the European part of the USSR, where the landscapes are varied, while their territories are enormous in the monotonous taiga zone

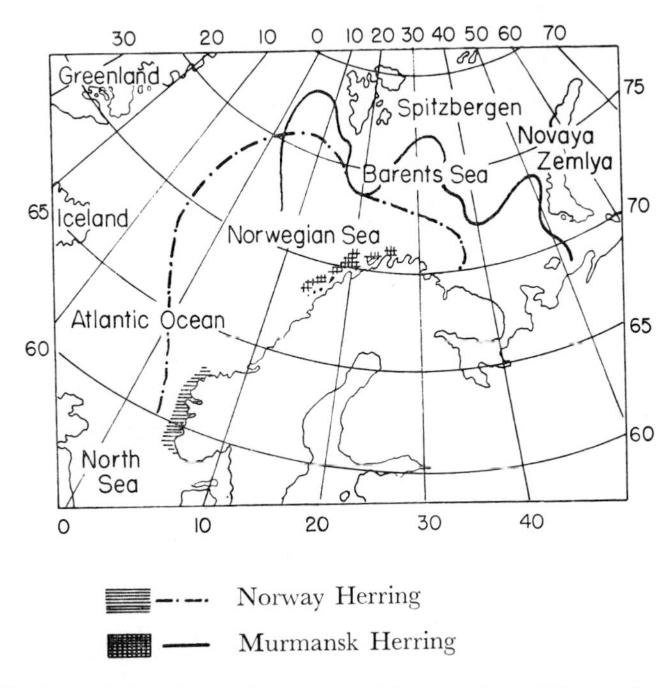

Fig. 154. Regions of migration and spawning of Murmansk and Norway herrings (after Tikhonov, 1939).

(Fig. 156). Large migrations of squirrels into the European USSR occur comparatively rarely and primarily in the north, but they are common in the Siberian taiga (Formozov, 1936). Therefore, it is impossible not to associate the magnitude of the habitat area of an individual subspecies of squirrel with the mobility of the creatures in the various parts of their range. The significance of this condition is even more convincing when we examine the geographic mutability of Arctic foxes. In the Eurasian tundras this species forms no geographic races, with the exception of the ocean-isolated populations on Spitzbergen and the Komandorsky Islands. The high mobility of foxes is well known; they make trips over the tundra and into the forest zone, wander to the shores of the Arctic Ocean, and make latitudinal migrations along the tundra zone. Foxes tagged on the Yamal Peninsula have been caught in the same season on the North Island of Novaya Zemlya some 1,200 km from the place they were tagged, 200 km away on the Taymyr Peninsula, and 150 km away on the Pechora River (Sdobnikov, 1947).

In addition to animal mobility, the magnitude of the area occupied by the subspecies may depend on the properties of intraspecies organization.

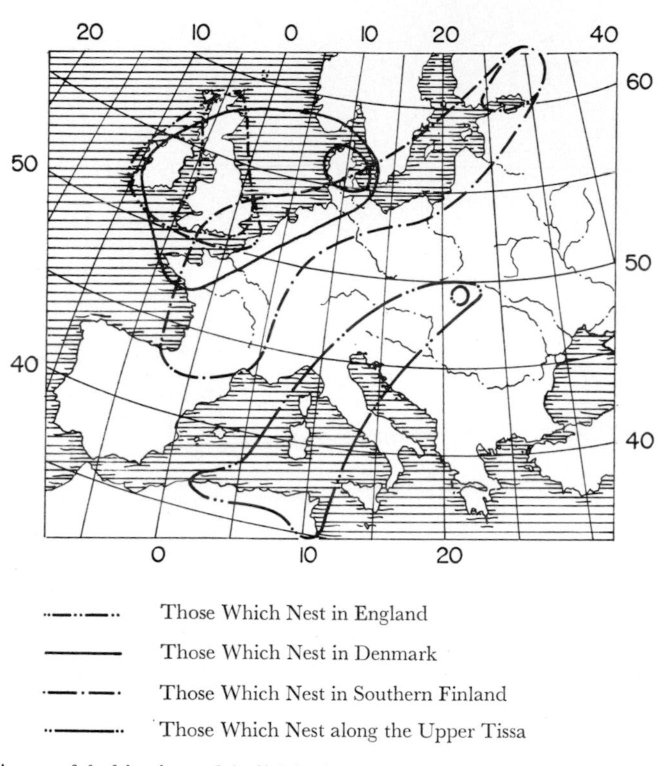

Fig. 155. Areas of habitation of individual populations of starlings (*Sturnus vulgaris*) (after Schüz and Weigold, 1931).

The bean geese in Eurasia are divided into three subspecies, while the broad-billed and gray ducks do not form separate geographic races either on this continent or in North America. This is explained by the fact that families of geese, unlike ducks, do not separate after maturation of the young and are preserved until the return to the breeding ground, where regrouping occurs and new pairs are formed. This, along with colonial nesting grounds, aids in the appearance of local races even without the presence of gradations or transitional populations (Mayr, 1947).

All these examples show convincingly that the formation of subspecies depends not so much on differences in place of habitation as on the ecological properties of the animals. Therefore, a series of subspecies may appear in one species and not be present in another on the same territory. This has been confirmed by so-called clinal geographic mutability, in which one feature or another changes gradually in space, making it impossible to distinguish subspecies or draw distinct boundaries between them (Huxley, 1939; Terent'yev, 1959). An example of clinal mutability

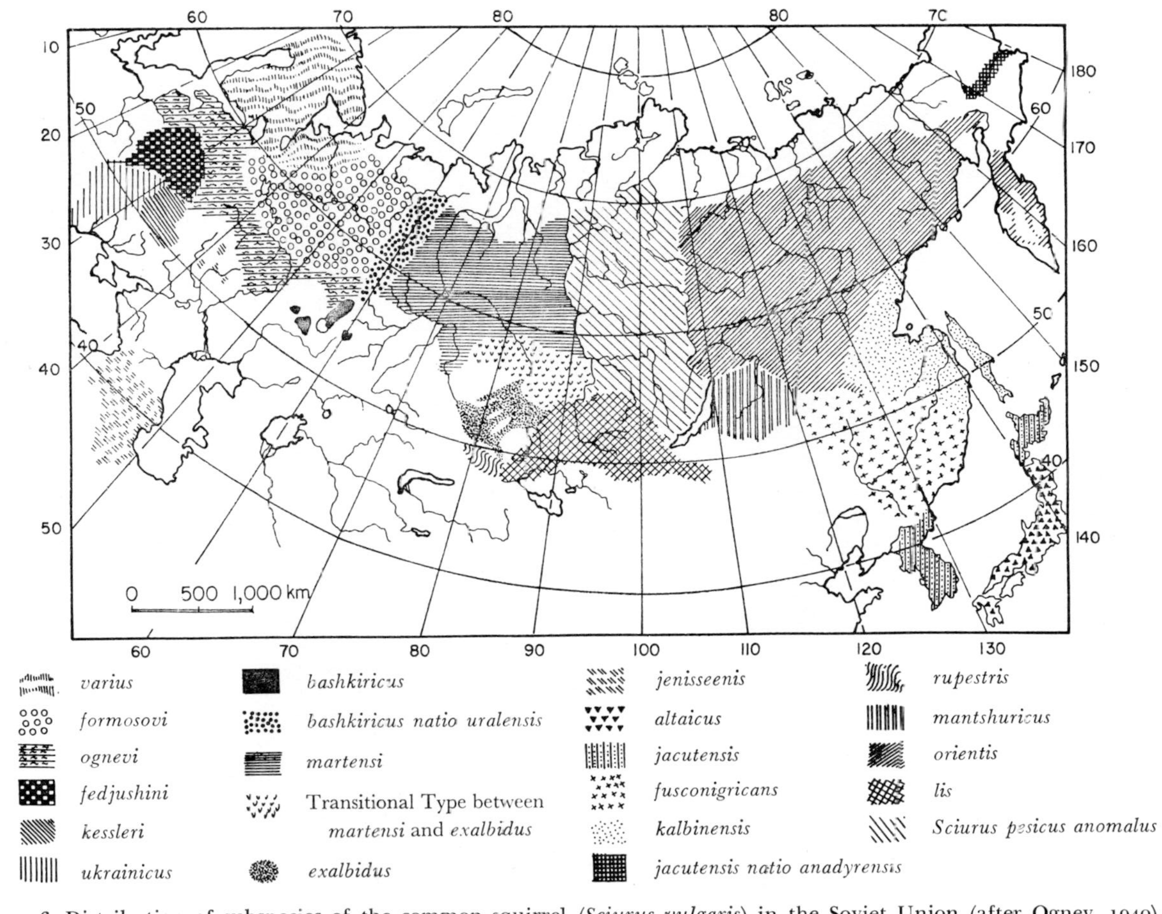

Fig. 156. Distribution of subspecies of the common squirrel (*Sciurus vulgaris*) in the Soviet Union (after Ognev, 1940).

would be the change in the number of different-colored forms in populations (among squirrels, some predatory birds, and others) (Table 40).

TABLE 40. *Changes in the percentage of colored forms of the red fox in British Columbia from north to south (after Cowan, 1938).*

	COLLECTION POINT NUMBER						
	1	2	3	4	5	6	7
Colored Form		*North*			*South*		
Black	19%	25%	14%	6%	8%	3%	1%
Gray	46	45	34	34	25	17	16
Red	35	30	52	60	67	80	83

It is usually accepted that clinal mutability reflects a gradient of changes in the environment. This is apparently confirmed by the fact that the various features of a species may change independently of one another. On the other hand, in several cases clinal mutability does not correspond to changes in the conditions of existence (the so-called nonadaptive cline), and it must obviously be connected with the animals' mobility and the amount of exchange of individuals between populations. As a result, a cline may arise even in the absence of gradual changes in the environment. V. N. Beklemishev (1960) has suggested calling such large populations "superterritorial"; they uninterruptedly settle enormous territories and are not limited to individual territorial groups. If we observe in the boundaries of a superpopulation a rather indistinct isolation of separate parts, such as alternation of densely and sparsely settled strips, Beklemishev proposed that we call them subpopulations. Inheritance factors may lie at the basis of clinal mutability.

GEOGRAPHIC POPULATIONS

In many cases subspecies are broken down into geographic populations or groups of individuals, each of which settles a territory with geographically uniform conditions of existence—"the zone of similar favorability." The territorial dimensions of geographic populations, as of subspecies, depend on the properties of the species and the territory inhabited. As an example one might introduce the three basic populations of the Pacific salmon. The first winters south of the Aleutian range and spawns in the rivers of Kamchatka and those which drain into the Sea of Okhotsk. The second lives south and southeast of the Kurile Islands and spawns in the rivers of eastern Sakhalin and the southern part of the Sea of Okhotsk. The third

winters south of the eastern Aleutian range and spawns in the rivers of Canada and Alaska. The territorial dimensions of squirrel populations are shown in Fig. 157.

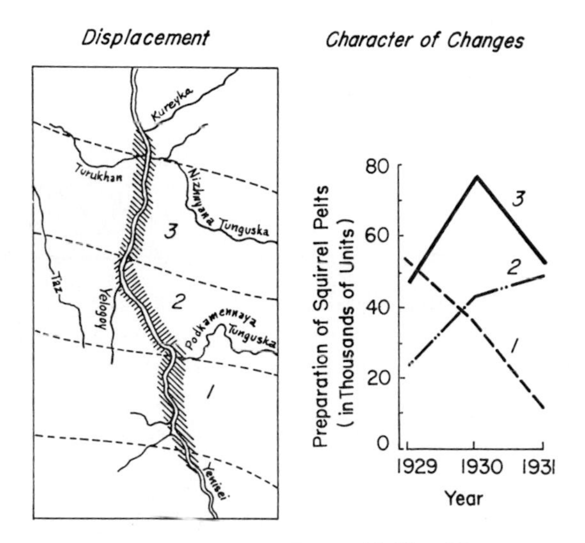

1 Population of Cedar-Fir Taiga with Silver Fir

2 Population of Leafy Taiga with Cedar

3 Population of Sparse Leafy Taiga of Northern Type

Fig. 157. Displacement and character of changes in abundance of a population of squirrels (*Sciurus vulgaris*) in the valley of the Yenisei River (after N. P. Naumov, 1934). Dashed lines represent yearly isotherms.

Cross-multiplication takes place within the geographic population. As a result, the individuals entering the population have the features of a general morpho-physiological type which constantly or temporarily[1] differs from the neighboring geographic populations merely in the rhythm of its vital phenomena (times and speeds of reproduction, hibernation, seasonal migrations, etc.), the total rhythm of population dynamics, and the resistance shown to unfavorable influences.[2] Various geographic populations may differ in their fecundity or sessility, or they may have different ecological and physiological properties, and these differences may be fixed by heredity. Thus the mountain populations of the Caucasian forest

[1] The latter is more common.

[2] Thus the eastern (Ural) populations of hares, roe deer, and several birds suffered less mortality than the western European populations in the unusually severe Siberian winters of 1928/29 and 1939/40.

mouse (*Apodemus sylvaticus ciscaucasicus*) have a stable, high hemoglobin level in the blood, while among the prairie populations this factor is lower and increases only as the mice are taken higher into the mountains (Kalabukhov, 1950).

Differences in feeding habits are of prime importance among the properties of geographic populations. Higher-calorie foods tend to predominate among the animals of the north, while meat eating prevails among predators. Among the forest martens of the Caucasus vegetable matter is found in 75 percent of the intestines, mammalian flesh in 29 percent, and birds in 7 percent. In the Pechoro-Ilych Preserve the occurrence of mammalian flesh increases to 70 percent and that of birds to 35 percent, while vegetable matter is found in only 13 percent of the intestines. The means of hunting are also different in the two populations. The Caucasus martens seek their food primarily on the ground, while the northern individuals hunt equally often on the ground and in the trees. Among the ducks on the Taymyr Peninsula there are species which are strictly carnivorous. On the Barabinsk steppes 50 percent of the total number of species and 32 percent of the individuals are of this type, while in central Kazakhstan they account for 43 percent of the species and only 23 percent of the individuals (Mikheyev, 1938). Seeds are found in the summertime in only 10.4 percent of intestines of the common voles of the southern Ukraine; in the Tula oblast' (forest steppe) they occur in 19 percent and in the Yaroslavl oblast' in 26 percent (N. P. Naumov, 1948).

Differences in the feeding habits of individual populations may depend on the abundance and availability of the various kinds of food in various geographic regions (Table 41). The prevalence of birds in the foods of

TABLE 41. *Properties of wintertime feeding habits of different populations of the sparrow owl* Glaucidium passerinum *(after Vorontsov, Ivanova, and Shemyakin, 1956).*

Food Group	*Suburban Moscow (251 individuals)*	*Scandinavia (445 individuals)*	*Central Europe (250 individuals)*
Mammals	92.0%	74.3%	38.4%
Birds	8.0	25.7	61.6
Shrews	6.8	37.3	14.4
Small rodents	85.2	37.3	24.4

Central European pygmy owls is associated with the fact that a number of passerine birds winter there. Owls of the same species in Moscow and Scandinavia feed primarily on mammals (rodents, insectivores) in the wintertime.

Among northern populations of animals one often observes a rise in food refusal, sometimes euryphagy, and often an increase in perennial and seasonal changes in feeding habits as an adaptation to a poor food base. In the forest steppes common voles eat approximately 70–80 species of plants (Tula oblast'). In the north (Yaroslavl oblast'), because of the multigrassed forest, the list of foods rises to 90–100 species of plants. The food of predatory birds on the tundra of the Yamal Peninsula changes a great deal (Osmolovskaya, 1948). There even the peregrine falcon, which in other regions feeds exclusively on birds, catches voles and lemmings when the number of white willow ptarmigan and shorebirds falls off. In other cases an increase in the total variety of feeding habits is reflected in marked local differences (Table 42). In this example the feeding habits of indi-

TABLE 42. *Variability of feeding habits of the whitefish* Coregonus lavaretus *in the north (specific weight of all basic food groups) (after Nikol'sky, 1944).*

River Basin	*Mollusks*	*Crustaceans*	*Chironomid Larvae*	*Other Insects*
Kara	0.0%	38.0%	61.0%	1.0%
Gyda	1.6	98.4	0	0
Upper Pechora	61.0	15.5	2.6	20.9

vidual populations appear specialized, while the population of the zone as a whole lives at the expense of a large number of foods.

Individual differences in feeding, which are usually determined by the local content and variety of foods, are often closely connected with geographic mutability. But in some cases they are associated with individual tastes and habits of food-getting; the latter are well known for predatory birds (Dement'yev, 1935). Differences in feeding habits among different families of birds and mammals have been observed by all investigators.

TABLE 43. *Feeding habits of individual broods of* Archibuteo lagopus *on the tundras of the Yamal Peninsula (encounterability of individual species of food) (after Osmolovskaya, 1948).*

	NEST LOCATION		
Food	*Sandy Watershed Tundra*	*Lakeshore (cliff)*	*Mossy Tundra*
Arctic lemming	61.7%	37.3%	11.4%
Ob' lemming	27.7	37.3	80.0
Middendorf's vole	8.5	15.7	2.1
Narrow-skulled vole	4.3	4.8	8.6
Voles (no precise determination)	2.1	2.4	11.4
Lapland plantain	–	6.0	–
Temminck's stint	–	1.2	–

They may be significant (Table 43), but usually they are not stable (Fig. 158).

In addition to feeding habits, different populations have different levels of gas exchange, chemical thermoregulation (Fig. 159), water exchange, and other ecologo-physiological properties (Sun Ju-yung, 1958; Shvarts, 1959). Among lake frogs from the Volga delta the permeability of the skin

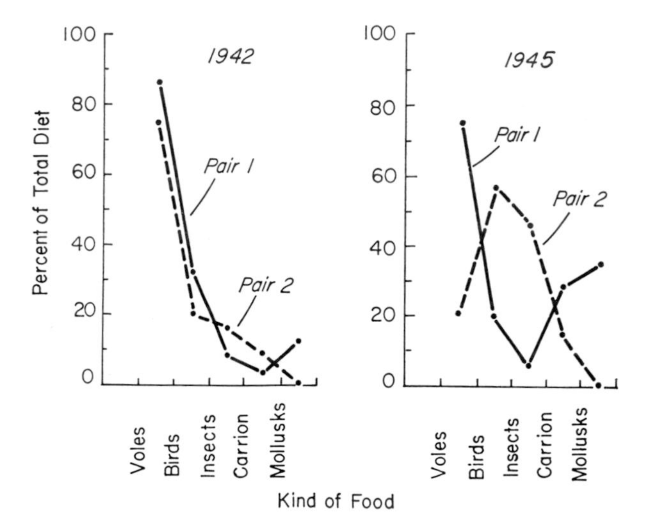

Fig. 158. Differences in feeding habits of two families of ravens (*Corvus corax*) in the Tula abatis during nesting (after Likhachev, 1951).

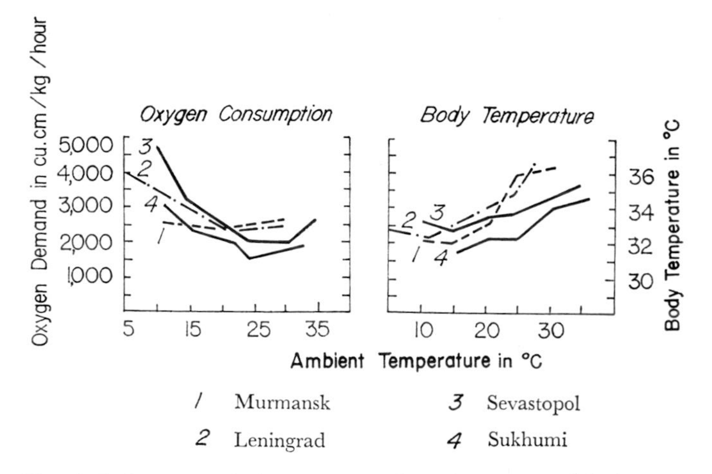

Fig. 159. Chemical thermoregulation: consumption of oxygen and body temperature of gray rats from various places (after Ruttenburg, 1953).

for water is 20.2 cu. mm per sq. cm per hour, while among frogs from around Astrakhan, where the air is dryer, it is 16.2. The resistance of erythrocytes in the blood to hypotonic solutions, which permits evaluation of the organism's ability to hold water, is higher in Astrakhan frogs (Dinesman, 1948). The permeability of the skin for water is also different for various populations of the grass frog (Table 44). Among two populations of the freshwater mollusk *Bithynia leachi* the ability to hold water in the body during heavy drought is different (Fig. 160). Inhabitants of perma-

TABLE 44. *Geographic variability of permeability of the skin to water in the grass frog* Rana temporaria *(after Dinesman, 1948).*

Time of Study	*Place*	*Amount of Precipitation in Nonfrozen Period in mm*	*Length of Frost-Free Period in months*	*Permeability of Skin in cu. mm/sq. cm/hour*
Apr.	near Prague	260	6.9	15.2
May	near Breslau	324	6.2	22.1
May	near Dresden	368	6.4	24.5
July	near Ternopol	340	6.2	21.6
Aug.	near Zsescuwa (Poland)	402	6.2	30.5

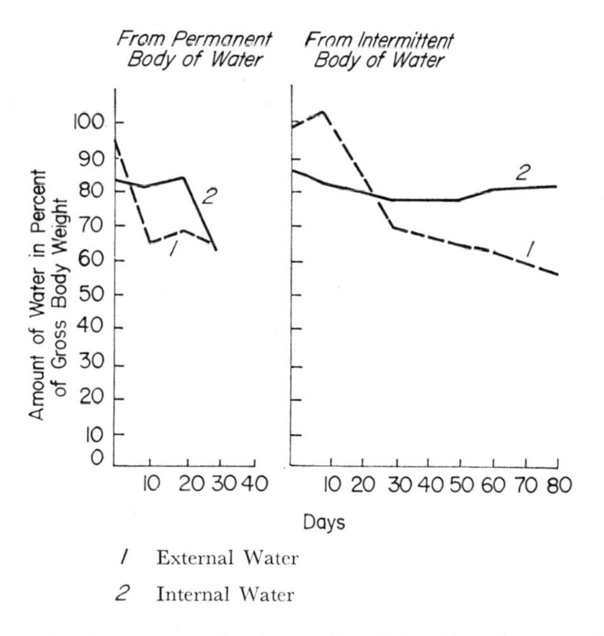

Fig. 160. Changes in the amount of water during desiccation of representatives of two populations of the freshwater mollusk *Bithynia leachi* (after Shkorbatov, 1953).

nent bodies of water quickly lose both their external and internal (tissue) water, while those which live in intermittent bodies of water can maintain the former and especially the latter (Shkorbatov, 1953).

We usually distinguish temperature constants, temperature resistance, and heat exchange among various geographic populations. As a rule, the sum of effective temperatures necessary to raise one generation is less in the northern than in the southern parts of a range. The optimal temperatures for activity in the housefly rise regularly from north to south (Lineva, 1950; Balashova, 1955). Among the American and European trichogrammas there are geographic forms which differ in their temperature preferences, speed of development, and other adaptive biochemical properties (Telenga, 1959). The chrysalis of *Exaereta ulmi* from the lower Volga region has greater cold resistance than that from the North Caucasus. The lower Volga variety can be supercooled down to —8° C (a maximum of —12° C), while the North Caucasus type can be cooled to only —5° C (maximum of —9° C) (Pantyukhov, 1956). Marine invertebrates of the Murmansk coast (*Mytilus edulis, Macoma baltica, Littorina saxatilis, L. obtusata, Balanus balanoides, Idothea baltica,* and others) tolerate temperatures below 0° C (down to —18–20° C). The same species along the coasts of Denmark and southwestern England died en masse in the winter of 1928/29, when the temperature went to —2° C (Polyansky, 1953).

Varying degrees of temperature resistance are associated with other physiological properties. Pulsation of the dorsal blood vessel in *Perinereis cultifera* (Polychaeta), which reflects the intensity of blood flow, is shown in Table 45. Among northern individuals the amount of blood flow proved to be higher, despite the significantly larger size of the individuals. The same thing has been found for echinoderms, crustaceans, and mollusks.

Finally, geographic populations usually differ in their fecundity. Among the marine invertebrates the index of fecundity (average number of off-

TABLE 45.

Place	*Month*	*Average Weight of Individual in g*	*Temperature in °C*	*Frequency of Pulsations*
Plymouth	Oct.	3.0–4.0	10.0	64
			14.5	92
			20.0	132
Tamarisse (southern France)	Aug.–Sept.	0.3	14.5	72
			20.0	90
			25.0	103

spring per g or ml of female body) changes in the following way (according to V. V. Kuznetsov, 1956):

Hyas arenaria haëki	
Coastal waters of eastern Murman (littoral and sublittoral)	386
Coastal waters of western part of White Sea bight (to 50 m deep)	364
Open portion of Barents Sea north of Cape Kanin Nos (50–120 m deep)	460
Onega Gulf on White Sea (1–20 m deep)	521
Balanus balanoides	
Littoral of open and semiopen coast of eastern Murman	5,611
Littoral of closed and semiopen coast of eastern Murman	3,633
Littoral of closed and semiopen coast of Kandalaksha Gulf	8,480

We also distinguish, of course, individual mortality rates, which leads to the discovery that the abundance dynamics of the various geographic populations are not identical. We distinguish both the mean level of abundance and its stability; both depend on the degree of favorability of the territory occupied by the population. In the widespread ladybird *Coccinella septempunctata* Bodenheimer (1958) distinguished four basic populations and gave them the following characterizations:

1. Europeo-Siberian. Development possible in May through September; winter hibernation present, none in summer; one generation per year possible; in some areas an extra one possible (in favorable years), but it is incomplete; active period corresponds in time with abundance of aphids; conditions of existence favorable.

2. Mediterranean. Up to 5 months of temperature below threshold of development, in summer above maximum, which determines torpor; sum of effective temperatures permits development of one full generation in spring and one in autumn; in addition, extra but aborted generation (in Israel), almost all of which die out before maturing and reproducing; conditions are entirely favorable for development.

3. Irano-Turan. Temperature during 6 months below level needed for development (winter) and 2 months above (summer); conditions in general unfavorable; combination of temperature and humidity in almost all months far from favorable (16–20° C, 65–80 percent relative humidity); especially unpleasant in autumn, while summer and winter are impossible for development; sum of effective temperatures permits one generation between May and June, but not all mature completely; this period corresponds with period of aphid abundance; sum of effective temperatures in autumn does not permit full development of one generation; spring generation migrates in summer to tops of hills, where some adults seek refuge in areas with a favorable microclimate (in torpor); the rest as a rule die; abundance unstable and low.

4. Sahara-Sindian. Dryness of air and lack of rainfall do not permit ladybirds to live in deserts; however, temperature and air humidity permit development of one or two generations on oases in winter, if there is vegetation and aphids at this time; even under comparatively favorable conditions (in Cairo) species cannot create a sufficiently stable population and is rare everywhere.

Temporary populations arise, but in parts of the range which are rather unfavorable for existence, these populations are not in a position to maintain their existence by reproduction. Instead they are maintained by newcomers, i.e., immigrants from the optimal parts of the range. Active or passive emigration plays an important role in the life of many populations. The eggs of *Chironomus batophilus* on Lake Ployen, which are laid over its entire surface in numbers up to 68 billion, end up with 16 billion on the bottom at a depth of 16 m, where part of them live; 50 billion at a depth of 16–36 m, where a few survive; and 2 million at great depths, where they all die. The population of the flounder *Platessa platessa* in the Baltic Sea cannot maintain itself because of an unfavorable hydrobiological regime and other conditions, but it exists there permanently, maintained by newcomers (Bodenheimer, 1958).

Thus the existence of some geographic populations may depend to a certain degree on others. These connections are usually set up by exchanging individuals through migration; they are sometimes mutual, while at other times they are one-sided. The latter is more common. This exchange may be constant, but in most cases it is periodic, when an increase in numbers of a population is accompanied by mass emigration. At the same time geographic populations are dissociated, with isolation so significant that it is accompanied by inherited differences between them. However, the degree of interdependence of geographic populations is not great; they are completely able to reproduce their kind and, like subspecies, deserve the name of independent populations (Beklemishev, 1960).

Geographic populations are separated from one another not only by distance and differences of habits but also by the existence of even the most temporary morpho-physiological differences caused by both factors, by noncoinciding rhythms of vital phenomena and different levels of progress in abundance fluctuations. At the same time and in the same space, place of habitation and morpho-physiological unity may connect the individuals of a population into one natural whole; special meaning devolves upon the times and places of reproduction and migration. The two populations of Scandinavian herring have different spawning grounds (see Fig. 154). A distinction is made between the wintering grounds and nesting areas of various populations of starlings (see Fig. 155) and seagulls (Fig. 161). The northern Aral population of bream is separate from the

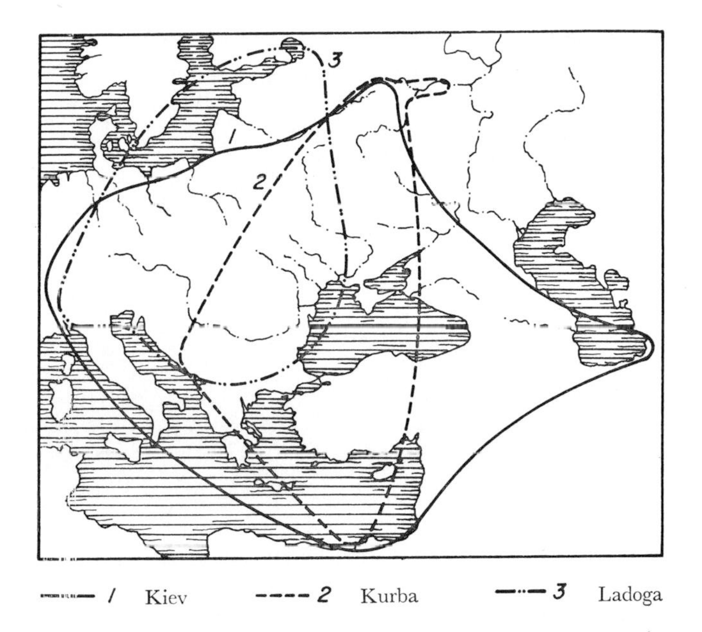

Fig. 161. "Vital area" of Kiev, Kurba, and Ladoga gulls (*Larus ridibundus*) according to data from banding (after Ptushenko, 1948). One can easily see the single "vital area" of all three groups of gulls, but differences in the way each one uses the area are also apparent.

southern one; it lays its eggs in the eastern part of the Syr Darya delta and fattens near Levushkin Mountain approximately 40–50 km away (Fig. 162).

An important condition for preservation of the isolation of populations among migratory animals is the so-called homing instinct, or nest conservatism. It is reflected in attachment to place of birth and the ability to find the road back to it when returning from a seasonal migration or even when accidentally removed. Its basis consists of nonconditioned reflexes (instincts) and individual experience (conditioned reflexes) acquired during postnidal development and during the "education" of the young by the parents. The existence of a homing instinct has been proved by banding birds and mammals. Of 31 flycatchers banded by V. V. Karpov and V. Ya. Parovshchikov (1941) near Moscow and caught repeatedly during succeeding years, 28 returned to the same place, one being found 1.5 km from its nest, one 5 km away, and one 400 km away near Smolensk. A pair of flycatchers was caught at the old place 4 years after banding. Of 11 marked gray flycatchers, 10 were repeatedly observed in summer at their previous location and 1 in its immediate proximity. In two cases birds were caught at the same place for 3 years in a row. Gray gulls banded as

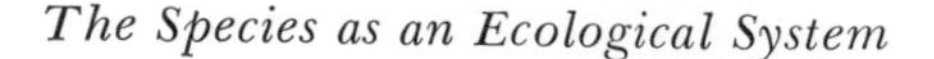

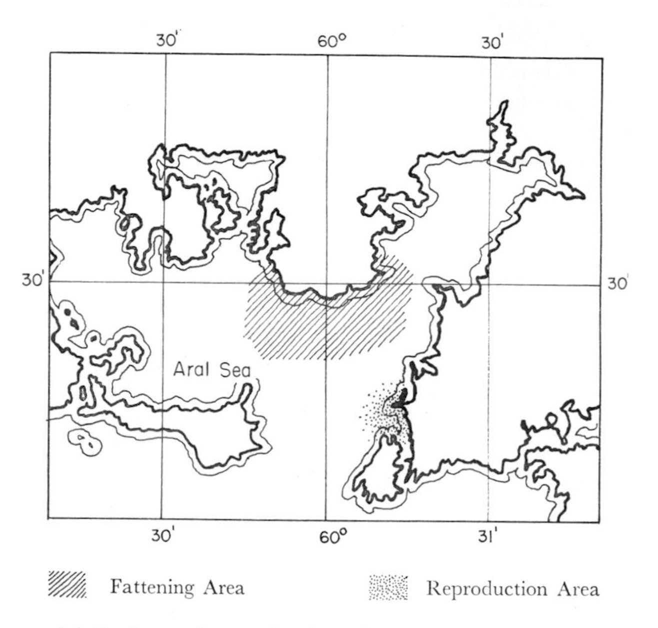

Fig. 162. Areas of fattening and reproduction of an eastern Syr Darya school of bream (after Nikol'sky, 1944).

fledglings on the lower Elbe had not returned to their colony the following year, but later one of the birds was caught in Holland and two were found at the place of banding. About the time of sexual maturation (the third year) all 11 of the birds encountered were found in their nest colony (N. Schulz, 1947). Redstarts have been observed in the same places for 5 or 6 years. A pair of cranes returned to the same swamp for 6 years (Kobylinsky, 1932). An eagle owl *Bubo virginianus* was found after 13 years 13 miles from its original nest, while a pink heron was found after the same number of years in the same colony where it had been banded. Of the adult birds banded and repeatedly caught during the reproduction period, a very high percentage (over 80 percent) returned to the old place (Schüz and Weigold, 1931).

The constancy of flight paths, stopping places, and wintering grounds of specific birds or groups of them has been observed for ducks and gulls. Individual gulls in the thousand-member colonies on Lake Kiyevo (near Moscow) used the same nesting place from year to year (Isakov, Raspopov, and Krumina, 1947). After filling up the Rybinsk Reservoir, the migratory birds first appeared on the same grounds where they had nested earlier. Skylarks have sung over open spaces of water, red-spotted bluethroats have uttered mating cries in the tops of flooded alder trees, and rooks have tried

to head back to their rookeries, although the nests had been covered a meter high by water (Isakov, 1948, 1949).

Starlings, terns, and other birds have been known to return to their nests from distances of 150 to 1,800 km after having been removed in darkened cages and even after having been chloroformed. In B. K. Fenyuk's (1936, 1941) experiments banded mice and voles returned to their burrows from a distance of 2–4 km after artificial removal. Their homing instinct was marked during the reproduction season and in winter, but it was suppressed in autumn, when wanderings take place. This ability is also known among the fishes. Pacific salmon return not only to the same river from which they entered the ocean, but they go to the same tributary and even stop at the same spawning ground where they were born several years before. This is associated with the delicate "feel" and "chemical" memory of the fishes (Schmidt, 1947). Of 13,004 marked fry and adults caught (red, Chinook, humpbacked, silver, and Pacific salmon), 34 percent were caught in the same rivers where they had hatched, 65 percent in nearby ones, and only under 1 percent in distant rivers.

ECOLOGICAL POPULATIONS

Geographic populations are broken down into even smaller spatial groupings, or ecological populations. The latter consist of inhabitants of one habitat (biotope), are less permanent in their composition, and are poorly isolated from one another. Ecological populations enter into the composition of local associations and form different populations. Among common voles in the Moscow oblast' two types of such settlement exist: inhabitants of sparse forests, clearings, and brushy meadows (floodplains) and, interspersed among them, populations of arable lands and small meadow and forest (cultivated land) plots. "Brush-forest" creatures multiply intensively from spring to the beginning of summer (April, May), but by September multiplication is usually over. Protected from predators and unfavorable weather by dense stands of grass, bushes, and an abundance of natural refuges (stumps, blown-down trees, etc.), these creatures suffer low mortality. As a result, there are few fluctuations in their abundance.

Among inhabitants of agricultural lands multiplication goes on almost year round. From spring on it is greatest on clover crops and meadows and in summer and autumn on fields of winter and spring grains. In winter it continues in haylofts, grain piles, and especially bins of fresh grain. Plowing, harvesting, hay-making, and other forms of farm work periodically deprive the creatures of cover and food, forcing them to migrate and thus

raising their mortality rate. As a result, a population's numbers may decrease 300-fold.

The desert rodent—the large gerbil in the Aral region—also forms two basic types of population. The first is accustomed to the clayey feather-grass and cereal-grass desert, where they live only along gorges, ravines, slopes of relict plateaus, and seashores. In this ribbon type of settlement the chain

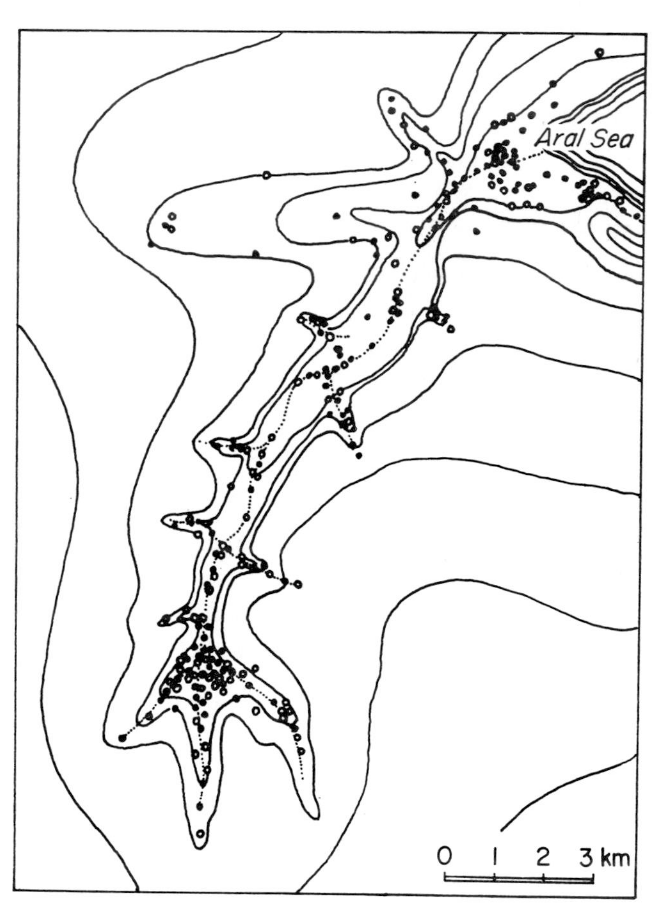

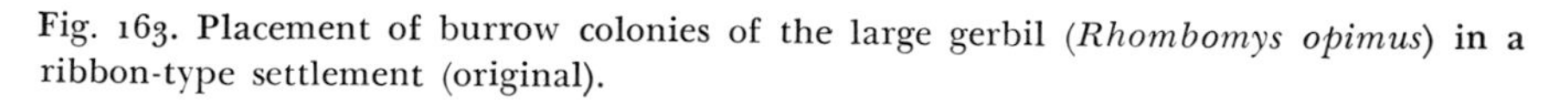

Fig. 163. Placement of burrow colonies of the large gerbil (*Rhombomys opimus*) in a ribbon-type settlement (original).

of colonies is often interrupted on plots of ground unsuitable for habitation (Fig. 163). Because of this, settlements break down into a series of islands or "pieces." When infections dangerous to the inhabitants spread through the settlement, the epizootic embraces only separate parts, and the abundance index in the suffering portion rises comparatively rapidly because of the arrival of newcomers from neighboring ones. Because of this, the total number of gerbils in the ribbon settlements changes within very small limits (Fig. 164).

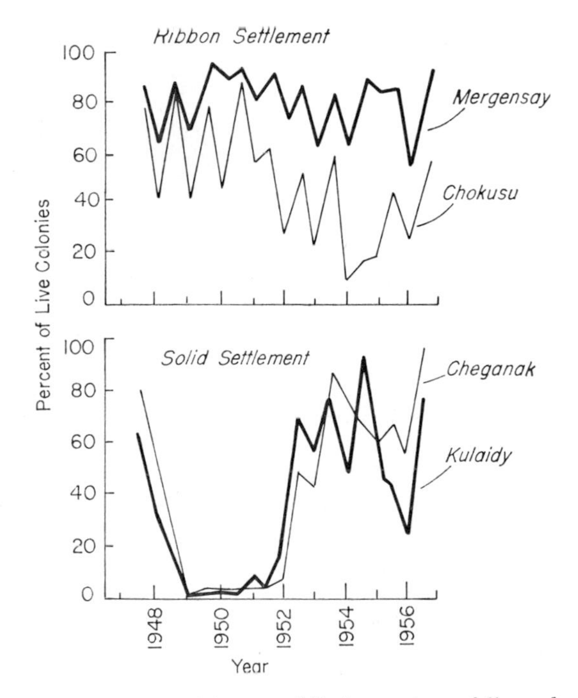

Fig. 164. Changes in abundance of large gerbils (percentage of live colonies) in ribbon-type and solid settlements of the northern Pre-Aral (after Naumov *et al.*, 1960).

The number of gerbils is especially stable in the mosaic landscape of the Pre-Aral Kara-Kum and the ancient plains of the Syr Darya and Amu Darya. Insular settlements, with a very dense population and a large number of creatures, arise there along the alluvial sands. Stable nidi of plague are acclimatized to these spots. In the pit-covered sands of the Kyzyl-Kum, along parts of the train of the Pre-Aral Kara-Kum, and in the Great Badgers and Little Badgers deserts they settle everywhere, and their colonies are scattered in checkerboard fashion, forming solid or diffuse settlements. When infection penetrates into such a settlement, epizootics spread unhin-

dered, embracing large territories and usually accompanied by high mortality rates. Because of this, changes in abundance have a high amplitude here, and whole populations are often killed off. Their re-establishment can come about only through immigration of animals from neighboring populations. Thus settlements of animals (ecological populations) can be divided into permanent and temporary ones. The former exist in favorable habitats, the latter in habitats favorable to existence but in which temporary unfavorable conditions sometimes occur (N. P. Naumov, 1948; Flint, 1958).

Ecological populations are also known among the fishes. The anatomic uniformity of individuals in such groups as the shoal or school has been observed by many investigators. According to N. V. Lebedev (1946), biologically separate groups are formed within the boundaries of a commercial school of fish (geographic population) which occupy definite living areas. Such populations are distinguished by the general rhythm of their vital phenomena, which are out of phase with those of the neighboring groups and are apparently stable, lifelong forms of behavior.[3] They are apparently formed at the hatching grounds but are not confined to one age. The Caspian roach forms schools in autumn which consist of fish of a single size. They do not usually mix with others and form a separate unit.

Among the insects ecological populations are no less distinct, differing clearly in the density of the population and in the times and rhythms of biological phenomena. The tick *Ixodes persulcatus* is significantly more abundant in the forest than on meadows, where cattle are infested with them two to five times less often. This is associated with the less stable temperature and lower humidity of the meadows, which lowers the viability of the ticks. Thus 95–100 percent of the unfed females survived in a forest from 16 June through 4 September, but only 62 percent survived on the meadow. The unfed females were more active in the forest than on the meadow (Fig. 165). Engorged females on the meadow died faster and produced fewer eggs, while their larvae hatched 7–8 days earlier than in the forest. In the forest larvae clustered quickly together to prepare for the winter, while in the meadow they crawled apart, and their mortality reached 75 percent, whereas it was only 25 percent in the forest.

Ecological populations of nearly immobile insects may develop as a result of dispersions even within the bounds of a small territory. In any gorge the population of dandelion weevils on the sunny slope multiplies 15–20 days earlier than the population on the northern slope, and crosses between them do not occur (Gilyarov, 1941). In the steppes populations of melon worms on the salty soils and sands are so different that their members are structurally distinguishable (Arnol'di, 1957). Many ecological populations are distinctly different in ecologo-physiological properties.

[3] Lebedev calls them "elementary populations."

The mollusks *Neptunea arthritica* and *Littorina squalida* on the southern Sakhalin coast in biotopes with changeable living conditions are more resistant to desiccation and to fresh water (Fig. 166) than the inhabitants of optimal waters (Golikov, 1959). The tempo of growth, speed of maturation, and size of fish from different lakes are usually not the same (Yurovitsky, 1958). The eggs of *Coregonus albulus* are distinguished by their size, amount of carotene and melanin pigmentation, and adhesiveness. Smaller size and increased pigmentation are adaptations to living in lakes poorer in oxygen. High adhesiveness and a long survival period of the eggs are

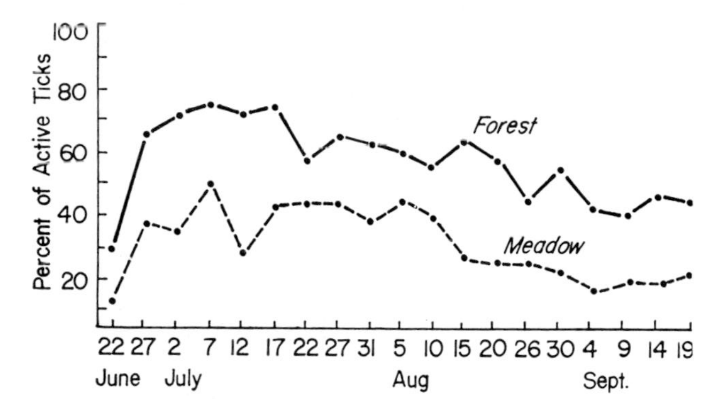

Fig. 165. Activity level of unfed female ticks (*Ixodes ricinus*) in a forest and on a meadow of the Karelian ASSR (after Lutta and Shul'man, 1958).

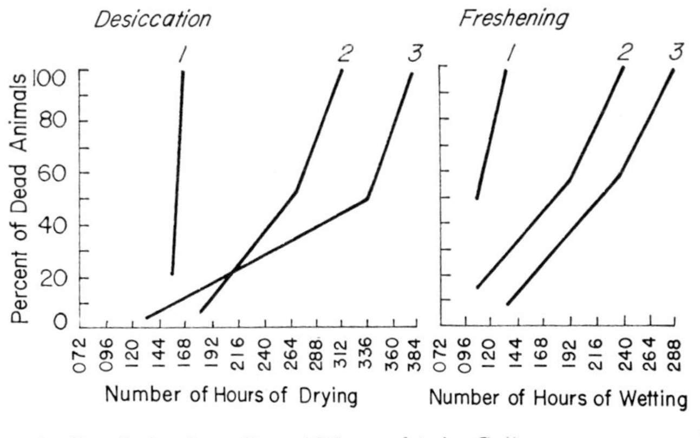

1 Population from Coastal Waters of Aniva Gulf

2 Population from Eastern Part of Busse Lagoon

3 Population from Undisturbed Part of Busse Lagoon

Fig. 166. The ability of a population of the mollusk *Littorina squalida* on the southern Sakhalin coast to tolerate desiccation and freshening of the water (after Golikov, 1959).

characteristic of populations living in lakes with faster-flowing water at the spawning grounds (Nikanorov, 1959).

Thus ecological populations do not occupy a geographic region as a whole but are accustomed to the various types of habitats within it. They are closely associated with one another and often exchange members, allowing the re-establishment of populations which for one reason or another have died out. Only some ecological populations are wholly capable of reproduction. Others either exist uninterruptedly because of constant replenishment by immigrants (semi-independent populations) or else die off entirely and are re-established by immigrants (dependent populations) (Beklemishev, 1960). The variety of ecological populations and the number of different types of biotopes they occupy may serve as an index of the ecological plasticity of the species. Adaptation to various conditions of existence and a variety of populations insure the species against extinction.

ELEMENTARY POPULATIONS

Elementary populations are the inhabitants of certain stages in nonuniform complex biotopes. They are known for birds (Isakov, 1948), for example, geese and ducks which nest together in combined flocks before migration and remain in stable form during flights and on the wintering grounds. Such groups not only nest together but also have definite flight routes, rest areas, and wintering grounds, which are often retained from year to year. The same stable flocks are also observed in the white-throated sparrow *Zonotrichia albicollis*, the American junco *Junco hyemalis*, *Spizella arborea*, *Melospiza melodia*, and other birds. They are united by the above-mentioned nesting conservatism; however, it is not impossible that individual birds or even whole flocks may separate themselves from such a group. Elementary populations usually have no stable structural differences but may be distinguished by their behavior. Among Japanese thrushes near Leningrad, A. S. Mal'chevsky (1958, 1959) found seven variants of spring songs among males, each of which was characteristic of a group of birds which lived in a definite area (the Old Peterhof Park, Gatchina Park, Hunter's Park, the park of the Forestry Academy, etc.). However, he also observed yearly changes in the character of the songs in the individual bird populations, which attests to a relatively weak isolation of such elementary populations and a broad exchange of individuals among them.

Among animals living in mosaic biotopes the only kinds of populations which can exist permanently are elementary ones occupying favorable

areas (survival stations). The remaining territories (dispersal stations) are settled temporarily, and in an unfavorable situation the animals there die or emigrate (temporary population). In dispersion part of the individuals may wind up in unfavorable habitats and may even maintain themselves more or less continuously. Populations of the polychaete *Nereis succinea* exist in the freshwater portions of the Caspian, for example, where it cannot reproduce. It is necessary to distinguish populations of various phases (stages of development) if they live in different biotopes, such as the amphibionts. V. N. Beklemishev (1960) has suggested calling them "hemipopulations." The different fate of each elementary population depends on the microclimate, the available food, protection from enemies, and the uneven dimensions of the mortality and birth rates at these stations. Continuity of existence and the stability of numbers of ecological populations usually depend on the number and quality of survival stations or permanent populations.

INTERACTING POPULATIONS

A species is an integral (single) formation, and its separate populations are connected with one another. These connections between populations of the same or different ranges are established with the aid of animal displacement (various types of wandering and dispersion), which are just as important to the survival and evolution of a species as reproduction. They provide for encounters between males and females and remove the danger of long-term, close-relative mating (inbreeding). Redistribution of animals comes about in conjunction with displacement of food supplies, more favorable microclimatic conditions are found, and, finally, animals disperse into new habitats or new regions; i.e., the range of the species is maintained and, if possible, expanded.

This is abetted by all types of migration, including those which are caused by changes in the surrounding environment and are represented by periodic (seasonal, daily) or nonperiodic wanderings arising out of seasonal changes in the weather, insufficient "harvests" of food, droughts, floods, steppe and forest fires, and many others (see Chapter 9). But a special place in the maintenance of species integrity belongs to the migrations associated with reproduction: regroupings of individuals at the beginning and end of reproduction and dispersal of the adolescent young. It is obvious that these movements, too, are associated with surrounding conditions, but not directly, only tangentially. They are caused by a change in the state of the animals and the intraspecies relations associated with them.

Independently of the distance of seasonal migrations and despite the preservation of a definite autonomy and integrity in the moving groups forming the migratory flock or cluster, individuals regroup during such migrations. Banding has shown that the individual birds marked at their nesting grounds are sometimes encountered in following reproduction seasons in places far from their "homeland." Ducks banded during the mating season in western Europe (England, Iceland) have been caught in the USSR, while those banded near Astrakhan have been captured in Yugoslavia, Czechoslovakia, and several other countries. Scissortails banded in February and March in Cairo wintered the next year in Daghestan (Mikheyev, 1948; Treus, 1957). Thus, despite the existence of nest conservatism, individuals do change places during seasonal migration and on a population's wintering ground, even though it is rare.

The multidirectional migrations of polar foxes over hundreds of kilometers are well known (see Fig. 99). These migrations produce such great population displacements that the formation of anatomically stable geographic forms is difficult or impossible. Migrations of wild northern reindeer are distinguished by their great regularity. Individual herds consisting usually of different sex and age groups (brood does with offspring, adult bulls) migrate. Observers have noted great consistency in migration routes, especially at the points where large rivers are crossed, but because of differences in amount of grass cover and changes in pasture location, the routes do change, leading to a mixing of different herds (Middendorf, 1869). Autumn-migrating flocks of blue titmice, goldfinches, siskins, crows, and daws occur while other groups of birds of the same species remain sessile, which opens possibilities for exchange of individuals. A mixed population of local and migrant *Parus atricapillus* exists during autumn and spring (see Fig. 149). Consequently, such migrations aid in regrouping the populations.

Wanderings associated with reproduction and dispersion of the young have a special meaning for intraspecies connections. So-called dispersion of the young is characteristic of all living organisms, since each species, regardless of whether its way of life is motile or sessile, exists in stages which move either actively or passively (larvae among animals, seeds and spores among plants). Among aquatic animals the young sometimes cover huge distances in dispersing. Such is the passive migration (with the current) of river-eel larvae, which multiply in the Sargasso Sea. They reach the coasts of America, Europe, and Africa, where the adult eels live, only in the third year of life (see Fig. 97). The route of the young Norwegian herring is complicated. From the spawning grounds along the southwestern coast of the Scandinavian peninsula the hatched larvae are carried by the current of the Gulf Stream to the coasts of central and northern Nor-

way. From there the young begin moving actively to the south, requiring 5 years. Every year schools of young fish approach the banks and then disappear back into the open sea. And each year the distance which the fish travel from the coast to the fattening grounds increases. At the age of 5 the now-mature fish come to the spawning grounds and then return there each year during the egg-laying period. The migrations of *Scomber scombrus* in the Black Sea are also of great magnitude (Fig. 167).

Distant resettlement of reptiles is also associated with dispersion. Cases are known in which the crocodile *Crocodilus porosus* swam across 900 km of ocean. *Natrix natrix*, the grass snake, has been encountered 5 km out to sea. Distant displacements (by wind) are known for many insects. Swarms of such good flyers as the dragonfly *Pantula flavescens* often appear on

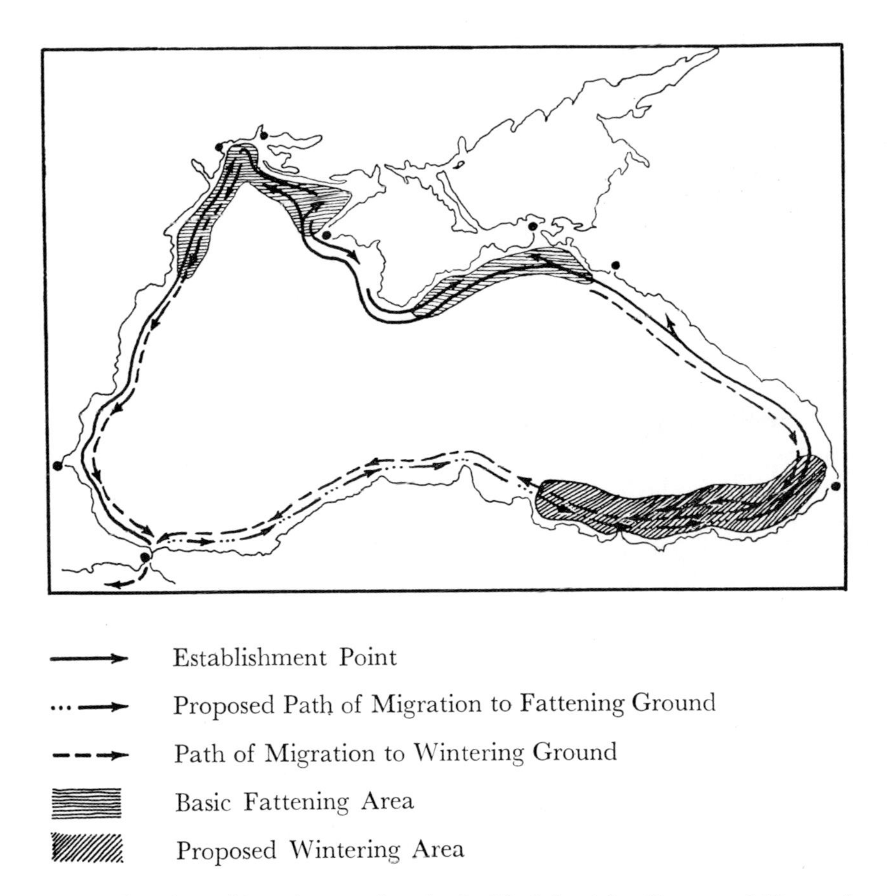

Fig. 167. Migrations of *Scomber scombrus* in the Black Sea (after Krotov and Vinogradov, 1940).

islands in the Indian or Pacific oceans, which they reach from the continents. Flights of several hundred kilometers are made by the *Schistocerca* (desert) locust, which flies from the Arabian Peninsula, India, and Iran into the borders of the USSR. It flies along definite routes, avoiding mountain regions with heights over 1,500 m above sea level (Fig. 168). In the majority of cases, however, insects cover small distances. The short flights of newly emerged malarial mosquitoes usually do not exceed 12 km and occur in all directions away from the hatching area (Fig. 169).

Among monocyclic birds and mammals (those which reproduce once a year) the family breaks up in autumn (moles, susliks) or spring after a common hibernation (marmots, many birds). Among species which reproduce several times during the year (polycyclic animals) dispersion of the young is a drawn-out affair. Nest conservatism in the small passerine birds is manifested distinctly among the adults but is practically absent among the young. The overwhelming majority of the latter do not return to the banding place but spread far into the environs, often over great distances.

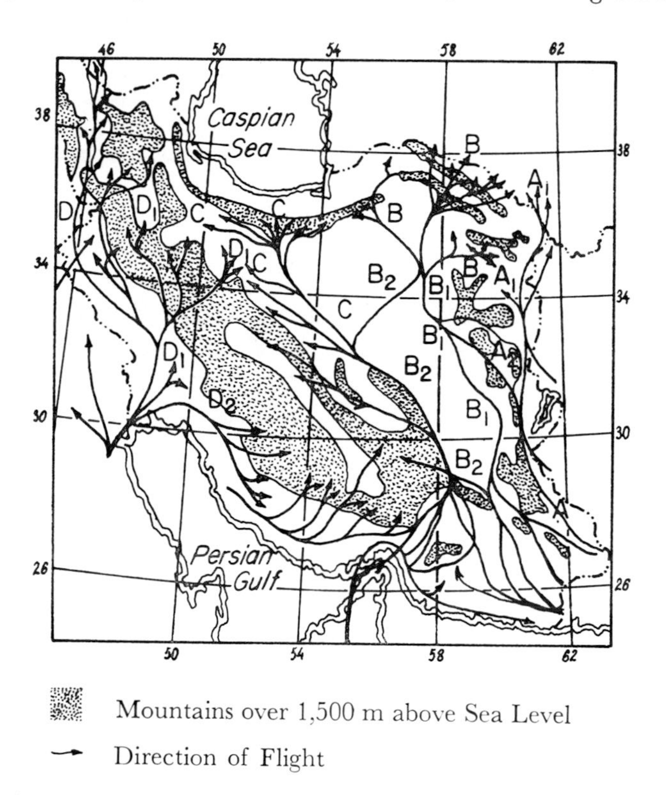

Fig. 168. Migratory routes of sexually mature desert locusts (*Schistocerca gregaria*) in Iran (after Predtechensky, 1935). Letters identify various migration routes.

Of eight adult birds banded at the Saval Forestry Farm (forest steppe), four were observed at the same spot during succeeding years, while only five of 1,229 banded fledglings returned (Mal'chevsky, 1958, 1959). Among

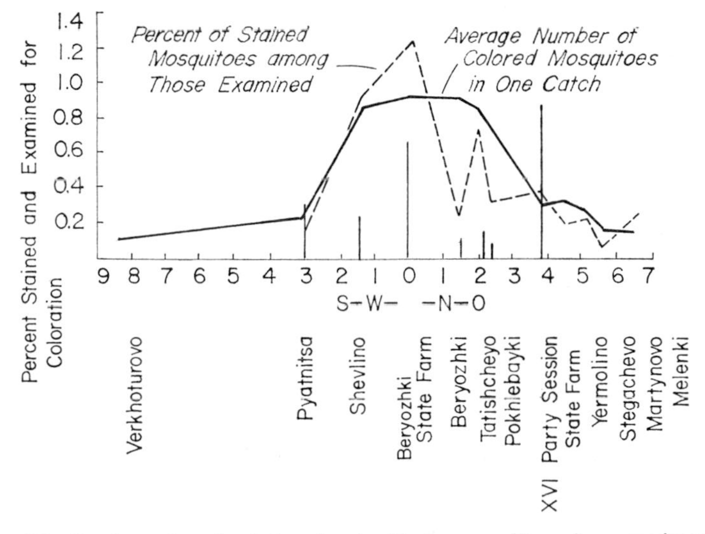

Fig. 169. Distribution of stained female *Anopheles maculipennis* mosquitoes around "The Little Birches" State Farm during daily sortics (after Ivanova, from Beklemishev, 1944). Columns represent herds of cattle in the villages.

TABLE 46. *Migrations of the common vole* (Microtus arvalis) *in the southern Moscow oblast' in relation to density of the animals and intensity of their reproduction (after N. P. Naumov, 1955).*

Time of Observation	*Average Number of Animals per hectare (Population Density)*	*Average Number of Migrant Animals Caught in Trapping Ditch in 10 Days*[a]	*Coefficient of Mobility*[b]	INTENSITY OF REPRODUCTION: *Percent of Females Participating in Migration*	INTENSITY OF REPRODUCTION: *Average Size of Litter*
June–Sept. 1946	11.5	3.8	0.33	66	5.6
June–Sept. 1947	4.9	2.1	0.43	60	6.4
June–Sept. 1948	45.2	6.4	0.14	57	5.9
June–Sept. 1949	2.7	1.1	0.41	70	6.3

[a] Catching ditches are trenches as wide and as deep as the head of a spade and run for 50 m. The bottom of the trench is dug out and filled with evenly spaced tin cans 70 cm deep and 30–35 cm in diameter.

[b] Coefficient of mobility is the ratio of catches in the ditches (third column) to number of animals, or the density of their settlement according to the data in the calculation (second column). The coefficient actually expresses the ratio of the mobile part of the population (migrants) to the sessile part.

the mammals there are no directed migrations in colonies of small susliks and southern gerbils (Fig. 170). Movements of small susliks and gerbils have been recorded over distances of 3–5 km (Kalabukhov and Rayevsky, 1935; Fenyuk, 1941), of long-tailed susliks (*Citellus undulatus*) over 180 km (Zverev, 1929), and of squirrels over 240 km (Salmin, 1938). Young moles make trips of up to 2 km, while adults go no farther than 700 m (Pavlinin, 1948). In uneven (insular or ribbon-type) animal colonies the wanderings are usually directed along the ribbons or from one island to another along routes favorable for movement (Fig. 171). On these wandering trails the intensity of movement is especially great wherever several "roads" come together, and therefore the burrows here are almost never empty; when their occupants die, they are quickly taken over by newcomers (Fig. 172).

Although weather has considerable influence on such movements (Fig. 173), they depend basically on the number of animals and the intensity of their reproduction. Movements of common voles during the reproduction season increase as their population grows sparser and their fecundity increases (Table 46). We can judge the connection between these wanderings and reproduction by the sex and age statistics of the migrants. In a group of sexually immature voles the numerical ratio of males to females is 1:1. In a group of mature and especially adult creatures the males predominate, their mobility being higher during the mating season than that of the females (Fig. 174), which remain close to their burrows at this time. The migrants consist largely of adolescent creatures seeking new shelters (Table 47). High mobility of the males during the mating season has also been observed among large gerbils in the northern Pre-Aral deserts.

TABLE 47. *Age makeup of sessile and migrant portions of a population of common voles* (Microtus arvalis) *in the southern Moscow oblast' (after N. P. Naumov, 1955).*

Method of Catching (character of material)	*Not Sexually Mature (under 12 g)*	*Adolescent (13–25 g)*	*Sexually Mature (above 25 g)*
Creatures caught in ditches (migrant)	25.7%	63.8%	10.5%
Creatures caught by plugging burrows (sessile)	60.0	20.2	19.8

The movement of mating adults and the dispersal of adolescent young are associated with changes in the relationships of individuals in the families and litters. These movements do not cease when populations are low but, in fact, seem very often to be more intense. The differences in mobility of individual sex and age groups make a good deal of biological sense. As a rule, the first stages of development—larvae and very young

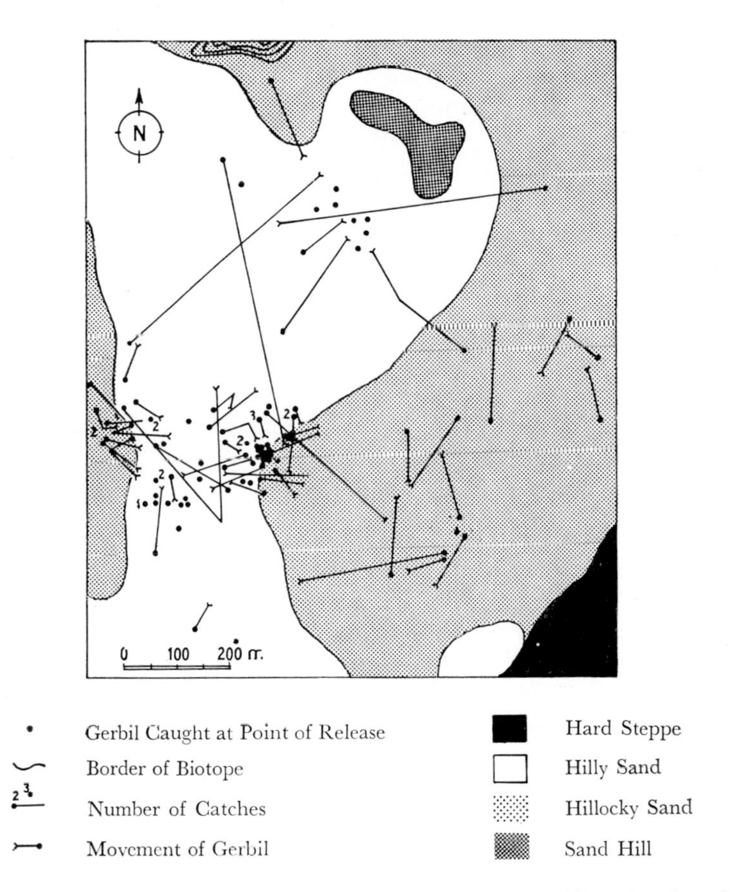

Fig. 170. Movement of banded southern gerbils (*Meriones meridianus*) released at a banding point in the Volga-Ural desert (after Fenyuk and Demyashev, 1936).

individuals which are not yet hardy—are not very mobile. Mobility increases as the animals approach maturity (dispersion). Dispersion and seizure of new territories are always undertaken by the most numerous group in the population, the adolescent young. They are very active, but they also have to bear the greatest losses (Fig. 175). After they have reestablished themselves following dispersion, both mobility and mortality drop sharply. They are minimal during the mating season among reproducing females and high among sexually mature males (N. P. Naumov, 1951).

Nonperiodic migrations arise when there are unfavorable changes in the conditions of existence and especially when there are increases in the numbers of animals. We are well acquainted with the mass wanderings of

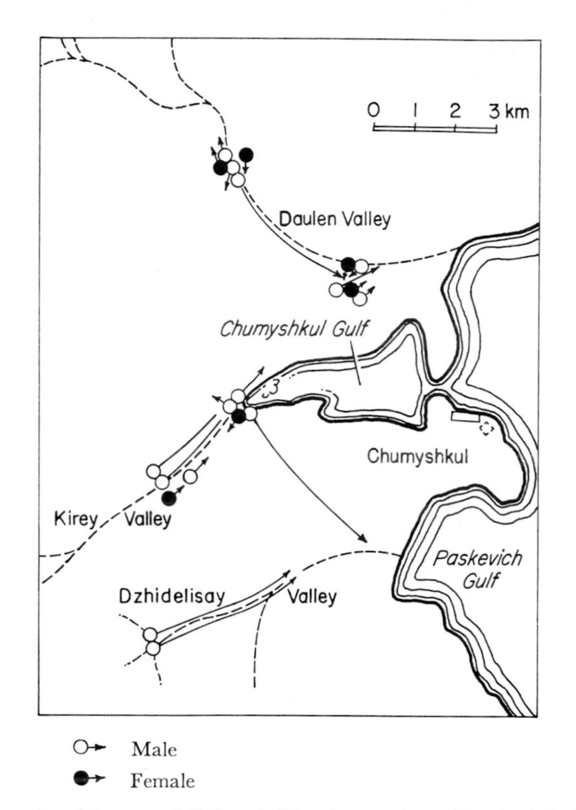

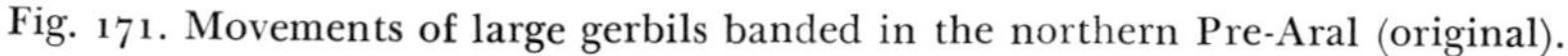
Fig. 171. Movements of large gerbils banded in the northern Pre-Aral (original).

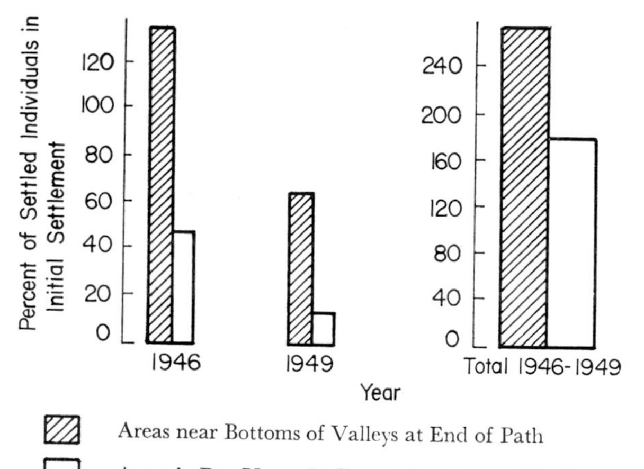

Fig. 172. Settlement by large gerbils of repeatedly rounded-up areas situated on various sections of the animals' migrational path (original).

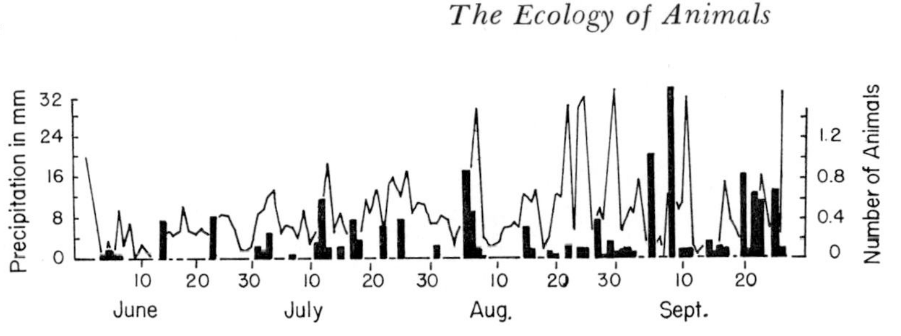

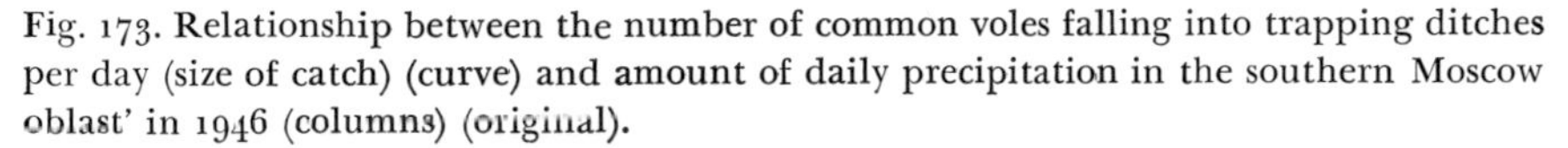
Fig. 173. Relationship between the number of common voles falling into trapping ditches per day (size of catch) (curve) and amount of daily precipitation in the southern Moscow oblast' in 1946 (columns) (original).

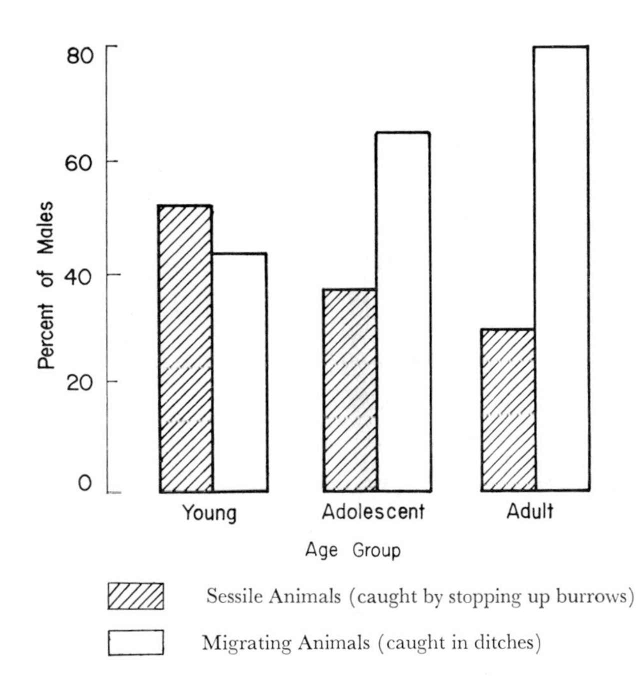

Fig. 174. Percentage of males in various age groups of migrating and sessile voles (original).

crossbills, squirrels, polar foxes, etc. when food is insufficient or unavailable. In droughts even the moles abandon their subterranean passages and move over the surface of the soil for long distances. The freezing of small streams and rivers during the cold, snowless winter of 1944/45 in the Urals was the cause of a mass out-migration of the usually sessile *Cinclus cinclus* (Kirikov, 1952). The often-encountered correlation between food shortage and the number of animals intensifies migrations of this type. Elemental deprivations in the form of fires, storms, floods, etc. also cause mass migrations of animals.

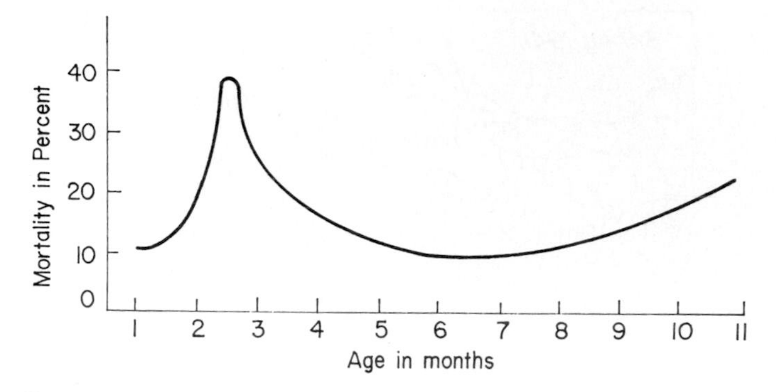

Fig. 175. Changes in mortality curve for common voles in the Moscow oblast' (original).

Nonperiodic migrations and resettlement of animals during rises in abundance expand, usually temporarily, the area of the species's habitation. Such "pulsations" in the borders of a range are known for a wide circle of species (Fig. 176). Resettlements sometimes lead to a stable expansion of a range. Such "jumps" in the borders have been described for the European hare, lemmings, and certain species of insects (dragonflies, locusts, mosquitoes, and others). Among the migratory locusts they occur in years of mass reproduction and when the solitary phase matures into the gregarious one. Among malarial mosquitoes mass migrations are also associated with "hyperproduction" (oversettlement) of nidi; these migrations are usually taken care of by the wind and are notable for their great distances. Their connection with external conditions has not been sufficiently explained (Beklemishev, 1944). The well-known mass emigrations of the sand grouse, Siberian nutcracker, crossbills, and certain other birds are caused by snowy winters, droughts, and crop failures. The blue kingfisher *Alcedo attis* of the Baltic coast increases in numbers after a series of warm winters but completely disappears during repeated severe ones. Subsequent settlement of the Baltic coast always comes from the south; "young" populations have one brood a year, while the "hardy" birds, i.e., those which have existed there for several years, breed twice a year.

By emigration from an optimal zone the population of the peripheral area of the range is maintained; this is the area where the high mortality rate of individuals is not balanced by reproduction. The stable existence of local populations is possible only when animals immigrate from other parts of the range, which is especially significant in years of a rise in abundance and mass wandering. This is the way populations of animals are maintained in circumstances which are far from optimal, and it is here that the species develops adaptations to new conditions. Actually, peripheral populations differ in several ways and in several ecologo-physiological

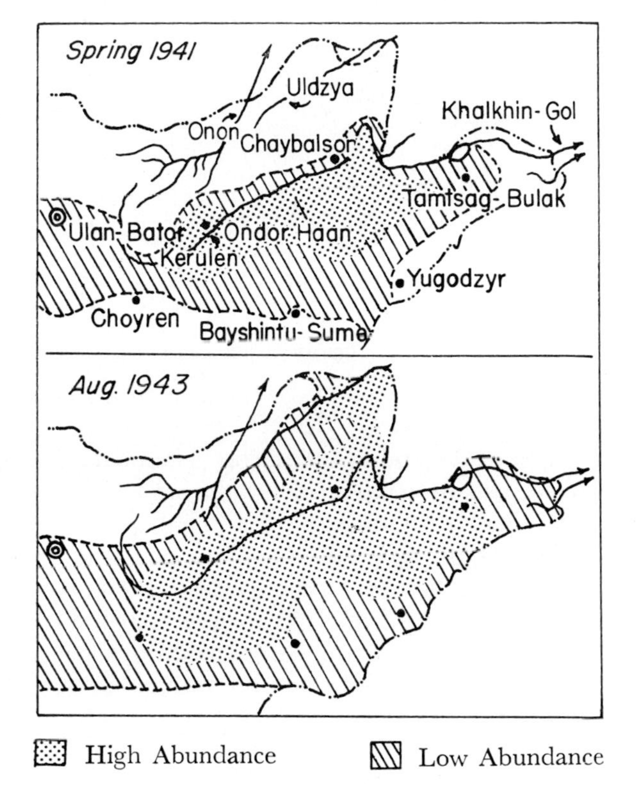

Fig. 176. Expansion of area of habitation of Brandt's vole (*Microtus brandti*) in Mongolia during mass mating (after Kucheruk and Dunayeva, 1948).

properties from populations living in the optimum range. It is known that their broods (among birds and mammals) are larger, their daily activity period longer, and their rest period shorter. There is also a growth in the size of individual areas and a change in feeding habits (increased variety). All of this constitutes an adaptation to areas where the death rate is high and reproduction is insufficient to overcome unfavorable conditions (N. P. Naumov, 1945, 1956). Mass migrations which maintain the population level at the periphery of the range aid in the adaptation of a species to conditions which are new for it; by the same token, they aid in expanding the range. Consequently, they are one of the important mechanisms providing for the preservation and integrity of the species.

Examination of the structure of the species allows us to conclude that territorial groupings of individuals, though autonomous, still share definite relationships with one another. These connections are established with the aid of migrations. Definite times, routes, and order of migrations

and similar rhythms for other biological phenomena connect individuals living on geographically similar territories into a major group, the geographic population. In such a group cross-reproduction is not only possible but actual, and it is motivated above all by animal movements. Each population occupies a "zone of similar favorability" and is to a known degree isolated from others. However, isolation is periodically disrupted, especially when mass migrations occur. These maintain settlement patterns or replenish extinct settlements in regions unfavorable to the life of the species. Even smaller is the isolation of ecological and especially elementary populations, among which only a part lives continuously. Thus a species population lives as an integral whole, having a definite, historically formed spatial structure and organization.

CHAPTER 12

The Content of Populations

SPECIES POLYMORPHISM

A population, especially one occupying a major territory (subspecies, geographic population), is not usually uniform. It is, rather, multiform, since it consists of various age groups (generations), seasonal phases (among cyclomorphic animals), males and females (among species which reproduce sexually), or forms with various functions (among the social insects). It may include phases which depend upon population density (among the migratory locusts). Finally, it may consist of so-called biological races, typified by character of feeding, reproduction, or other biological properties.

Some of these groups may change from one into another (age phases, seasonal phases, and phases dependent on population density), but all of them are biologically different, since they have more or less stable morphophysiological and ecological properties which give them a definite autonomy and sometimes a separate existence. Simultaneously, all of them are closely connected with one another, not only by common origin but also by numerous mutual adaptations, which S. A. Severtsov called congruents. These latter unite such different individuals into an integral ecological system, the species. They include anatomic properties providing for encounter, reproduction, and raising of young (sex organs, sense organs, fluorescence organs, secondary sex characteristics, mammary glands, marsupial pouches on mammals, etc.); the mutual adaptations of multiform individuals among colonial animals; adaptations providing for the dispersion or association of individuals in a flock, brood, or any other group (the white flash of reindeer and hares, signal coloration of many animals, migrational instincts, etc.; various odorous substances excreted by animals

and serving for mutual attraction, sometimes for demarcation, or for a warning to stay out of an occupied refuge or portion of territory). These "telegrons" are widely used among animals, and they occur among plants and microorganisms (synergistic substances or antibiotics) (Kirshenblat, 1956).[1] Finally, the means of mutual notification, communication, or dispersion include ordinary sound, ultrasound, light, and magnetic waves which the animals perceive.

BIOLOGICAL RACES

We define as biological races those groups of individuals of one species which are distinguished by feeding habits, i.e., which occupy different food niches or which, living together, still reproduce separately (usually at different times), such as the winter- and springtime races of fish. All of them, as a rule, are not isolated in space but differ in their biological cycles. In several cases forms which are widely separated in the ecologo-physiological sense cannot be distinguished anatomically and are referred to as one species merely because of this feature. Actually, they do not cross; they are "biological species," such as N. A. Kholodkovsky (1908, 1910) described for the aphids *Chermes* and *Adelges*, or "twin species," like the malarial mosquito *Anopheles maculipennis* and others.

The rise of biological races through the splitting off of groups of individuals into different ecological niches is one means of species divergence and species formation, a means still studied very little. It differs from the geographic form of species formation in that the individuals forming the different biological races may live right next to one another, separated only by their association with different foods and by their different periods for biological phenomena (especially mating). It is the latter which especially hinders their cross-breeding.

One example of biological forms differentiated by their mating seasons is the herring *Clupea harengus* of the northeastern Atlantic, whose races are easily distinguished morphologically. The so-called banok herring living in the southern part of the North Sea spawns in August and September, while the "spring herring" mates from March through May. The spawning grounds of both races are often the same, and outside the mating season these herrings even form mixed schools on the feeding grounds. But the impossibility of cross-breeding preserves the differences between the populations.

[1] All these substances play an important role in not only intraspecies but also interspecies relationships, where their significance is especially great.

It is well known that there are winter and spring races of salmons, sturgeons, and several other fishes. The former need a rest period (diapause) at low temperatures for maturation of the sexual products; after leaving the sea in autumn, they pass the winter in an inactive state in rivers. Only in spring do they continue their progress to the spawning grounds along the upper course. Spring races leave the sea with their sexual products almost mature and go directly to the spawning grounds, completing their spawning in the same year (Berg, 1935, 1953). The initial phases of such a division may be observed among seagulls, pelicans, herons, and other colonial "shoreline" birds whose colonial grounds are not free of meltwater or floodwaters at the same time, forcing the birds to nest at different times. The differences may amount to a month or more, leading to the rise of separate flocks. A similar picture occurs in the reproductive patterns of certain groups of insects living on plants which grow in sunlight or shade and which bloom at different times.

Among the *Chermes* aphids described by N. A. Kholodkovsky the biological species *Chermes abietis* and *Adelges lapponicus* arose through changes in the cycles of development and reproduction. Unlike the usual cycle of development, theirs occurs without a shift to other plants. Only one food source (fir) is used, and multiplication is parthenogenetic, a distinct anatomic and biological difference. The absence of cross-breeding between individuals has isolated these populations from the closely related species *Chermes viridis* and *Adelges strobilobius*, from which they obviously descended. Forms of aphids living on secondary food plants may also become single representatives of the species. The aphid *Thecabius* lives on poplar and ranunculi, but in areas where poplar does not exist, it forms a type with an incomplete cycle only on ranunculi. This form cannot be distinguished from the phase of full-cycle *Thecabius* on ranunculi in areas where it can run a full cycle. Similar cases are known for the aphids *Eriosoma*, *Tetraneura*, *Myzodes*, and certain others (Mordvilko, 1926, 1939). Often man causes the appearance of incomplete-cycle forms of aphids by destroying certain of their food plants (Kozhanchikov, 1956).

The widespread ichneumon fly *Aphitis maculicornis* is represented by a complex of forms, the adult phases of which are indistinguishable. But the "Persian" and "Indian" races of this species multiply sexually, while the "Spanish" and "Egyptian" varieties are parthenogenetic; they also differ in other biological properties. The same picture of complex intraspecies differentiation and the existence of biological races can also be seen in other well-studied species of entomophagous insects (Rubtsov, 1960).

Differences in mating often send different races into different ecological niches and make their cross-breeding impossible. But the initial stages usually consist in a difference in feeding habits. Many biological races

differ in this feature; among insects they are referred to as "food races" and among parasites as "host races." I. V. Kozhanchikov (1956) recognized four types of intraspecies biological differentiation among plant-eating insects:

1. Feeding adaptations, which arise easily but also disappear easily (common among omnivorous species).
2. Intraspecies biological forms differing in their feeding habits and place of residence. They are usually separated (often territorially) in nature, but under experimental conditions they can often be combined. They are very different anatomically (insects, internal parasites of plants).
3. Intraspecies biological forms similar to the previous ones but not well differentiated anatomically and markedly different biologically (characteristic of oligophages).
4. Biological species (after N. A. Kholodkovsky).

Over the course of several generations the plant preferences of polyphagic insects may be reconstructed through "nutritional rearrangement." However, such reconstructions are usually reversible physiological adaptations and quickly disappear when continuous training is dropped. But the forms which have a stable preference for the new food (stenophages, especially monophages) appear as a result of the long-term action of natural selection; new mechanisms of digestion and other ecologo-physiological properties appear (Kozhanchikov, 1958; Rubtsov, 1960). The willow-leaf eater *Lochmaea capreae* has two forms—the willow type, which feeds on willows, and the birch form, which feeds on the leaves of the downy birch (*Betula pubescens*). They are practically indistinguishable anatomically, but the willow variety cannot live on birch and vice versa. Moreover, they differ in their temperature preferences and do not willingly cross-breed, even under experimental conditions (Kozhanchikov, 1946). The rice (small) and wheat (large) forms of the beetle *Calandra oryzae* do not interbreed and apparently belong to different species (Birch, 1944).

It has usually been felt that the impetus toward formation of food races was a forced switch to an unaccustomed food. Observations have shown that the matter is not quite so simple. The basis of the appearance of a new form with a different food preference is the adaptation of separate individuals stemming from a reconstruction of digestion which leads to a new food demand. Among insects it is common for the new food specialization to arise with greater difficulty as the species becomes more omnivorous (Kozhanchikov, 1956).

Food races of a species are usually anatomically different and are described as different species. The armored acacia beetle *Eulecanium corni* lives on acacia, pear, almond, blackthorn, etc. The acacia population differs markedly from the others in its size and in many of its structural fea-

tures. Moving such an insect from a pear tree to an acacia makes the offspring completely similar to the acacia individuals, but moving acacia beetles to the pear tree is not successful, since they die. The variety of the willow beetle *Chionaspis salicis* which lives on mountain ash and which differs from the typical variety can be easily transferred to the willow tree *Salix nigricans*. There the offspring acquire the characteristics of the willow population, but their offspring cannot be returned to the mountain ash (Rubtsov, 1952).

The apple aphid *Lepidosaphis ulmi* exists in two biological forms, sexual and parthenogenetic, differing in their manner of reproduction. They prefer various food plants, although in experiments one form can be forced to develop on the preferred plant of the other. The bisexual form is more plastic and viable. In certain zones the biological forms of this insect even have different yearly cycles. Copulation between males of the bisexual form and parthenogenetic females is possible but very difficult (Dantsig, 1959). Both forms are an example of almost complete differentiation and transformation into separate species. Differences in feeding habits have been supplemented here by the parthenogenetic behavior of one of the forms, and this provides reproductive isolation for both of them. Attempts to transform both of them by altering their food plants and isolating the females of the bisexual form have not been successful; the forms have proved to be completely resistant.

Plant nematodes, as a rule, prefer to feed and live on the same species of plant host on which their parents lived. This tendency is strengthened over several generations, and finally the specialization reaches an extreme degree of narrow specialization. Sometimes a food race is formed quickly. Man has witnessed the penetration of *Carpocapsa pomonella* into the California walnut groves, where it adapted to its new host. Now each of the forms (walnut and apple tree) of this parasite is unable to exist on the other host.

The same types of food races are known for parasites under the name of "host races." The human body and head lice are well known. They are reversible: the head form of the human louse will turn into a body louse in several generations if allowed to live on open human skin. It will then differ in size, fecundity, and behavior; the instincts of the head louse change when this happens, becoming the instincts of the body form. Individual populations of *Trichogramma evanescens* are specialized for definite hosts, and each develops primarily on one of them. These forms are differentiated physiologically and biologically rather than anatomically. They prefer different hosts, temperatures, and humidities and have different numerical relationships between the sexes in their populations and different levels of fecundity (Rubtsov, 1960). The common ascarid of man

and domestic animals (swine) has two types which cannot be differentiated anatomically but which are incapable of cross-breeding.

Biological races are known among the nest-stealing parasite birds. The cuckoo *Cuculus canorus canorus* has biological races associated with its various hosts. In the European USSR there is a numerous and stable race which parasitizes white wagtails. The cuckoos which lay light blue eggs parasitize *Phoenicurus ochruros* and *Saxicola rubetra.* In the south European USSR there is a common group of cuckoos which lays its eggs in the nests of *Acrocephalus arundinaceus.* Cuckoos associated with the mating habits of the Sylviidae and *Lanius cristatus* are common in western Europe but rare here. *Fringilla montifringilla* and *F. coelebs* are the basis for races of cuckoos and are acclimatized to the northwestern regions of the USSR; their numbers diminish toward the south. Cuckoos parasitizing the nest of the gray flycatcher have so far been recorded mainly in the Baltic area. In addition to the above-mentioned nurseries of young cuckoos, we know of several dozen (48, including the above) species of birds in whose nests cuckoos lay their eggs, but these are all rare or accidental hosts.

The individual biological races of the common cuckoo apparently have different centers of establishment and different routes of distribution. The ranges of cuckoo populations belonging to the different races usually overlap one another quite widely. In areas where the ranges coincide, the different groups of cuckoos usually stay in different biotopes and concentrate around the different species of birds. This maintains the necessary separation of the various racial populations (Mal'chevsky, 1958).

GENERATIONS

Age-based populations (generations), which live apart in some species, are of quite a different order of intraspecies grouping. Such are the migratory fishes, especially those which spawn once in a lifetime (the Far Eastern salmon: red, Chinook, and humpbacked, or pink). The pink salmon is known to alternate between its strong and weak populations (Fig. 177), which is apparently the characteristic feature of this species' abundance dynamics (Nikol'sky, 1949, 1952). The different age groups of the Far Eastern herring *Clupea harengus pallassi* mix with one another very little; they enter Peter the Great Gulf at different times (Fig. 178). The adult herrings begin to enter in February, and the flow is ended by the year-old group in June.

The sea flounder is highly isolated by age groups. The hatching pelagic larvae of this species move over to the shore, covering 100 km or more.

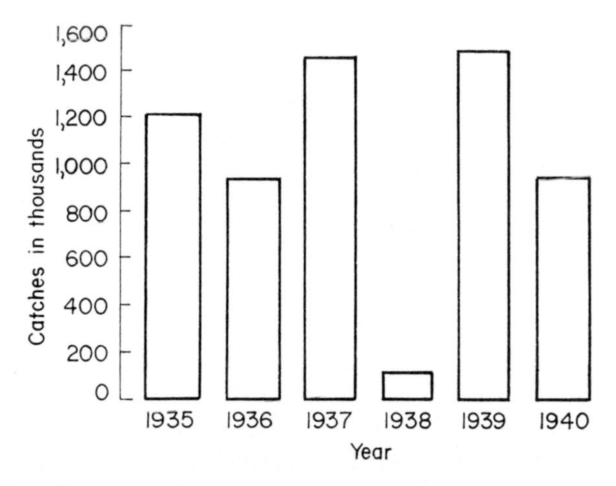

Fig. 177. Catches of *Oncorhynchus gorbuscha* in the Far East (after Nikol'sky, 1950).

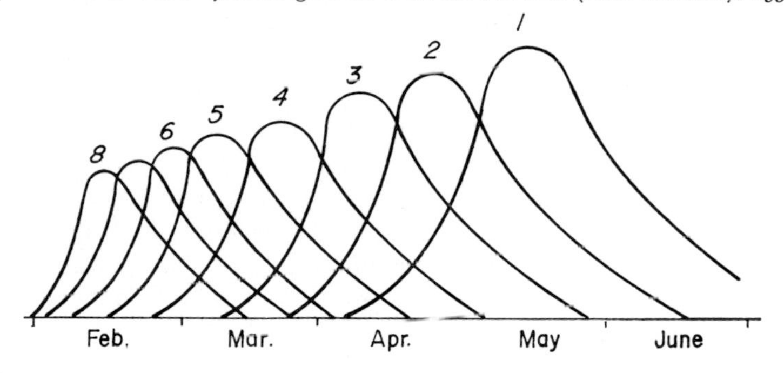

Fig. 178. Entrance of Far Eastern herring of various ages into Peter the Great Gulf during migrations (from Shmidt, 1947). Numbers represent ages in years.

During this time the symmetrical larvae turn into asymmetrical adults; these drop to the bottom in the shallow areas along the coast (no deeper than 10–20 m) and begin to lead a benthic way of life. When 1 year old, the young flounders migrate into deeper areas, and they ultimately repeat this move each autumn. Therefore, the separate age groups occupy definite marine zones, free or nearly free of flounders of other ages (Fig. 179).

Among most species there is no such spatial displacement of the individual age groups; they are mixed, living in the same area. But even in this latter case they are often biologically isolated, since they occupy different ecological niches. This is especially true among animals with complex developmental cycles, i.e., with metamorphosis accompanied by a shift in the surroundings (habitation in water or soil by the larvae of many insects and transition to a terrestrial existence in the adult phase). Com-

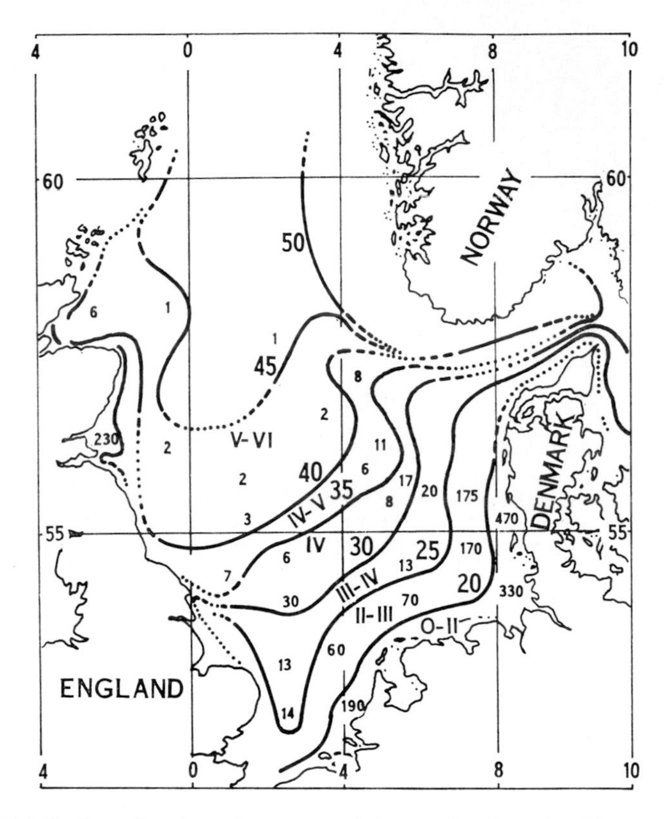

Fig. 179. Distribution of various size groups of the marine flounder *Platessa platessa* in the North Sea (after Garstang, from Shmidt, 1947). Curves correspond to the average length of the flounder in cm (Arabic numerals above curves); in each zone the area between the curves is dominated by flounders with a length intermediate between the two figures above the line. Roman numerals indicate the age classes of the flounder; Arabic numerals in parentheses represent the relative densities of the population (number of flounders caught in an hour of trawling).

plex developmental cycles with alternation between free-living and parasitic stages are known for the cestodes and trematodes. Among the insects with complete metamorphosis (Holometabola) development from the egg to the adult form includes the intermediate phases of wormlike larva and pupa. In the so-called hypermetamorphosis the number of stages increases to six or more (Fig. 180).

In all cases each stage is a special life form, and its replacement is accompanied by a change in the character of its association with the environment and often by a change in its way of life as well (Fig. 181). Actually, the wormlike caterpillar of a butterfly or the free-swimming larva of the

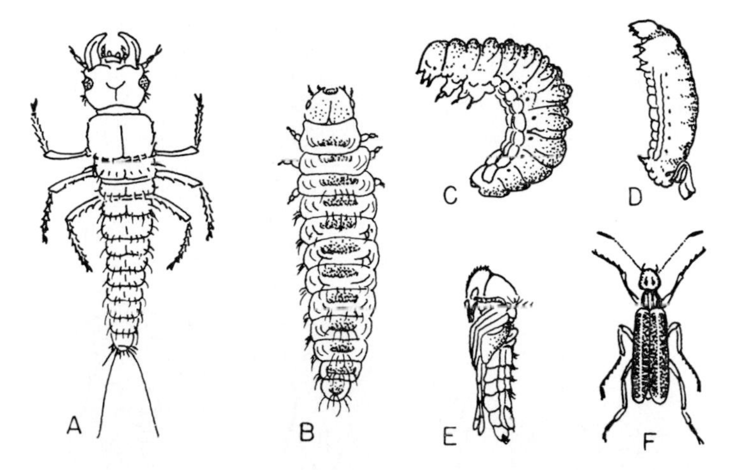

Fig. 180. Hypermetamorphosis of the *Epicauta* beetle (after Folsom, from N. Ya. Kuznetsov, 1951). A: first phase of campodeo larva ("triungulin"); B: second phase; C: last form of second phase; D: shriveled and quiescent larva; E: pupa; F: adult beetle.

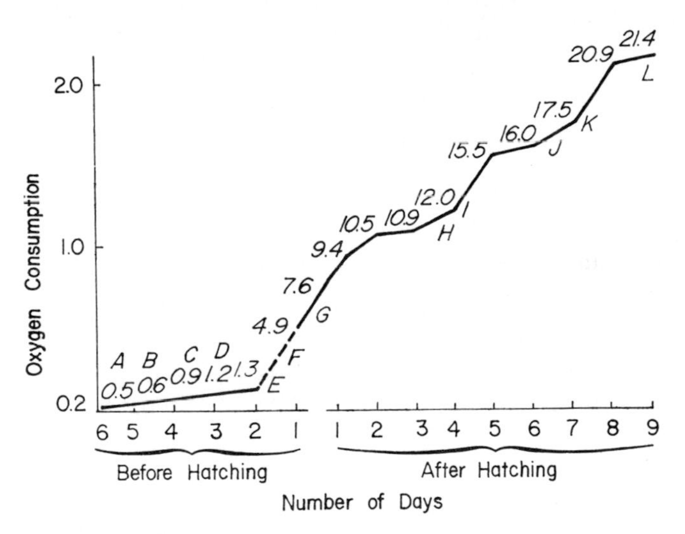

Fig. 181. Consumption of oxygen by eggs and larvae of the sturgeon (after Korzhuyev, 1941). A: birth; B: gastrulation; C: nerve tube; D: heartbeat; E: motile larva; F: hatching; G: abdominal breathing; H: opercular gill; I: gill petals; J: movable gill coverlet; K: active feeding; L: dorsal fin.

Antedonidae differs from the adult stage just as greatly as do representatives of different orders and even classes. There is a difference in the demands of each stage for conditions of existence and a different level of toleration (Table 48). Resistance to cooling, for example, changes. The

caterpillars of the first phases of the malvella moth *Gelechia malvella* (an important cotton pest) die off completely if kept at a temperature of —4° C for 6 hours, while at —10° C in the same period 44 percent of the older stages survive (Babayan, 1950). The differing resistance of separate stages of insects to DDT and hexachlorethane depends on differing amounts of fat in their bodies and differing levels of gas exchange, which regularly change at each stage of development. As the food regime of the caterpillar grows and develops, the animal lives on full-valued and relatively protein-rich foods. In the pupal stage the insects do not feed, while adult butterflies either feed on nectar, obtaining chiefly carbohydrates, or do not feed at all.

TABLE 48. *Relationship of the beet webworm and the turnip moth to the temperature of the environment (after Kozhanchikov, 1937).*

Phase of Development	BEET WEBWORM		TURNIP MOTH	
	Optimum	*Limits*	*Optimum*	*Limits*
Egg	28.0° C	30.0–26.0° C	25.3° C	30.0–20.0° C
Caterpillar (stage I)	25.0	26.9–23.3	25.6	28.8–21.1
Caterpillar (stage III)	32.7	36.0–20.0	21.0	30.0–11.7
Pupa	28.1	32.0–20.0	19.0	29.0–18.0

The development of fishes of the carp family goes through several stages demarcated by quick organizational restructuring, which sometimes occurs in the space of several hours. Anatomic changes are inseparably associated with changes in ecology (Vasnetsov, 1948, 1953). The newly hatched larvae are not very mobile and feed on the yolk; this is their first stage of development, and it lasts 2 days at a temperature of 18° C. In stage 2 they feed not only on yolk but on external food as well (small, nonmotile plankton organisms); a movable mouth, tracheal teeth on the gill arches, intestinal cavity, and swim bladder also form in this stage. In stage 3 (6–7 days) the fish have passed over to external foods entirely (larger and more mobile plankton organisms). In stage 4 (7–8 days) the bony rays appear in the fins, and the fish are capable of making turns; they begin hunting for more motile Entomostraca, and the mouth acquires the ability to move. The intestine lengthens and bends and the olfactory aperture splits in stage 5 (3–4 days). In stage 6 (10–12 days) the body shape typical of pelagic plankton eaters is formed, and the scales appear. In stage 7 the terminal mouth is transformed to a semiventral mouth, the entrails lengthen and add another bend, the lateral-line organs begin to pass from the dorsal to the ventral side, and the fish goes from eating plankton to tiny benthon. In stages 8, 9, and 10 the process of transformation from a rather quick-moving plankton eater to a slower benthic fish

is completed. This is accompanied by a transition from the school to an almost solitary existence. And the changes in feeding and behavior continue. The role of different crustaceans changes significantly with age in the case of the Gobiidae (Table 49). The types of age-related changes in food content among the fishes of the northern Caspian are shown in Fig. 182.

After appearance of the mouth, frog tadpoles eat plant detritus, diatoms, and green algae. Later their ration includes small animals. Silt, crayfish,

TABLE 49. *Change in significance of various components of the food of* Gobius fluviatilis *in connection with its growth (after Kinalev, 1937).*

	SIZE OF FISH IN CM				
Food	*5–7*	*7–9*	*9–11*	*11–13*	*13–15*
Amphipoda	16.8%	32.0%	35.1%	67.1%	74.4%
Mysids	24.3	11.8	3.0	6.0	0
Others	58.9	56.2	61.9	27.9	25.5

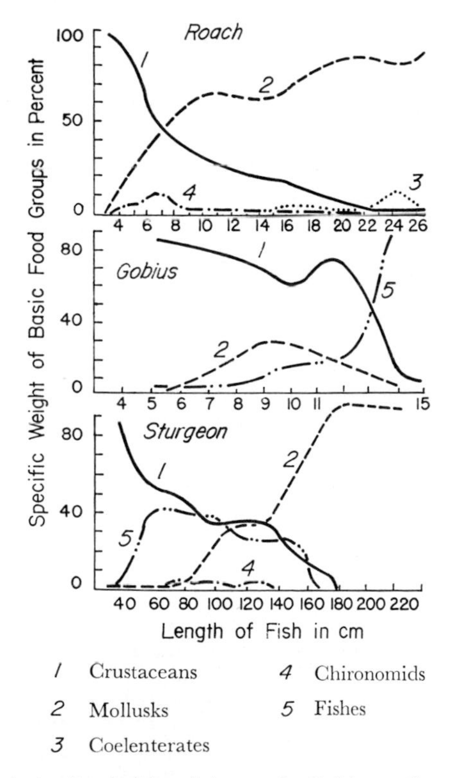

Fig. 182. Age changes in feeding habits of the roach, *Gobius*, and sturgeon in the northern Caspian (after Shorygin, 1952).

and plant remains have been found in the intestines of more mature tadpoles, and the adult forms are wholly carnivorous.

Among birds and mammals the young are incapable of maintaining the body temperature at a constant level; they are poikilothermic. The body temperature of fledgling murres, seagulls, guillemots, puffins, and others may drop without harm to 7.3–13.6° C and even to 4.5° C. With the growth in feathers and increased body size, a constant body temperature—homoiothermy—is established. Among open-nesting Sylviidae, thrushes, and other small birds this occurs in 7–9 days, while among those which nest in tree holes (flycatchers) it occurs in 10–12 days. Chemical thermoregulation among passerine birds is developed in jumps, being established over the course of 24 hours at most (in individual fledglings), and it corresponds to the moment when the feathers are developed (Fig. 183). But the rise in

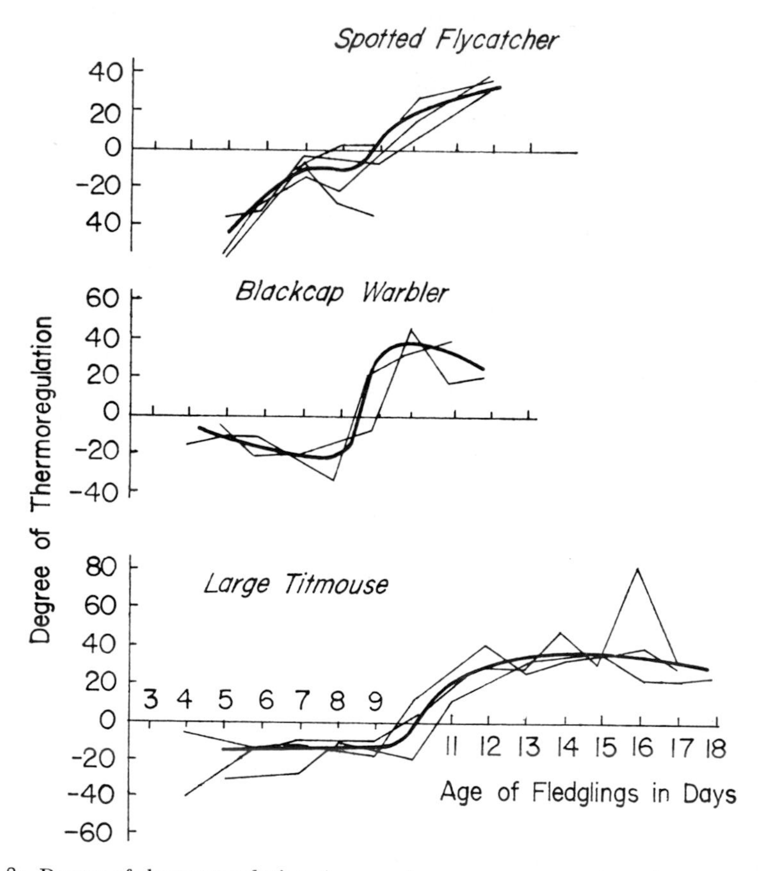

Fig. 183. Degree of thermoregulation (increase in oxygen consumption when temperature is lowered by 1° C) among fledglings of various ages (after Shilov, 1957).

resistance to external temperature increases gradually with age, since this is a function of increased body size and decreased surface area (Fig. 184).

There are two stages in the development of small passerine fledglings, differing in the character of their relationships with the environment and the behavior of the parents. In stage 1 the fledglings are naked or covered with down. Thermoregulation is absent, since fledglings are able to tolerate rather low temperatures without danger. The parents spend most of the time warming the fledglings, giving them a high temperature and rapid growth rate. In stage 2 plumage develops and chemical thermoregulation appears, guaranteeing the fledglings survival at low temperatures but only within comparatively narrow bounds. Parents spend most of the time feeding the young; they warm them irregularly (Fig. 185). Warming only aids the fledglings in preserving the normal body temperature during marked deviations of external temperature from the optimum (Shilov, 1957).

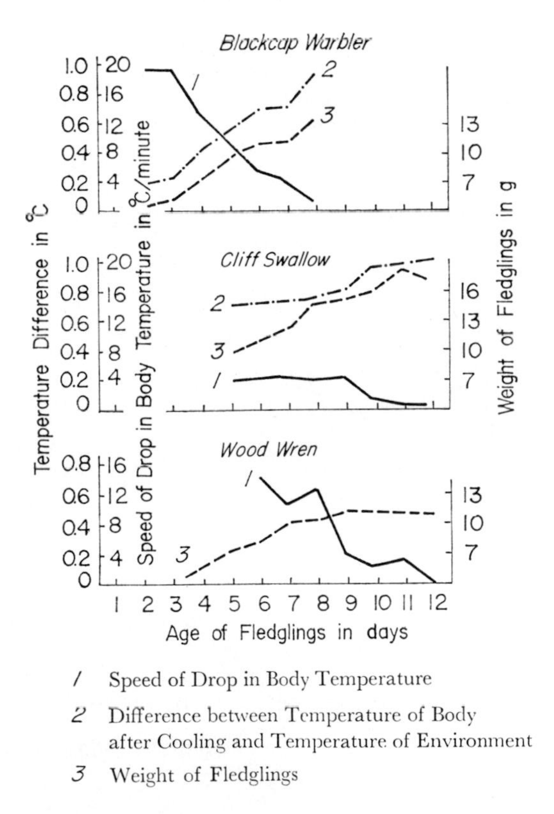

Fig. 184. Changes in stability of body temperature of the fledgling blackcap warbler, cliff swallow, and wood wren (after Shilov, 1957).

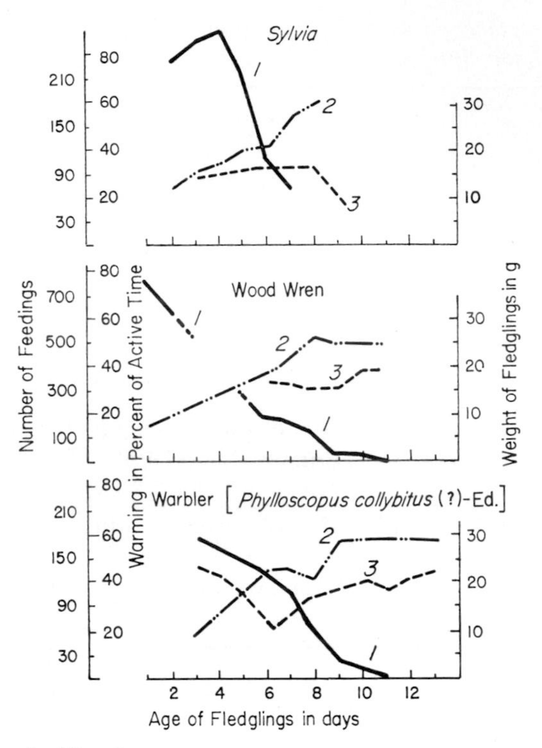

/ Warming

2 Feeding (number of flights to nest with food by parents)

3 Air Temperature

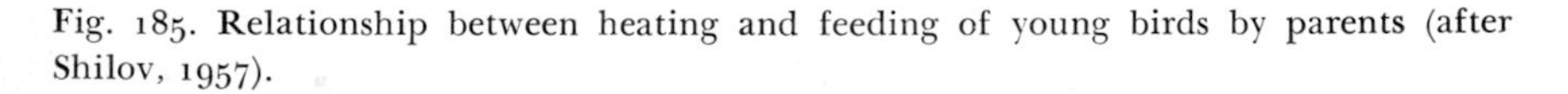

Fig. 185. Relationship between heating and feeding of young birds by parents (after Shilov, 1957).

Young fledglings are resistant not only to cooling but also to a drop in the partial pressure of oxygen. By "raising" birds to different "altitudes" in a barochamber (i.e., by decreasing the atmospheric pressure within it), Stroganova (1950) obtained the results shown in Table 50. Fledglings and young birds feed on animal food after leaving the nest. Only later in their lives does the amount of vegetable matter in their diet rise considerably (Fig. 186). During the nesting season the food of adult field sparrows consists of 81 percent seeds and 19 percent insects; among young fledglings at the same period seeds make up 29 percent and insects 71 percent (Mal'chevsky, 1959).

Thermoregulation in small mammals is usually established by 10–15 days of life and accompanies the opening of the eyes. At this time young

TABLE 50. *Tolerance of fledgling rooks* (Corvus frugilegus) *to a drop in barometric pressure (after Stroganova, 1950).*

Air Pressure in mm of Mercury	*Partial Pressure of Oxygen in mm*	*"Altitude" in m*	STATE OF EXPERIMENTAL BIRDS OF VARIOUS AGES				
			Adult	*30–40 Days*	*15–25 Days*	*3 Days*	*1–15 Days*
634.14	130.20	1,500	alive	alive	alive	alive	alive
286.74	58.78	7,500	alive	alive	alive	alive	alive
183.38	36.76	10,500	died	alive	alive	alive	alive
169.60	34.00	11,000	died	died	alive	alive	alive
–	–	11,300	died	died	died	alive	alive
140.40	28.00	12,200	died	died	died	died	alive
105.60	21.20	14,000	died	died	died	died	part died

common and society voles have an increased demand for heat, while the temperature optimum of the adults is lower (Polyakov and Pegel'man, 1950). The levels of matter exchange and chemical thermoregulation are higher in young animals and noticeably lower in adults (Table 51). Speed

TABLE 51. *Degree of chemical thermoregulation at different ages among mammals (after Slonim, 1952).*

Species	*Age*	*Percent of Change in Exchange at 1° C*
Guinea pig	1 day	4.7
	adult	1.7
Jackal	36 days	5.5
	adult	1.9
Raccoon dog	15 days	2.7
	23 days	3.0
	34 days	2.6
	88 days	1.8
	adult	0.2
Domestic cat	15 days	6.3
	adult	4.1

of development and growth in young mammals is associated with feeding on milk, especially on the amount of protein in the milk (Table 52).

Thus we have seen that feeding habits change over the course of life in almost all animals. Heat exchange and water exchange with the environment also change, and, as a result, the way of life and other biological properties change. The changes coincide with turning points in the development of the animal and take place in short intervals of time; they alternate with longer periods of relative constancy.

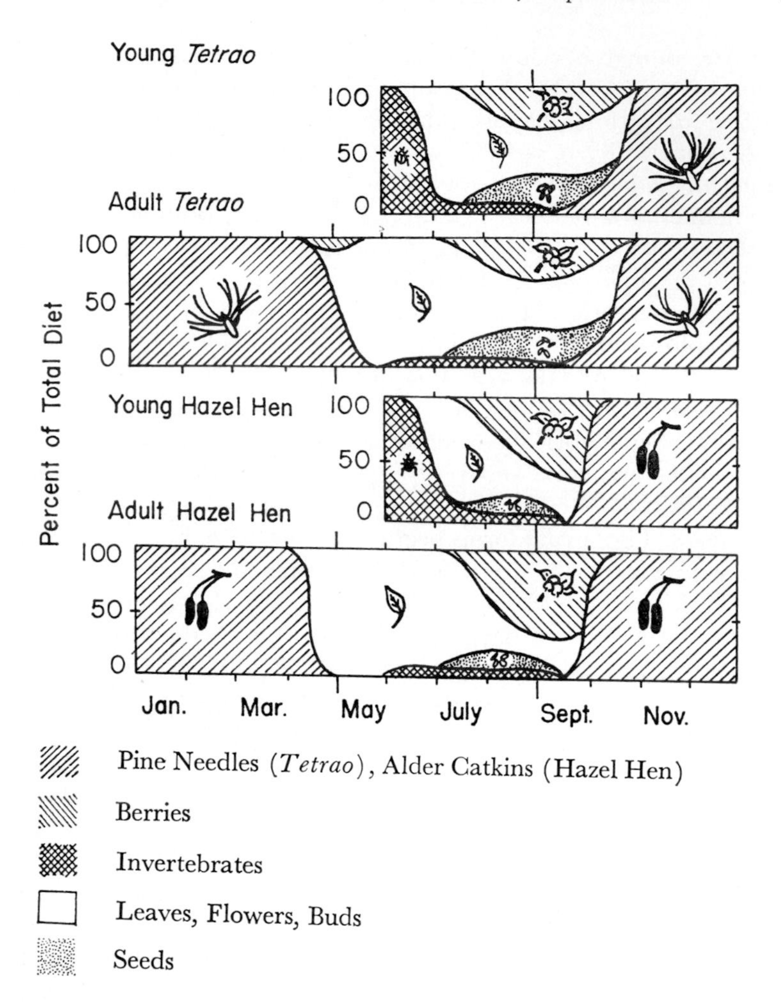

Fig. 186. Relationship (in capacity percentages) of basic species of food among *Tetrao urogallus* and hazel hens (after Oliger, 1957).

TABLE 52. *Relationship between speed of animal growth and amount of protein in the mother's milk.*

	Man	*Horse*	*Cow*	*Pig*	*Sheep*	*Dog*	*Cat*	*Rabbit*	*Greenland Seal*
Number of days needed to double weight	180	60	47	18	10	8	9	6	5
Amount of protein in milk in g/l	19	20	33	37	70	97	95	104	119

In many animals development is accompanied by so-called cyclomorphosis, when definite stages of development or whole generations are timed according to various seasons. The cladoceran *Hyalodaphnia* multiplies parthenogenetically in spring and sexually in summer. Among the generations of Cladocera which appear in the warm part of the year, the size of the helmet and the length of the spike increase, leading to an increase in surface area. This is an adaptation to the decrease in viscosity and density of the environment during warm weather.

Seasonal alternation of generations is known for the aphids. In spring the hop aphid *Phorodon humili* lives on plum trees. The winged generations appearing in May and June fly off to the hops, where they give birth to several generations of wingless aphids on the young sprigs. In August and September the winged mothers of sexual females and the winged males are born and return to plants of the genus *Prunus*. In *Chermes, Gnaphalodes,* and other aphids living on conifers the life cycle takes 2 years. The sexual forms begin it on the main host, making a start for the wingless female "founders." The latter give birth to winged migrants which fly to the secondary plants (from fir to larch, pine, silver pine, etc.). Here, after several generations of "virgins," the winged mothers appear, producing either females alone or males and females. The mothers fly back to the basic food plant, and the cycle is complete.

The developmental cycles of other aphids may differ in the schedule of generation alternations. As was said above, the individual phases of a cycle (such as sexual reproduction, change of host, etc.) may disappear, often leading to the appearance of noncyclic forms along with the cyclic ones and thus to collapse of the species. The complex cyclomorphosis of aphids is an adaptation to a climatic factor which changed in the Quaternary period. Then cooling and the shortening of the growing season for food plants made the existence of several generations of bisexual aphids on one host impossible (Mordvilko, 1926, 1936, 1939).

SEXUAL GROUPS

Males and females often differ in their ecologies. The ecological differences between the sexes are expressed first of all in feeding habits. As a rule, adult males of the ixodid ticks, unlike the females, do not feed. Among mosquitoes only adult females suck blood; the males suck only the sweet juices of plants. Among the social insects there is a marked difference in the feeding habits of females, workers, and drones. The in-

complete development of the sex organs of the worker-bees is caused by insufficient feeding during the larval phase.

Male sturgeons in the Caspian Sea catch a great variety of fish, while females eat only fishes of the families Gobiidae and Bentophilidae. The total amount of fish eaten by both sexes is close. Gammaridae predominate among males (19.7 percent of all food) and are rarer among females (8.9 percent); on the other hand, Corophiidae account for 29.5 percent among males and 48.3 percent among females (Shorygin, 1952). The different feeding habits of male and female martens are shown in Table 53. The smaller-sized females catch fewer of the large prey. Their skulls are lighter and the masticatory musculature is weaker, weighing 22 percent less than that of the male. Similar differences are known for the ermine and the solongoy [*Mustela altacia*—Ed.] (Grigor'yev and Teplov, 1939; Gusev, 1953). On the other hand, the female is larger among the predatory birds; the lighter, more flight-adapted, and more maneuverable male bears the burden of caring for the setting female and later the young.

TABLE 53. *Feeding characteristics of male and female forest martens* (Martes martes) *in the Pechoro-Ilych Preserve (after Yurgenson, 1947).*

Species Encountered	*Males*	*Females*
Squirrels	48.8%	54.2%
Forest voles	13.9	33.3
Shrews	4.6	4.7
Birds (total)	30.2	54.7
Gallinaceous birds	11.6	19.0
Tetrao urogallus	4.6	–
Willow ptarmigans	10.3	–
Hazel hens	4.5	19.0
Black grouse	11.6	–
Small birds	–	14.3
Bird eggs	–	9.5
Wasps	4.6	23.8
Cedar "nuts"	4.6	28.6

Females and males differ in their morpho-physiological properties as well as in their ways of life. Among male gray owls the level of oxygen consumption is higher than that of females (Segal', 1958). The primarily sessile males of the Central European finches have a relative heart weight of 1.43 percent, while females, which fly south for the winter, have a relative weight of 1.28 percent. The minimum temperature for torpor among female houseflies in the summer months is 2.5–5.1° C; the minimum for males is 4.2–6.4° C (Balashov, 1955).

In many cases males and females lead separate lives. Male *Callorhinus*

ursinus seals live apart from the females for the greater part of the year and make relatively small migrations, while the females and the young migrate far to the south (the Japanese islands and California). Male and female reindeer live apart for a considerable portion of the year. Many of the migratory birds (snow buntings and finches) winter apart. Among species which do not form permanently paired families, males and females occupy different biotopes for most of the year.

MEANING OF POLYMORPHISM OF POPULATIONS

The differences between various generations (age populations) have enormous meaning for the abundance of the species. Depending on the conditions of birth and development, generations may be powerful (abundant) or weak (sparse), and they may also differ by being more or less viable. This circumstance is of especial importance for species with a short developmental cycle and monocyclic (once in each generation) reproduction. When several generations exist simultaneously in the population, the age makeup may be a good index of the relative strength of a generation (Fig. 187); this index is commonly used to predict catches of game fish and other animals. Among monocyclic species with a developmental cycle of several years the appearance of a strong generation may have an effect over a period of years. Generations of the Siberian *Bombyx*, which requires 2 years to develop, are now distinguished in East Siberia and Mongolia by the fact that the strong generations come in the odd years (Grechkin, 1960). The same regular alternation of fat and lean years is known for the Pacific salmon (Soldatov, 1912).

The biological significance of age and sexual differences for structure, function, and behavior is great, since they provide for the qualitatively differing demands of the individual stages of development. Simultaneously, the age and sexual variations in food rations and ecologo-physiological characteristics aid in increasing the abundance of the species. Individual groups using different foods and sometimes inhabiting different areas do not occupy the same ecological niches and hardly even compete with one another.

The species is thus not only a complex system of subordinated territorial groupings but also a no less complex system of biological forms. This polymorphism complicates and enriches the connections between organisms and the environment, opening up the possibility of a wider use of the latter. The complex structure of the species population does not per-

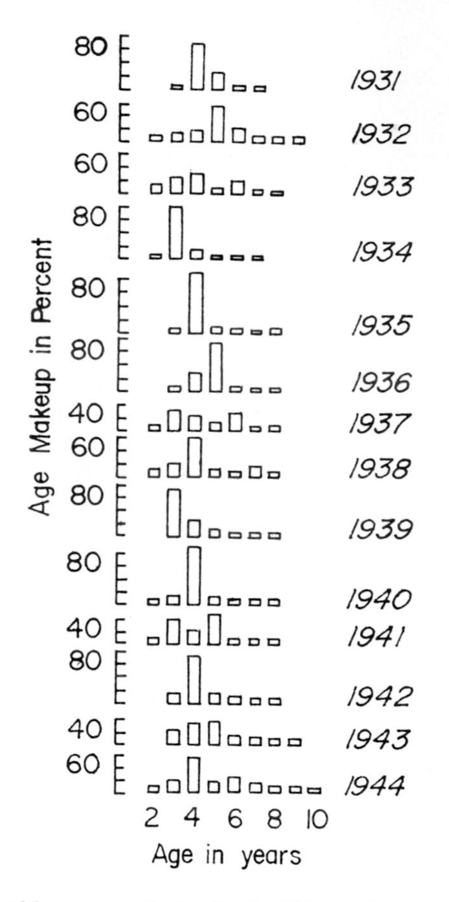

Fig. 187. Age makeup of bream population in the Volga delta (after Monastyrsky, 1952).

mit us to speak of a single ecological niche for the species, since its individual biological races and sex and age groups (phases) almost always occupy several and sometimes a great many niches. Spatial (geographic) differentiation of a species population and its biological races provides paths of divergence capable of leading to the formation of new species if the interactions of a given population and its environment take on a qualitatively different character.

CHAPTER 13

The Population Structure: Way of Life and Use of Territory

WAY OF LIFE AND DEMAND FOR TERRITORY

Individuals of one population or another live in isolation and lead solitary lives, or they unite into a flock, herd, or colony and lead a group life. Combination in groups or dispersion of individuals or families is determined by the demands of the animals and their relationships with one another. Different ways of life are characteristic of different species. The way of life may change even in the same species in various parts of its range, at different seasons, or in different developmental phases if the phases belong to different biological types (life forms).

In order to satisfy its demands, the animal must have the kind of territory on which it can find enough food, refuges from enemies and inclement weather, sources of water (in the case of those which need flowing water), minerals, and gases, and a favorable microclimate. Finally, it is necessary to remove metabolic products which contaminate the environment. The complexity and variety of the animals' demands for territory explain the fact that the amount of territory required for the needs of the individual or group is usually of considerable size and is always larger than the amount needed to fill a single need at a given moment. As a result, animals have worked out during the course of evolution certain demands for a definite amount of territory, regardless of the vital resources existing thereon at any given moment. Because this biologically important demand is not always taken into consideration, certain animals cannot be raised in captivity. It is known, for example, that the insect *Calpodes ethlius* reproduces only in large ponds. If the individuals are forced into a cluster, they

will not reproduce and will turn to cannibalism. The same phenomenon has been observed for rodents. Sparrows will multiply successfully in large nurseries but will not raise offspring in closed cages.

The demands for territory and the means of satisfying them are associated with the way of life of the animals, a way of life dependent on the morpho-physiological and ecological properties of the species. The basic significance comes from character of feeding habits and means of catching food, relationships of competition or symbiosis between individuals and their groups within the species, relationships with other species, manner of reproduction, and demand for refuges from enemies and the weather. There are two basic ways of life—solitary (often transformed during the mating season into solitary-family) and group. At the basis of both lie the properties of individual behavior and the mutual relationships of individuals. These reactions may be positive (individuals are attracted to one another) or negative (individuals avoid one another) and may even change over the course of the year.

In order to communicate with one another, animals use visual, olfactory, and aural signals. Recognition marks are also found in bright spots, sexual properties, the "flag" of reindeer and rodents, the plumage of birds, various types of odoriferous substances excreted by glands, the voice (oral calls, danger signals, threats, and cautionary signals), and properties of behavior (threatening poses, manners of nursing, etc). These means of communication are widespread in both solitary and group animals, but they serve different goals. In herds, flocks, and colonies they furnish an agreed-upon behavior, while among solitary animals they serve as repellents and lead to dispersal.

Various types of groupings in the lives of individual species do not have the same significance and differ in the duration of their existence. Being associated, like all relationships within the species, with the conditions of existence, they change as the latter change.

SOLITARY OR SOLITARY-FAMILY WAY OF LIFE

By the solitary or solitary-family way of life we mean a separated animal existence, with each animal on its own individual territory. During the mating season each pair of many species occupies the same individual piece of territory, on which their offspring dwell until maturity and the breakdown of the family. In some cases the piece of territory occupied by an individual or family is actively defended against invasion by pretenders,

leading to collisions between individuals. In other cases only a part of the area is defended, that part immediately adjacent to the nest or burrow. Under a third set of circumstances the animals keep apart but do not have specific pieces of territory isolated from others.

A special case occurs among the sessile aquatic animals, both those which live immediately next to one another and those separated by small distances. The possibility of such existence in an aquatic environment is insured by the small demands of sessile forms and by the fact that food and oxygen are brought to them by currents of water and their metabolic products removed in the same manner. The near-immobile inhabitants of the bottom (benthos) prevail over individual territories usually no greater than several dozen square meters (burrowing crabs, *Onchidium* and *Acmea* mollusks).

Separate use of territory is common among the insects. The caterpillars of many butterflies need so little food when vegetable matter is abundant that the need for defending their tiny territories never arises. However, in experiments with the flour beetle it has been found that when the density of the population is greater than 44 individuals per g of flour, reproduction is sharply curtailed, not so much because of a lack of food but more because the newly laid eggs are eaten along with the flour. The same has been confirmed in the case of the coffee beetle (Table 54). However, under natural conditions these beetles will not remain in the same spot when the population becomes dense but, instead, will fly off.

TABLE 54. *Multiplication and survival of the coffee beetle* Stephanoderes hampei *at various population densities (after Frideriks, 1932).*

Number of Coffee Beans	*Original Number of Beetles*	OVER SPAN OF 1 MONTH				
		Eggs	*Larvae*	*Pupae*	*Total Offspring*	*Number of Beetles*
30	60	3	208	95	306	195
30	450	4	98	45	157	109

The aspen hawkmoths lay their eggs a certain distance from one another, and the caterpillars of the butterfly *Nymphalis antiopa* disperse before going into their cocoons. Certain fishes (carps, pikes, *Aspius*, and others) keep apart when in the adult state, feeding in their own individual territories. During spawning females of the Pacific salmon spread their eggs, cover them with earth at a distance from one another, and then defend the spots where the eggs are laid (Fig. 188). The same type of nesting ground is known among sticklebacks.

Among reptiles the phrynocephalids, agamas, and others have com-

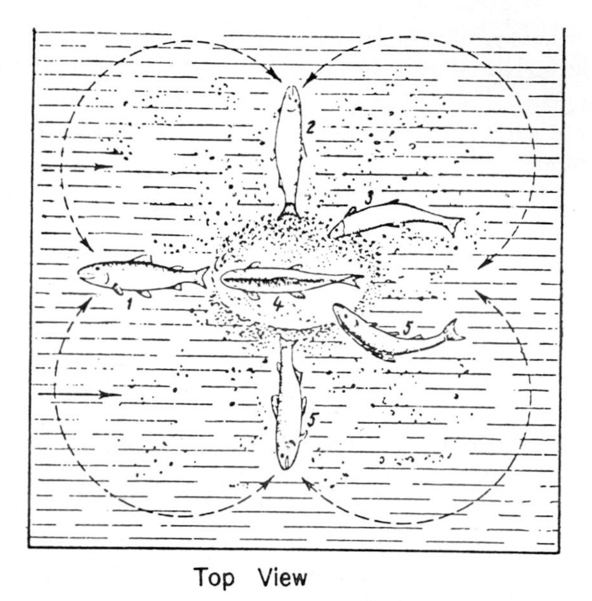

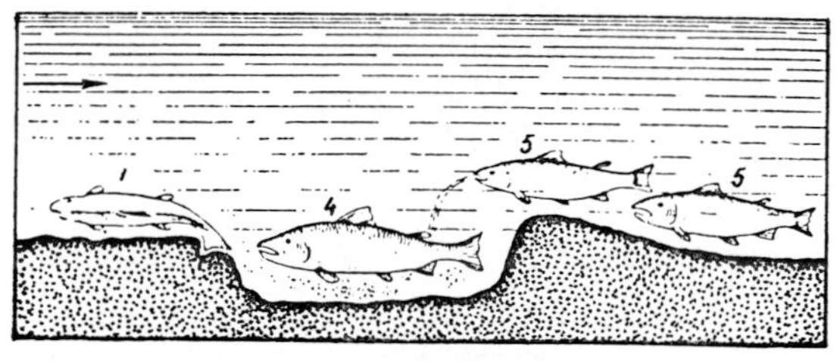

Cross Section

1-3 Position of Female While Digging Out Nest

4 Principal Male

5 Secondary Male

Fig. 188. Construction of nest and egg-laying scheme of the red salmon.

pletely or partially isolated individual pieces of territory. Female jumping lizards have individual territories only several meters in diameter in the spring, while the males are highly mobile and wander widely, covering dozens of meters in a short time (Fig. 189). Among the North American *Sceloporus occidentalis* the radius of the individual piece of territory does not exceed 500 feet. The California species *S. graciosus* has been known to live for 3 years in the same place. The territory of *Anolis sagrei* is 400 sq.

feet. The territory occupied by the turtle *Terrapene carolina* is 220 sq. yards, while for the desert-dwelling *Gepherus agassizii* it varies between 10 and 100 acres. The steppe tortoise (*Testudo horsefieldi*), on the other hand, has no isolated territory, and several individuals may feed on the same plot of ground.

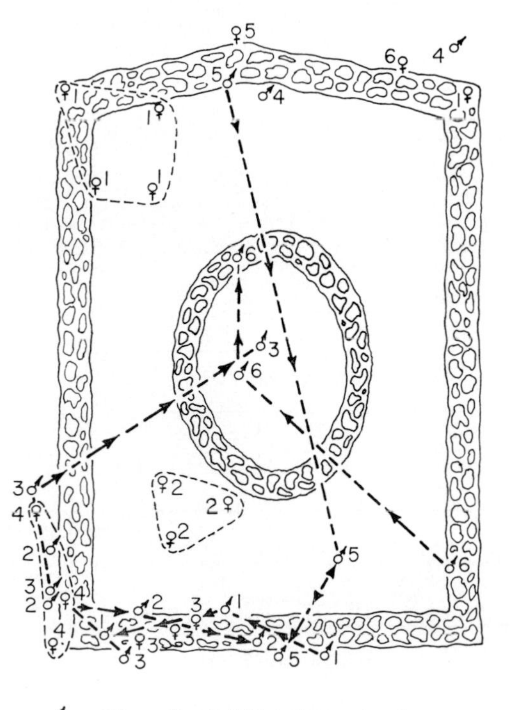

♂ Where Banded Male Encountered

♀ Where Banded Female Encountered

--- Border of Individual Female Plot

-→ Line Joining Areas Where Males Repeatedly Caught

Forest Planting

Fig. 189. Individual plots of jumping lizards (*Lacerta exigua*) in forest plantings in the Dzhanybek region of the western Kazakhstan oblast' (after Dinesman and Kaletskaya, 1952). Numbers indicate the number-designation given to each animal.

Many birds, especially such poor flyers as finches, *Phoenicurus ochruros*, European robins, the Sylviidae, yellow buntings, and certain others, lead a solitary-family way of life during the mating season (Howard, 1922). Each family's need for a separate piece of territory is determined not only

by the relatively poor ability of animals to move around but also by their tremendous demands for food. According to observations made by A. N. Promptov (1940) near Leningrad, a pair of blue titmice which had been feeding their young made up to 332 flights (an average of 250–300) per day (18.4 hours) to bring food back to the nest, usually returning with several insects each time. In one day (20.2 hours) *Phoenicurus ochruros* made 469 sorties, while a spotted flycatcher made an average of 350–450 sorties in 19.3 hours, with a maximum of 561. In addition to food brought for the brood, the adults also need food for themselves. For such birds the possibility of satisfying food demands can be met only by ruling over a separate piece of territory which can be reached by flight. It is energetically defended by both parents or by the male alone, which flies in ahead of the female in spring. Defense of the nesting ground, building of the nest, and the singing of the male nearby all have significance for the normal establishment of the sexual cycle in the female, and such species mate only after they have established and begun to defend a nesting territory.

The size of an individual or family plot usually depends on the abundance of food and is not the same in each biotope. The hunting grounds of a spotted flycatcher near Moscow had a radius of about 100 m; those of a redstart occupied 300 m, while those of the hoopoe in the Ukraine and Kazakhstan were 400–800 m (Formozov, Osmolovskaya, and Blagosklonov, 1950). Different parts of the territory are generally used at different hours of the day or in different types of weather (Fig. 190). Different individuals often occupy the same piece of ground for several years in a row or even for life.

Among many species of birds the area immediately adjacent to the nest is the only one defended; newcomers of the same, and sometimes of other, species are actively driven off. This is achieved by dispersal of nesting pairs, a factor important not only for the elimination of competition in food-getting but also for greater protection against enemies, which have a more difficult time finding isolated nests (Lek, 1957). Among such species the feeding grounds are often commonly shared. The snipe and the lapwing nest only in dispersion but feed on the same areas and even warn one another of danger.

One proof of the fact that the individual preserve of small passerine birds cannot always be protected from food competitors is the experimental switching of tree-hole nests between redstart and titmouse offspring. If the nests are not moved too far away, the parents do not abandon them but quickly find the new location and continue to feed the offspring. In sparse settlements the nests were removed on one occasion up to 60 m and for one day as much as 150 m away, while for 15 days (120 removals) they were removed a distance of 1.5 km. Even in these cases the parents

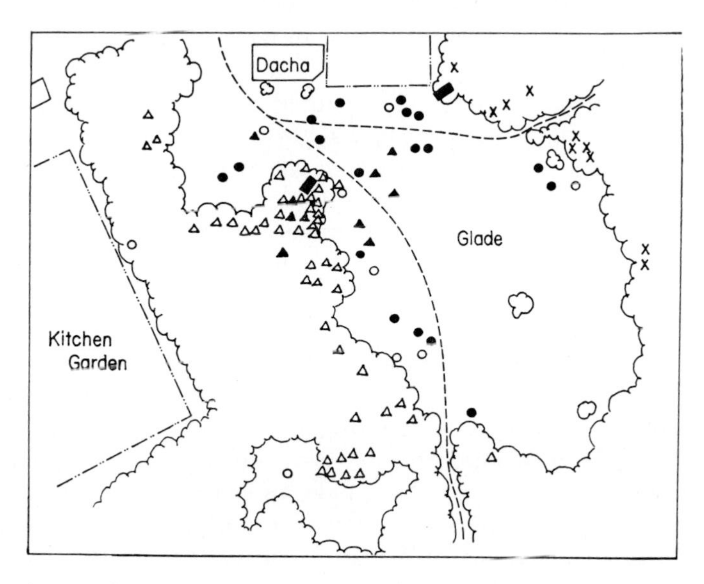

● Hunting Ground of Flycatcher from Tree Hole 6 during Cold, Windy Weather (insect taken from ground)

▲ The Same (insect taken in flight or from tree branch)

△ Hunting Ground of Same Flycatcher during Warm Weather (insect taken in flight or from tree branch)

○ The Same (insect taken from ground)

× Hunting Ground of Flycatcher from Tree Hole 5

▬ Tree Hole

Fig. 190. Hunting (nesting) plots of spotted flycatchers (*Musicapa atricapilla*) in a mixed forest at the Zvenigorod Biostation of Moscow State University (after Formozov, Osmolovskaya, and Blagasklonov, 1950).

did not abandon their offspring and continued to feed them at the new location. When one nest was moved into an area already occupied by another pair, the hosts and the new settlers did not fight but spelled one another in feeding the fledglings, obtaining the food from the same plot of ground (Kadochnikov, 1960). This again shows that food does not directly determine the feeding ground and its use, although the role of the food factor cannot be doubted. As was pointed out before, its importance is underlined by changes in the magnitude of the individual feeding ground in food-rich or food-poor areas.

During the mating season families of predatory birds have hunting grounds, but they only defend actively those parts immediately adjacent

to their nests. On the tundras of the Yamal Peninsula the peregrine falcons and rough-legged buzzards panic when alien birds of their own species or their enemies approach closer than 200–300 m from the nest. However, they hunt over a territory 4–6 km in diameter (Osmolovskaya, 1948).

The magnitude and character of the use of hunting grounds by *Buteo buteo* near Moscow change along with changes in the surroundings. In spring and early summer the birds hunt primarily over clearings and to some extent over meadows, where small rodents and other vertebrates are numerous at this time. The hunting grounds of individual pairs are isolated from one another, although their nests are sometimes situated in close proximity (Fig. 191). After grain crops have been harvested, the buzzards hunt primarily on cut fields and mown meadows; when this happens, different families make use of the same fields. The transition to co-utilization of hunting grounds coincides with the increase in all species of foods in nature.

A division of grounds during the first half of the nesting season is observed among steppe eagles and *Accipiter* hawks in the deserts of the northern Aral region. It exists among nesting pairs of the same species, while grounds belonging to families of different species often adjoin one another (Fig. 192). The eagle owls also have individual, actively defended nesting grounds. The nests of the steppe eagle in the Astrakhan and Volgograd oblasts are scattered about a relatively even area (Fig. 193). They are especially numerous in areas where small susliks are abundant, and the distance between broods here fluctuates between 0.3 and 3.8 km; in areas where there are not many susliks, the nests are 2–6 km apart (Agafonov *et al.*, 1955).

The nesting regions of individual pairs of ravens are varied, as is the case with predators (Fig. 194). *Milvus korschun* nests rather close together in the Tula abatis, sometimes almost colonially, and hunts over very wide territories. But even these birds have their own nesting grounds, although they are quite narrow (Fig. 195). Among the large spotted woodpeckers the individual grounds are retained even in winter. Successful wintering among these comparatively poor flyers is associated with the presence of shelter (hollow trees) and a food supply (seeds of coniferous trees). Therefore, it becomes biologically necessary for the woodpecker to rule an individual feeding ground in winter. Toward spring the males increase the size of their feeding grounds in the direction of the female feeding grounds until both territories overlap (Dergunov, 1928).

As D. Lek (1957) has pointed out, the individual territory of a bird cannot be regarded merely as a feeding ground. It is also a means of regulating the density of the population, as E. Howard, the first to establish its existence, insisted. This factor, too, is relative, since clusters of nests

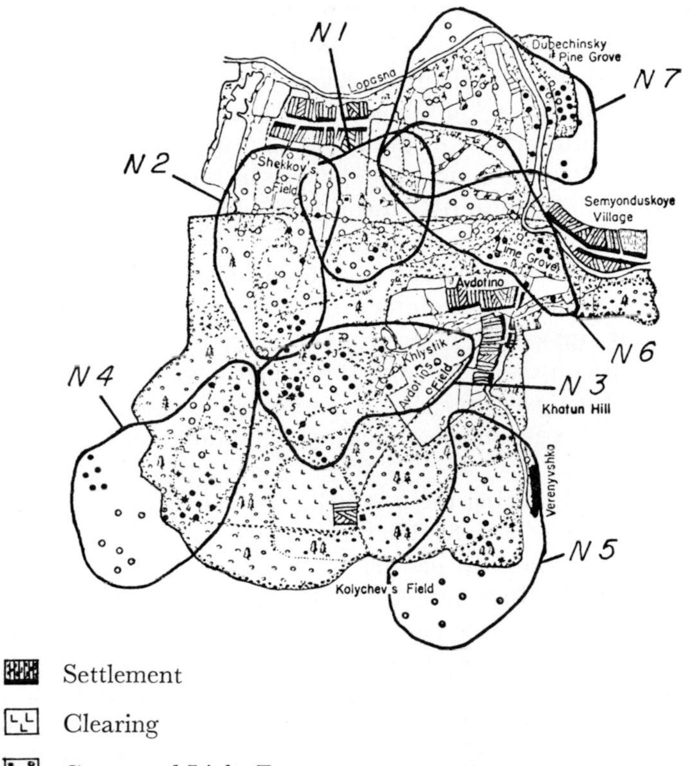

Settlement

Clearing

Copse and Light Forest

Forest

Meadow

* *Buteo* Nest

• Where *Buteo* Was Observed in June and July

○ Where *Buteo* Was Observed in Aug. and Sept.

N1,2 etc. Border of Hunting Ground of Individual Pair in May, June, and July

Fig. 191. Hunting ground of *Buteo buteo* at the beginning and end of summer in the southern Moscow oblast' (original).

are known also for solitary species. There are many cases in which birds nest apart but feed together (snipe, magpie, lapwing, etc.). Likewise, dispersion of nests, like dispersed egg-laying among insects, insures a lower mortality rate from predators. It is no wonder that seagulls, which usually nest colonially, begin to nest apart as soon as foxes or other predators gain

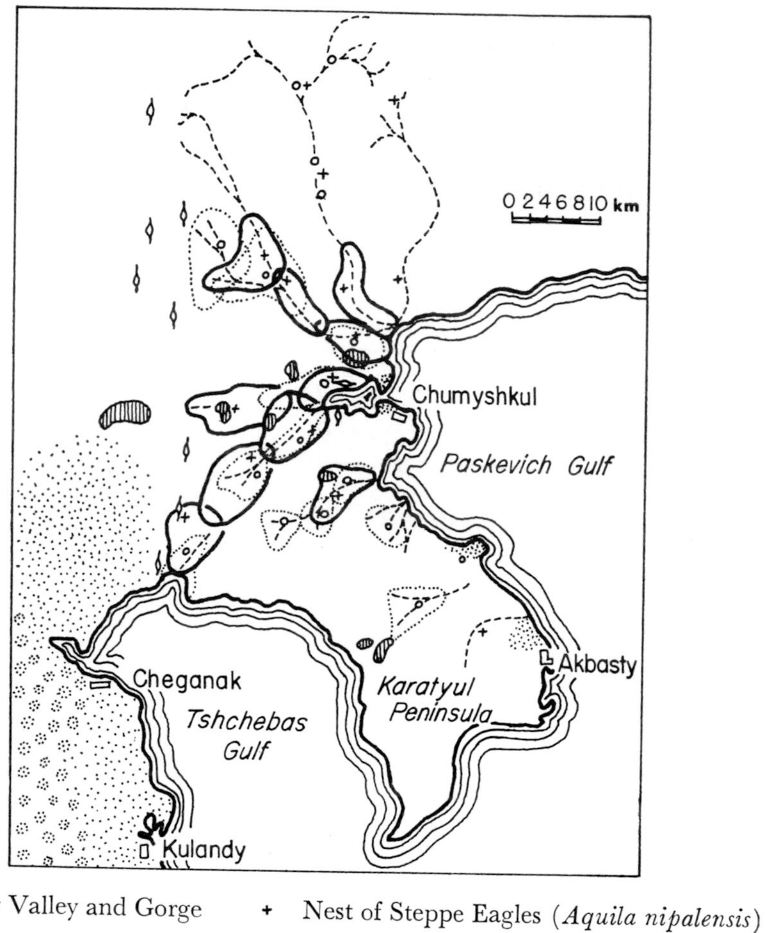

--- Dry Valley and Gorge
Tableland
Salt Flat
Hillocky Sand
Hilly Sand
+ Nest of Steppe Eagles (*Aquila nipalensis*)
— Their Hunting Ground
o Nest of *Buteo ferox*
.... Their Hunting Ground

Fig. 192. Placement of nests and hunting grounds of predatory birds in the northern Pre-Aral (original).

admittance to a colony and plunder it regularly. The nesting grounds of migratory birds are first occupied, as a rule, by the earlier-arriving males. Their singing over the occupied grounds attracts the females. At the same time the birds defend the area against newcomers. In this instance singing

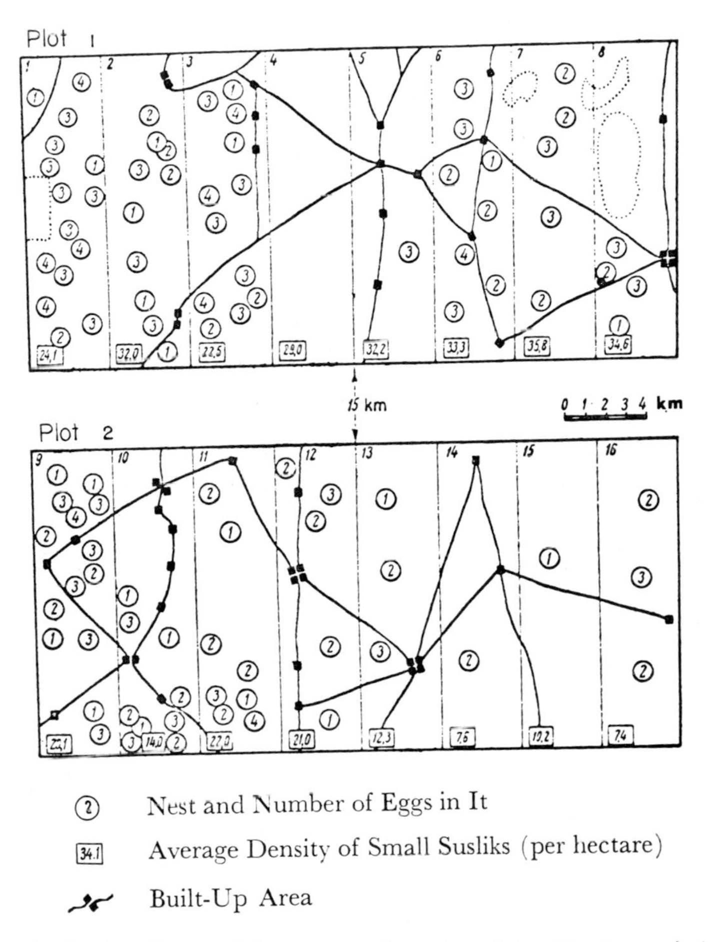

Fig. 193. Distribution of nests of the steppe eagle on two plots of the Sarpa virgin steppe in the Volgograd and Astrakhan oblasts (after Agafonov *et al.*, 1957).

plays the role of a threat signal to frighten foreign individuals away from the occupied nest. The same purpose is served by the threatening pose (Mal'chevsky, 1952).

Among mammals individual or family grounds are preserved for life. Among sables these grounds have a definite stability, and their sizes depend on the amount of food in the territory. According to V. V. Rayevsky (1947), the average area of an individual hunting ground on the upper Konda and Little Sos'va rivers is 10.3 km in winter (Fig. 196). In the better feeding grounds the territory is a little smaller and two grounds may partially intersect; i.e., they are not wholly isolated. After mating and the

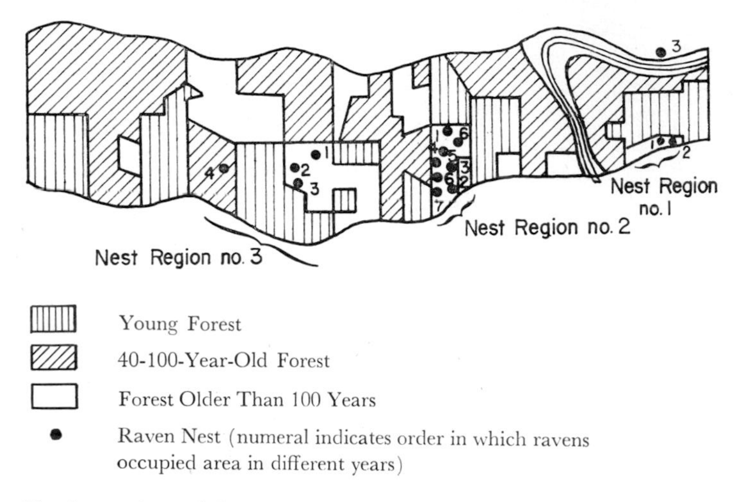

Fig. 194. Nesting regions of *Corvus corax* in the Tula Preserve (after Likhachev, 1951).

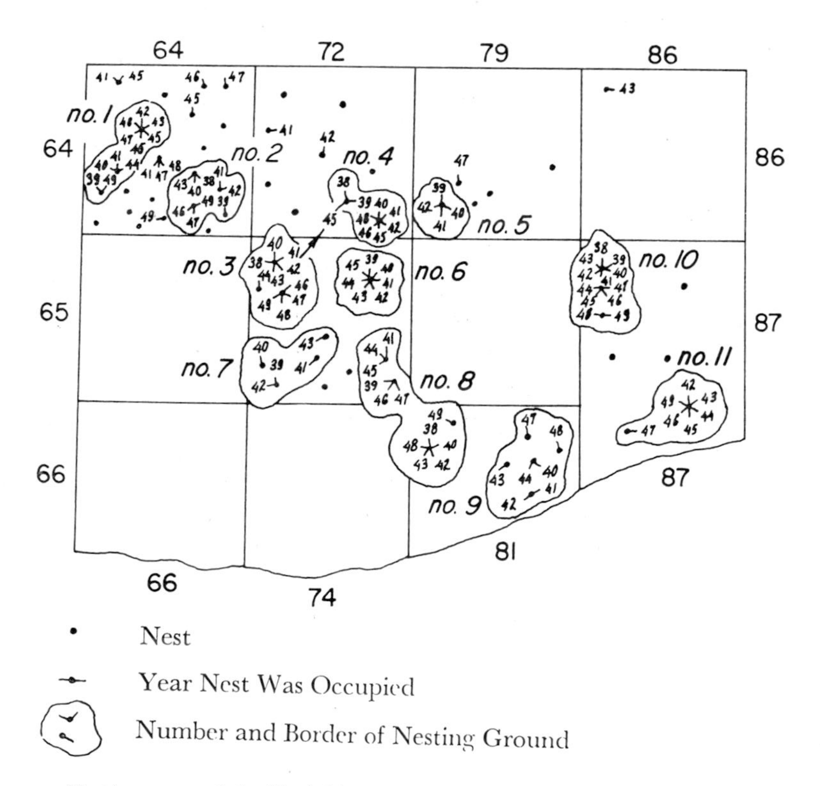

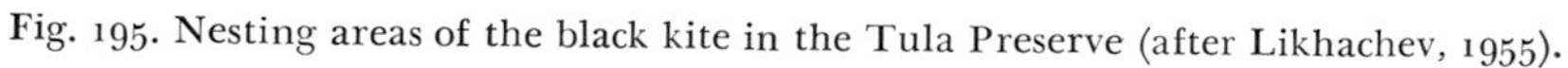

Fig. 195. Nesting areas of the black kite in the Tula Preserve (after Likhachev, 1955).

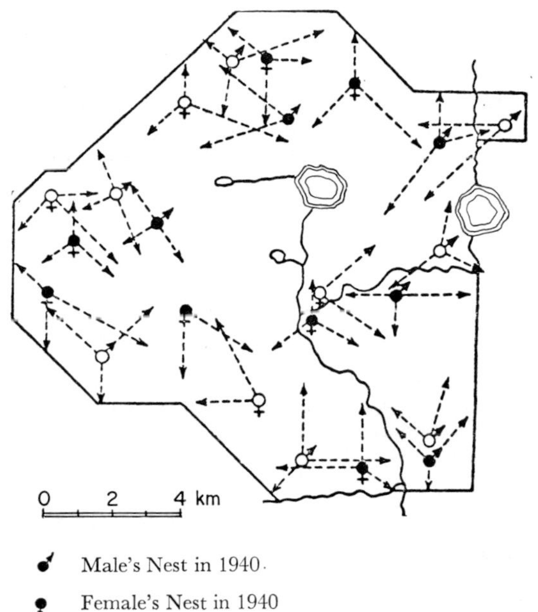

Male's Nest in 1940

Female's Nest in 1940

Male's Nest in 1941

Female's Nest in 1941

Path of Sables (used to determine their hunting grounds)

Fig. 196. Individual plots of sables (*Martes zibellina*) in the Kondo-Sos'va Preserve in different years (after Rayevsky, 1947).

maturation of the young the latter resettle beyond the borders of the territory occupied by their parents. If all the area suitable for habitation is already settled, the animals do not increase, since the young die off. When animals become more abundant, the area available for each is significantly decreased.

Among ermines the hunting grounds are large in autumn before the first snowfall. At this time the hunting grounds belonging to some animals often adjoin one another, while in winter they do not usually overlap. The daily hunting areas of the males are larger than those of the females; in Tataria the latter average 7.1 hectares (0.2–42.3 hectares) and the former average 20.5 hectares (4.2–49.0 hectares).

Relatively stable feeding grounds exist among the wolves in the central oblasts of the USSR, where these beasts lead a sedentary life. During the mating season pairs and their offspring remain no closer than 10–12 km from one another (Fig. 197). In winter they also keep to their hunting grounds, and the entire life of this group proceeds within their bounds

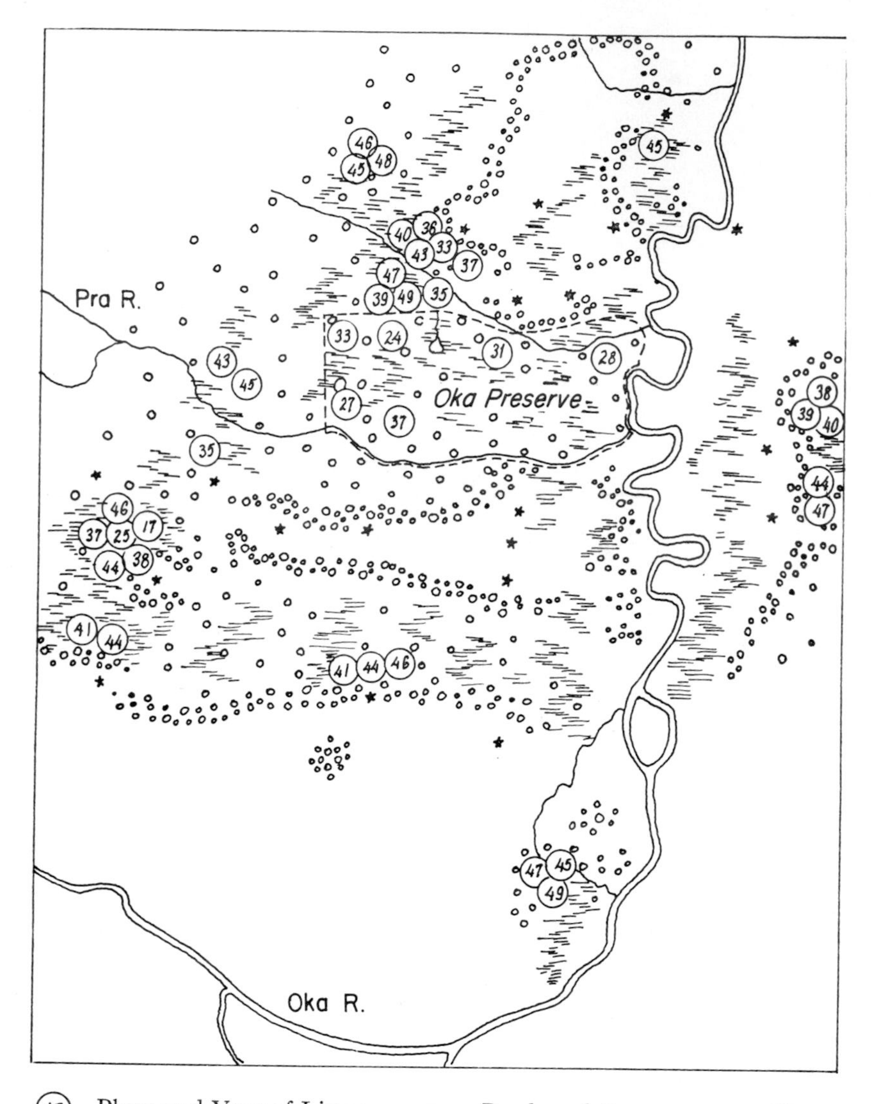

(47) Place and Year of Litter --- Border of Preserve ★ Town

Fig. 197. Location of wolf offspring over a series of years in the region of the Oka Preserve (after Kozlov, 1952).

during the cold part of the year (Fig. 198). The group is usually a family (parents and young) to which solitary animals attach themselves for the winter. It is seldom that such a "pack" of wolves contains more than ten animals (Kozlov, 1952).

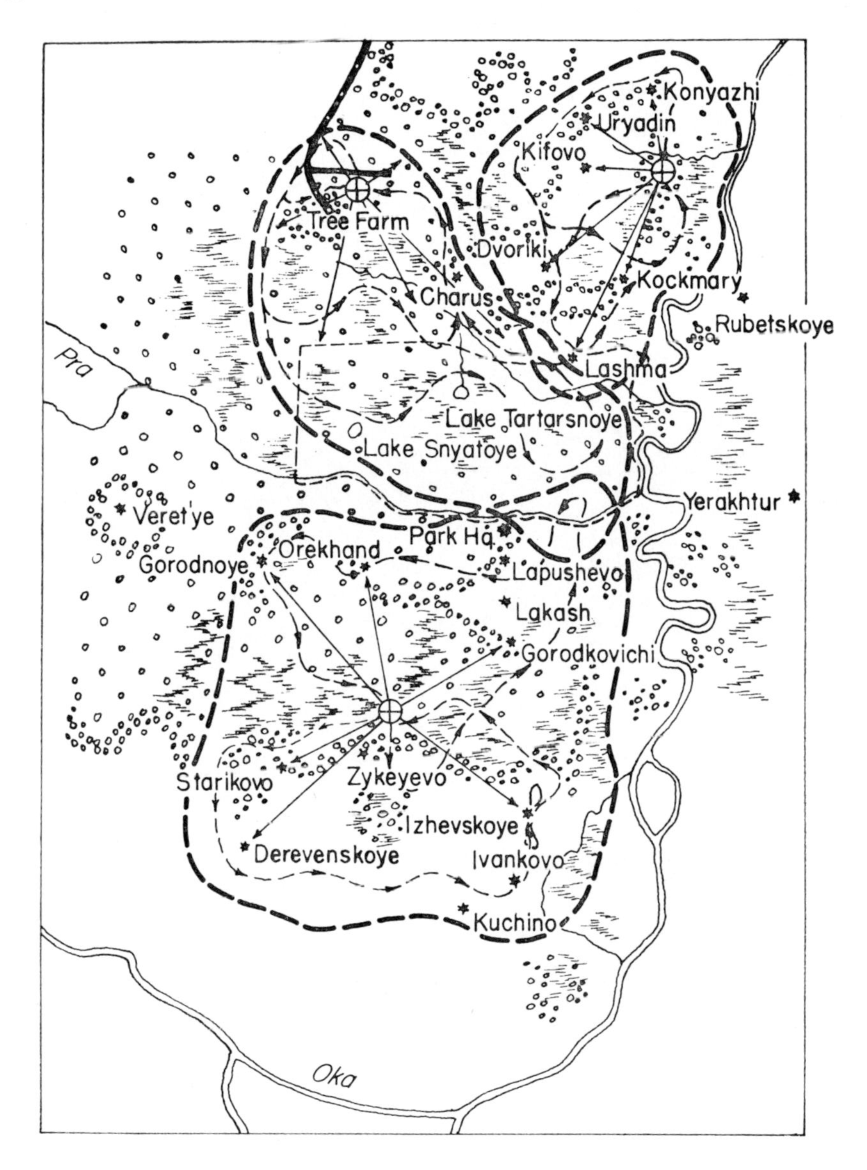

⊕ Place of Litter

-- Border of Hunting Ground

-► Permanent Movement of Wolf

--- Boundary of Preserve

→ Trip of Adult Wolf for Food in Summer

Fig. 198. Hunting grounds of wolf litters in the region of the Oka Preserve (after Kozlov, 1952).

Individual or family grounds are typical of many species of rodents. Our squirrels maintain them not only during the mating season but also in winter. Among these creatures pairs often start forming in autumn. By winter the pair builds several depots on the feeding ground and collects food supplies. During years of poor harvest a significant amount of territory is necessary for this, and during years when the creatures are abundant, they compete for the necessary space. Therefore, in such years the squirrels begin to migrate early, even before they have completely used up the local food supplies and begun to starve. As a result, the migrating animals appear so well fed that they seem to contradict speculations about a famine's being the cause of the migration. Actually, the migrations arise as a result of difficulties in selecting winter feeding grounds and collecting food. The migrations attract more and more creatures until they finally take on a mass character. In such years an early beginning of migration is desirable from a biological point of view, since the participating animals will not yet be emaciated and the trip will be accomplished while the weather is still favorable. Migrations are sharply interrupted when frost sets in (Formozov, 1936). During the trip a considerable number of the creatures die, but those which survive settle in biotopes provided with sufficient food.

Among mice and their ecologically close relatives, the American deer mice (*Peromyscus*), each sessile creature lives in a definite territory, but the plots belonging to separate individuals, although permanent, are not isolated and often overlap (Fig. 199). This overlapping decreases in winter. Among yellow-throated mice the plots belonging to adult females participating in reproduction are isolated from one another during the mating season (Fig. 200), but they overlap considerably with the plots of barren adult and young females and males of all ages (Merkova, 1955). The motility of mice is not great during the winter and autumn and at the beginning of summer, and their individual plots are limited to dozens or hundreds of square meters. Their areas increase (to 2,000–3,000 sq. m) only as the basic supply of food seeds begins to ripen. However, in spring, summer, and autumn the individual plots are used for a short time, after which the family (or individual solitary animal) resettles in a new refuge, sometimes moving 1 km or more, later returning to the old place.

Among field mice in the Volga delta the dimensions of the individual plots of males participating in reproduction changed from 3,200 (1,800–5,300) sq. m in June to 1,500 (1,000–2,700) sq. m in September. Among nonmating males the plots were one-third as large (500 sq. m), and among adult females they averaged 1,300 sq. m in June and 600 sq. m in September. Among young animals they fluctuated between 100 and 900 sq. m, depending on the age of the animal (Nikitina, 1958). Mating female red

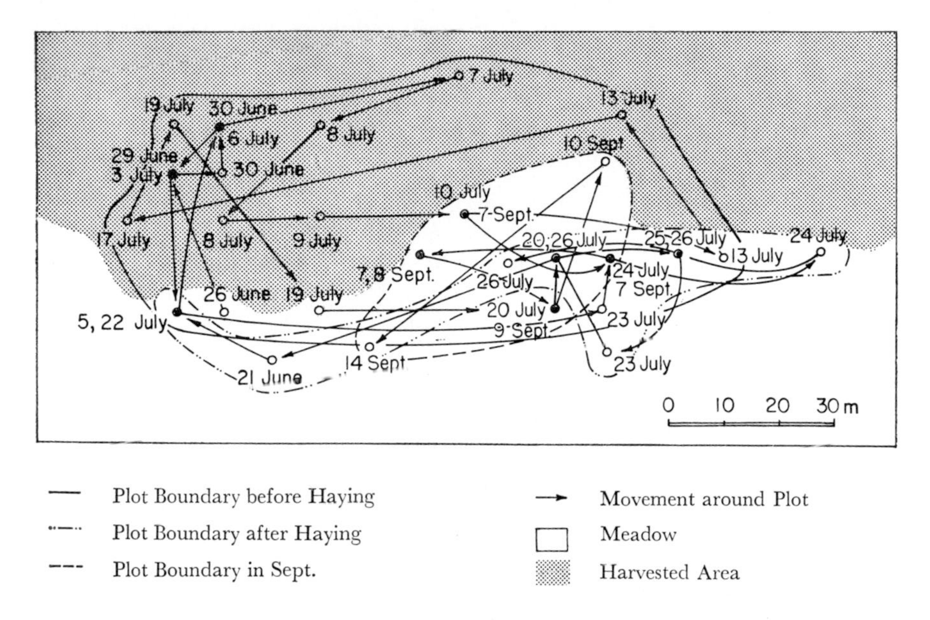

Fig. 199. Individual plot of an adult male field mouse (after Nikitina, 1958).

voles have plots isolated only from one another; the individual plots of the young voles and of adult males overlap and coincide with the plots of mating females (Fig. 201). The plots of sexually mature males are especially large during the mating season (1,000–4,500 sq. m, with an average of 2,700 sq. m). Among the nonmating females and immature males their dimensions change within limits of 100–1,100 sq. m (average 600 sq. m). The plots of the young animals are the smallest of all. After the end of mating (autumn) the difference disappears, and the plots of adult males and females are about equal in size (N. P. Naumov, 1951).

The individual plots of many of the jerboas are not isolated either and consist of two parts. The jerboa feeds in the "basic" part, where other jerboas do not usually live, and his burrow nest is found not far from this "pasture." When feeding is finished—and it occupies only the first part of the night—the jerboa wanders widely, running across plots belonging to other jerboas. During this period it does not feed but merely "tests" food, apparently seeking new "pastures." Actually, within a few days or in 2 or 3 weeks the creatures abandon their old burrows and build new ones near the new places which they have discovered to be abundant in food (Orlov, 1957). In deserts, with their quick shifts in vegetation and invertebrate animal populations, this type of territorial exploitation is biologically

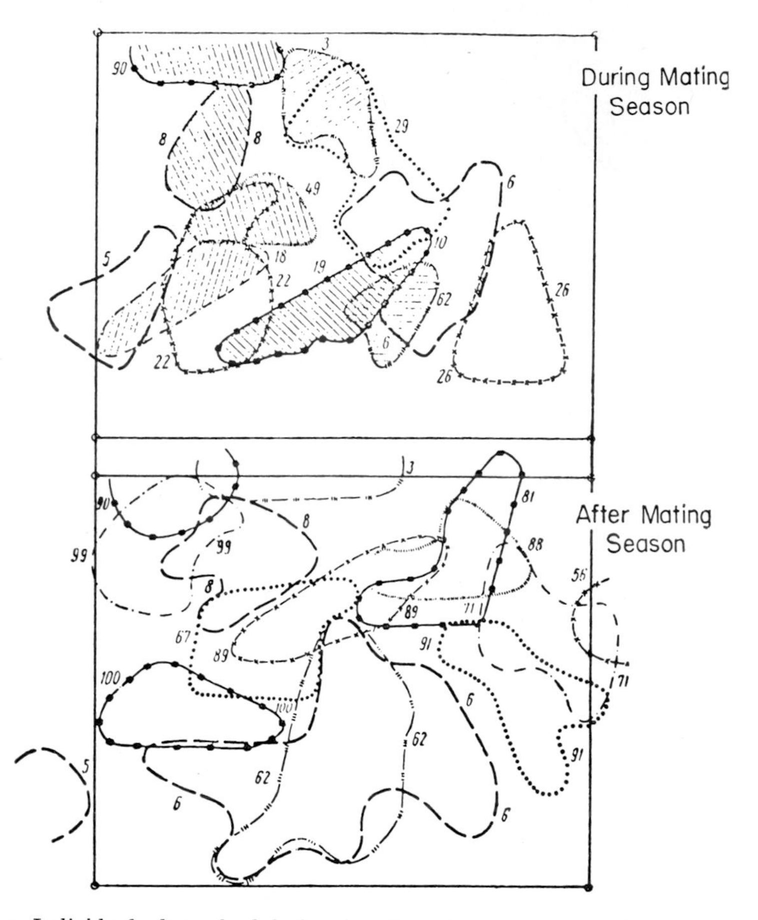

Fig. 200. Individual plots of adult female yellow-throated mice during mating season (July and August) and after its completion (September) in 1950 (after Merkova, 1955). Numbers are those assigned to banded mice.

desirable, but on superficial observation it creates an impression of "anarchy" in the use of the land (Fig. 202).

Among common voles the burrows are occupied by families or single animals for a comparatively short time, after which the voles resettle in a neighboring, unoccupied set of burrows or build new ones. But new inhabitants sometimes settle in the old burrows after a short time. Each family periodically changes its refuges within the boundaries of a definite plot, usually not exceeding 1,000–2,000 sq. m (Karaseva and Kucheruk, 1954; Karaseva, 1957). The plots of individual families partially adjoin one another. When the plots are small and supplies of food limited, an "inter-

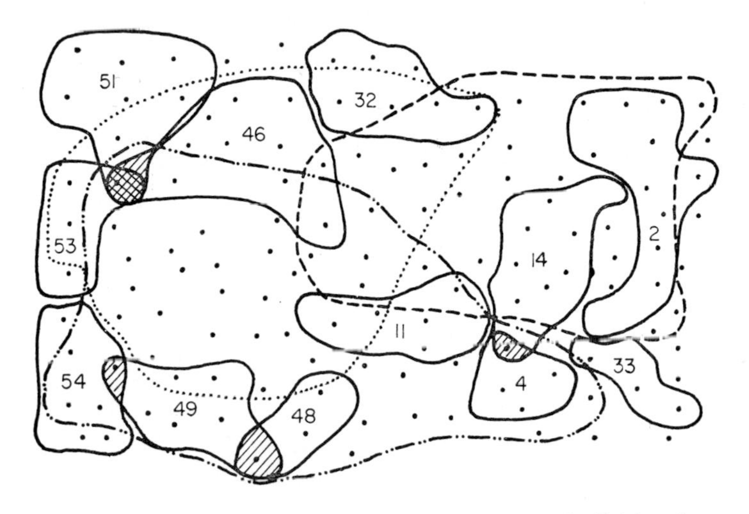

Plot of Mating Female (overlapping parts of adjoining plots are cross-hatched)

Plots of Three Adult Males

Live Trap

Fig. 201. Individual plots of red voles (*Clethrionomys glareolus*) in a park forest in the southern Moscow oblast' in May and June 1948 (after N. P. Naumov, 1951). Numbers are those assigned to animals.

locking" system of territorial use arises in which a family of animals lives in a burrow and uses the plot adjoining it until the food on it is exhausted, after which the animals resettle on a new plot (Formozov and Voronov, 1939).

A type of land use similar to that of the voles has been observed for the large gerbil (*Rhombomys opimus*), which, like the voles, feeds primarily on the growing portions of plants (Fig. 203). When settling in a new burrow, the gerbils drag their young along with them, even when they are still blind and naked. Frequent and regular exchange of refuges, in which the females drag along even the babies which are unable to walk by themselves, has been observed for squirrels, susliks, other gerbils, voles, mice, and hamsters (Stakhrovsky, 1932; Nicholson, 1941; Sinichkina, 1950). This exchange is apparently explained by the multiplication of ectoparasites in the old burrows and the soiling of passages and chambers in the burrow, and the nest itself, with excrement. Polar foxes, red foxes, and wolves also carry their young from one burrow to another, sometimes over a distance of several kilometers, in years of famine.

Thus the existence of isolated food and nest plots is associated with

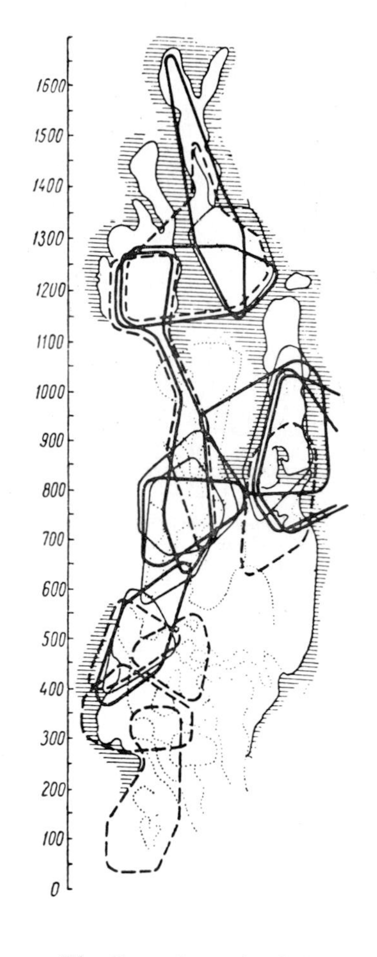

—— Plot Boundary of Adult Male
- - - Plot Boundary of Adult Female
—— Plot Boundary of Young Jerboa

Fig. 202. Mutual position of individual plots of shaggy-legged jerboas in the northern Kyzyl-Kum (after Orlov, 1957).

different characteristics of the way of life. Among insectivorous birds it enables the parents to acquire enough food to feed the young, especially when the parents have poor flying abilities, by regulating the population density in keeping with the ability of the biotope to furnish food. In other cases dispersal and a solitary way of life diminish the danger of attack by predators. Among mice and mating female voles the presence of natural

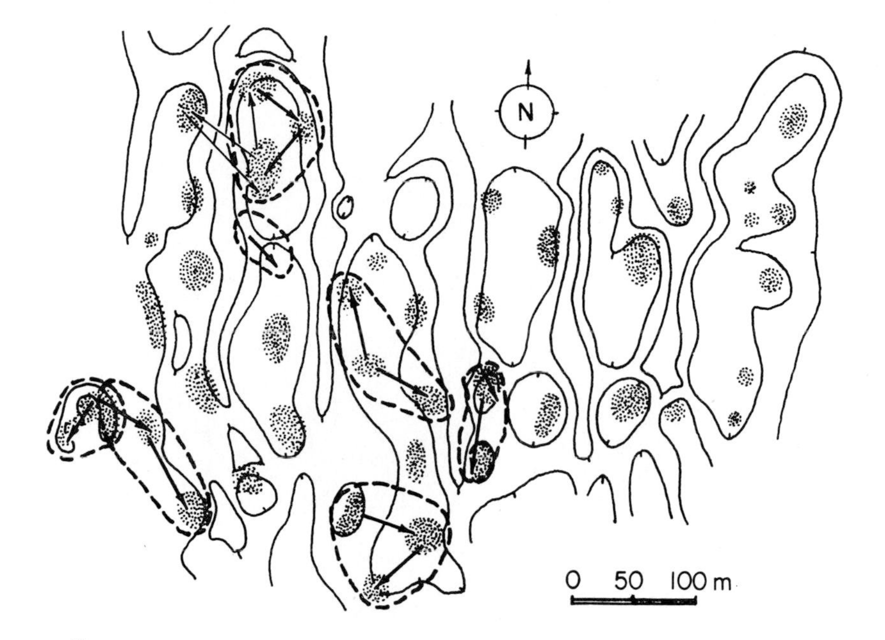

Gerbil Colony (burrow)

Trip Made by Whole Litter

Boundary of Plot Used by Litter

Displacement of Individual Animals after Breakdown of Litter

Contour Line (every 2 m)

Fig. 203. Trips made by litters of large gerbils in the northern Kyzyl-Kum in May and June 1953 (after Kulik, 1955).

refuges takes on a deeper meaning, since these animals do little of their own digging. The necessity of periodically transporting the brood makes these refuges a factor limiting the dispersion of the multiplying individuals, and their absence acts as a brake on reproduction (N. P. Naumov, 1948). Refuges and the dynamics of the food supply determine the character of the jerboa's use of the land.

In many cases refuges limit the number and dispersion of the animals. The possibilities for housing a population of birds which nest in earthen burrows depend on the presence of outcroppings into which holes can be cut. The rise and dimensions of colonies of bats are associated with the presence and capacity of suitable dwellings. Trees or natural hollows are necessary for the construction of suitable residences for the hollow-

dwelling nest builders (woodpeckers, titmice, and many others). They are often insufficient; of the old woodpecker holes, 70 percent in the forests of the Kostroma oblast' and 48 percent in the Moscow oblast' are resettled by redstarts, flycatchers, titmice, nuthatches, martins, and owls (Osmolovskaya and Formozov, 1950). The abundance of hole-dwelling ducks (golden-eye, *Mergus*, and others) depends not so much on the abundance of food as on the presence of suitable holes for nesting near the water. The same is true for the inhabitants of similar burrows or other refuges. Even in the case of open-nesting birds a shortage of suitable sites limits the population. The abundance of wild ducks is in some places determined not by the ability of the land to feed them but by the presence on the banks or in swamps of well-covered and nonflooding elevations about 60 cm in diameter. Therefore, the construction of artificial nesting spots raises the population of the tract. The number of flycatchers, titmice, redstarts, and wrynecks in Latvia in the presence of two to three tree holes per hectare of land grew from 7–10 to 80–120 pairs per sq. km (Taurin'sh and Mikhelson, 1950).

The magnitude and degree of isolation of individual or family plots change in areas with different amounts of available food, refuges, water holes, etc. and in different geographic regions. Thus the average radius of the sable's hunting grounds in the northern taiga (Zhigansk) is 15 km, in the Altay 3 km, on the Pechora 2–3 km, in the Barguzin taiga 2 km, in the Kondo-Sos'va Preserve 1.0–1.5 km, and on the Shantar Islands 1 km. The hunting radius of the forest marten is less than that of the sable. In Lapland (Kola Peninsula) it reaches 15–20 sq. km, in the Pechora taiga 5–6 sq. km, in the forests of the Smolensk oblast' 6 sq. km, in the Tambov and Penza oblasts 3–4 sq. km, and in the forests of the Caucasus Preserve only 1 sq. km (Yurgenson, 1950). Among the deer mice (*Peromyscus maniculatus* and *P. leucopus*) the average size of individual plots is minimal in the ecologically optimal area (the southern Michigan meadows), being 2,550 sq. km for females and 3,125 sq. km for males. In the maple and beech forests of Upper Michigan this size increases by two or three times, while to the south, in southern New Mexico, it increases six to seven times (Blair, 1940, 1943, 1951). Growth of the area on the periphery of the inhabited region (which is necessary for satisfaction of the animal's vital demands) limits the possibilities of the species's distribution.

Among animals with a solitary way of life the need usually arises for a way of marking occupied areas. Among mammals this is done by the scent glands; these exist in solitary species attached to a definite type of biotope (stenotopic) and distinguished by marked competition for territory and, hence, by fighting. The leaving of odorous trails and tracks on the ground and on noticeable objects is a means of demarcating territory occupied by

a given individual as a warning against occupation by a competitor. The other functions of these glands (sexual functions and protection from enemies owing to unpleasant smell) are clearly a secondary phenomenon. In addition to the secretions of the scent glands, single animals, pairs, and families mark territories by other means as well. Wolves lay out "urinals" on all noticeable objects in the area where they hunt and leave their droppings in easily seen areas. Birds use their voices to warn against occupation of the area (singing, cawing, screeching), and woodpeckers use their drumming for the same purpose (P. P. Tarasov, 1960).

The total biological significance of the solitary or solitary-family way of life consists in the fact that separation of territories among animals attached to definite refuges or nests guarantees feeding, protection of the young, or survival by the elders of unfavorable seasons. This method of using an area of habitation is indispensable for species with limited means of movement, but it is possible only in places where the food supply is distributed relatively evenly and where its level on the individual plot is sufficient for feeding the animals living thereon and sometimes for the creation of supplies. This situation actually exists in summer for many small insectivorous birds. Among forms which are less mobile and which use food in the immediate proximity of the refuge (voles, gerbils), the need arises for a periodic change of residence, a need which is augmented by soiling of the nests and multiplication of parasites—hence, the "fallow" system of land use.

The solitary way of life is very important as a means of protection against parasites. We have already noted the significance of dispersion, which precludes the possibility of mass destruction of animals by a predator. Colonial nesting grounds of birds (spotted thrushes, rooks, gulls, sea birds) can be found only in areas which cannot be reached by predators. Good knowledge of one's own grounds, feeding areas, paths of movement, shelters, and location of protective burrows (among mammals) or other refuges is very important as a means of defense. The stereotyped behavior of the area's host is also very important in this connection.

The existence of individual plots and the driving out of pretenders from the already-occupied territory lead to the rise of migration and especially to wide dispersion of the young. This assures a more even and complete distribution of the population suitable for the area and causes a mixing of the individuals, leading to cross-breeding. This eliminates the danger of long-term inbreeding, cases of which are a frequent phenomenon in nature. At the same time such movements cause the animals to become overdistributed, and local overpopulation is eliminated.[1] However, when

[1] Overpopulation is understood to mean a number of animals too great to have sufficient food supplies, number of refuges, or even places of existence.

numbers increase and the proper territory is filled, the solitary way of life leads to more frequent collisions, an increase in intraspecies contacts (which aids in spreading illness), the appearance of mass emigration, a general rise in mortality, and usually a decrease in the population.

GROUP (COLONIAL OR FLOCK) WAY OF LIFE

The group way of life exists equally with the solitary way. Under certain circumstances the combining of animals into groups gives them some advantages and aids them in more fully conquering their environment. These advantages consist of the creation of a favorable microclimate and of teamwork in catching food, controlling competitors (especially the closely related species), and defending against enemies. The working out of such a way of life is aided by the uneven distribution of food; it is possible when sedentary animals are highly mobile and are associated with permanent refuges (colonial species) or live the so-called nomadic way of life.

Many species which lead a solitary-family way of life in the mating season turn to group living in autumn and spring. For others the group way of life is permanent, although the dimensions and makeup of the groups usually change from one season to another. The group way of life is widespread among the various classes of animals.

Flocks and herds

Flocks and herds are encountered among many species of insects (the migratory locusts, such butterflies as *Pyrameis cardui*, and others), fishes, birds, and mammals. Flocks and herds lead a nomadic life, moving constantly from one part of the territory to another. But nomadism is not mere wandering, since each flock (herd) usually has its own region of habitation and its own migration routes different from those of neighboring groups. This brings about an even mixture of population throughout the territory suited for habitation. Sometimes such regions adjoin, often mixing with one another. Distinct separation of pasture lands may occur among the ruminants (Dmitriyev, 1938; O. I. Semënov-Tyan-Shansky, 1948). The size of the herd changes from season to season, and the comparatively large winter groups break down into smaller ones in the summertime (Table 55).

TABLE 55. *Seasonal changes in herding behavior of the northern reindeer* Rangifer tarandus *in the Lapland Preserve (after O. I. Semënov-Tyan-Shansky, 1948).*

Season	*Herd of 1 to 10 Head in percent*	*Herd of More Than 10 Head in percent*	*Average Number of Reindeer in Herd*
First half of winter (Nov. to Jan.)	1	99	52
Second half of winter (Feb. to Apr.)	0	100	86
Period of calving (May)	16	84	16
Summer (July to first half of Sept.)	63	37	3
Calling period (second half of Sept. and Oct.)	7	93	20

Flocks and herds are distinguished first of all by their tight unity and the common behavior of their individuals, but they are not clusterings of animals which are uniform within themselves. Among northern reindeer the vanguard of the herd is usually made up of the more restless and wild individuals—the ones which finish their feeding and lie down to rest before the others but which do not rest for long and are the first to set out on the migration. The rear guard is made up of laggards which spend most of their time on the pastures. The quiet animals which have fed better and suffered fewer losses to predators are usually found in the middle of the herd (Sdobnikov, 1935).

Birds which feed on carrion (carrion crows, buzzards, vultures, and others) are an interesting example of the transition from a solitary to a flock type of existence. They are sedentary, but they form unique frangible flocks. When spotting food from the air during a solitary hunt, they follow one another in an unbroken line and quickly gather wherever any single member of the group observes carrion. In doing this, they often make use of vocal signals (cawing) emitted not only by their own but even by different species.

Colonies

A colony is a group of cohabiting sedentary animals and consists of free individuals sharing a general shelter (the social insects, bats) or having nests situated right next to one another (colonial nesting grounds of birds).[2] The complex families of insects (bees, ants, etc.) are similar to

[2] We are not concerned here with colonial organisms which form a single complex aggregation in which the individuals closely adjoin one another, have a common matter exchange, and often have functions allocated to and exchanged by the different individuals (sponges, corals, polyps, etc.).

colonies but have their own unique system in which separate groups of individuals are morpho-physiologically distinguished and have different functions. These family-colonies are distinguished not only by division of labor but also by their high degree of integration of the individual, as expressed in the intracolonial organization and the agreed-upon behavior of these individuals. Complex instincts lie at the basis of both traits.

In colonies of higher animals the connections between individuals are less tight. Among bats the size of the colony depends upon the ability of the species to move around. Among the long-winged and fast-flying species the colonies consist of several tens of thousands of individuals. They are adapted to large caves and are separated from one another by great distances; in the Crimea two or three such colonies are known, and there are five or six in the Caucasus.[3] The less mobile broad-winged bats form smaller colonies. The number of individuals in them seldom exceeds 100 and more often consists of several dozen. Colonies of these species are usually smaller in the northern regions, but the number of colonies per unit of area is greater. In the south the actual colonies are larger, but they are situated more sparsely. The hunting grounds of the northern colonies are, as a rule, isolated; those of the neighboring large colonies in the south usually adjoin one another, at least partially (Kuzyakin, 1950).

Colonial settlements are aggregations of individuals leading solitary and group ways of life. Among birds they are represented by the colonial nesting grounds of rooks and gulls, the avian bazaars of sea birds, etc. Each such colonial settlement or group of colonial nesting spots has its own hunting area, the size and feeding capacity of which govern the size of the colony (Fig. 204). The so-called bird bazaars are a colonial nesting ground of several species: gulls, auks, guillemots, puffins, and certain others (Kaftanovsky, 1951; Belopol'sky, 1957). Usually the nests of the different species occupy different parts of such a colony, but sometimes they are situated in strips. After mating and birth the number of guillemots on a nest cornice reaches 80–100 per sq. m. The adults energetically defend the nest area and the nest proper against their neighbors. However, they themselves try to steal food brought by other birds (Modestov, 1939). Such attacks and combats are sometimes accompanied by the wounding and even killing of young birds, but this seldom occurs when nests are sparsely scattered (Skrebitsky, 1940).

The negative significance of these collisions is beyond doubt, but as compared with the advantages of colonial nesting it is not great, especially in terms of protection against predators. In a dense nesting area of *So-*

[3] In Bakharden Cave on the north slope of the Kopet Dagh (Turkmenia) live some 40,000 long-winged bats (*Myniopterus schreibersi*), two species of horseshoe-nosed bats, and several others.

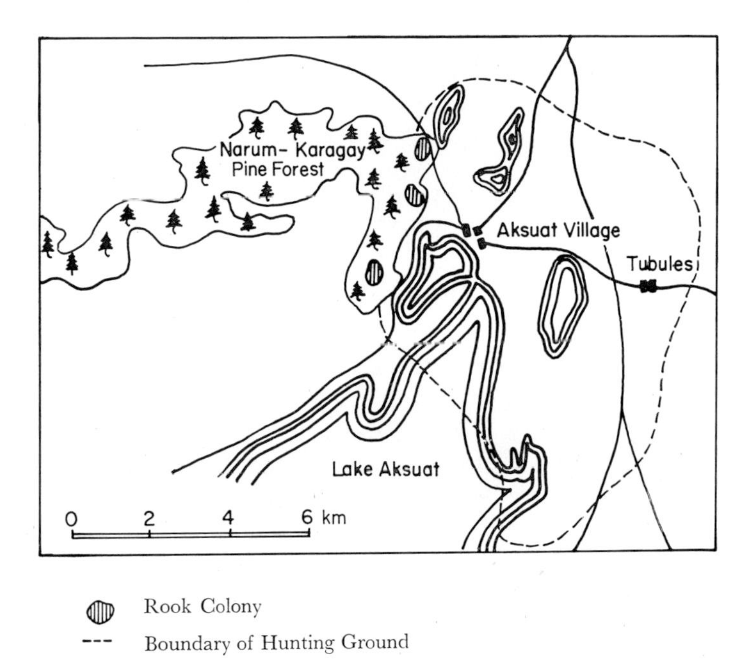

Fig. 204. Hunting plot of three colonies of rooks (*Corvus frugilegus*) in the Naurzum Preserve (after Formozov, 1952).

materia mollissima situated in the open (on Novaya Zemlya), it was mainly the seagulls which ravaged 5.0–7.5 percent of the nests, while in a scattered colony 22–28 percent of the nests were plundered. At the same time the average number of eggs per nest was 3.2 in the former case and 2.6 in the latter. *Somateria mollissima* nests are scattered only in areas with a large amount of shelter under which they can be built. But these same birds form large and dense nesting colonies in open areas. Predators energetically attack small and scattered guillemot, kittiwake, and polar-gull colonies, but they cannot cause any significant loss to the dense populations of these birds, which stoutly defend their nesting grounds. In this way predators actually augment the coloniality of such birds. However, under other circumstances the same predatory activity may lead to different results (Belopol'sky, 1957).

A unique type of group combination is characteristic of the susliks, marmots, large gerbils, some voles, and apparently the pikas living in the sparse-grass steppes and deserts. Their colonial settlements consist of individual plots of separate individuals (or families) with borders adjoining one another. Because of the comparatively small size of each plot, a single

settlement is formed in which the animals see each other and can signal danger by shrieking. As in the colonial nesting grounds of birds, penetration of foreign individuals is not allowed, but the distant feeding areas are often used conjointly. However, each individual or family feeds primarily on its own little plot (Fig. 205).

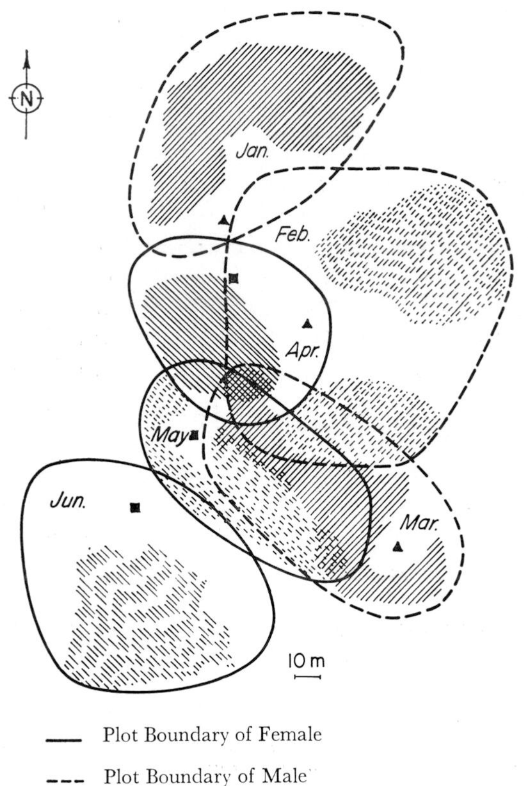

Fig. 205. Individual plots of small susliks (*Citellus pygmaeus*) in the lower Volga region (Dzhanybek Station) in the period 10–30 May 1952 (after Soldatova, 1955). Cross-hatching shows the feeding area preferred by each suslik.

The biological significance of group combinations of animals is multifaceted. They aid in creating a favorable microclimate and sometimes a favorable environmental chemistry, influence metabolism, and change the character of growth and anatomic and behavioral properties of the indi-

viduals. The creation of favorable microclimatic conditions plays an important role in the lives of animals and plants. The varying density of growth in plants creates optimal illumination and maintains a favorable temperature-humidity regime. Animals utilize the ecoclimate created by the plant association, but often they create their own.

In families of social insects the clustering of many individuals permits an almost constant temperature to be maintained. Despite the poikilothermic qualities of its individual members, a bee colony can maintain an almost constant temperature within the hive. In summer the worker-bees beat their wings, ventilating the hive, increasing the evaporation of water from the fresh honey, and lowering the temperature in the hive; in winter they collect in large clusters, thus decreasing heat dissipation. A drop in the temperature of the hive to 13° C causes the bees to become restless, and they increase their movements and thus their heat production so as to raise the temperature once again. When 26° C is reached, the bees cease moving and again collect in bunches (Réaumur, 1740). The rudiments of microclimate regulation may be observed even in such aggregations of insects as cultures of flour worms (larvae of the beetle *Tenebrio molitor*), in which the temperature is higher and more stable than outside the aggregation (Fig. 206). In anthills and clusters of termites the temperature and humidity are very stable. In these clusters the animals easily tolerate large doses of light energy.

The density of the population in a culture of *Drosophila* affects the life span as shown below. Both very high and very low densities have a negative effect on fruit flies.

Water is changed by the organisms living in it; substances are excreted

Initial Density (number of flies in test tube)	*Average Density (number of flies)*	*Average Life Span in days*
2	1.8	27.3
4	3.3	29.3
6	5.0	34.5
8	6.7	34.2
10	8.2	36.2
15	12.4	37.9
25	20.7	37.5
35	28.9	39.4
55	44.7	40.0
75	59.7	32.3
95	74.5	27.2
105	80.4	24.2
125	94.4	19.6
150	111.9	16.2
200	144.5	11.9

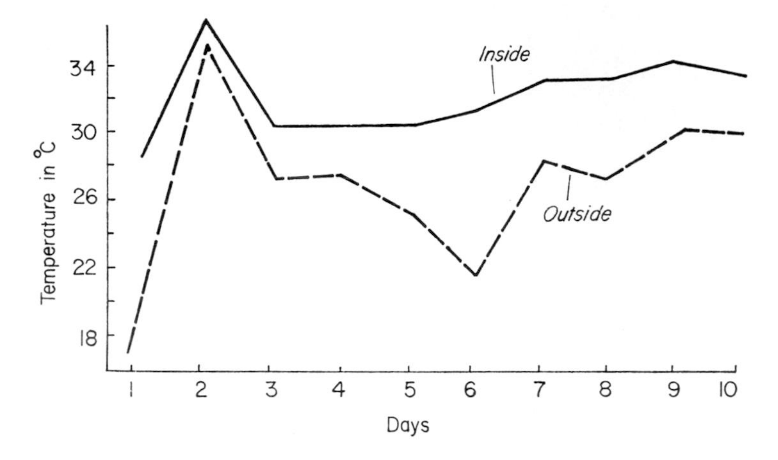

Fig. 206. Changes in temperature inside and outside a "culture" of flour worms (larvae of the beetle *Tenebrio molitor*) (after Michal, 1931).

into it which change its chemistry and physical properties. These changes are favorable to a known degree for growth and development, but if they exceed that degree, they serve as a brake. The water in which fish or aquatic invertebrates have been found acquires the ability to fluoresce when illuminated with ultraviolet light, and the fish in it grow faster. When the excretions of water animals accumulate too quickly in a water environment, they poison it. Therefore, in creating high-production bodies of water, the excretions are ameliorated, lowering the harmful effects of contamination and poisoning by metabolic products.

Among certain species of animals clustering has a favorable effect upon metabolism. Thus oxygen consumption by the oligochaete *Enchytraeus albidus* in atmospheric air increases when the worms are in either a high-density or a low-density situation (Fig. 207). In the ground the worms use oxygen less than in the air, and here the effect of population density on oxygen consumption is not very noticeable. Lower oxygen consumption and more economical oxidation processes occur in water, with an increase at low population densities (Ivleva, 1960). Goldfish which have been placed alone in a bowl will consume noticeably more oxygen than those which are kept together. It has been shown that lowering the basic exchange acts on neighboring individuals (Allee, 1934).

Among the "peaceful" fishes which form winter clusters (the carps, certain sturgeons) the level of matter exchange drops whenever several individuals get together. The ocular receptor is the leading indicator; blindness accompanies loss of the ability to lower matter exchange in groups (Shtefan, 1958). Among species with imperfect vision and a highly developed

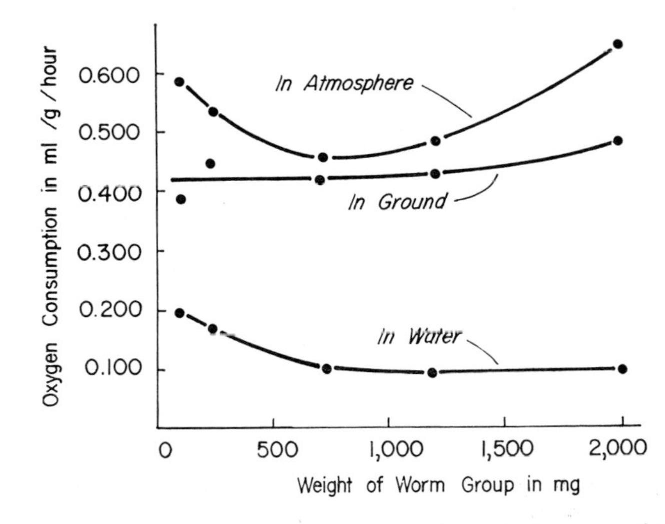

Fig. 207. Intensity of respiration at a temperature of 20° C of the aquatic oligochaete *Enchytraeus albidus* in different media, depending on density of population (after Ivleva, 1960).

sense of smell, the latter is the basic means of sensing the presence of neighbors. A minnow put into water in which other minnows had been a few moments before ("school" water) lowered its level of oxygen consumption almost the same as if it had been confined with a group (Fig. 208). Among predatory fish and other lone types there is no ability to lower the level of exchange in a group, or this ability is poor. Lowering the level of exchange during clustering is achieved by a more economical expenditure of reserve matter, which is important for survival during the winter and in the following mating season. The same holds true for the gregarious (in winter) amphibians and reptiles.

Because of the combined habitation in winter of small rodents (mice,

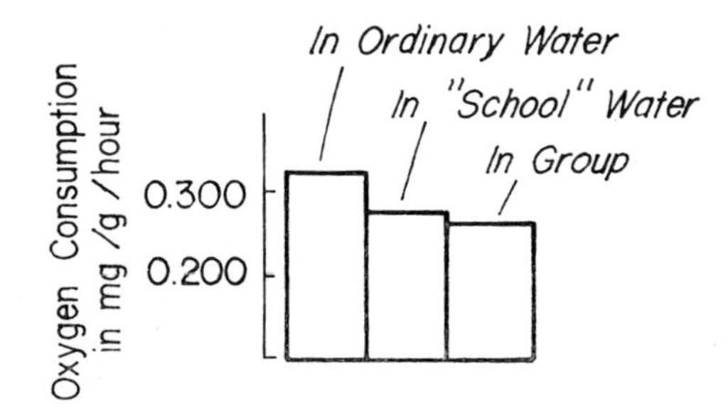

Fig. 208. Consumption of oxygen by the minnow in ordinary water, in "school" water, and in a group (after Shtefan, 1958).

voles, gerbils, etc.), the temperature in the burrows and especially in the nesting chambers, which contain five to ten or even more animals, exceeds the temperature of the surrounding environment by 10–20° C, usually going no lower than 12–15° C (Fig. 209). Conjoint habitation has a great effect on animals' heat exchange. Young common voles living alone at an air temperature of 15° C had a body temperature of 15–21° C, while in groups it rose to 31–34° C (Strel'nikov, 1950).

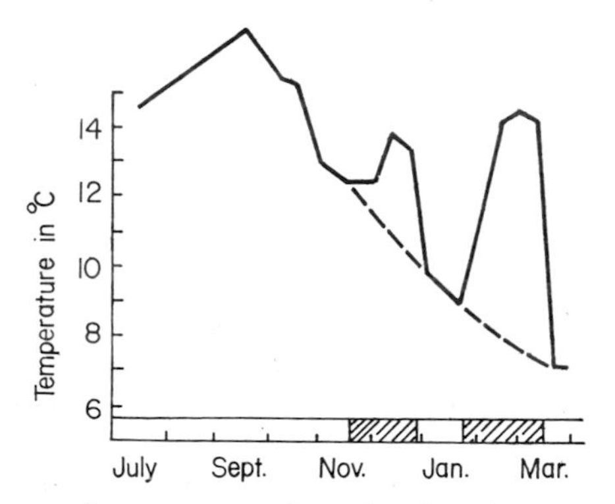

Fig. 209. Seasonal changes in temperature in a chamber of a burrow of southern gerbils situated at a depth of 285 cm (after Rall', 1939). Cross-hatched area represents the time when five animals were in the burrow; the dashed line represents the probable course of the temperature in an empty chamber during this period.

As is true among the peaceful fishes, clustering mammals can also depress their metabolism automatically when they cluster (Fig. 210). The decrease reaches 38 percent among the frequently clustered society voles and only 22 percent among common voles. It has also been found experimentally for the small suslik, the garden dormouse, and several species of bats, but it is absent among long-eared hedgehogs, thin-fingered susliks, plate-toothed rats, white rats, and red-tailed gerbils (Ponugayeva, 1953). The existence of flocks among gallinaceous birds permits them to survive even severe frosts. The importance of mutual warming has been shown in experiments with bobwhite quail at low temperatures (Table 56).

The formation of nesting colonies of birds and herds of mammals is

TABLE 56. *Death rate of bobwhite quail at a temperature of −18° C (after Gerstell, 1939).*

Death Rate	NUMBER OF BIRDS IN GROUP			
	10	*5*	*3*	*1*
Percent dead within 13 hours	10	20	100	100
Percent dead within 20 hours	10	60	–	–

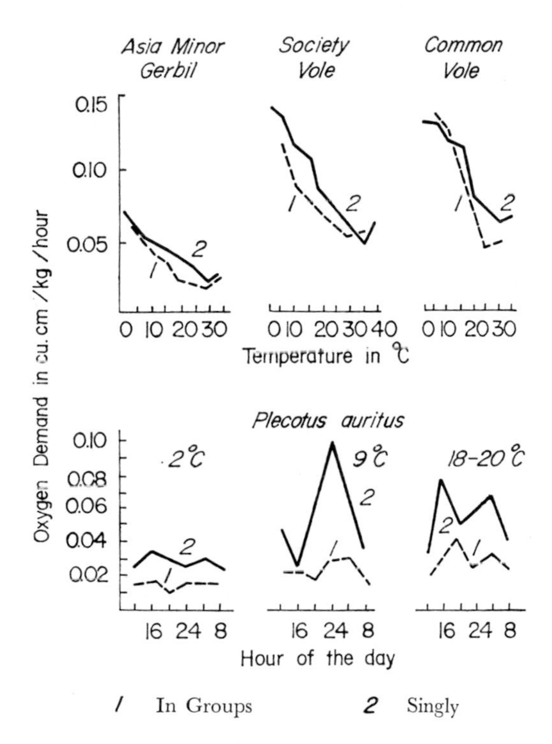

Fig. 210. Oxygen consumption level of gas exchange at various external temperatures and according to confinement in groups or singly of the Asia Minor gerbil, society vole, and common vole (after Ponugayeva, 1953), and daily changes in the consumption of oxygen by *Plecotus auritus* at various environmental temperatures (after Ponugayeva and Slonim, 1949).

associated with mutual defense against predators and group hunting of food. Many colonial and semicolonial birds (rooks, starlings, thrushes, and others) not only defend the nesting grounds against enemies together but also seek their food together. Rooks seek their food this way on a meadow. A row of birds proceeds in a line like a chain of beaters on a hunt, moving along and scaring up the catch (large insects), to be harvested either by the beaters themselves or by the birds following them.

Conjunctive storing of food (gerbils, sometimes voles, mice, and others) is also a basis for animal unity during the hungry part of the year. Large stores, sometimes reaching dozens of kilograms, are collected through the efforts of many creatures. Such stores are especially necessary for species living in regions with a severe windy and snowless winter which does not permit them to feed on the surface of the soil (the large gerbil, Brandt's vole). On the other hand, the digging species which feed in winter on

subterranean plant parts do not form aggregations, since this would be biologically unprofitable. Semicolonial animals which cluster in the summertime and feed during this period on bast and sedge roots, like the water voles, change their solitary way of life in the northern part of the range. But in winter each creature lives in its own burrow with its long, surface food passages, during the digging of which it obtains root hairs, roots, and bulbs (Vishnyakov, 1957). In the Volga delta the semicolonial way of life of the water vole is maintained all year long among the great bast and cane patches (Kondrashkin, 1957).

The preservation of family groups of wolves for the winter, and even an increase in their sizes by adoption of solitary individuals, is associated with the particular way in which the animals hunt their catch. In summer families and lone wolves catch the small animals which are abundant and easily obtained during this season. In winter they live on large prey (ruminants), which are best hunted by packs. Conjunctive food-seeking plays an important role among ruminants in winter, and it is then that their herd tendencies are most strongly developed. The larger animals scrape away the snow to aid the younger ones in getting food. When the numbers of the herd diminish and the size of the group contracts, feeding is more difficult and the number of animals lost to predators rises considerably. Colonial settlements of digging rodents (marmots, susliks, large gerbils are associated only with protection from predators. Similar clusters are formed by some insects. The caterpillars of some butterflies begin to go into mass twisting movements, like the sweep of a pendulum, whenever danger appears. When danger threatens, the *Macropsis* beetle warns its neighbors by squeaking.

Finally, many groups are associated with mating. Such are spawning schools of migratory fishes, harems of ruminant animals or some pinnipeds (seals), combinations of birds on their mating grounds, and similar formations encountered among vertebrates and invertebrates alike. The colonial nesting grounds of birds are also associated with mating. Progressive diminution of their dimensions among silver gulls, fulmars, and gannets is generally accompanied by a decrease in the amount of mating, a slowing down and extension of the egg-laying period, and a lower rate of success in bearing and raising the young. However, among the penguins *Pygoscelis papua* and *Eudytes cristatus* no such relationship between reproduction and the size of the flock can be found.

A special phenomenon is to be found in the phases of the metamorphic locusts (migratory, desert, Moroccan, and Italian types) which arise when the populations reach different densities. Depending on whether they live in groups or alone, not only their external appearance and physiological properties but also their behavior will change. The usual state of these in-

sects is the solitary phase, which is distinguished by a green or yellow outer covering. Formation of the flocking phase occurs as the result of an increase in abundance and forced concentration of nymphs in a limited area with a density of hundreds of individuals per sq. m. Such groups are caused by uneven drying out of vegetation (during a drought) or unusual moistening (during rainy weather). In the former case the multiplying insects concentrate in the depressions where green vegetation has been preserved, while in the latter they accumulate in high, unflooded places. In both cases they live in such close contact that they are constantly touching one another. The result is that grouping becomes their normal way of life. A conditioned reflex of gregariousness is developed, as a result of which the insects not only fail to disperse but even try to stay together if they are artificially separated. This is how swarms of locusts are formed. The transformation of the solitary phase of the desert locust *Schistocerca gregaria* into the swarming phase takes place in three ways: (a) concentration—an absolute increase in the number of locusts on a given area of land; (b) aggregation—active grouping into swarms, in which the behavior of neighbors more and more determines the behavior of each individual; (c) gregarization—formation of even greater swarms with rapid attraction of all individuals encountered. These are very active and undertake large-scale migrations.

Great importance in the rise of the swarming phase and the consequently huge swarms of locusts is now attached to gregarization as it occurs among adults. Because of the flocking way of life, the excitability of the nervous system is increased, and along with this the activities of metabolism and motility. Part of the metabolic products which are produced are deposited in the cuticle, giving the body its characteristic coloration, with dark and red spots. This increases the absorption of solar rays, which further increases the intensity of the metabolic processes and the locusts' mobility. The temperature of clustered locusts often exceeds that of the surrounding air by 10–15° C. The increased work of movement creates conditions for a change in the proportions and structure of the individual parts of the body (Fig. 211). A reverse formation of the solitary phase takes place when locusts die off and the density of the population is decreased. In the northern part of the range of the Asiatic locust, where population density is never very great, only the solitary phase exists (*Locusta migratoria rossica* in the European USSR and *L. m. danica* in the Baltic area) (Kozhanchikov, 1956).

The swarming phase of locusts has a biological meaning, since they are capable of undertaking distant migrations. These are a means of dispersing the locusts, causing new areas to be settled and old nesting grounds to be re-established if they have begun to die out for any reason. This method not only maintains the area of settlement (the range) but also produces

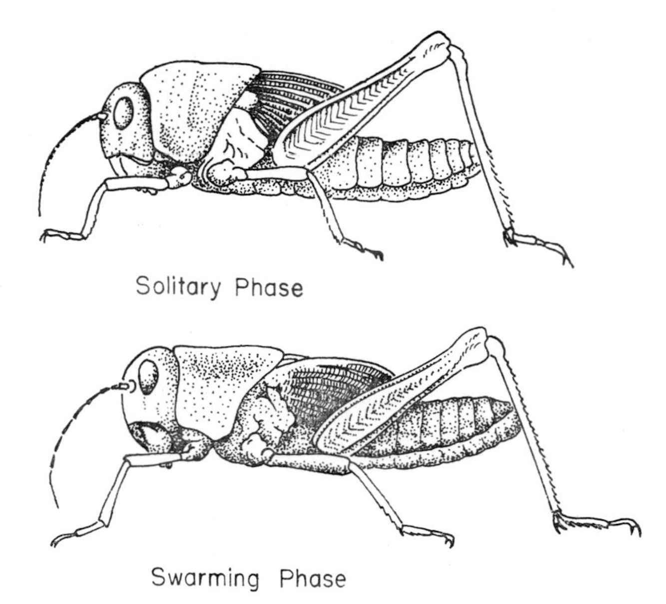

Fig. 211. Nymphs of the Asiatic locust *Locusta migratoria* of stage 5 (from Bey-Biyenko and Mishchenko, 1951).

continuing attempts at expanding it. Thus a phase is a special form of existence of the species in a definite environment, while the specific form of existence depends on the abundance (density) of the animals themselves.

Unlike the other flocking locusts, whose swarming grounds are formed in areas with a mosaic of biotopes and nesting areas which are more or less constant, the desert locust has no permanent nesting area. Therefore, succeeding generations as a rule are born in different places, and the nests of the offspring will not be those of their parents. The desert locust wanders over the great expanse of its range whenever an increase in density causes it to seek the most favorable areas. In the solitary phase it is as immobile as the other locusts, but in the swarming phase it may fly from the Sudan to Río de Oro or Morocco, East Africa, India, Afghanistan, or the USSR (Shcherbinovsky, 1952; Udarov, 1957; Bodenheimer, 1958).

SEASONAL CHANGES IN WAY OF LIFE AND ORGANIZATION OF POPULATION

Seasonal and other changes in conditions of existence are accompanied by changes in intraspecies organization. The solitary-family way of life

characteristic of many species during the mating season is replaced by a group form of life in autumn and winter. The group way of life at this time has several advantages associated primarily with protection against unfavorable temperatures and with searches for food. Shield bugs, ladybugs, and other insects cluster in groups toward winter (Arnol'di, 1947). Many fish, amphibian, and reptile species which are solitary during the warm part of the year winter in groups. Winter clusterings are known for freshwater and semi-migratory fishes. The carps and sturgeons approach the mouths of rivers after fattening and settle in holes in groups of several thousand. In winter they do not feed and remain immobile. In the Volga and Ural rivers fishes of several species will often settle down together, but each species will stay apart from the others. Among the Far Eastern flounders *Liopsetta obscura* and *L. glacialis* the wintering clusters are even formed in small bodies of water where the fish gather in large numbers to lie immobile on the bottom.

Similar clusters are formed in the temperate latitudes by many amphibians (grass, pond, lake, Siberian, and sharp-nosed frogs), which collect on the bottom after a drop in temperature. Among grass frogs the average number of individuals per wintering spot is 24; areas with 100 or more individuals are rarely encountered. The viper *Vipera berus* also winters in groups, clustering in the empty spaces under tree roots and in other refuges (Raspopov, 1935). Heat-exchange conditions are improved in such a den of snakes, and the animals are protected from freezing with a minimal expenditure of reserve matter, thus raising considerably their chances of surviving the winter.

In the postnidal period the broods of many birds combine to form larger flocks and lead a wandering or sedentary form of life. When migrating during the winter, some species cluster, and the separate individuals warm each other. Food consumption usually rises during a temperature drop among small wintering birds. According to A. N. Formozov (1950), the Lapp titmouse searched 280 branches in an hour in order to collect the necessary amount of food (insect larvae), and in the same length of time a long-tailed titmouse visited 144 trees and inspected some 1,120 branches. The number of branches which one bird can inspect during an autumn day must therefore approach 10,000. This necessitates a transition to a wandering way of life, such as is characteristic of most small winter birds at this time of year. The group way of life arising at this time is profitable as a protection against predators (warnings of danger). Group searches for food and exchange of information about food, which are characteristic of many species, have their advantages. The cries of the Siberian nutcracker when it finds a harvest of cedar "nuts" attracts all the neighboring birds. When harmful insects appear en masse, the birds

easily concentrate in the area; not only their own but other species as well respond to the signal. Such relationships aid in the formation of mixed flocks of several species.

Small mammals form winter groups which provide a favorable temperature in burrows and refuges. The last autumn litters of the large gerbils do not, as a rule, break up; in fact, they often pick up the loners which broke off earlier. As a result, the burrows of these animals have a larger number of inhabitants in winter than in summer. Among the Brandt's voles living in the Mongolian steppes, 10–20 animals usually live in one burrow in winter. The entrances to the burrows are stopped up with earth from the inside at the start of winter, and for 2.0–2.5 months the animals live on food supplies stored in large warehouse chambers (Kucheruk and Dunayeva, 1948). Southern gerbils, which lead a solitary-family way of life in summer, combine into groups of considerable size in winter and change to a colonial way of life (Rall', 1938, 1939). Mice and voles also cluster in their burrows in winter. In the Volgo-Aktyubinsk floodplain up to 29 house mice, 15 common voles, and 17 gray rats per burrow were unearthed in the winter of 1937/38 (Fenyuk, 1941).

PROPERTIES OF INTRASPECIES RELATIONSHIPS

The organization (structure) of a species is a special group adaptation providing for population reproduction, increase in numbers, maintenance of the space already occupied, and settlement of as yet unpopulated areas. The population structure has developed during evolution by natural selection, but this certainly does not mean that it is free of contradictions.

The external forms of relationships between individuals of a species are similar and are sometimes difficult to distinguish from the relationships of different species. Thus among many species we find frequent cases of aggression in which individuals of the same species kill and eat one another. This intraspecies predation is widespread among fishes. Cannibalism is regular and even biologically desirable among cod, burbot, and Balkhash perch. The adult fish, by eating their own young, may assimilate a type of food (such as plankton or benthos) which is unavailable to them directly. Cannibalism explains the fact that only one species (perch, pike) may be able to exist in some bodies of water. In addition to feeding of adults on the young of their own species, we also observe the reverse, as in the case of the young Pacific salmon which eat the bodies of their elders, which have died during spawning. In the fish-processing plants of the

Soviet Far East the bodies of old salmon are often used as food for the young (Klyuchareva, 1956).

Cannibalism among large-mouthed perch is caused by the difference in growth rates of individuals of the same age which fatten on feeding grounds of differing value. When this happens, the large ones eat the small ones. Cannibalism is also widespread among other groups of animals.

Birds, mammals, reptiles, and many invertebrates fight actively for shelter or nesting grounds; this is associated with insufficient territory, food, or shelter. True parasitization of males on females or vice versa occurs among several invertebrates and in fishes among the Ceratioidea. Passive competition and active struggle for food, shelter, and other conditions are widespread among individuals and their groups, but cases of mutual aid are also frequent, and so is parasitism. This similarity of external forms and intraspecies relationships has been the cause of many false interpretations of the nature of intraspecies connections.

Darwin himself saw that within the relationships of individuals of a species, which use the same vital resources and which contaminate and poison the surroundings with their excretions, there always lies a hidden danger of marked competition—a danger which increases as the population density of a species increases. However, Darwin did not realize that the prevalence (sometimes a very great prevalence) of offspring over survivors is an adaptation compensating for the high mortality rate of the individuals in their various stages of development. Therefore, fecundity corresponds to a known degree to the hardiness of the individuals of the species (N. A. Severtsov, 1855; S. A. Severtsov, 1941) (Fig. 212).

Although the character of reproduction is also an adaptation to the conditions of the species' existence, it, like every adaptation, is only a relative response to the changing surroundings. It does not exclude the possibility of either surplus multiplication, and overpopulation, or underpopulation. The danger of both, though controlled by several intraspecies adaptations, is still not eliminated and is constantly expressed in fluctuations in species populations.

Despite external resemblances, inter- and intraspecies relationships are different in principle. Coadaptation, or mutual adaptation of individuals of different species, is the evolutionary result of interspecies adaptations. Examples can be found in the coadaptation of predators and prey, parasites and hosts, pollinators and pollinating plants, and many others. In all these cases individuals of various species win out because of the survival of individuals of one species or of both together. A consequence of interspecies relationships is the formation of associations differing only in the relative stability of their species contents, the possibility of one species' easy replacement by another, and the simplification or complication of the

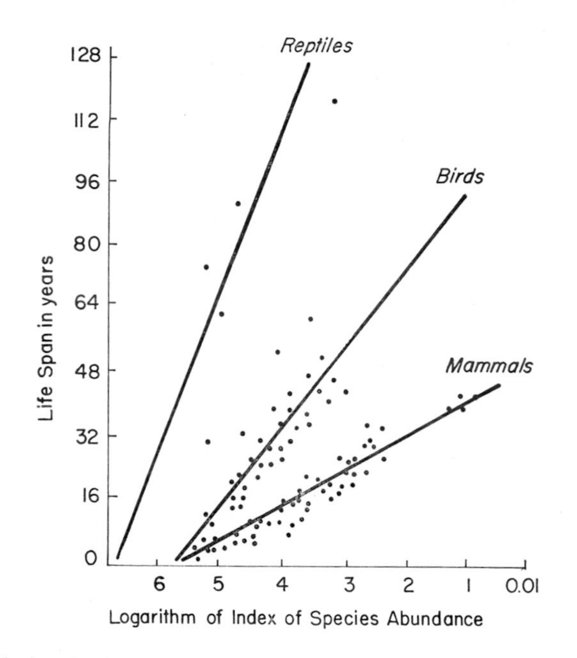

Fig. 212. Correlational table of life span and fecundity of mammals, birds, and reptiles (after S. A. Severtsov, 1936). Each mark represents a separate species.

species present in an association by the inclusion of new or removal of older species from the association. When this happens, the associations change but do not disappear or die off. Hence they are highly motile systems of distinct species.

As was pointed out above, the species is also a system of biologically different forms. But the connection between them is tighter and differs in principle, since the species is a solid, integral system with subordinate and closely associated groups. At the basis of this connection lie the congruences described by S. A. Severtsov (1951): the mutual adaptations of individuals of the same species. These are quite different from coadaptation, since they are primarily directed toward maintaining the integrity and unity of the species. Congruences often appear "harmful" for the individual, diminishing the chances of its survival and subordinating its life to the survival of the species.

Such are all the properties of reproduction and the various adaptations associated with it, an example of which is the slowing down of defense reflexes of parents protecting their offspring, bringing with it the danger that the parent will lose its own life. The horns of reindeer are of little or no use to the individuals, but they are a form of tournament armament

and assure that those allowed to participate in mating will be the strongest, thus maintaining the possibility of healthier offspring (Fig. 213). Another and similar role is played by the various types of signaling devices of mammals; these increase the animal's ability to be seen and hence its danger (Petrusevich, 1960).

The biocenose is an aggregation of populations of various species and thus differs from the species, which is a system of intraspecies forms. In the former case the mutually related populations of different species form a very mobile and frangible system in which the basic role is played by contradictory and sometimes antagonistic relationships among the various species. In the latter case we have a very tightly integrated system of mutually dependent populations of a single species, the existence of which is mutually governed and is possible only as an integrated totality. The breakup of a species population or the appearance of a new species in the biocenose leads to a recombination of relationships, sometimes a recombination which is not even significant. However, the appearance of new forms within the species, and their isolation from one another, lead to destruction of the integrity of the species and to its divergence, i.e., to the appearance of new species.

Fig. 213. Fighting male reindeer (from S. A. Severtsov, 1951).

CHAPTER 14

The Dynamics of Species Abundance

VARIABILITY OF ABUNDANCE

The abundance of individuals of any species never remains constant. It changes with the seasons, increasing during the reproductive season and decreasing after its completion. Nor is it identical in different years, for the latter usually differ in the degree of favorable conditions they offer (supplies of food, good weather, etc.). Finally, abundance may increase or decrease over the course of several years, as a result of which the species either settles new areas, thus expanding its range, or loses territory it had conquered earlier. In some cases increase in abundance is replaced by decrease, as a result of which "pulsations" arise in the boundaries of the range. In other cases the changes prove to be irreversible, and the previous abundance level cannot be re-established.

The immediate causes of changes in abundance may be of the most varied type, but their general character is specific for the species concerned. Differences may involve the age groups in the population, which are governed by the birth and death rates of the individuals, the speed at which they mature, the ratio of the sexes, and the amplitude and frequency of changes in abundance. Thus among the fishes, rodents, insects, and several other animals there are species with very significant fluctuations in abundance and others which, except for the effects of human beings, change only slightly from year to year. The amplitude of fluctuations in the abundance of voles and mice, for example, reaches a magnitude of 500 on the steppes, while among hamsters it only varies tenfold or less. Changes in the abundance of insect pests are enormous, while they are comparatively small among bees. Among the ocean herrings and cods the new generation

may exceed the old by more than 50 times, while among the flounders and sturgeons the ratio never goes higher than ten.

The character of a species' abundance dynamics is determined by the species' morpho-physiological and ecological properties, i.e., by its ability to adapt to constantly changing conditions of life. A species' abundance dynamics reflects the character of its interrelations with the environment. These interrelations include the speed of growth, development, and reproduction of animals which have adapted to definite conditions and have therefore survived, and also their own interrelations, which determine the structure of the population and its way of life. The mutual relations of species and environment are accompanied by both reversible and irreversible changes in the individuals and by changes in the environment which the individuals have caused. The latter have to do not only with the way the species utilizes the environment's vital resources and the way in which it contaminates the surroundings, but they also involve the way in which the environment responds to these demands, such as the way in which it creates a favorable microclimate (ecoclimate).

Quantitative changes in populations are the sum of three phenomena—reproduction, death, and animal migration (immigration and emigration). A known stability of abundance and range can be explained by the fact that the species uses all suitable places of habitation, while the surplus population settles in unsuitable habitats and dies off. Dispersion of the species and expansion of its range testify not so much to the species' fecundity as to the absence of obstacles to its resettlement. Finally, progressive decrease in abundance and contraction of the habitat's dimensions may be the result either of an increase in mortality which is not compensated for by reproduction or of an unfavorable change in the conditions of life (food supply, climate, number and strength of enemies, etc.). All this is reflected in changes in the abundance and biomass of the population, the movement (dynamics) of which is thus the quantative expression of the character of the interrelations between species and environment or of the results of the struggle for existence in its widest sense. The dynamics of abundance thus expresses the mechanism of natural selection. Seasonal and perennial changes in the physiological state and structural appearance of a population are well known for insects, rodents, and other animals. They may be either reversible or irreversible, and it is not easy to draw the line dividing them. These qualitative changes in individuals are expressed in changes of speed in population reproduction (age of maturity, fecundity), its variability, character of survival, mortality, and other indices of population dynamics. When abundance decreases and populations divide into separate groups, conditions are created for fixing different, sometimes accidental, deviations.

E. B. Ford (1945) studied the influence of such fluctuations in abundance on the variability of properties of butterflies in nature. In England an isolated population of *Melittea aurinia* had two long peaks of abundance in 49 years. Variability occurred in this species at the time when abundance was low—right between the two peaks—especially in the period when numbers were being re-established after the depression. At that time it was almost impossible to find two butterflies identical in size, coloration, and form. During the peak abundance period this variability disappeared and a definite type of butterfly prevailed, but the prevailing type was different in each period of high abundance. Among small rodents albinos and other color deviations are also observed more often during a rise in the number of animals (Kalabukhov, 1940; Bodenheimer, 1958). Observations have been made of an increase in immunity among rodents after a widespread epidemic in their populations (Tinker, 1940), of an increase in cold resistance after freezing winters, and many others. It is interesting that the character of such changes is often similar to seasonal shifts in the state of populations.

Changes in abundance reflect the quantitative side of the relationships between populations and environment and are the sum of reproduction, survival, mortality, and displacement of individuals compounded under the influence of the conditions of life. These reactions of animals to external forces are not a simple, one-sided effect of external factors (or combinations of them) on organisms accepting them passively. These influences, established according to general physico-chemical principles in living systems (such as organisms taken individually or, even more so, in populations), with their definite organization, are complicated by regulatory or compensatory adaptive reactions (S. A. Severtsov, 1941; Errington, 1945, 1951; Bodenheimer, 1955; Nikol'sky, 1955).

These reactions affect all basic biological phenomena. Thus, when abundance and population density increase, the tempo of growth slows down as food conditions worsen and other conditions of life grow more difficult (contamination of the environment by products of metabolism, insufficient oxygen, etc.). In many cases the speed of development also decreases, as a result of which the animals are smaller in size, allowing more of them to occupy the same amount of territory. Such phenomena are well known among the parasitic worms and have been observed among many fishes and several other animals. A slowing of sexual maturation has been observed among small rodents under such circumstances. Usually voles become sexually mature at the age of 1.5–2.0 months and mice at 2–3 months, but in years when they are very abundant they do not mature for 8–10 months, i.e., in the following spring.

Along with the decrease in growth tempo and the slowdown in sexual

maturation, the intensity of reproduction also decreases under conditions of dense population. This is expressed most distinctly among species with a more stable population density, such as the flounders, but it has also been observed among animals with highly variable abundance indices, such as the small rodents. In these cases a decrease in a population's fecundity is not associated with a simple lack of food but reflects the external stimulus (signal) of population density. Among species with a distinctly expressed separate use of territory and a more or less distinct set of individual plots for single animals, the animals simply will not mate unless they can find free territory for themselves. Thus a decrease in the fecundity of a population during high abundance is not so much due to direct suppression of reproduction by unfavorable conditions of life, such as hunger, as it is to refusal to reproduce. There is no question about the regulatory character of these phenomena. Thus voles and mice, which do not mate when abundance is high, do not appear emaciated under these conditions; they are well fed and often have fat deposits in their bodies. A study of their food in such cases usually shows that its level is sufficient for the existence of the population at the given moment but that it is insufficient to allow collection of the necessary winter supplies (N. P. Naumov, 1948, 1956).

When abundance rises, there is a marked lack of territory and the number of homeless, wandering animals rises, even though the size of individual plots has decreased. This increases the motility of the population, and in extreme cases it leads to broad and often mass migrations (emigration). Among species which do not have large rises in abundance the same increase in mobility is not so marked and is not so attractive to others. As was said earlier, such migrations are most important in the interpopulational sense, but they doubtless have great compensatory significance as well. Local resettlement is eliminated and the abundance level is reduced in the biotopes where the conditions of life are insufficient for animals which have reproduced.

When abundance decreases, we observe reverse phenomena—a speeding up of growth and development and an increase in reproduction. When this happens among rodents, there is usually a preponderance of females in the population; high density usually results in a preponderance of males. When abundance drops sharply, some of the animals which have survived and dispersed to less desirable living space migrate back to the optimal spots, or "survival stations." These phenomena also have compensatory significance, since they bring about quicker re-establishment of the population after it drops.

Tempo of growth, age groups, and sexual makeup of the population change correspondingly, as do intensity of reproduction, mortality, char-

acter of territorial utilization, and animal motility. In both cases the ecological significance of the changes is identical, and the response reactions of the populations are basically not simple or direct but instead form a complicated reflex mechanism. In this case the strength and sometimes the specifics of the exciting factor may not correspond absolutely to the strength and character of the response reaction (just as the weak pressure of a finger on a rifle's trigger does not correspond to the force of the shot).

Regulatory or compensatory mechanisms are of great importance in population dynamics, since they help stabilize the population at a definite level. The latter is an important condition of the species's existence. Through these mechanisms the population is able to adapt to changing conditions of existence. But the meaning of such mechanisms in the dynamics of abundance, like the role of any adaptation, is not the same for all species and is relative in each. They do not furnish complete stabilization of abundance and biomass of the population primarily because the relationships between the species and the conditions of its existence (which give rise to its unity with the environment) are contradictory. The chief contradiction lies in the species' one-sided use of the environment—its exhaustion and contamination of the environment with its own excretions. It cannot live for any length of time alone, out of contact with other species. Therefore, the unity of the species and its environment is in this sense one-sided and undervalued.

The species and the environment are not merely interconnected. They also stand in opposition to one another as separate entities, changing not only in the course of their mutual relations with one another but also independently under the influence of cosmic and other factors. The scales of such changes sometimes exceed the capacities of the species' regulatory mechanisms, thus rendering the species' abundance variable. Among species which are better adapted to their places of habitation, or among populations occupying optimal biotopes, the regulatory abilities are greater than the external forces, and abundance is stable. Consequently, its level and degree of stabilization may be regarded as a measure of the species' ability to adapt to its place of habitation. On the other hand, the high variability of abundance reveals the weakness of the population's compensating mechanisms. In this case the abundance of the population may be limited by climatic factors or (with the aid of interspecies relationships) by predators, the rise of diseases, etc.

The most important condition of abundance dynamics in a separate population, or in an aggregation of populations making up an integral species, is the exchange carried on when individuals from one set of biotopes resettle in another. The basic cause of such migrations is the maturation and dispersion of the new generation, a phenomenon strength-

ened by an imbalance between fecundity and mortality. In separate populations uneven reproduction and mortality are the rule, not the exception. This unevenness is also the basis of emigration or immigration of animals, which connects the species as an aggregation of various populations into an integral whole. It manifests itself primarily in the alternation of sexually active and sexually quiescent periods, different for different regions and even for neighboring biotopes. Therefore, the seasonal peaks—rises and decreases—in abundance seen in many populations come at different times. The seasonal migrations of animals are partially associated with this cause.

Even more important are the dissimilar magnitudes of fecundity and norms of mortality within one species in the different geographic and ecological populations formed by it. In the more favorable zones and biotopes the increase in population (reproduction) is greater than the loss (mortality), and the dangers of insufficient food, shelter, and even living space often arise. The regulatory adaptations mentioned above, as the history of the mass reproduction of insect pests, small rodents, and other fecund species shows, are in no condition to eliminate these dangers, and massive out-migration begins. As was shown above, the possibility of such a migration is established by the existence of suitable, but less favorable, zones and biotopes in which the death rate is often higher than the birth rate. The development of migrations associated with the abundance dynamics of a population, which are always accompanied by a higher mortality rate, may also be regarded as a compensatory phenomenon. In the final account migrations eliminate the accumulated surplus of individuals but have no single regulatory significance, as have fluctuations in fecundity, speed of maturation, and other phenomena dependent on population density.

When the total character of the conditions of existence is preserved unchanged, a more or less balanced scheme of reproduction and death is established among species which preserve their level of abundance and the borders of their ranges. In the opposite case either the abundance of the species increases and the habitation area is enlarged, or the range is cut back. Examples of both phenomena are equally frequent, as attested by the fact that an even balance between reproduction and death is only a partial case or else constitutes the total of reproduction, death, and migrations, each occurring in its own way in individual populations. One reflection of the regular imbalance among these three phenomena can be found in the seasonal and perennial fluctuations of animal abundance. They are specific for each species and depend on its properties of reproduction, death, and mobility.

Fluctuation in abundance which covers several years' time is the summation of seasonal population fluctuations. These latter are expressed es-

pecially distinctly in the high and temperate latitudes and relatively poorly in the tropics and subtropics. By the same token the total variability of abundance among animals living in the temperate-zone tundra, forest, and steppes is high, while populations of animals in the tropical forests are very stable.

The general character of population movement among separate species is closely associated with the seasonal dynamics of populations. Among animals which reproduce once a year or strictly seasonally the peak of abundance comes at the time when the young are born; the nadir is reached just before birth. Among those which bear several litters per year or which multiply uninterruptedly the time for the rise or fall in abundance depends on the course of reproduction in a given year. Since fluctuations in abundance over a series of years are governed by the uneven course of the abundance index of each year, a study of the seasonal population dynamics of different years is of primary importance in understanding the specifics of the population movements of a given species and its special properties in the different areas of habitation.

FECUNDITY OF SPECIES

The average magnitude of fecundity of a species has been construed in evolution as an adaptation furnishing a means of overcoming population losses. Among species which are not resistant to unfavorable conditions it compensates for high infant mortality or a short adult life span.

The range of fluctuations in fecundity is enormous: from the hundreds of millions of eggs during the life span of a parasitic worm, an insect, or certain fishes down to the small litters of mammals. Care of the young to curtail infant mortality plays an especially important role here. The moonfish *Mola mola,* with its pelagic roe and unprotected larvae, lays up to 300 million eggs at one time, while several of the viviparous sharks produce only a few eggs. Among the amphibians the green toad *Bufo viridis* lays 8,000–12,000 eggs per year, the toad *Alytes obstetricans* lays about 120, *Nototrema marsupiatum* about 200, the South American *Pipa americana* lays only 100 eggs, which develop in brood sacs on the mother's back, and the live-egg-bearing black salamander gives birth to a total of two offspring per year.

It used to be an accepted fact that fecundity corresponded to the normal average death rate of individuals (N. A. Severtsov, 1855; S. A. Severtsov, 1930, 1941). Support for this position is evident in the fact that in areas more favorable for life the fecundity of the species is usually lower, while

in places with poorer conditions of existence it is higher. The predominance of more fecund species in the temperate zone as compared with the inhabitants of tropical forests is well known for warm-blooded animals, in particular for birds (Table 57). An increase in fecundity in northern

TABLE 57.

	NUMBER OF SPECIES OF BIRDS	
Size of Brood	*Southern Wisconsin*	*Borneo*
1–2 eggs	5 (4%)	38 (60%)
3–4 eggs	52 (46)	15 (24)
5 or more eggs	56 (50)	10 (16)
Total	113	63

zones as compared with southern ones and on the borders of a range as compared with its central regions has been observed among mice, voles, hares, and many other mammals.

However, D. Lek (1957) found that the magnitude of fecundity of birds (observations on *Apus apus, A. melba,* starlings, and others) and apparently of other animals is determined not by mortality but by the possibility of raising offspring. Lek found that populations are dominated by those individuals of whose litters the maximum number attain sexual maturity. Hence selection takes place for the most effective fecundity, leading to the greatest development of the adult part of the population and, consequently, to maximal growth (Table 58). It is apparent from the table that

TABLE 58. *Survival of young starlings* (Sturnus vulgaris) *in Switzerland after leaving the nest (after Lek, 1957).*

		CAUGHT MORE THAN 3 MONTHS AFTER LEAVING	
Number of Young in Brood	*Number of Young Banded*	*Per 100 Banded Birds*	*Per 100 Banded Broods*
1	65	0	0
2	328	1.8	3.7
3	1,278	2.0	6.1
4	3,956	2.1	8.3
5	6,175	2.1	10.4
6	3,155	1.7	10.1
7	651	1.5	10.2
8	120	0.8	–
9, 10	28	0	0

in nature litters of average size (four to six fledglings) prevail; they also yield the maximum number of surviving individuals (the last two columns). Selection for the most effective fecundity should be regarded as a means of handling competitors of other species.

The differing character of fecundity of the individual species depends on inherited features: speed of sexual maturation, size of the litters born during the year (in one or several parturitions), and the ratio of males to females in the population. S. A. Severtsov (1930) called these indices "species constants of reproduction," while R. N. Chapman (1931) called them "species multiplication potential."

Animals are divided according to the character of their reproduction into those which reproduce once in a lifetime, or monocyclic animals (many invertebrates, certain fishes), and those which reproduce repeatedly, or polycyclic animals. Monocyclicity is typical primarily of the lower species and groups and usually characterizes forms in which the adult phase is rather short-lived and the larval phase or other immature stages are quite long-lived. The latter may be associated with the strong press of predators and especially the large number of enemies of the young in places where the adults live (spawning in streams and rivers like the Pacific salmon, open-dwelling insects whose larvae live in the wood of trees, in the soil, or in bodies of water, etc.). In extreme cases the function of the adult phase is limited only to reproduction. The moths of the winter looping caterpillar do not feed when they lay their eggs in autumn; there is no food, but neither are there many enemies.

The way in which many insects and fishes lay their eggs in batches may be considered a transition toward polycyclic reproduction. It has the advantage of guaranteeing that when eggs or live offspring are deposited in small groups throughout the year, the offspring as a whole have a greater chance of surviving. Polycyclic reproduction is either uninterrupted (among species with short-lived individuals) or periodic (seasonal).

Among species with monocyclic reproduction the adult group (the producers) consists basically of one generation, with only a small admixture of fast-maturing or late-maturing individuals from other age groups. Among polycyclic species the sexually mature phase comprises several generations, which aids in stabilizing the fecundity of populations (Monastyrsky, 1952). Special cases include reproduction at different stages of metamorphosis (life cycle), found in the coelenterates, protozoa, and other invertebrates (schizogonia, polyembryonia, etc.). These special forms of polycyclic reproduction increase the fecundity of the population and aid in its resistance.

The intensity with which a population reproduces changes with changes in the environment and in the state and content of the population. It was shown in Chapters 2–8 that environmental factors which change the state of individuals influence the times and intensity of their reproduction. These fluctuations are of an adaptive character. Under favorable circumstances reproduction increases and its results rise too, while under un-

favorable circumstances reproduction is cut back or curtailed altogether. However, the dimensions of such regulation, as already pointed out, are insufficient and do not provide a stable level of numbers or a population consonant with the state of the environment (food supply, favorable weather, etc.). In particular, they cannot prevent rises in abundance and emigration, nor can they provide for quick utilization of newly arising conditions favorable to reproduction.

The reason for this is a unique biological inertia associated with the complexity of biological systems (organisms, populations). Reaction to new influences takes time, sometimes a significant amount of time. As a result, there exists an almost constant disproportion between the demands of the population and the state of the natural resources on which the population lives. When, for example, a great harvest of tree seeds is available in the second half of summer and in autumn, the available number of seed eaters is usually so low that a considerable part of the harvest remains unused. But when in the following year the number of seed eaters rises as a result of intense reproduction, the supplies of the old food are either used or spoiled. As a result, the expanded population of squirrels, forest mice, and other seed eaters is faced with famine, causing migration and mass death among the animals.

Another group of factors influencing fecundity is associated with fluctuations in the density and structure of the population itself. Fecundity changes in response to increases in numbers and density of the population and to the degree to which all its members are furnished with necessities. Therefore, the action of these factors is also adaptive.

Fluctuation in the intensity of reproduction during a rise or fall in population density is well known for fishes. Such fluctuation has also been observed among birds, and almost all investigators have noted it in the case of rodents. It has been induced experimentally in laboratory cultures of animals: the flour beetle *Tribolium confusum*, the fruit fly *Drosophila*, plankton shrimp, protozoa, and many other animals. It has been found, however, that a rise in fecundity during a decrease in population density occurs only within definite limits. Below these limits the probability of males encountering females is limited, and among the colonially nesting species the advantages created by a group way of life are lost.

The causes of fecundity curtailment during a rise in population density are not the same in all cases. Among flour beetles the decisive factor is cannibalism (the beetles eat their own eggs); among protozoa it is poisoning of the environment by metabolic wastes; among a large number of species, from *Drosophila* to the fishes and rodents, it is lack of food. In all these cases the direct cause of the curtailment of fecundity is unfavorable changes in the environment caused by an increase in population density. But a

direct role in the regulation of reproduction is also played by those features of nervous activity which are associated with the intrapopulational relationships of the individuals, their use of territory, and the structure of their population (Chitty, 1955, 1958). Chapter 13 showed how important the possession of individual plots can be as a condition for reproduction in many species; the absence of such plots is usually accompanied by a refusal to mate and by emigration. Fecundity depends upon the dimensions and quality of these plots. The role of nesting grounds and refuges has been convincingly revealed by experiments with artificial nesting grounds for birds. In these cases the intensity of reproduction changes only in relation to changes in the structure and density of the population, regardless of other factors. However, even these fluctuations in fecundity, though they may attain considerable dimensions, are insufficient to bring numbers into agreement with the state of the population's vital resources.

The quality of offspring has a great deal to do with their survival. Fecundity alone, i.e., the number of eggs laid or the number of live births, cannot completely rebuild the population even under the most favorable circumstances if the offspring are not hardy. The hardiness of the offspring depends to a great extent on the feeding conditions which obtained among the parents. When feeding conditions are proper among fishes, the amount of fat in the white of the eggs is quite a bit larger than when feeding conditions have been poor, and this is associated with the survival of the offspring. Thus "weak" and "strong" generations are characterized not only by size (number of animals born or number of eggs laid) but also by the survival rate which follows birth.

Individual growth is important in population dynamics. Rapid growth is associated with more rapid recovery from the action of predators, which, for example, destroy primarily the young of many fish species. Fast growth and development bring quicker sexual maturity and thus aid in maintaining the population. Finally, the different qualities of growth among individuals in a single age group of fish enable the population to utilize a broad spectrum of foods and thus to have more food available to it at a given time.

MORTALITY

An animal's death may be caused by a thousand different external or internal factors, many of which may be entirely accidental for a given individual. But within the population, mortality has a regular character and is not identical for the different sexes or age phases. The causes and extent

of mortality change regularly as the environment and the population change. The magnitude of mortality determines the life span, the sequence of death among the different generations, and the character of the age groups in the population.

Among most species early infant mortality is higher than adult mortality. Among species of fish which do not protect their roe and which have pelagic larvae, only 1–2 percent of the eggs laid actually reach maturity. Among the salmon, which bury their eggs in the ground, protect them, and spawn in rivers containing few natural enemies of young fish, there is a larger survival rate. The death rate of birds and mammals before reaching sexual maturity usually reaches 70–80 percent. Of the small susliks born at a given time, only 15–20 percent will reach sexual maturity, and the survival rate of the various species of voles is no more than 20–30 percent. The death rate for small rodents is not usually very great while they are still in the nest, but it rises sharply when they make the transition to independent life, especially when families break down and the young disperse. During this period a generation suffers 60–70 percent mortality. In the adult state, after dispersion and the transition to a sedentary life, the mortality rate drops. Even in the wild the animals sometimes live to the limit of their life span (2 years), although only about one to three of every thousand creatures born ever succeed in doing this (N. P. Naumov, 1951; Merkova, 1955).

The death rate is different for adult males and females because, as was pointed out before, the mortality of the more mobile males is higher during the mating season than that of the less mobile females, while in winter it is the other way around. By the end of the mating season the females actually predominate in numbers among the adult population; in winter, however, the ratio evens out. The inequality of death rates for males and females is compensated for in some degree by the birth of a greater number of individuals of that sex which ultimately will have a higher mortality. Since the males are usually more mobile and their mortality is higher on the average than that of the females, the "male" species (those in whose litters males predominate) include most animals (almost all the rodents, ruminants, insectivores), while the "female" species are noticeably fewer (wolves, martens, and some others) (Teplov, 1954). The ratio of males to females in the litter changes in different years; under unfavorable conditions (famines) more females are usually born, while under favorable conditions more males are born. In many cases the development of polygamy has its apparent evolutionary basis in an increased male mortality rate.

In most species the death rate is lower among the adults than among the young, but in many insects it is higher. The larvae of such insects live

in the soil, where there is a favorable microclimate and relative freedom from predatory enemies. But the adults live on the surface, where they are subject not only to the full range of unstable weather conditions but also to the depredations of various enemies, from mammals and birds to predatory insects. In the evolutionary sense this has led to a curtailment of the life span and often to a limitation of the functions of the adult phase; it has also aided in the growth of the species' fecundity. Increase in mortality among adults and the cutting back of the productive part of the population create conditions for selection for higher fecundity and transition to larval reproduction (neoteny). Not only the amphibians but also the small rodents (voles) have proceeded along this path.

The general character of population mortality depends on the death rate of individuals during their various stages of development. It may be expressed by three basic types of curves (Fig. 214). The first is an exponential curve (diminishing geometric progression) and is characterized by identical norms of mortality at all ages. It is seldom encountered and is characteristic of animals with a permanently optimal set of living conditions for individuals of all ages.

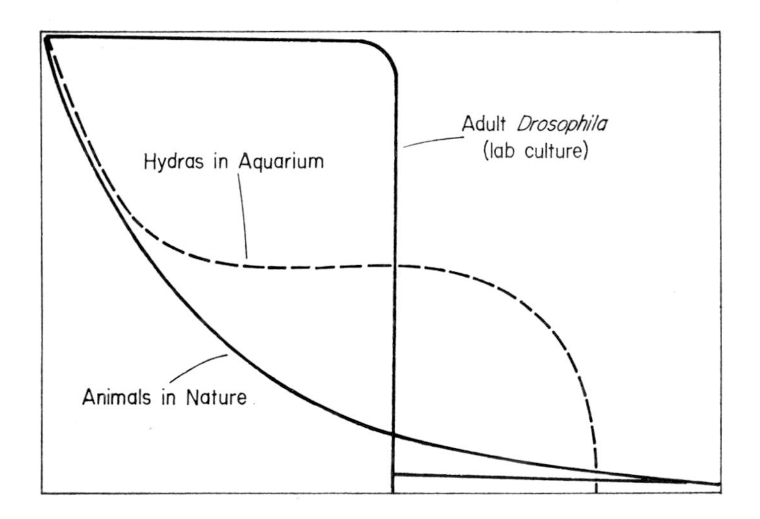

Fig. 214. Mortality curves of various populations (after Pearl, 1934).

The second type is distinguished by a higher mortality rate for the young and relative stability among the elders. This is the type encountered among most species of animals. It is interesting that this type was characteristic of man in the earlier periods of his development. The life span of a gen-

eration (life expectancy of individuals) depends on the adult mortality rate.

The third type of curve is characterized by stable existence among the young individuals but higher mortality among the adults, especially among the old individuals. This type of curve distinguishes the already-mentioned insects whose larval stages live in the ground, bodies of water, and other such places. To a lesser degree it is typical of animals which care for their young a good deal.

At the present time, as a result of raising small animals in the laboratory, analyzing the age groups and fecundity of samples from natural populations, calculating the abundance of animals in the wild, and marking them, we have accumulated accurate data on the death rates of individuals and the character of population abundance dynamics. Numerous and partially successful attempts at mathematical integration of the data have been made, and several quantitative principles of abundance dynamics have been discovered (Volterra, 1926; Pearl, 1930; Nicholson, 1933, 1947, 1958; Gause, 1934; Lotka, 1934). These principles have been useful, beyond doubt. However, the inevitable simplification of complex biological phenomena has led to a situation in which the results are most often quite far from adequate depictions of reality (Bodenheimer, 1958). Still, they have thrown some light on the numerical relationships between animals and their prey, between parasites and hosts, and between competitors.

The mathematical characterizations and analyses of empirically obtained curves of age groups, expressing both reproductive properties and individual death rates in the population, have proven especially useful. These data enable us to compile so-called life tables, which characterize quantitatively the basic population parameters and dynamics in concrete cases. Such parameters include norms of survival to a given age (l_x), mortality in each age group per 1,000 animals born (d_x), the death norm (per 1,000 animals) in each age group (q_x), and the expected (mean probable) life span of individuals in each group (e_x). Examples of such calculations are given in Tables 59 and 60 and Figs. 215 and 216.

AGE GROUPS IN A POPULATION

The age distribution of a population is an expression of the general character (type) of its abundance dynamics. At the same time it clearly reflects its properties in different years and reacts keenly to an increase in reproduction, survival, and other changes. In favorable years, when in-

TABLE 59. *Life table of the mountain sheep* Ovis dallii dallii *(data from an age analysis of 608 skulls collected in Mt. McKinley National Park in Alaska) (after Deevey, 1947).*

Age in years	*Number Which Died in Given Interval out of Those Born* (d_x)	*Number Which Survived to Beginning of Interval out of Those Born* (l_x)	*Mortality Norm (per 1,000) in Given Interval* (q_x)	*Expected Life Span* (e_x)
0–0.5	64	1,000	54	7.1
0.5–1.0	145	946	153	–
1–2	12	801	15	7.7
2–3	13	789	17	6.8
3–4	12	776	16	5.9
4–5	30	764	39	5.0
5–6	46	734	63	4.2
6–7	48	688	70	3.4
7–8	69	640	108	2.6
8–9	132	571	231	1.9
9–10	187	439	426	1.3
10–11	156	252	619	0.9
11–12	90	96	937	0.6
12–13	3	6	500	1.2
13–14	3	3	1,000	0.7

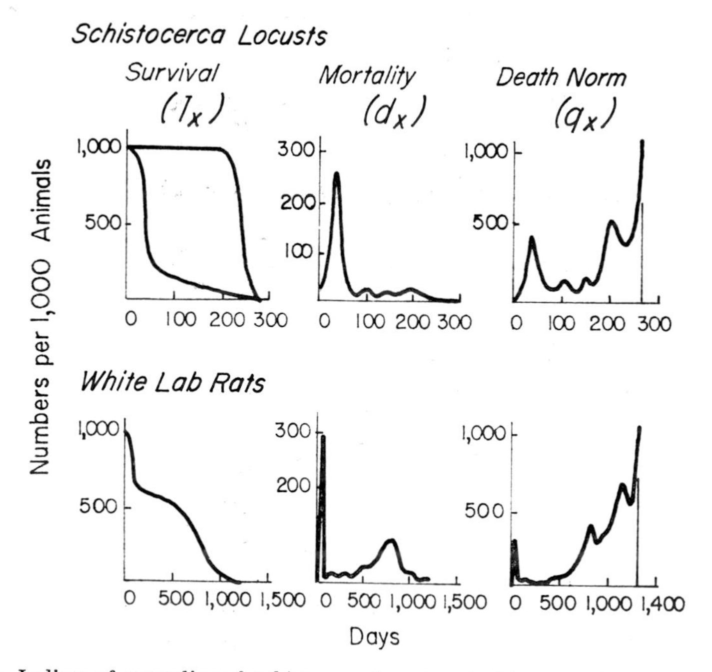

Fig. 215. Indices of mortality of *Schistocerca* locusts raised in cages in the open and of white lab rats (from Bodenheimer, 1958).

TABLE 60. *Life table of* Balanus balanoides *which settled on a clean-swept rock in 1933 (after Murie, 1944).*

Age in Months (x)	*Number Surviving at Start of Interval out of 1,000 Born* (l_x)	*Mortality Norm (per 1,000) in Given Interval* (q_x)	*Life Expectancy* (e_x)
0–2	1,000	90	12.1
2–4	910	110	11.3
4–6	810	62	10.5
6–8	760	79	9.1
8–10	700	114	7.8
10–12	620	258	6.7
12–14	460	174	6.7
14–16	380	203	5.9
16–18	280	179	5.7
18–20	230	174	4.7
20–22	190	526	2.4
22–24	90	667	1.9
24–26	30	667	1.8
26–28	10	800	1.4
28–30	2	1,000	1.0

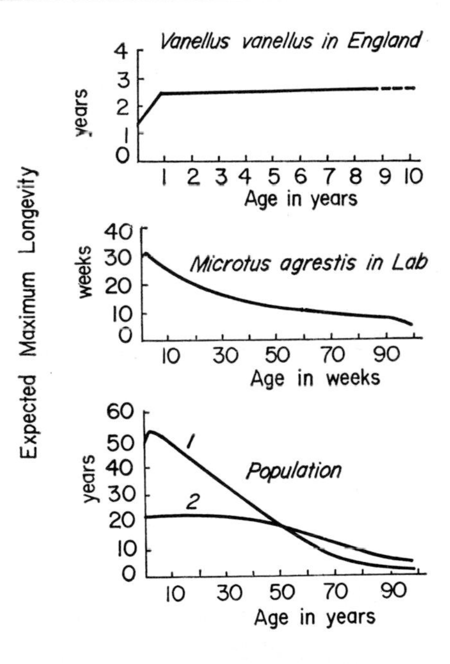

1 England and Wales from 1891-1900 Data

2 Ancient Rome in First-Fifth Centuries A.D.

Fig. 216. Probable life expectancies at various ages (after MacDonel, 1913; Leslie and Renson, 1940; from Lek, 1957).

tensity of reproduction is great and mortality is low, it is easy to trace the appearance of "strong" generations which occupy the dominant position in the population (Fig. 217). When multiplication is intense and

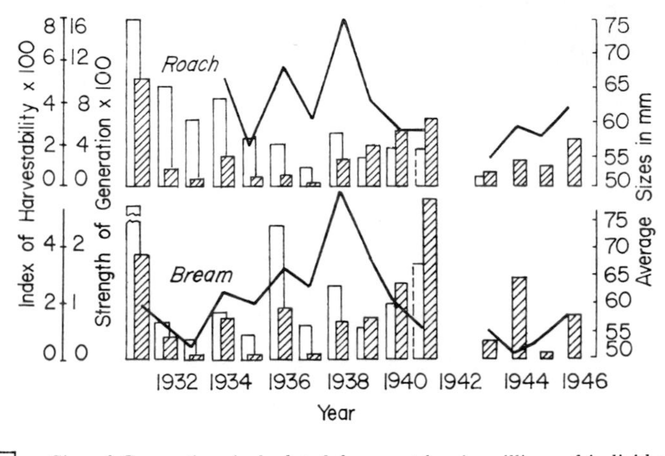

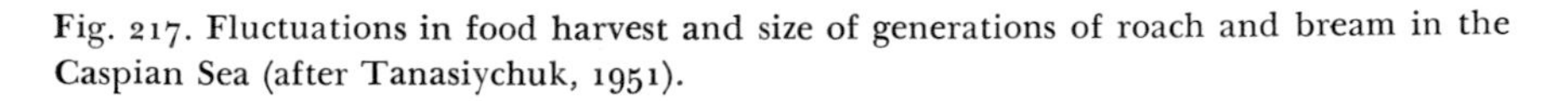

Fig. 217. Fluctuations in food harvest and size of generations of roach and bream in the Caspian Sea (after Tanasiychuk, 1951).

abundance increases, the age distribution of the population is distinguished by the prevalence of the young over the adult individuals. The reverse occurs when abundance is on the decrease. Curtailment of young age groups attests to a decrease in fecundity. Increased mortality among adults is reflected in the disappearance of the older age groups and a great decrease in the productive part of the population, represented primarily by the sexually immature animals (Fig. 218).

W. Ricker (1954) analyzed the abundance dynamics of game fish and formulated the valuable concept of a "supply," or the sexually mature portion of the population participating in reproduction, and a "complement," or those young individuals which mature in the course of the year and enter into the "supply." Both the general character of population dynamics and its particular properties stemming from concrete circumstances in a given year are determined by the ratio of "supply" to "complement." Using these features, G. N. Monastyrsky (1952) distinguished three basic types of spawning populations (the sexually mature part of

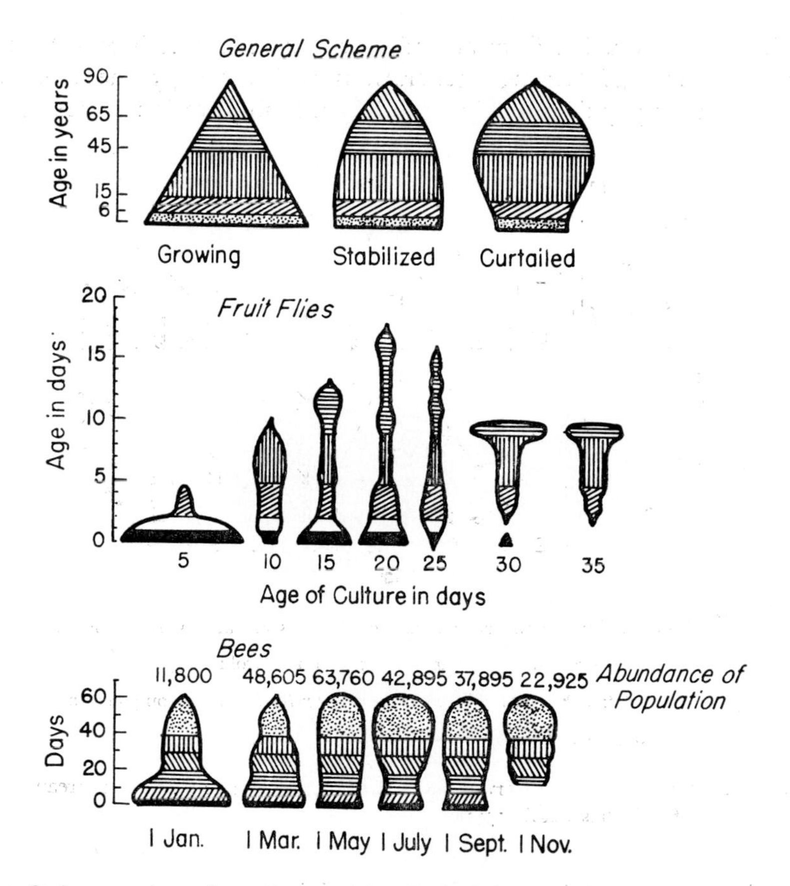

Fig. 218. Age structure of populations (after Bodenheimer, 1958).

the population) of fish. Species of the first type reproduce once in a lifetime, and except during the mating season the entire population consists of sexually immature animals. Their spawning population includes only individuals which are reproducing for the first time ("complement"). This is the situation with the Pacific salmon (Siberian, humpbacked, Chinook, and red) and certain Gobiidae. It is also true for most of the insects and other invertebrates with monocyclic reproduction.

Among the Pacific salmon the spawning group often consists of several ages, but with a sharp predominance of one of them, since in each generation some individuals undergo either faster or slower development (Fig. 219). This fact is of great biological importance, since it insures the species against extinction. With strictly synchronous development of in-

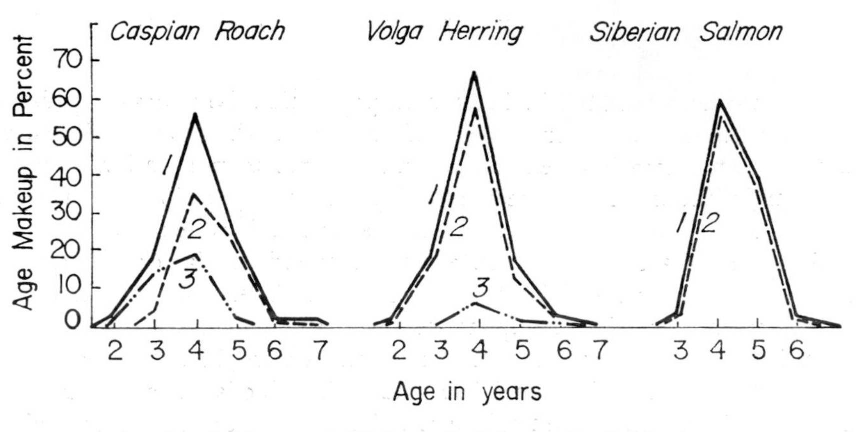

Fig. 219. Types of spawning populations among fishes (after Monastyrsky, 1949, 1952).

dividuals, the danger of such a population's extinction would be great even if there were 1, or especially 2, years in a row with no reproduction. The abundance of species in which the sexually mature part of the population consists of one generation is very unstable and fluctuates within very broad limits.

The second type includes species with repeated reproduction but whose adult individuals have a short life expectancy. Therefore, the "complement" (individuals multiplying for the first time) prevails over the "remainders" (individuals which have reproduced before) in the reproductive population. This group includes populations of short-lived mammals, birds, and other animals distinguished by their unstable abundance levels. The third type includes long-lived species of fish and other animals. In these populations the "remainder" prevails over the "complement." As a rule, the abundance of such species is stable.

Abundance may fluctuate in any of the three groups distinguished by the structure of the productive herd or flock but within boundaries typical of the given type of group. Stability of the population depends on stability of the reproductive and mortality indices. We have seen that they change uninterruptedly under the influence of the external environment and changes in population density, the state of the population, and its structure.

TYPES OF POPULATION DYNAMICS

The significance of individual factors in population dynamics depends on the strength and character of their effects. We usually distinguish exogenous factors (those independent of and external to population density) from endogenous ones (those which depend on population density and are factors of automatic regulation). The terminology here is imprecise and in many ways incorrect, since the "independent" factors are closely related to the "dependent" ones and constantly interreact with them.

One example of the "independent" factors to which a species has no special adaptations consists of major, often catastrophic, changes in the environment, such as harvest failures, droughts, floods, severe frosts, blizzards, etc. They cause a general, usually large, and nonselective mortality among the animals. The factors "independent" of population density strengthen their braking effect when the population grows and weaken or curtail it when there is a decrease in abundance. A well-known example of the action of such factors is the growth of laboratory cultures of animals and other organisms. At first they grow slowly, then increase according to the laws of geometric progression, but after a definite point is reached their growth diminishes and finally stops altogether (Fig. 220). Growth stoppage is due to unfavorable changes in the environment caused by the

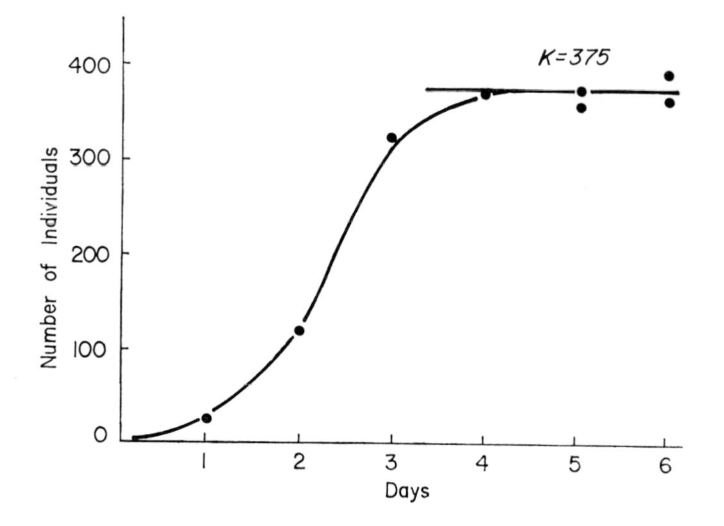

Fig. 220. Growth in abundance of a laboratory population of *Paramecium* according to a logistic curve (after Gause, 1934). The circles represent calculations of abundance (factual).

increase in number of animals and the density of their population. These changes may be insufficient food, development of cannibalism, poisoning of the environment with metabolic wastes, changes in pH or salt content, insufficient oxygen, or any one of a number of other factors. Usually one factor is the leading cause, but it combines with others. When this happens, mortality rises and fecundity falls (Fig. 221). The latter phenomenon often has the character of reflexive braking (Chitty, 1957). If unfavorable changes pile up on one another, the result may be a decrease in population abundance, sometimes profound (Fig. 222).

Despite differences in causes leading to increase in mortality and decrease in fecundity as revealed in the growth of various populations of laboratory animals, the one thing common to all these examples is the fact that increase in the abundance of a population leads to accumulation of unfavorable phenomena which ultimately cause the population to become smaller. The course of some events in nature resembles this scheme closely. Large increases in the abundance of insects and small rodents end usually with a precipitous drop and sometimes with a profound and lasting depression. The growing food deficit, multiplication of parasites, development of diseases, and growth in number of enemies play a large role here.

The effect of meteorological factors on reproduction and death is usually regarded as independent of population density. But it is closely associated

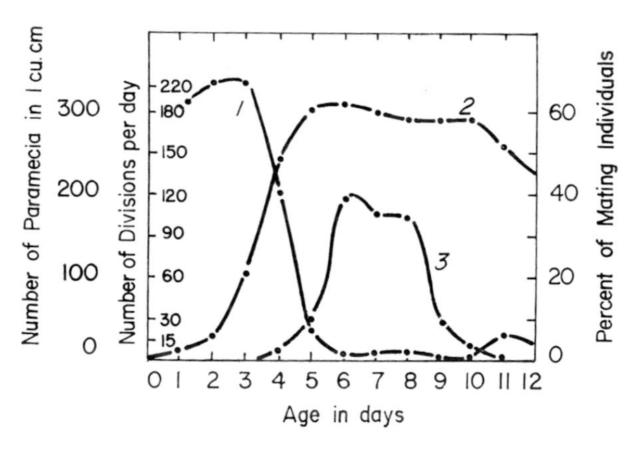

1 Average Number of Cells Dividing per day

2 Abundance

3 Average Percent of Conjugating Individuals in Culture

Fig. 221. Changes in intensity of reproduction in a *Paramecium* culture (after Bodenheimer, 1958, after Dudorov).

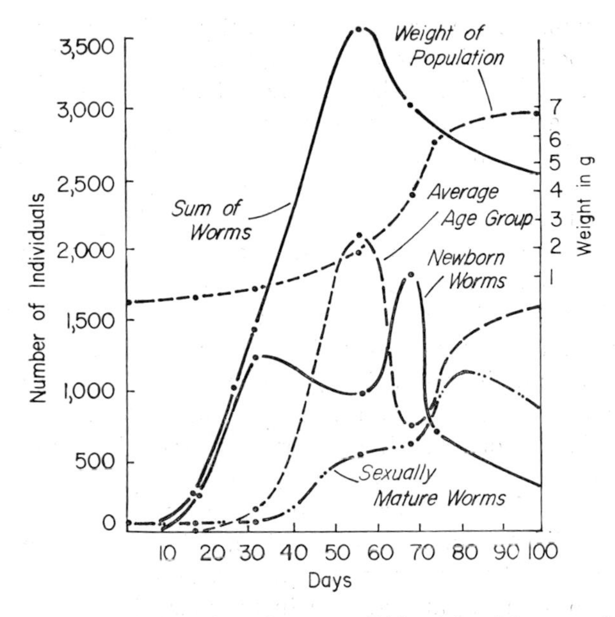

Fig. 222. Growth in population of *Enchytraeus albidus* (after Ivleva, 1953).

with the supply of food and shelter. Any unfavorable change in circumstances which occurs independently of the animals is more likely to have an adverse effect on dense populations, which inhabit all the available refuges, than it is on sparse populations, which live only in the most favorable ones. On the other hand, it is undoubted that the spread of diseases in the population, which depends on density of settlement, is closely associated with the mobility of the animals and the conditions for preserving the agent in the external environment. In many cases these conditions are determined by the weather and the characteristics of the place of habitation.

All of these complexly related factors determine the character of population dynamics. For each individual species it is the sum of interactions between the population and the environment, the result of its adaptations to the concrete circumstances of existence. This interrelationship is determined by the biological properties of the species. Therefore, the character of a species' population dynamics is specific for it to the same degree as its unique structure, functions, behavior, etc. distinguish it from all other species.

Populations of some species depend greatly on the direct or indirect influence of climatic and other factors. The abundance of others may be limited by supplies of food, competition with other species, number of

enemies, or, finally, by the development of diseases, which often cause a decrease in abundance. Regardless of the causes which limit abundance, reproduction prevails over death in some populations, and the population grows as long as this growth is not interrupted by an increase in the action of predators, diseases, or other causes or as a result of individuals migrating away. This is the way in which populations that occupy favorable areas (optimum range) behave.

In other cases population abundance is more or less permanently limited by one or several factors. As a result, the density of the population is not high and never brings about the appearance of the automatic regulating system. This is what we observe in the case of squirrels, whose population dynamics are controlled by supplies of their basic food (the seeds of hardwood trees) (Fig. 223), gallinaceous birds and small rodents in the north, where abundance is limited by the warmth of the growing season (see Figs. 93 and 94), and many insects. This kind of population dynamics is characteristic of species or individual populations which dwell in the less favorable habitation areas. Here reproduction merely compensates for death and is sometimes lower than mortality, there are no conditions for emigration, and, on the other hand, the limited reproduction favors immigration. This type of abundance dynamics is often en-

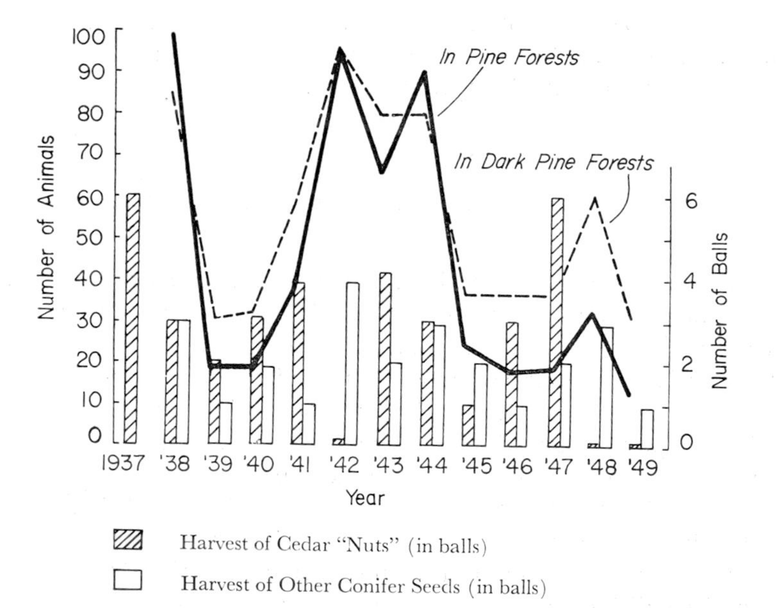

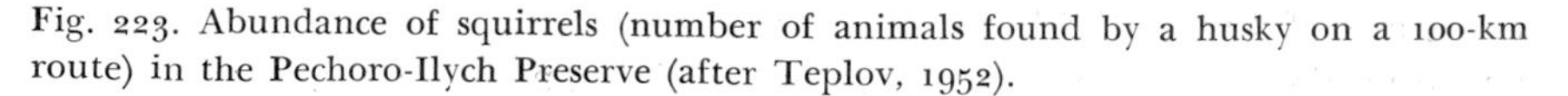

Fig. 223. Abundance of squirrels (number of animals found by a husky on a 100-km route) in the Pechoro-Ilych Preserve (after Teplov, 1952).

countered in the high and temperate latitudes or on the margins of a species's range (N. P. Naumov, 1958).

The amplitude of fluctuations in abundance or the general stability of abundance can be an important feature of population dynamics. This amplitude is different not only for different species but also for different populations of one species. As a rule, it can be associated with the breadth of the ecological niche occupied by the species or the population, and with fluctuations in its magnitude. The concept of the breadth of the niche should be understood in terms not only of spatial dimensions but also of the possibility of using the vital resources to be found within the territory as governed by abundance of competitors, number of enemies, and many other conditions. In the complex tropical rain forests, with their rich and varied population, the niche of each species is limited, so that the abundance of each species changes within narrow boundaries. On the other hand, in uniform landscapes—tundras, the taiga, the steppes, the deserts, and sometimes in fields with uncomplicated biocenoses—population abundance fluctuates sharply, while ecological niches of species are broad and variable. Harvests or famines, the possibility of using food, favorable or unfavorable conditions—all may expand or contract these niches. Common voles inhabiting uniform stretches of farmland have large increases in abundance, while populations of animals living in a mosaic landscape of forest steppe in areas with live hedgerows, copses, boundaries, and small fields do not have such outbreaks of abundance (Frank, 1950, 1954; Van Wijngarden, 1957).

ROLE OF INDIVIDUAL FACTORS

Successful development, reproduction, and survival of individuals, and consequently the abundance of a population, all depend on an aggregation of conditions of existence. Changes in abundance are determined by the relationships between populations and the leading environmental factors. The latter are not identical either for different species or even for different populations of a single species. Their significance may change depending on the state of the population and the constantly changing cluster of environmental factors, so that a condition which is the critical one at a given time and place may become unimportant under another set of circumstances. The important and constantly acting factors include food and conditions of gas, heat, and water exchange. They determine the state of the individuals, and their deviation from the optimum is accompanied by a decrease in reproduction and survival. Density and

structure of the population itself, as well as its influence on other species—competitors, enemies, parasites, etc.—may acquire importance.

The effect of the food supply on population abundance has been well shown by A. J. Nicholson (1955) in the case of the carrion fly *Lucilia cuprina.* In his first series of experiments the food of the larvae was limited, while that allowed the adults was abundant. As a result, the survival of the adults depended on competition with the larvae for food. Adult survival was greater in those cultures where 99 percent of the hatched flies were destroyed artificially, thus reducing the magnitude of the "complement" of the population and hence its offspring. But when this was not done, there was practically no increase in adults, while the newly hatched larvae were distinguished by a short life span and very low fecundity. In a second series of experiments, in which the food given to the adults was limited and that given to the larvae was in surplus, distinct cyclic fluctuations of flies arose which depended on periodically repeated emaciation of adults owing to insufficient food (Fig. 224).

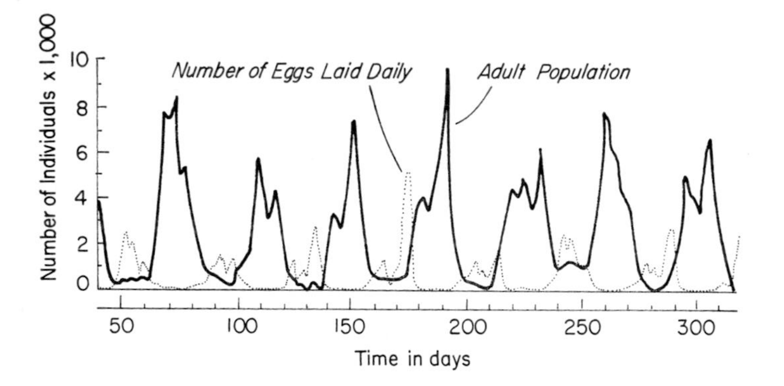

Fig. 224. Changes in abundance of a laboratory population of carrion flies (*Lucilia cuprina*) when the adults were given 5 g of food daily and the young were given a surplus (after Nicholson, 1955).

The commanding significance of food has been observed many times in the wild, especially in stenophagous species. The quality of the food plays a great role. The basic cause of decrease in the abundance of brown-tailed moths during years of their mass reproduction was the feeding pattern before hibernation. The food was not the autumn oak leaves, which are rich in sugars and cellular matter, but young leaves which secondarily cover the tree in summer after the spring extermination of foliage by pests. These new leaves are rich in proteins and water but poor in sugars. When the caterpillar eats them during mass-reproduction years, it grows

rapidly right up until autumn and therefore goes into its cocoon unprepared physiologically. Its frost resistance is noticeably lower than that of individuals which have fed on the proper leaves. The survival rate of caterpillars from trees which had leafed out twice is 12.1 percent by the end of winter, while those which feed on leaves not attacked by pests have a survival rate of 60.5 percent (R. V. Naumov, 1959).

The climatic factors determining the physical aspect of metabolism are just as important. They are the energetic background on which an animal's entire life unfolds. Naturally, almost all the vital phenomena and their sum total—population dynamics—have a close correlation with changes in the weather. This is explained by the varied (direct and indirect) influence of these factors on all aspects of the development and existence of individuals and populations. Recognition of the character and significance of this influence makes possible a sufficiently accurate prognosis even of long-term changes in abundance according to regularly repeated synoptic situations. The biological bases of such prognoses have been worked out for many species of insect pests both here at home and abroad. It has been made quite clear that the resolving factor in abundance is the state of the weather in the more important (the so-called "sensitive") periods, different for each species, when the success of reproduction and viability of populations are determined. Thus the abundance index of the cotton beetle *Anthonomus grandis* in Texas is determined by the relative humidity of the air in June and July, the minimum temperature during the preceding winter, the humidity of the air in July and August, and the cloudiness of the sky during the June-September period of the preceding summer. Reproduction of the beet aphid *Apis fabae* on its primary host (bushes) is favored by dry, warm weather in April and May, when this pest lives on beet plantings in large numbers. Further development of the pest is aided by warm, moist weather in June and July. The ladybug enemies of aphids multiply right after their prey and reduce their abundance even more (Paly, 1960).

In Tataria a correlation has been observed between the furs collected from foxes—reflecting the variation in abundance of these creatures in the wild—and the level of spring flooding of fields and the mean temperature during the period from 1 April through 10 May. Similar correlations for the ermine-fur catch are associated with the amount of flooding and the mean air temperature in the period 10 April–20 June. For the European hare the correlation is with the mean temperature from 10 April to 10 May, the relative humidity of the air from 10 July through 10 September, and the time when snow cover appears (Tikhvinsky, 1938). For small rodents, hares, and grouse in the north the critical periods usually prove to be the mating seasons (O. I. Semënov-Tyan-Shansky, 1938; Kalela, 1944; Siivonen, 1948, 1950, 1954; Bashenina, 1951; Shelford,

1954; Shelford and Petter, 1955). For northern populations of nocturnal animals coincidence of short critical periods with the half-moon seems to be especially favorable, and thus there is quite a good correlation between abundance dynamics and the lunar phases (Fig. 225).

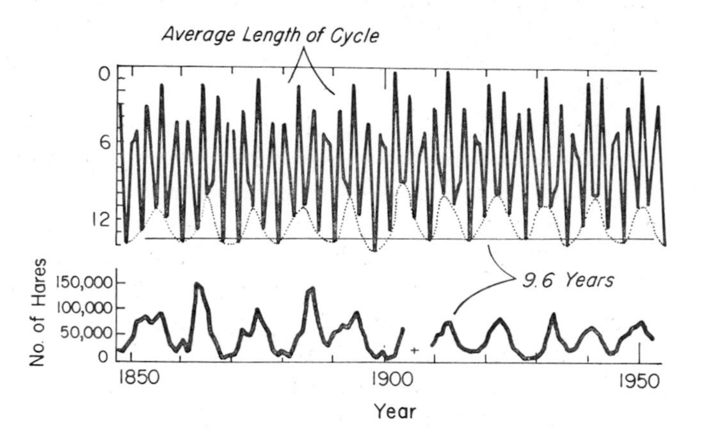

Fig. 225. Lunar 9.6 year cycle (dotted line) from 6 April as a critical period (upper curves), contrasted with fluctuations in abundance of the varying hare in Canada (lower curve) (after MacLulich, 1937; Chitty, 1933–1948; Rowan, 1955; from Siivonen and Koskimies, 1955).

However, in all or most of these cases we are speaking not of direct causal relationships but, rather, of coincidences which attest to tangential connections. This may be the influence of climatic factors on the supply and quality of food or on partners in the association. Or these factors may act as signals. The action of relationships with other species (enemies, competitors, parasites, symbionts, etc.) is less permanent in its effects on the abundance of the population. But even these animals can acquire critical importance from time to time.

Finally, the density of the population and its structure, which change at various seasons of the year in relation to a rise or fall in the number of animals, are important conditions of population dynamics. The action of factors "dependent" on population density is associated with this (Petrusevich, 1957, 1958). Usually when numbers increase, the exterminating activity of enemies and the effects of competition also increase (Volterra's law), as does the possibility of the spread of pathogenic parasites (Farr's law). The danger of starvation rises, the possibility of intraspecies competition increases and the adaptations serving to limit reproduction come into play. Finally, mobility increases, and in many species it leads to mass emigration if the number of animals is high enough. In nature all of this,

acting in conjunction with the physical and biotic factors in the environment, leads inevitably to a decrease in abundance after the rise. After this, other phenomena begin to develop. When abundance is low, the animals' mobility increases, leading to their concentration in favorable places (survival stations), and multiplication is greatly increased, with the mortality rate being very low.[1] This is the time when populations emerge from depression, and the period is marked by especially rapid replenishment of the population.

The role of population density is especially great for species with variable abundance (small rodents and other mass pests). As the population grows, the space devoted to individual plots diminishes, the size of the individual animal often decreases, conjoint use of territory increases, and fecundity usually decreases. When population density is high, resettlement increases and takes the form of mass migration, or mass shock diseases develop, not all of them infectious in nature (Chitty, 1958). K. Petrusevich (1957, 1958) found that there is an increase in reproduction after recombination of individuals into populations (changes in its structure).

In these examples the mutual action of physical, biotic, and population factors is seen very clearly, as are the mutual dependence of reproduction, mortality, and migration. The extreme example of such dependence is the desert locust *Schistocerca,* which leads a primarily nomadic life and has no permanent nesting place. The desert vegetation which appears for short periods leads to increase in numbers of this species at the place and time when food is abundant. The existence of vegetation forces the locusts to congregate in depressions. The resultant group, when united into flocks, migrates on the wind into new areas where, if it finds favorable conditions, it begins to found new generations to start the whole cycle over again. The entire existence of the locust population is thus a series of uninterrupted wanderings about temporarily favorable areas, the individual stages of which are established by different generations (Shcherbinovsky, 1952; Uvarov, 1957; Bodenheimer, 1958). This special form of species existence arose under the influence of natural selection as an adaptation to ephemeral food resources which arise repeatedly for short periods of time in different parts of the range. From time immemorial frequent increases and decreases in the abundance of small rodents, accompanied by mass emigrations of the reproducing creatures, have been known. Both phenomena are typical of species with low life expectancy, variable abundance, and many strong enemies.

The complex mutual relationships between factors in the environment and their effects on the population force us to carry on constant observa-

[1] After this, mobility drops to a minimum (N. P. Naumov, 1955).

tions of populations and the circumstances of their existence. At the basis of changes in abundance lie not the standard periodic combinations but, rather, the always unique combinations of external conditions, which the population perceives differently depending on its state and its contents. This perception depends not only on existing conditions but also on the circumstances of development. Survival and reproduction, even under identical environmental conditions, are different among animals which have been born and have developed under different sets of conditions and which have different levels of hardiness and viability. V. N. Stark (1955) found that the placement of eggs laid by wood-eating insects depends on the position of the trees in the plantation and on the weather during the time the females are flying. Therefore, in certain years the eggs are laid on different trees, and this is associated with success of development and the number of individuals in the new generation. It is also known that the prime impetus for reproduction of forest pests is usually not favorable weather but the destruction of stands of trees and the breakdown in condition of the trees, which changes the fauna of the planting and often reduces the number of enemies and parasites of the destructive species (Rudnev, 1959).

The causes of a drop in abundance of the mass species after their reproduction in the same place are, as a rule, different in different cases. When the pine caterpillar *Panolis flammea* reproduced in Germany, the first outbreak was liquidated mainly under the influence of climatic factors, while in the second mortality from enemies and predators was 632 times greater than mortality owing to unfavorable weather (Schwerdtfeger, 1936, 1941). Among reproducing *Dendrolimus pini* moths primarily climatic factors lowered the population by 98 percent in the period from hatching until winter torpor, while in the succeeding period enemies and diseases destroyed 89 percent and climate destroyed only 10 percent. Only 0.2 percent of the eggs laid were preserved. In all carefully studied cases of rise and fall in numbers it cannot be said that any single factor was the cause.

A definite conservativism in population dynamics and a constancy of its leading factors may be observed among fishes and other animals which inhabit comparatively stable biotopes where conditions change little with time (Dement'yeva, 1961). But even here the repetition of changes in abundance is relative, since each rise or fall is distinguished by its level, the causes which created it, and their consequences.

The complexity of population dynamics is inevitable and natural, since it reflects and expresses quantitatively the entire aggregation of mutual relations of population and environment, in the course of which the pop-

ulation itself must change as it adapts to the circumstances involved. Therefore, any attempt to assign changes in population levels to an effect of one or several "resolving" external factors passively perceived by the population is incorrect; it simplifies and impoverishes the most complex biological phenomenon. Two such attempts are known, both of which tie population dynamics of organic forms to the principle of a "dynamic balance" in nature as formulated by Spencer (1852). The theory of "biological cycles" seeks the causes of abundance fluctuations only in the relations existing between food and its consumers (plants and animals, predators and victims, parasites and hosts) or in self-regulation of reproduction and mortality when population density increases (Howard and Fiske, 1911; Lotka, 1925; Nicholson, 1933). The environment in this case is regarded merely as an inert background, influencing the dynamics of abundance but not determining its character or causes.

Just as one-sided is the theory of "climatic control," which ties the population dynamics of animals to the direct or indirect effects of climatic factors. It was developed by R. N. Chapman (1928) in his theory of "biotic potential," by which is understood the species' native tendency and ability to multiply and survive as limited by the "resistance of the environment." According to these considerations, multiplication, death, and population level should generally correspond to the state of the environment. But, as was pointed out above, because of the braking effect of many factors ("biological inertia" of the population), the existing factors do not, as a rule, correspond to the level of the population. In this disagreement we find the puzzle of all the causes of migrations and adaptations: a balance toward which the population is striving as it adapts to its environment is suddenly disturbed.

GEOGRAPHIC FEATURES

The species exists as an aggregation of geographic and ecological populations, distinguished from one another by the rhythms of their biological phenomena. The favorability of the environment governs the density and character of the species population's distribution. The species is "common," "frequent," or "numerous" in its optimum range and becomes "rare" or "accidental" on the periphery of the range. Broad settlement throughout all or many of the biotopes in the optimum part of the range is replaced by a tendency toward a few places of habitation on the periphery. Solid populations become mosaic. Stenotopic behavior grows. The charac-

ter of the population's dynamics also changes. The less favorable regions and biotopes are sometimes entirely free of populations and are settled only when there is a geometric rise in numbers and they can be filled by emigration from the more favorable centers. In the less favorable zone the population may attain a high level only under a favorable combination of factors, which makes an increase here a rare occurrence; the amplitude of fluctuations here can have enormous scope. Finally, in the optimal zone abundance may be either relatively constant or subject to more or less regular fluctuations, but the general level is relatively high (Kozhanchikov, 1937, 1950; N. P. Naumov, 1945). In agricultural entomology this geographic regularity has found expression in the concept of "zones of destructiveness" (Fig. 226).

In the optimal zone of the range and in biotopes favorable to the species, reproduction exceeds mortality on the average. The consequent increase in numbers creates unfavorable situations (shortage of food, refuges, and space, increased intra- and interspecies competition, higher death rate owing to predators, spread of disease, etc.) which are usually resolved by more or less mass emigration to less favorable biotopes. Along with the growing death rate, this leads to a decrease in abundance. Abundance is regulated here primarily by the very course of its changes, which are determined by the fact that reproduction does not encounter substantial hindrances and thus exceeds the mortality rate. This type of dynamics might be called "regulatable."

In the pessimum range and in all biotopes which are not wholly favorable, reproduction does not, on the average, equal mortality, and populations there may die out entirely, being re-established (or constantly maintained) by a trickle of immigrants from the optimal habitation regions. The abundance dynamics of such populations has a more complex character. Their reproduction and abundance are limited on the spot by several or even by one existence-limiting environmental factor distinguished for its great effect. An important role is played by the inflow of migrants, which does not always correspond to fluctuations in the local conditions. This type of dynamics is usually called "limitable" (Lek, 1957). Both types are mutually connected and together serve to preserve the species' range and enable it to use suitable areas, if only for a short time. In addition, migrations and population exchanges knit the species into an integral whole, while the variety of conditions and the character of abundance dynamics in the various biotopes open up the possibility of divergence (radiation) of the species in different directions (Fig. 227). Such divergence is preceded by wider and wider use of more and more varied forms of environmental resources.

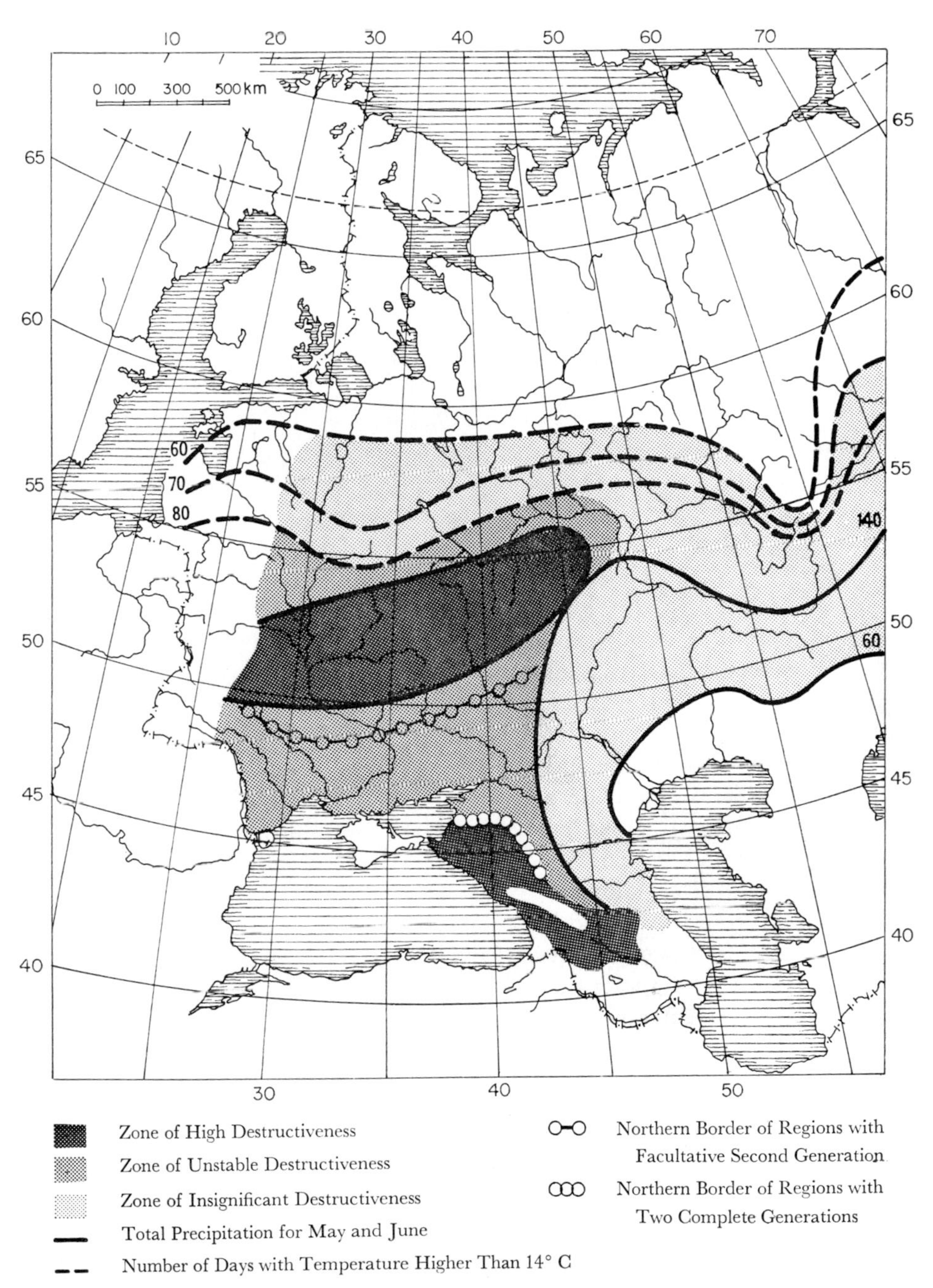

Fig. 226. Territory of distribution, zones of destructiveness, and boundaries of regions with differing numbers of generations of the corn borer (*Pyrausta nubilalis*) (after Shchegolev, Znamensky, and Bey-Biyenko, 1937).

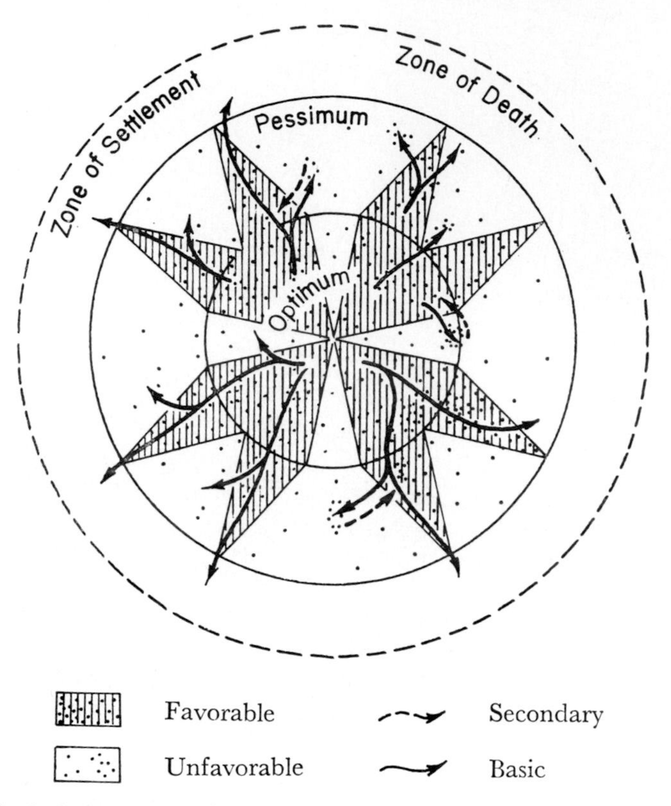

Fig. 227. Ecological structure of a range, showing displacement and interrelations in it of individual populations (original). Individual sectors are those suitable or unsuitable for species; points indicate density (abundance) of the populations in the biotopes; arrows show direction of emigrations.

PART III

THE ECOLOGY OF ASSOCIATIONS

CHAPTER 15

Plants and Animals

BASIC FORMS OF INTERSPECIES RELATIONSHIPS

Populations of various species of plants, microorganisms, and animals form natural associations based on the biological turnover of matter, which grows more complex as the association develops. The basic form of connection between species in the association is the relationship between food and its consumers; the consumers form a food chain in which each link is the food of the following one. In schematic form the chain consists of the following basic links:

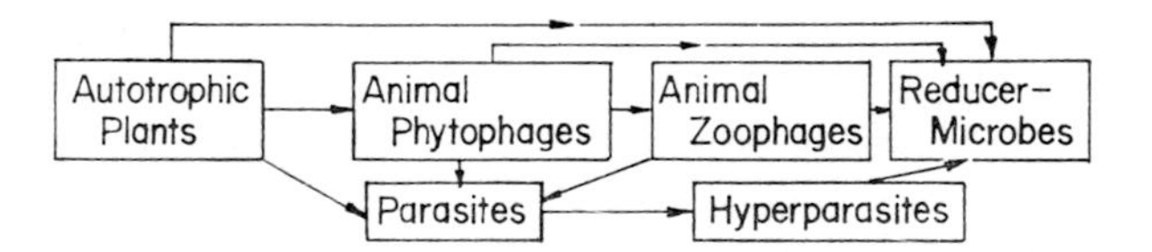

The biological turnover of matter can be set in motion only with the help of food chains consisting of interconnected food relationships of all three of the living kingdoms (microbes, animals, and plants). Not only the organisms themselves (their bodies) but also their excreted and expired metabolic products are used.

Besides food relationships certain spatial relationships must of necessity arise among the component organisms of associations. Individuals of some species may use others as an environment (endoparasites) or as a substrate and place of attachment (ectoparasites and free-living, sessile, or nearby immobile species). Associations containing various species inhabit caves, rocks, outcroppings, cracks in the earth, and other natural refuges.

Many species may live as tenants in the burrows, nests, or dens of other species, finding there a defense against enemies and a favorable microclimate. The cover created by plant associations or groupings of sessile animals attracts numerous animals as a refuge and means of protection against enemies and sometimes as a defense against the harmful effect of abiotic factors.

Three basic types of relationship arise on the basis of food and spatial connections between species: cohabitation (symbiosis), incompatibility (antibiosis), and indifferent toleration (neutralism).[1] In the broad sense symbiosis is defined as any case of close spatial association of two or more species from which at least one derives benefit. The historical result of such species cohabitation usually proves to be a mutual adaptation (coadaptation) of the symbionts which enables them to exist together.

Symbiosis takes various forms. In the mutual relationships of food and consumer (relationships of plants and animal phytophages, predation, and parasitism using the host as a medium of existence and a source of food) an interdependence arises. Not only the existence of the consumer but also that of the consumed organism depends on the relationship; the food source cannot exist without the species which consumes it. The stable existence of meadow and steppe vegetation is based on the presence of plant-eating animals (insects, ruminants, rodents, etc.) which keep perennials trimmed, limit the growth of their leafy parts, pollinate flowers, distribute seeds, etc. In the same way predators are important for the existence of their victims, since they lessen the danger of mass illness spreading among them. Many species of parasites are directly useful to their hosts, while others slow down their reproduction or stop the penetration of organisms destructive to the host. Thus, when antibiotics are immoderately applied, the destruction of the microbe population of an organism sometimes leads to a flood of ordinarily nonpathogenic fungi and the consequent death of the animal. We know that animals infected with the protozoon *Lamblia intestinalis* suffer less from dysentery. Food relationships sometimes take on the character of boarding, or commensalism (epiphytes, commensals, carrion eaters, coprophages, etc.).

Mutually beneficial cohabitation (mutualism) may be very close. Symbionts can form systems with a total exchange of matter in which some species use the metabolites of the others (lichens, mycorhizae, and higher plants, cellular zoochlorellae and zooxanthellae, intestinal microfloras and microfaunas of the higher animals, and many others). Or they may be less closely bound together (plants and pollinators of their flowers or distributors of their seeds and spores, mushroom "nurseries," the "farms" of

[1] This classification is a somewhat modified version of the scheme suggested by P. R. Burkholder (1952) and G. L. Clarke (1954).

aphids raised by ants, hermit crabs, actiniae, and many others). Spatial symbiosis takes the form of tenancy: habitation in the refuges of another species, on its body (epoikia), or within its body (enthoikia). A close relationship and some benefit to at least one partner are also characteristic of these cases.

Antibiosis is any form of antagonism which limits or precludes the existence of species. Antagonism, in which the presence of one species precludes the possibility of another's being present, is widespread among the microorganisms (fungi and bacteria), in which mutual exclusion is accomplished by allowing special antibiotic substances to get into the medium. Many similar examples may be found among the higher plants and animals; they are associated with the presence of individual or colonial plots and are especially frequent in closely related species. Biological antagonism explains the demarcation of their places of residence. Another form of antibiosis is competition, which limits the possiblities for existence. Food competition is often accompanied by a struggle for space, refuges, and other means of existence. Formation of solid, single-species clusters of plants and sessile animals apparently serves as such a means of struggling for space between different species.

Neutralism, or indifferent relationship, is often observed in an association between two species which are not connected by spatial or food relationships. In many cases the division of interspecies relationships into symbiosis, antibiosis, and neutralism is relative. Close cohabitation also has its negative aspects (poisoning or contamination of the environment), while antibiosis sometimes takes on mutually beneficial forms. The three most important consumers of pine-cone seeds in our forests—the great spotted woodpecker, the squirrel, and the crossbill—are competitors, since they all live on the same food. But the crossbill, unlike the woodpecker, throws away the core of the seed, which is then used by the squirrel. This is useful in that it maintains supplies of seeds which would otherwise be lost (Formozov, 1934).

MEANING OF PLANTS IN THE LIFE OF ANIMALS

Plants are the necessary food of herbivorous animals; geographic distribution, displacement by habitat, and abundance of the animals depend on them. Special microclimates are created in different tiers of the plant association which are favorable for different animals. Vegetation supplies refuges from enemies and complicates or, on the other hand, lightens the

work of predators. Many mollusks, fishes, insects, and other animals use water and land plants as a substrate, a place to lay their eggs and raise their larvae. The lower fungi, bacteria, and viruses cause numerous and often fatal diseases among animals, but many of them are an important food. The evolutionarily formed food and spatial relations connecting animals and plants have determined the basic properties of structure in many groups of animals. Today the animals' abundance dynamics are also associated with these features.

The relationship between the number and distribution of animal phytophages and the abundance and distribution of plants is manifested very distinctly among animals with specialized feeding habits. Monophages occur only where their food plant lives. The animal's range is usually smaller than the food plant's, since food is not the only condition of life. The so-called lacework of the range, or the concrete placement of the species within the area of habitation, corresponds to the distribution of food, but usually it is narrower, since it is also limited by the microclimate, the chemistry of the environment, and the presence of refuges, enemies, and competitors. Considerably more often, the area of habitation of an animal corresponds to the ranges not of individual species but of complexes of plants. Such is the penetration of Siberian taiga animals and taiga complexes of plants into the European plain, along with broad-leafed forests and the fauna typical of them.

Changes in the amount and content of plant foods lead to changes in the geographic distribution and abundance of the animals associated with them. Cutting of forests, breaking and seeding of the earth, pasturing of cattle on virgin territory, plowing, and cultured meadow-growing have led to a reconstruction of pre-agricultural plant cover. This has proved favorable for some animals and unfavorable for others.

Variations in the supply of vegetable matter in nature govern changes in the abundance of plant-eating animals. In some years there are variations in the harvest of tree seeds, the basic food of many vertebrate and invertebrate animals. Years of average or good seed production alternate with years of shortage. Different tree species bear their seeds at different times and with differing frequency. Bearing is stable and abundant among trees which are in the proper circumstances for bearing, while those in shade or on poor soils yield a sparser harvest. In the southern regions seed years come more often and the harvest is greater. The pines near Moscow yield a good harvest every 3–5 years. In Lapland the pines bear well about once a decade, at latitude 68°30′ they bear no oftener than once every 15–20 years, and on the margins of their distributional area the harvest is even rarer. In a year of average harvest near Archangel about half a kilo of seeds falls per hectare, while in the southern pine forests of the Bryansk

Forest Preserve (Buzuluk Pine Grove) it is 15 kg per hectare. In the Danubian countries the oaks produce acorns practically every year. In southern Germany acorns are abundant every 5 or 6 years, while in the northern part they come every 8–10 years.

When insect pests or fungus diseases attack tree seeds on a mass scale, the birds and mammals have that much less to eat, even in a year of good harvest. In the central forest belt in 1936, at an average harvest of 16,800 fir cones per hectare, the supply of seed food was 10.04 kg per hectare. In 1937 there was a good harvest, and the average number of cones per hectare rose to 31,720, but the seed supply dropped to 2.06 kg per hectare because of the depredations of pests and diseases (D. N. Danilov, 1941, 1944, 1950).

The variability of fruit and monocotyledon, shrub, and semi-shrub seed production is well known, as is the great variability of fungus production. The mass and quality of the growing parts of grassy vegetation are variable. F. Nansen and B. Helland-Nansen (1909) turned their attention to the relationship between hay production in Norway and the flow of the Gulf Stream, which has a tremendous effect on the weather of northwestern Europe. When temperature is depressed and precipitation abundant, fruits, seeds, bulbs, tubers, and stems all contain large supplies of starch, but when the air is very dry, the plants are filled with protein, fats, and sugars.

The role of changes in the supply and quality of vegetable foods is especially great in the population dynamics of those stenophages which have variable food sources. In years of poor pine- and fir-seed harvests the crossbills do not produce and lead a primarily nomadic life; when pine cones are abundant, egg-laying, setting, and raising of the young take place in winter. Increases in the number of squirrels take place in the year after a good evergreen-cone harvest (see Fig. 223). Because the reproduction season is over before these seeds are ripe, they can be used effectively only in the succeeding year. The food situation is very rarely so good that the squirrels bring forth a litter in late October and November of the same year.

A relationship is known to exist between hazel, oak, lime, and maple bearing in the oak forests of the European USSR and the Caucasus, and changes in the numbers of forest, yellow-throated, and field mice, which use the seeds of these trees as their basic food. Mass multiplications of voles (common, field, and society) and of the steppe lemming usually take place after a warm, moist autumn. In such years seeds are dropped in great masses, abundant young green vegetation remains under the snow, and the animals frequently multiply during the winter (under the snow) in fields, meadows, and even the tundra.

In all cases the primary cause of increased reproduction is the abundance

of food and its high nutritional value, in particular, a high vitamin-E content. In years when the seed harvest is low, not only does reproduction drop but migration and mortality rise as well. Food plays a less direct role as a regulator of abundance among species which feed on different foods or on foods which are not subject to such great quantitative and qualitative fluctuations. This is especially true in the southern regions, where snow cover does not hinder the search for food.

ROLE OF ANIMALS IN THE LIFE AND EVOLUTION OF PLANTS

Both evolution and the contemporary existence of plants are closely associated with the activities of animals. This is confirmed by the many anatomic and physiological properties of plants, most of all by their protective adaptations. Their mechanical defenses consist of tough leaves, hairs, thorns, hard bark, bristles, burrs, sticky secretions, and several others.[2] Among the species of *Prunus* and the wild pear tree, spines exist only on the lower branches and are absent on the upper part of the tree, which even large animals cannot reach. The sap of many plants contains bitter and poisonous substances (ether oils, glucosides, and alkaloids). Even the flowering portions can be protective. The toxicity of the narcissus, tradescantia, orchid, and many others; the unpleasant taste or the presence in the leaves of needle-like crystals of calcium oxalate, which cause a painful burning sensation when chewed; poison organs and hooks—all give relative protection to the species which possess them.

Atropine, which is found in belladonna, protects the flower reliably against ruminants, but the beetle *Epithrix atropae* feeds on its leaves. Fly agaric, which is poisonous to flies, is readily eaten by certain slugs. The euphorbia is toxic for cattle, but it is readily eaten by common voles and is the basic food of the caterpillar of the euphorbia butterfly *Deilephila euphorbiae*. Many plants containing ether oils, alkaloids, acids, and milky juice are eaten by caterpillars. The latter render the poisonous substances harmless with a special gland situated in the fork of the two little horns at the back of the head. This gland absorbs the poisonous substances from the blood and excretes them into the cuticle of the dorsal area. The skin glands of *Melasoma* beetles, which live on willows, apparently perform the same function by secreting salicylic acid. The poisonous berries and seeds of many plants are eaten with no harm by passerine birds, snipes,

[2] In addition, these adaptations cut down on evaporation and protect the leaves and other organs against injury.

seagulls, and chickens. The berry *Empetrum nigrum* is the favorite food of the hazel grouse, black cock, and white and tundra ptarmigans.

Even incomplete protection against extinction is important and explains the wide distribution on heavily grazed cattle pastures of such "inedible" plants as thistle, euphorbia, spurge, greater celandine, hellebore, and many others. The protective properties of plants (immunity) against the penetration and development of parasites are very important. They are noticeably weakened by unfavorable conditions. This explains the higher rate of damage caused by pests among plants growing on poorer soils. Aspens growing on podzols, for example, suffer from insects more than trees living on marls. There is a vast group of "secondary pests" which attack weakened plants when their defenses are down; these include the bark beetles, the oak pest *Cerambyx cerdo,* and many others.

One defense against destruction is a plant's ability to regrow lost parts and capacity for vegetative reproduction. The ephemeral desert cereal *Poa bulbosa* var. *vivipara,* after a short growing season (in spring), forms bulbs with solid walls on the root parts of the stem and in the florets; they are carried freely by the wind from place to place. In autumn the bulb takes root and puts out the shoot of the new plant. Many trees and shrubs which bear juicy fruits (mountain ash, bird cherry, buckthorn, raspberries, sweetbriar, and others), and which are often broken off and eaten by animals, easily restore the broken shoots. The ability to produce a new organism by putting out shoots from stumps is a property of almost all leafy trees and shrubs, but it is absent among the conifers, those ancient gymnosperms which developed in the Paleozoic and middle Mesozoic, when the modern seed destroyers did not yet exist.

The development of this ability among the leafy trees was apparently favored by the destruction of seeds by insects, birds, mammals, and other animals. Seed eaters are a serious hindrance to seed reproduction among several species. In most places it is impossible for the oak to reproduce by means of acorns because the latter are eaten up by small forest rodents and insects. When these pests multiply in mass numbers, the acorn crop is almost entirely destroyed. The above-mentioned periodicity of bearing among woody trees may be regarded in this sense as an adaptation. Abundant seed years usually coincide with low abundance among the seed-eating animals, since a good crop year is generally preceded by a poor one. Because of this, a large part of the seeds which fall during an abundant year actually take root (Table 61).

The maple crops in the years 1929–1939 changed very little. Large crops came in 1939, 1935, 1931, and 1929. The poor harvests were in 1936 and 1932. Few forest rodents were observed in 1939 and 1935, while the years when these pests reproduced en masse were 1938 and 1933/34. In these

TABLE 61. *Age groups of young maples in the Tula Preserve in 1939 (after N. P. Naumov, 1948).*

	YEAR OF HARVEST										
	1939	*1938*	*1937*	*1936*	*1935*	*1934*	*1933*	*1932*	*1931*	*1930*	*1929*
Average number of shoots of a given age per 10 sq. m	323.0	5.4	16.2	3.3	3.3	1.9	0.8	1.3	1.5	1.2	1.8

years, too, the lowest numbers of maple seedlings survived. In the year when large numbers of rodents were wiped out (1939), the number of seedlings surviving rose enormously. Mass reproduction of forest rodents affects similarly natural renewal of the beech in the forests of the Caucasus Preserve (Zharkov, 1938).

Animals aid reproduction (pollination) and distribution (spreading of seeds and spores) of plants. The basic pollinating animals are the insects. A smaller amount of responsibility lies with the birds, which pollinate the so-called ornithophilic plants (orchids and certain others). Many cases are known of surprising coadaptations (mutual adjustments) which permit the flower to be penetrated only by certain pollinators attracted by bright coloration, smell, or radiation from the flowers. Plants pollinated by nocturnal animals often have white, strongly fragrant flowers. Diurnal pollinators are attracted by the flower's colors. In the steppe and desert, with their strong winds, flowers with no fragrance predominate. Those plants which are pollinated mainly by Hymenoptera are distinguished by their hidden nectar and require a very involved mechanism for getting it out. Usually they are red or violet and have special spots for the arriving insect to land on (primroses, the Labiatae, orchids, papilionaceous plants). Plants pollinated by Lepidoptera have the nectar in deep, tube-shaped receptacles and the pollen in the open (lilies, phlox, certain carnations). Flowers pollinated by Diptera are often white or dark blue; they are open, with easily reached nectar. The trapping flowers (*Arum, Aristolochia, Piguicula,* Asclepidaceae) hold onto the flies which settle on them until pollination has taken place. Flowers which give off the odor of carrion are pollinated by the carrion and meat flies they attract.

Among the insects which visit flowers we distinguish four kinds. Dystropic insects are incapable of pollinating and usually destroy the stamens and pistils of the flower (various larvae, beetles, thrips, ants). Allotropic ones are the buzzing and lapping insects whose mouthparts are poorly adapted to pollination (most of the flower-visiting beetles, bugs, Orthoptera, long-nosed flies and flies of the families Stratiomyiidae, Tabani-

dae, Leptidae, Empididae, Dolichopodidae, and Muscidae, the social wasps, the short-nosed cuckoo wasps). Hemitropic insects have lapping or lapping-sucking mouthparts, dense down on the body, and agile movements while in the flower (solitary wasps, long-nosed cuckoo wasps, short-nosed bees, Orthoptera of the families Bombyliidae, Nemestrinidae, Syrphidae, Conopidae, most of the Lepidoptera). Finally, eutropic insects have dense down on the body, a long proboscis, and a series of very highly coordinated special movements which aid in pollination (the hawkmoths among the Lepidoptera and the long-nosed bees). Among the pollinating birds the most important are the hummingbirds, the Trichoglossidae, the Meliphagidae, and several others. There are some 1,580–1,600 species of birds which feed primarily on nectar and the insects found in flowers. Ornithophilic plants usually use a red or, less often, a white or yellow color scheme to attract birds. They generally have no odor.

Zoochoria—the carrying off of seeds and spores by animals—plays an important role in the distribution of plants and is caused by several adaptations. There are two types of spore and seed dispersion—ectozoochoria, or transmission on the surface of the animal, and endozoochoria, or transmission in the digestive tract. In the first case the seeds either attach themselves to the body surface by means of special adaptations (hooks, burrs, spines, sticky substances, etc.) or, being very tiny (spores and the seeds of many swamp and water plants), become mixed with mud and stick to the extremities or other parts of the body. The stinkhorn (*Phallus*) gives off a strong rotting smell, attracting flies which carry away the spores on their feet. The distribution of many steppe and meadow plants, and especially of the seeds of littoral and aquatic species, is accomplished primarily by animals.

"Active ectozoochoria" is the name given to the transport by birds, beasts, and certain insects of relatively large seeds which do not have special adaptations for attachment to the body of the animal. This method is associated with the feeding habits of the animals; during feeding part of the seeds collected do not find a host or for some reason are not used. Oaks are distributed mainly by jays, which hide acorns under forest litter, logs, and tree roots. In the Voronezh forests jays have been observed to hide acorns in a pine forest several kilometers from the oak grove; on 1 hectare of a 27-year-old pine forest 522 young oaks were found which had grown from such acorns. In the Caucasus transport of acorns from a valley forest 2–3 km up into the hills has been observed. The spread of cedars by the Siberian cedar bird and the spread of hazels by European nutcrackers take place in the same way. In the southern Cis-Baykal the cedar birds carry a considerable number of cedar "nuts," from 43,000 to 8,500,000 (2.5 kg)

per hectare per year, into areas where cedars have died from silkworms, lumbering, or forest fires. The loss of cedar seeds as food is more than compensated for by the extra distribution the birds create. At the same time mouselike rodents threaten the distribution of cedar "nuts" by storing them away (Reymers, 1956).

In anthills and termite colonies the hosts of these associations construct their own "fungus nurseries." Ants of the genus *Atta* cultivate the cap mushroom *Rhyozites*, termites raise *Monilia*, the *Hyloroetus* beetle (Lumetilidae) raises the *Dipodoscus* mushroom. In all these cases the fungi, by breaking down the medium on which they are raised, serve as a means by which the ants can use it.

Seeds and spores of many plants preserve their germinating powers and their ability to develop after passing through an animals's digestive tract. Endozoochoria is characteristic of plants which attract animals with bright colors and tasty (sugary) fruits. These are the grassy plants with berries, the shrubs, and the fruit trees, whose soft fruits are eaten by animals, while the seeds pass through the digestive tract unharmed. The germinative powers of such seeds are actually increased by the acid in the digestive juice.

The ornithochoric distributors of seeds include many of the small passerine birds. The mistle thrush is responsible for the wide spread of mistletoe, a plant parasite. The distribution of the Mexican lantana, a pasture weed, in the Hawaiian Islands, took place only after the Indian myna bird became acclimatized there. This bird willingly eats the berries of lantana. Birds which eat strawberries, elderberries, guelder-rose berries, buckthorn berries, spurge laurel, honeysuckle, and other plants spread their seeds widely during the autumn migrations. This is the reason for the rise of "aspen bushes" in the steppes—small woody patches addicted to karst or other depressions with closer groundwaters and moister soils. When a well-watered spot turns up, the first plants to appear are willow and aspen; their light seeds are brought from afar by the wind. The migratory birds stop in these young patches and seed them with such plants as sweetbriar, buckthorn, and many others (I. I. Popov, 1914).

However, not all species of birds which feed on fruit spread seeds. Bullfinches and certain others crack them with their beaks, and further breakdown is assured by the presence of tiny stones in the gizzard. Seeds are digested in the intestines of grouse and ducks, although in the latter this process is slow and incomplete. The spores of the higher fungi, protected by their indigestible coverings, pass through the animal digestive tract in a still-viable state. Thus animals which willingly eat the reproductive parts of fungi (reindeer, hares, squirrels, small rodents, and others) turn out to be important spreaders of these plants.

ROLE OF ANIMALS IN THE LIFE OF PLANT ASSOCIATIONS (PHYTOCENOSES)

By eating plants, aiding their reproduction and distribution, stealing their fruits, etc., animals influence the content and structure of plant associations. Selective destruction of various species of plants changes their relationships within the association. Extermination of plant pests and action directed at the soil (plowing, solidification or improvement of soil to change its structure, hydrothermic regime, or chemistry) also have a tremendous, though indirect, effect on the life of the phytocenose. Since individual animals are associated with various groups of plants, the activity of each of them has a different effect on the association.

In the steppes close associations connect the turf-grass cereal grains and the ruminant animals. Weakening or curtailment of pasturing leads to the overgrowth and domination of feather-grass turfs and the driving off of other grasses (fescue, hairgrass, and *raznotrav'ye*). Dead plant remains accumulate; they break down very slowly in the dry climate of the steppes, leading to the formation of a solid layer of plant compost which hinders the growth of young shoots. So the curtailment of pasturing for long periods of time leads to the death of cereal turfs. Their places are taken by weeds, and the re-establishment of the previous steppe association led by cereal turfs takes a long time (15 to 20 years) and proceeds in stages ("weeds" of root cereals and steppe multigrasses) (Pachossky, 1915, 1927). Consequently, the normal existence of cereal turfs is impossible without the constant mowing of the superterranean portions of plants by animals, trampling of plant remains, improvement of the soil, and trampling of seeds into the soil. In this way the pasturing of ruminants maintains the relative stability of the steppe's plant association: pasturing regulates the relations between various species of plants and assures that a variety of species will exist in the plant cover.

However, intensification of pasturing above and beyond known limits, different in each set of circumstances, destroys the stability of steppe associations. Oversolidification of the soil endangers the hydrothermic regime and increases salinity.[3] Trampling of the cereals leads to death of the turfs they have formed, which, along with salinization, aids in the growth of feather grass, salt grasses, etc. Artificial desertization of the plant association takes place, leaving the previous turf choked with tall weeds.

Similar phenomena unfold on sandy deserts and desert feather-grass

[3] As a result of the predominance of rising currents of soil solutions in unstructured soils and of capillary action.

pastures. Lack of periodic "weeding" by hoofed animals leads to "aging" and extermination of plants. On the other hand, overintensified pasturing is accompanied by trampling of these plants and spreading of inedible salt grasses. In the northern Pre-Aral area, near built-up areas where pasture is used with great intensity, the food-rich feather grasses are replaced on saline soils (soils containing more than 1 percent salt) by *Anabasis salsa* and on light soils by *Peganum harmala,* which cannot be eaten by ruminants. In the sandy desert overpasturing leads to destruction of the grass cover and shrubs and to the blowing away of the previously fastened soils, so that near wells, settlements, and cities there are considerable areas of moving sand dunes.

The activities of rodents affect the existence of many species of dicotyledonous plants (salt weeds). The eating of perennial cereals and multigrasses by voles, susliks, and large gerbils often leads to the death of the individual bush or turf patch. Vegetation-free patches form around the opening of the burrow; because of this, the creatures abandon their old burrows and settle in new ones. The bare, ripped-up patches, fertilized with urine and dung and the inedible remains of plants, are favorable for the development of weeds which need friable soils containing humus (Fig. 228). Along with these grow annual ephemerals and, somewhat later, plants valuable as food. Among the monotonous cover of steppe turfs or meadow cereals, spots with a different vegetation spring up. Thus a micro-

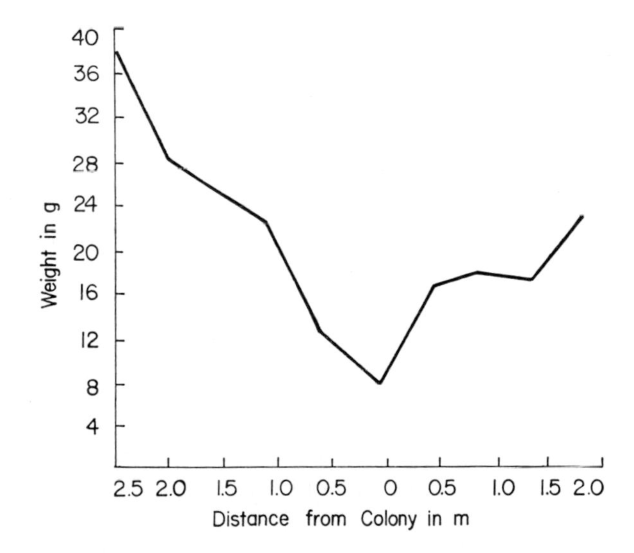

Fig. 228. Weight of green mass of vegetation from plots of 0.25 sq. m inside and outside a colony of steppe lemmings (*Lagurus lagurus*) (after Formozov and Voronov, 1939).

mosaic of plant cover is created and maintained. An increase in the number of rodents is usually accompanied by the spread of weeds and a slowdown in the establishment of the natural steppe vegetation on fallow lands. Many species which now clutter our fields existed before cultivation and inhabited plots which the animals themselves (moles, susliks, etc.) had turned over, from which they spread to cultivated areas when the soil was plowed up.

Digging burrows and throwing friable soil onto the surface lead to the rise of new plants on the broken soils and also to an increase in the variety of the plant cover. When susliks build mounds, the desalinization of the soil which results causes growth of the less halophilic plants (white feather grass, cereals), which attain larger sizes and better development than on unbroken soils. Marmot burrows are usually distinguished for their dense growth and for a greater number of plant species than are to be found on neighboring plots. In the tundra the vegetation near the lairs of Arctic foxes is markedly different from that nearby because of its relative richness and splendor, which issue from the fertilization of friable soils by garbage, dung, and urine from the animals. These green spots have received the name "fox gardens." The vegetation of anthills, termite mounds, and the nests of other insects differs greatly from that surrounding them. In the desert, where colonies of large gerbils are found, the initial vegetation is cut over (Fig. 229), and annual ephemera and inedible salt grasses appear, creating a mosaic plant cover. Destruction of plants, cutting of vegetation, and increased wind erosion create here an odd combination—an increase in species of plants and an enrichment of pastures with valuable food species.

Destruction of bulbs and other subterranean plant parts by insects and

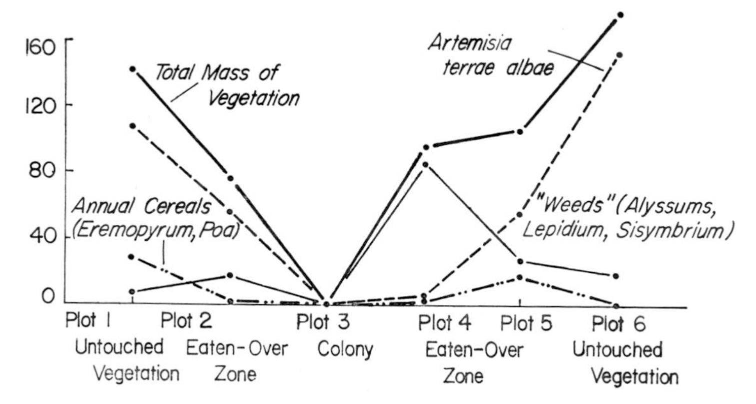

Fig. 229. Consumption of vegetation by large gerbils (*Rhombomys opimus*) in a colony, showing change in size of the mass of grassy vegetation from a plot of 1 sq. cm (original).

rodents causes considerable losses among the early-spring foods (viviparous *myatlik* (*Poa arctica*), annual couch grass, onions, and tulips). In 1933 in northwestern Kazakhstan, even under conditions of low population among small susliks and jerboas, these animals had already destroyed 16,400 bulbs per hectare on the feather-grass plots and up to 19,300 in the wormwood associations by 24 June. These animals accumulated 221 and 457 bulbs a day respectively (Formozov and Voronov, 1939). In the Cis-Aral deserts up to 100 holes of dug-up tulips and goose onions were found in 1 sq. m in some places.

Lemmings aid in the development of the tiny-hillock micro-relief of the tundra and augment the spotty mosaic character of the plant cover. Mass destruction by tundra grouse of flower buds and the tops of *Dryas* and *Salix* shoots diminishes the seed production of these plants and causes their characteristic structure. Geese do a great deal of harm to the plant cover of the cotton grass–moss tundra, where the percentage of damaged cotton-grass shoots runs from 50 to 83 percent. In some places they destroy the plant cover, intensifying the heating and aeration of such plots, which increases thawing and cracking and causes the soil to creep down the sides of slopes. As a result, the cotton grass–moss tundras gradually turn into sedge–cotton grass–moss groupings and then into sedge-moss spotty associations (Tikhomirov, 1955).

Thus the plant cover of the steppe, semidesert, desert, meadow, and tundra was not only formed evolutionarily under the influence of animals but is continually subject to their influence even today. Changes in the number and species content of the animals lead to replacements in the plant cover, and the total absence of animals is accompanied by the death of the association. Therefore, the idea of a "steppe," a "desert," a "tundra," a "meadow," etc. must of necessity include not only the vegetation but also the animal population, not merely because they live together but because they are mutually dependent.

The same thing is true of forest associations. The obstacles encountered in the reproduction of woody trees and shrubs have different significance for different species. Thus the species content of a naturally self-renewing forest depends both on the conditions for germination and growth—on the ability of the individual species to bear fruit, multiply from their growing parts, or regenerate—and on their means of protecting themselves against pests. Animals which harm individual species selectively (unevenly) change their ratio in the plant association.

The influence of pests is retarded by their extermination by species of animals, especially insectivorous birds and predatory birds, mammals, and insects (as well as such parasites as dung beetles, parasitic flies, etc.), which are useful in forestry management techniques. Impoverishment of this

useful fauna, formerly a consequence of predatory use of forests and the single-crop techniques popular in the West, has led to increased pest activity and to the death of large plantings (C. Elton, 1960). Mass reproduction of pests has also become more frequent under such conditions. "The history of forestry is full of examples showing that lack of knowledge of the role of fauna or inability to foresee the consequences caused by one technical method or another have led to the sorriest results for forest management" (Tkachenko, 1939). In the protection of timber against harmful insects a special role is played by insectivorous birds. Their significance is determined by their extraordinary voraciousness, their ability to find swarms of insects quickly, and their capacity for concentrating in these areas with great speed (Formozov, 1950).

When pasturing is not restricted, both wild and domestic hoofed animals destroy and trample the grassy vegetation, pack down the surface layers of the soil, and chew up and then completely destroy the young shoots and bushes. As a result, the hierarchical structure of the forest plant association is simplified and its species content impoverished. All this hinders or even precludes the possibility of natural renewal, is accompanied by a dwindling or even complete disappearance of many insectivorous birds (Sylviidae, thrushes, etc.), and gives rise to an increase in the abundance of forest pests. Impoverishment of the bird population is not compensated for by the slight growth in the number of birds nesting in hollow trees or on the ground, which feed primarily on Diptera. Thus the existence of a forest without animals is impossible, just as the existence of other plant associations without animals is impossible. Here, too, animals influence the content and character of forest plantations.

SPATIAL CONNECTIONS BETWEEN ANIMALS AND PLANTS

Spatial relationships play an important role in associations by determining their stratification (hierarchical structure) and other structural properties. At the basis of this connection lies the animals' use of plants as protection from enemies and inclement weather, as a medium of habitation or a substrate, as refuges for reproduction, and as agents of a favorable microclimate. The last is of special importance for animals, since in the plant association and its individual parts (tiers) a microclimate is created which differs substantially from the climate of the open air and soil.

"Microclimate" is the name given to the properties of the regime of meteorological elements (temperature, humidity, air movements, etc.),

which depend on the character of the surface relief of a given area and especially on the character of its living cover. The character and magnitude of daily, seasonal, and nonperiodic changes in temperature, humidity, wind, and pressure are different in the various air layers, on horizontal ground and flat slopes, on slopes of different exposure, in the various tiers of the plant cover, in the soil, in caves, burrows, hollow trees, etc. The climatic characteristics of the individual parts of a body of water also differ (Fig. 230).

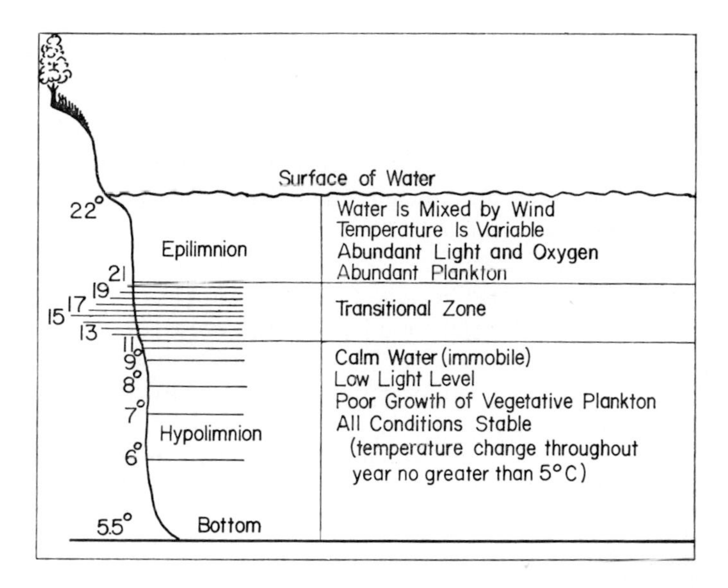

Fig. 230. Summer distribution of temperatures in a lake (after Needham, from Allee *et al.*, 1949).

The properties of climate on plots covered with various types of living growth are called the ecoclimate, since they are determined by the living cover of the plot of land. Such are the climatic characteristics of forests and meadow plots, for example; these characteristics depend on the character of absorption and reflection of light energy by plants and on the movements of the air. In a dense forest about 3–5 percent of the light energy penetrates down to the moss cover. Beneath the beech forest of the Crimean Preserve the average intensity of radiation in summer is only 0.05 calories per sq. m, or about 20–25 times less than on an open space. The forest substantially alters the moisture turnover of the soil and the moisture of the air. Forest plantations in the steppes, though they may cause a drop in groundwater levels, still provide a means of holding snow;

they slow down the melting rate of the snow to such an extent that in the final reckoning the supply of moisture in the soil is increased. However, in the sandy portions of the steppe zone the cultivation of forests has sometimes been accompanied by a drop in the supply of groundwater and consequent drying out of the forest (Table 62).

TABLE 62. *Soil moisture (in percent of its weight) on plots with different types of plant cover (after Vysotsky, 1930).*

Character of Plot	DEPTH IN CM						
	0	*10*	*25*	*50*	*100*	*150*	*200*
Leafy forest	15.0	18.3	18.4	17.8	14.8	14.8	13.1
Virgin steppe	5.9	12.4	17.2	17.5	16.0	16.8	17.6
Black fallow	3.6	21.8	24.2	24.4	24.4	20.8	19.4

A forest aids in preserving moisture in the upper soil horizons and dries out the deeper soil layers. In the forest and forest-steppe zones the dehydrating effect of the forest on the soil is slight and cannot even be observed in many types of forests and soils. In the northern parts of the forest zone marshes are formed after a forest fire or a lumbering operation. By withholding the runoff from rains and meltwaters, forests play an important part in controlling water losses.

In the forest massifs the humidity of the air is noticeably higher than in neighboring open spots. Temperatures here are distinguished by their lesser fluctuations and their lower yearly averages. Thus the forest tends to shift the area it occupies into a cooler, moister climatic zone (in the European USSR, into the northwest). The relative humidity of the air is different in the separate tiers of the forest and the grassy cover of the meadow. On a sunny, windless day in July, with the temperature of the air at 29° C, the relative humidity was 57 percent in the open air 100 cm above the surface of the soil, 78 percent at a height of 13 cm above clover shoots, and 96 percent at a height of 2 cm above the soil and in the midst of stalks and leaves of cereals and shoots of meadow tea (Geyger, 1931). Forests, especially the field-protecting forest strips, which break the strength of the wind by five to ten times, markedly alter the distribution of snow. In summer, because of their ability to screen, they minimize the temperature and evaporating strength of the air (Fig. 231).

At the same time the temperature regime of the soil is changed. The plant cover, especially the forest, reduces the possibility of freezing at the soil surface and lessens the scope of temperature fluctuations in the soil itself. In summer the soil under the forest is colder and in winter hotter than on neighboring unforested plots. Therefore, soil freezing is usually

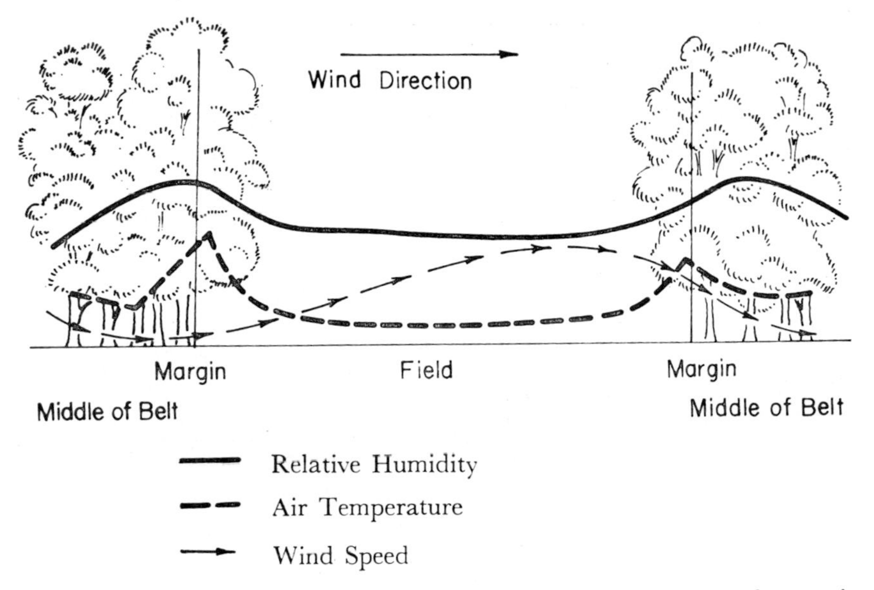

Fig. 231. Changes in basic climatic elements in a system of field-protecting forest strips (after Mel'nichenko, 1949).

kept to a minimum in the forest. After a snowfall the soil, which may have been frozen since autumn, thaws again in many places (Fig. 232). The soil freezes most deeply in places where it is totally devoid of plant cover. Here the soil hibernators and digging animals burrow the deepest; the population density of soil-dwelling animals is usually minimal in such areas.

Thus the forest climate is distinguished by a greater resistance to meteorological elements, although average yearly temperature and amount of precipitation in forested and unforested areas may differ very little (Berg, 1938). It is understandable that, in any of the natural zones, forests provide greater possibilities for the existence of climatic stenobionts than areas of open landscape do.

The relief and exposure of a plot have a great effect on its climatic characteristics. A horizontal surface receives only slightly more calories than a southerly vertical surface, almost twice as many as an easterly and westerly surface, and almost 50 times as many as a northerly exposure. And the soil temperature is different on slopes of varying exposure. Unequal amounts of snow pile up on different slopes depending on the prevailing winds. This is expressed in the moisture turnover and the duration of the snowless period of each slope and makes it possible for heat-loving plants and animals to penetrate far into the mountains along the slopes of southern exposures. In the central Tyan'-Shan' the sunlit slopes at 3,000–4,000

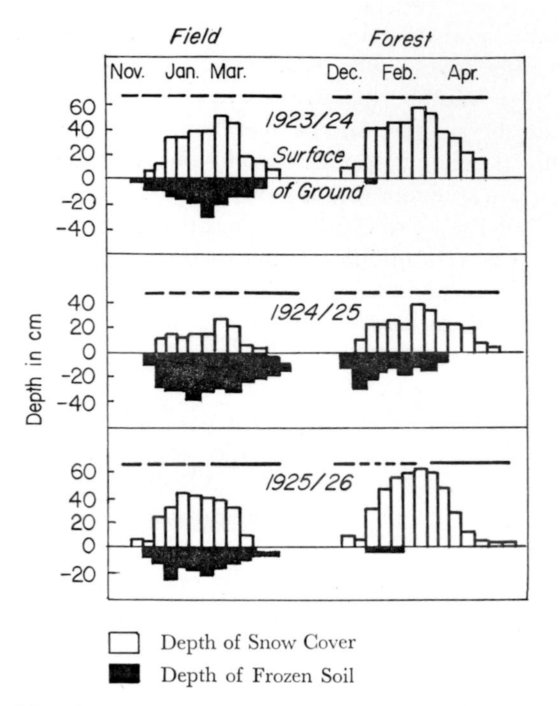

Fig. 232. Course of freezing and melting of soil in a field and a forest near Moscow (after Alisov, Drozdov, and Rubinshteyn, 1952).

m are occupied by steppe or even semidesert groupings of plants and their accompanying animals (locusts, desert snake-eyed lizards, viviparous *Eremias* lizards, least skylark, gray hamster, jumping jerboa). The northern slopes, situated nearby, are covered with fir forests or cobresia barrens, with a boreal fauna of insects, birds (nutcrackers, willow tits, three-toed woodpecker, horned skylark), and mammals (rust-colored, Tyan'-Shan', narrow-skulled, and economist voles). The times for the periodic phenomena of animals and plants on such plots may differ quite a bit, which isolates even populations of the same species living right next to one another.

In the lowlands changes in temperature have a greater range than in the highlands. In valleys, ravines, and gullies the nighttime temperatures are lower and the daytime temperatures higher than on elevated portions. As a result of the flow of cold air into hollows and valleys, the winter temperatures in Trans-Yenisei Siberia are very low, while the very nadir of cold is reached in the Verkhoyansk hollow.

Cave climates are unique. They depend on the cave's overall size, the dimensions of its entrance, and the ground in which it is found. Humidity

is high and stable, and the temperature is constant in the depths of large, deeply penetrating caves. Even yearly fluctuations do not exceed a few degrees (Fig. 233). Winter in a deep cave is warmer and summer cooler than the air at the surface. In the Karakamysh Caves (near Tashkent), where horseshoe-nosed bats were wintering on 24 January 1934, there was frost for almost a whole month, with outside temperatures reaching —25° C. On the day of the observation the temperature was —7.6° C in the morning and —1.5° C at midday, but at the cave entrance it was —0.6° C, in the first reaches of the cave it was 3° C, and at the back of the cave it was 8° C. In the Sablin Caves near Leningrad on 4 February 1937, with the outside air temperature at —4.7° C, the temperature in the cave at the point where bats were wintering fluctuated from 2.5 to 6° C (Kuzyakin, 1950). In caves near Berlin in which bats were wintering the temperature in winter varied from 8 to 2.7° C, while outside it varied from —3 to —8° C (Eisentraut, 1934). In the famous Bakharden Cave in the foothills of the Kopet Dagh (Turkmenia) an underground lake keeps the humidity very high (80–95 percent) and the temperature stable; even in winter it never drops below 25° C. In summer the temperature never rises above 31° C, and its daily fluctuations are insignificant. The seasonal temperature maximum in the cave occurs in autumn and the minimum in spring. The lag is associated with the retarded penetration of temperature changes into the depths of the rocks. Air movement in caves is ordinarily hard to feel.

The microclimate of burrows is comparatively stable. Its differences depend on the diameter of the entrances and their length, shape, and

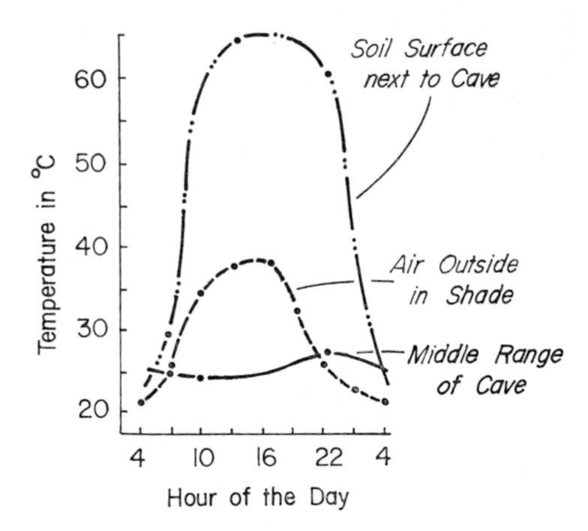

Fig. 233. Daily changes in air temperature inside and outside a cave near Stalinabad, 3–4 August 1958 (after Kuzyakin, 1950).

depth. A burrow with no inhabitants has a temperature and humidity very close to those of the surrounding ground. The air inside is practically motionless. The cooling rate in the nesting chamber of a vole burrow was 2.4 m-calories per sq. cm of surface per second, which corresponds to the cooling rate in still air at a humidity of 80 percent and a temperature of 21° C; the cooling rate in the same air 0.5 m above the soil was ten times as much—24 m-calories (Strel'nikov, 1950). In an inhabited burrow the temperature and humidity rise in proportion to the number of inhabitants (see Fig. 210). A high, stable air humidity in the burrow favors the penetration of hygrophilic animals into drier zones. By using the burrows as a daytime refuge, these species hunt their food at night, when the danger of dehydration and overheating is not so great. The same type of microclimatic properties are found in cracks in the soil or rocks (Fig. 234). These cracks are also used widely by hygromesophilic animals. On this basis the

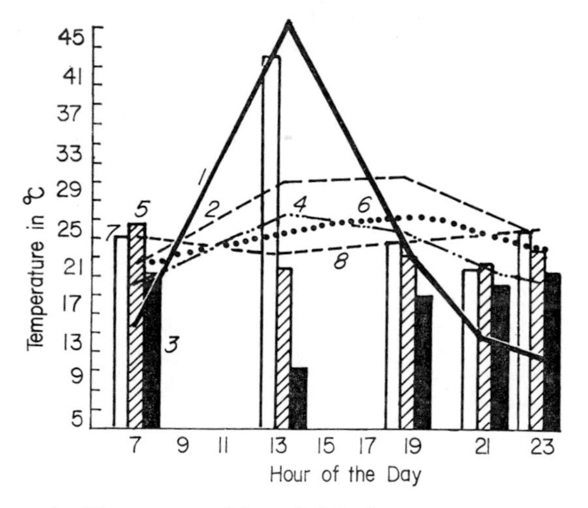

1 Temperature of Open Soil Surface

2 Temperature of Soil at Depth of 10 cm

3 Humidity at Soil Surface in Thick Grass

4 Temperature of Soil Surface in Thick Grass

5 Humidity in Cracks

6 Temperature in Cracks

7 Humidity in Burrow

8 Temperature in Burrow

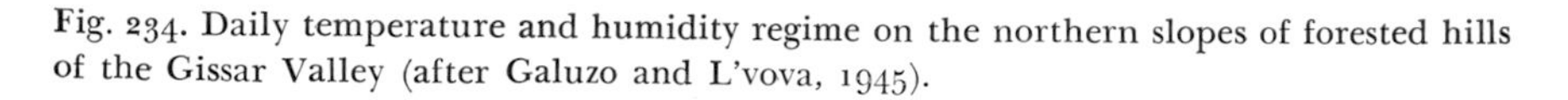

Fig. 234. Daily temperature and humidity regime on the northern slopes of forested hills of the Gissar Valley (after Galuzo and L'vova, 1945).

"burrow biocenoses" arise, some of them quite complex. There are at least 200 species of animals which live in their own unique crack burrows as tenants—cohabitants—of burrows along with their hosts. Many of them (parasites, predators) are associated by food relationships as well as spatial ones.

Although they do so to a lesser degree, the hollows of trees also offer a stable microclimate (Fig. 235). The degree of difference depends on the size of the hole, the thickness of its walls, the size of the opening to the outside, the species of tree, the state of its wood, and many other factors. Tree holes are more subject to cooling in winter than they are to overheating in summer. Having relatively poor thermo-isolating properties, wooden holes maintain a relatively stable humidity; this accounts for the relatively delicate and porous shells of eggs of hole-dwelling birds as compared with those which nest in the open.

The microclimate of the place of reproduction also accounts for other structural and developmental properties in the young. Hares and hoofed animals give birth to offspring already covered with hair and able to see; from the first days of life they are capable of getting around by themselves. This is associated with the fact that they have no permanent refuges. Rabbits, mice, voles, squirrels, susliks, and marmots are born naked, blind,

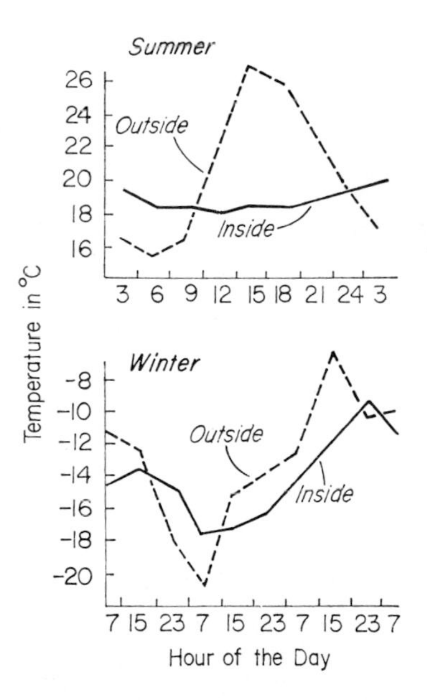

Fig. 235. Daily changes in air temperature inside and outside a lime-tree hole in summer and winter (after Kuzyakin, 1936).

and helpless. The broods of nesting birds can be distinguished in the same way: those small birds which are born in tree holes develop more slowly, and the establishment of plumage and a constant body temperature is retarded as compared with the fledglings of open-nesting birds. The latter are usually 1–3 days ahead of the former in developing these features.

The existence of many microclimates with different degrees of variability in one locality permits many different species with differing demands to exist side by side. At the same time the presence of different microclimates aids in the formation of several relatively independent unions of a nonterritorial character within the biocenose associations of the second and third orders.

CHAPTER 16

Predators and Prey

ADAPTATIONS OF PREDATORS

The relationships between flesh-eating animals and their victims are just as close and mutual as those between plant-eating animals and plants. They find their expression in mutual adaptation of structure, physiological properties, behavior, and way of life. The digestive tract of a carnivore, especially the intestine, is shorter than that of an herbivore, and the chemistry of its digestive juices is distinguished by the abundance and activity of its enzymes—the chemicals needed to break down animal proteins and, in some species (birds), bones. In predators the central nervous system is highly developed, the sense organs are more perfected, and nervous activity is more complex. Predators are usually very mobile. These progressive features of construction and vital activity are associated with activation of the relationship between predator and environment and do much to explain their outstanding role in evolution. Actually, almost every new evolutionary group, especially those of the higher taxa (classes, orders), had as its ancestors animals which led a predatory life (hoofed animals, primates, arthropods, and many others).

Reproduction of predators is closely associated with the feeding habits and nutritional dynamics of the food base. The mating season occurs during the time of greatest food supply. The fecundity of predators corresponds to the variability of the food base. Predatory birds whose food supplies change little from year to year are not very fecund, and the size of their broods does not change much from one year to another. The buzzard lays two eggs, the steppe eagle one to four (usually two or three), the smaller osprey two or three, the peregrine falcon two to four, and the even smaller honey buzzard one to four (usually two). All of them are more or less specialized in their feeding habits.

Among species living on variable food supplies (locusts, small rodents, etc.) fecundity is higher and varies according to the state of the food supply at a given time. The *Buteo* hawks lay five or six eggs, but in years of famine they do not mate at all. In good food years falcons lay as many as 8 eggs, merlins 10 or 12. The hawk owl has been known to lay as many as 13 eggs, the polar owl as many as 12, the long-eared owl up to 9, and the marsh owl as many as 14. When mouselike rodents are in short supply, these owls either do not nest or else only a small part of them mate, laying one to three eggs. Even then feeding of the young is not always successful, and if starvation sets in, the adults will even eat the young (Sharleman, 1958). On the other hand, in years when food is abundant (usually in autumn) a second nesting takes place in which greater numbers of eggs are laid than in summer. In 1929 the marsh owls of the North Caucasus were observed laying eggs in the wintertime. This same species had a second nesting at the start of winter in Bessarabia in 1906, and fledglings appeared in December and January. In the period from 1896 through 1927, when small rodents were abundant, the grass owls near Leipzig laid eggs two and even three times a year in 11 different years, the late broods containing up to six or eight eggs, as against the usual four or five. The fledglings left the nests only in January.

The small insectivorous birds also do not multiply when food is lacking or is in short supply. In the cold, rainy summer of 1939 cliff-swallow nesting was retarded in the shoreline banks of the Upa River. Sometimes the birds turned up near the nesting colonies, and sometimes they left them for a week or more (observations made in the Tula Preserve). The same is known for mammals. The most fecund are the polar foxes, whose basic food (small rodents, willow ptarmigan) is subject to sharp fluctuations from year to year. When these foxes are raised in captivity, they produce litters of up to 25, and B. M. Zhitkov (1904) found 16 embryos in a female on the Yamal Peninsula. In famine years reproduction drops, the survival rate of the young decreases, and the fatness of the adults diminishes as well. The low multiplication intensity is maintained into the following year. Red foxes, which have a richer and more stable food base than the Arctic fox, give birth to five or six pups on the average. In only a few cases does the litter rise to 9 or, rarely, to 12.

PROTECTIVE ADAPTATIONS OF PREY

Protective adaptations are methods of individual defense displayed by sessile or relatively immobile animals (armor, spines, hooks, stinging or-

gans, poisonous or odorous secretions, various types of protective coloration, and several others). They also include group protection in the form of animal-created refuges and shelters, flocking tendencies, and other behavioral characteristics.

Stinging organs are widespread among the lower animals. They exist, among the protozoa, in the Dinoflagellata and the infusorian *Frontonia*. They are represented by finger-shaped trichocysts which turn into resilient threads whenever a physical or chemical disturbance causes them to be launched from the body. Among the cnidarian coelenterates there are species of polypoid or medusoid coelenterates with numerous trichocysts in the body. The external epithelium of certain ciliated worms (such as the rhabdocoele *Microstomus*) contains stinging cells acquired from ingested hydroids. Some of the opisthobranch mollusks also have these "stolen" stinging capsules.

The protective adaptations of insects include hard covers and glands which excrete strongly odorous or poisonous substances (Hymenoptera, blister beetles, and many others). The hairs which cover many caterpillars and butterflies are dangerous. The stinging Hymenoptera (bees, hornets, wasps, and others) protect themselves with poison glands whose presence is associated with a special warning color scheme. Poison glands exist in certain fishes and amphibians. The cephalopod mollusks are protected by the secretion of a special gland, the ink sac, which contains an opaque, dark, sepia pigment, or India ink. The electric organs of the eel and skate are differentiated muscles. Many species have armorlike covers (shells of the gasteropod mollusks, bony plates of the slow-swimming fishes, bony plates and scales of reptiles, and armored coverings of mammals) or quills (the benthic fishes, hedgehogs, porcupines, and many others).

Like all adaptations, these defensive structures of the organism do not guarantee security. Many predatory birds and beasts eat hedgehogs, and the larger ones even eat porcupines. Fuzzy caterpillars are not eaten by most of the insectivorous birds, but cuckoos destroy them successfully. The great titmouse "prepares" such caterpillars by eating only the internal organs, while nightingales "butcher" the caterpillar by pounding it against the ground or some object and separating the hairs from the surface of the body. But animals protected by armor plate, needles, poison glands, electric organs, unpleasant tastes, or a repulsive smell do die in smaller numbers.

Hunted animals are often distinguished by their protective coloration. There are two types of protective coloration—camouflage, or cryptic coloration, and warning, or aposematic, coloration. A frequent case of the latter is mimicry, in which an unprotected animal uses its body form and coloration to imitate an inedible species or an object. Camouflage is also

characteristic of stalkers; it enables a predator to creep up on the prey unnoticed or to watch it.

In the case of camouflage many species have various degrees of coloration which enable them to blend into the background in which they ordinarily live. The "sandy" coloration (a yellowish-gray shade) of many desert animals is well known. Arctic species which do not hibernate turn white or grow much lighter for the winter (polar foxes, varying hares, Arctic lemmings, northern reindeer). The inhabitants of treetops or grassy vegetation are often green. In caterpillars this is governed by the amount of light-resistant chlorophyll contained in the body and extracted from food. In the spotted sphinx *Smerinthus ocellatus* the green pigment of the egg is obtained from the chrysalis and the butterfly, and the young caterpillars acquire their coloration even before they begin feeding on green vegetation. In other cases green coloration has nothing to do with chlorophyll. Among adult insects, amphibians, and reptiles the green color is often created by a combination of yellow pigment and structural properties of the skin, which are light or dark blue. The emerald-green color of tree frogs comes from a combination of three types of pigment cells—melanophores, lipophores, and guanophores—which act as carriers of black or brown, yellow or red, and white pigments respectively. The same thing apparently exists among many of the snakes and lizards. The green coloration of the hair of the South American sloths *Bradipus tridactylus* and *Choloepus didactylus* is actually caused by symbiotic protococcus algae which live in it. The coloration of the big-eyed sphinx caterpillar, the tree frogs, and certain other animals corresponds to that of the background in relation not only to visible light but also to the infrared part of the spectrum.

In industrial areas, where soot darkens the background, the earlier-existing light-colored populations have been forced out by dark forms. This "industrial melanism" has been observed for such moths as *Biston betularia, Boarmia extersaria,* and *Boarmia punctinalis* in England and other western countries. This phenomenon can be explained by the activities of insectivorous birds, which quickly exterminate light-colored moths when they appear against the new dark background. However, it was later shown that there is a relationship between the color of the moths and the tendency of their caterpillars to feed on one plant or another.

There are many examples of local races with their own special camouflage among the insects, birds, mammals, reptiles, and other animals. The butterfly *Camptogramma bilineata,* in addition to its light-colored yellow form, is also known in a dark variation distributed on the dark rocks of the Isle of Jersey, in the adjacent parts of the Continent, and in the dark peat bogs of Connemara, Ireland. The light-colored form lives on the

light gray limestones of the neighboring regions. Skylarks are usually colored the same as the soil on which they live. This is not associated with moisture or temperature, since the regions settled by the variously colored races do not always differ in climate but only in the color of the substrate. The color of the eggs of *Lobipluvia malabarica*, which nests on the open soil, is the same as the color of the ground. On dark soils they are of a dark earthen color, while on the reddish laterites they are yellowish red with brown speckles.

The colored forms of the deer mouse *Peromyscus* sometimes occupy quite small areas of land with sharply differing ground colors. On light-colored sands and dark lavas in North America almost all rodent species have corresponding colorations, which are apparently associated both with the direct effect of the physical circumstances and with increased extermination by predators of individuals which do not have a color scheme that agrees with the background (Sumner, 1921).

The direct effect of physical conditions is undoubted in the case of seasonal dimorphism in insect coloration. The wet and dry seasonal varieties differ not only in color but also in form. Camouflage colors and shapes are better expressed in generations whose offspring are born during the dry season. The insects are less mobile then, and the numerous insectivorous animals hunt their prey more intensively. Under such circumstances cryptic coloration acquires special importance.

Camouflage is achieved by various means. Among inhabitants of the water partial or complete transparency of the body is widespread, giving rise to the name "glass animals." Such are the medusae, salpae, worms, crustaceans, and the larvae of insects (such as *Corethra*). Individual translucent organs do such a poor job of camouflaging, however, that this method is less important than a slight darkening of the body, which permits its general contours to be seen. Among the terrestrial animals this kind of camouflage is rarely observed. Among the South American Sthomiinae butterflies the scales on the wings are so reduced that they are as transparent as glass, and the very delicate body completes the illusion of no flesh.

Color which changes in response to the background is associated with mobile pigment cells in the skin, or chromatophores. This type of camouflage is characteristic primarily of animals with a complex nervous system (cephalopods, crustaceans, certain fishes, many amphibians and reptiles). It is absent in birds and mammals, the skins of which are covered with feathers or hair. The change in color may occur relatively slowly and may be stable; in this case it is associated with a rise or fall in the number of chromatophores. In other cases a change in color occurs quickly and is associated with a change in the form and position of the chromatophores.

The complete transition of a chromatophore from contraction to expansion in the cephalopod *Sepia* takes place in two-thirds of a second. The acquisition of a series of colors when an animal is placed on a multicolored background is especially surprising. Certain fishes are by turns monocolored, then longitudinally or transversely striped. The longitudinal striping is usually characteristic of the animal as it moves; while at rest, the color scheme is a transverse stripe.

Permanent camouflage (cryptic coloring) is of two types: concealing antishadow and dismembering coloration. In the former case the brightly illuminated parts of the body are dark in color, while the parts in shadow are light-colored or white. This eliminates "shadow" on the body of the animal, so that it is perceived visually as a flat object rather than a voluminous one (Fig. 236). This type of coloration is profitable to animals which usually remain in the same position with regard to the light source. It is widespread among fishes, birds, mammals, reptiles, and amphibians, is encountered among insects and other invertebrates, but is absent in animals whose way of life causes them to change their body position fre-

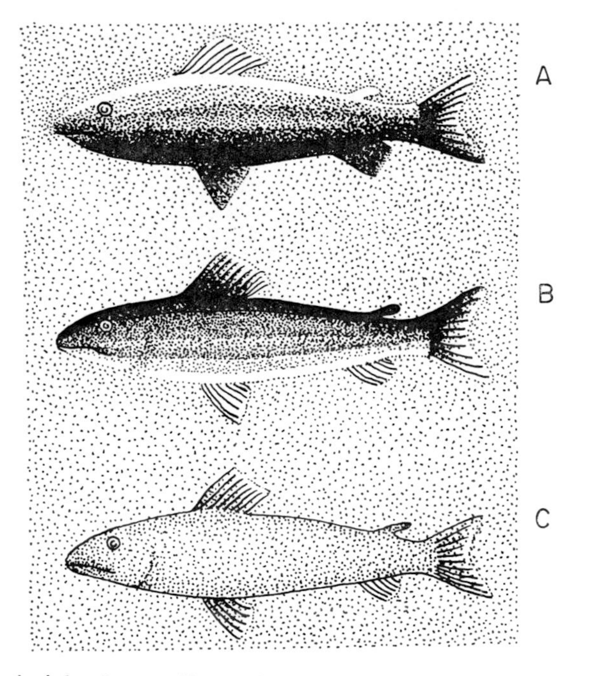

Fig. 236. The principle of concealing anti-shadow (from Kott, 1950). A: distribution of light and shadow on a monochromatic body illuminated from above; B: external appearance of a fish which has a color scheme founded on the principle of anti-shadow under uniform illumination; C: the same fish illuminated from above.

quently (such as the sticking fish *Echeneis naucrates*) and among species which live in unilluminated locations. On the other hand, in the Nile fish *Synodontis batensoda,* the pelagic mollusk *Glaucus atlanticus,* and the caterpillar of the spotted sphinx *Smerinthus ocellatus* the abdomen is dark and the back is light, since all these animals hang upside down.

Anti-shadow is an effective camouflage only against a monotonous background; it is easily recognized against a mosaic background from the contours of the colored spots. Dismembering coloration destroys the contour picture of the body and may retard or withhold recognition of the animal against a multicolored background (Fig. 237). Dismembering color schemes are a feature of many insects, peaceful fishes, the eggs of open-nesting birds, and many young ruminants; it is also common among predators (tiger, marbled polecat *Vormela peregusna,* many snakes, lizards, frogs). Dismemberment of the contour is best achieved by contrasting the adjacent colored spots on the body, so the figures of the color pattern are often contoured in white or bright-colored strips.

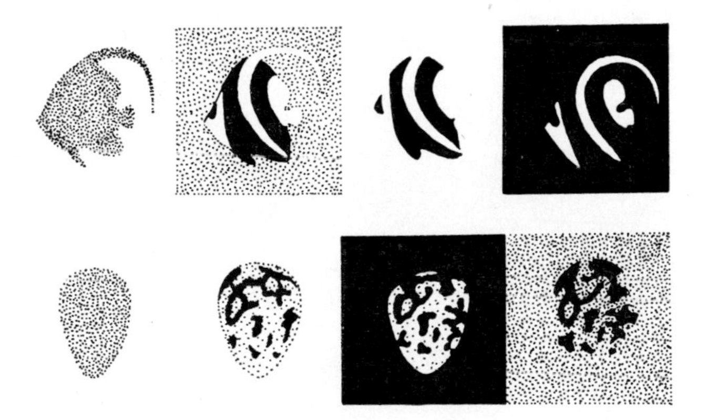

Fig. 237. Dismembering design and its camouflaging properties as displayed against different backgrounds (from Kott, 1950).

It is more difficult, however, to use dismembering coloration to mask the component parts of the body, i.e., such especially noticeable ones as the extremities and the eyes. Here the methods are more complex, as demonstrated by the color patterns of locusts: the sides of the body, the extremities, and the antennae are colored black or some dark color, while the center of the body is light. Irregularity is also used to show the familiar contour of the body in an unrecognizable form. Lateral blades on the tail and sides of an animal which rests on the ground break up the shadow which would ordinarily unmask the animal.

Cryptic coloration never gives the animal complete security, even when it is perfected to a high degree. Movement can give even a highly camouflaged animal away to its predator. Therefore, protective coloration is, as a rule, combined with some behavior which heightens its effect. Cryptic behavior is expressed in the ability to conceal oneself against a suitable background. The rapid movements of desert lizards alternate with complete immobility when the animals literally "fall" into their protective background. The bittern is almost invisible in a patch of rushes when it stretches its neck out and stands motionless. The ability of the hazel grouse to maintain immobility even under conditions of manifest danger serves as a means of defense against feathered predators which rely mainly on their vision. When hiding, animals cover the brightly colored parts of the body which serve for recognition during movement (the white tails of rabbits and hares, the bright feathers of the plover, houbara bustard, etc.).

Many animals are known to mask themselves with the aid of inedible objects. *Hydropsyche* larvae construct little houses which protect the soft respiratory surface of the abdomen; these are built from material unsuited for use as food (grains of sand, stones, plant remains). The crab *Dromia vulgaris* masks itself with the inedible sponge *Suberites*, which it holds above itself with its legs. Female Majidae crabs cover their back plates with pieces of algae, colonies of sponge, moss, or polyps. In other cases the animals themselves may take on the appearance of inedible objects. The larvae of the praying mantis are amazingly similar to the flowers on which they rest. Butterflies often imitate the trembling of leaves or plant shoots (*Callima*; the caterpillars of butterflies which resemble the twigs or branches of the trees on which they live; the tropical orthopteran *Phyllium*, which resembles a blowing leaf; etc.). A resemblance to tree bark distinguishes many of the inhabitants of trunks and sometimes achieves amazing perfection. The plant-eating bugs look like the leaves, flowers, or seeds of the plants on which they feed. Other animals imitate rushes, clods of earth, etc. The sickle-wing butterfly *Drepana lacertinaria* looks like a bird dropping in the caterpillar stage, while the older caterpillar looks like a dried-out birch catkin, and the sessile butterfly resembles a withered leaf of the same tree. In its shape and color the sea needle fish *Syngnathus* resembles a blade of the eel grass *Zostera*.

The adaptive significance of camouflage color and shape is often subject to doubt. The doubts usually consist of the assertion that "unprotected" animals are just as common in nature as the protectively colored ones; both often move from one background to another. Studies of the biological role of protective coloration have clarified its undoubted capacity to protect. Observing the destruction by birds of praying mantids

on various backgrounds, Cesnola (1904) established that over a span of 17 days no brown insects died on cut grass, nor did any green ones die on green grass, while of 70 mantids set out on a background which disagreed with their coloring, only 10 survived to the end of the experiment. When placed against a similar background, insects were destroyed by birds two to three times slower than those which were placed against a contrasting background (M. M. Belyayev, 1927). In experiments with differently colored *Gambusia* fish the selectivity of the catch depended on the exterminators. Penguins destroyed 70 percent of the "nonadapted" fish and only 30 percent of the "adapted" ones, herons destroyed 28 percent of the "adapted" and 45 percent of the "nonadapted," and the predatory fish *Apomotis* destroyed 25 percent of the "adapted" and 53 percent of the "nonadapted" fish (Sumner, 1934).

The selectivity of the catch depends on the means of hunting. For instance, flycatchers catch primarily flying insects, and camouflaged animals easily crawl away from them. The robin redbreast catches sessile and flying insects; it is less likely to miss a protectively colored insect. Finally, titmice, which carefully search the branches of trees and bushes, seldom miss even insects which are sitting motionless, and only a prey with a highly perfected color scheme resting against its natural background can fool this bird.

As a result of selective extermination, a good correlation between animal distribution and background is observed. Of 314 praying mantids found in nature, 77 percent were observed on plants which completely corresponded to their coloring and only 23 percent on "uncharacteristic" backgrounds (M. M. Belyayev, 1947). In Yorkshire 96 percent of the *Oporabia autumnata* found in the relatively dark pine forests belonged to the dark race, while in a neighboring birch grove 85 percent were light. This type of distribution may be a consequence of selective elimination of the wrong-colored individuals, but it may also depend on the background which is actually preferred by the insects. The actions of predators thus aid not only the development of protective coloration but the behavior associated with it as well. The latter is especially important as a condition for directing the development of variability and selection.

Another means of lessening the danger of death is the so-called warning signal or fright coloration—aposematic coloring or body form. It is usually associated with active defense against enemies or with inedibility. Some animals puff up the body (spiny-bellied poisonous fishes, Didontidae, *Phrynosoma* lizards, chameleons, *Dispolidus* snakes, cobras, many frogs and toads). The apparent increase in size of some birds and animals is achieved by ruffling the hair, spines, or feathers.

Poisonous, unpleasant-smelling, and inedible animals are usually bright-

colored. Sometimes warning spots of color appear momentarily during the danger period; the suddenness of the warning heightens the action. At the moment of danger the frog *Bombina* bends and exposes bright black and red (or yellow) spots on its breast. When threatened, the long-eared *Phrynocephalus* lizards living on the sandy desert soils of Central Asia usually run and hide or dig into the sand. But if they are caught in a back-to-the-wall situation, they open their mouths, stretching out the sides of the mouth into great flaps of skin which quickly turn bright red from the blood flowing into them, and use this as a threatening pose (Fig. 238). The caterpillar of the sphinx *Leucorhampha ornata* shows the bright spots on its abdomen by suddenly raising the anterior half of its body.

Fig. 238. Long-eared *Phrynocephalus mystaceus* lizard in a threatening pose (original).

Along with coloration and warning movements, sounds are sometimes used. The hissing of snakes, the rattlesnake's customary sound, the chirring and buzzing of many insects, the panic screams of birds and mammals, and many others are of this type. The white-toothed musk shrew, which lives in India and has a sharp smell, hides in the twilight and usually squeaks rather loudly, as if warning of its presence.

Aposematically colored animals are usually slow-moving and often form clusters. The reef fish *Plotosus anguillaris*, of the sheat family, forms clusters consisting of hundreds of individuals, creating what is almost a true sphere which then swims about the reefs. Apparently none of the predatory fishes will risk attacking such a sphere. Animals with warning coloration are often diurnal. Most of the cicadas with camouflage are nocturnal, while the black and scarlet *Huechys sanguinea* (from Malaya) flies in the daytime. Most of the Sphingidae have a modest cryptic color scheme and are active at night, but the wasp sphinxes *Hemaris fuciformis* and *H. tytius*

are similar in color to stinging Hymenoptera and are active during the day.

Aposematic coloration and the behavior associated with it are strengthened by the poisons and unpleasant smells which some animals employ. The poison of the bright-colored blister beetles will kill rabbits even in very small concentrations. Hedgehogs and bustards, however, can eat these insects without endangering themselves. Unpleasant tastes and smells given off by the body fluids and tissues are a means of protecting chinch bugs. Among North American skunks a smelly substance is secreted by glands situated at the base of the tail; the skunks also have typical demonstrative threatening poses.

Studies of the stomach contents of predators, observations in the wild, and special experiments have shown that there is a manifest and sometimes nonconditioned refusal to eat representatives of the defiantly colored species. Of a total of 11,585 insects found in 993 stomachs of eight species of tree frogs, only 20 (0.17 percent) were colored aposematically (Kott, 1950). Experiments with mammals, birds, reptiles, amphibians, and fishes have shown that avoidance of aposematically colored prey is the usual result of personal experience (Fig. 239). In an experiment the gray toad *Bufo bufo* was fed on bees; in the beginning the toads ate them willingly, but on the seventh day not one of the toads would select bees. After 3 weeks the new habit was changed only to a slight degree, and only a few toads had "forgotten" the unhappy experience.

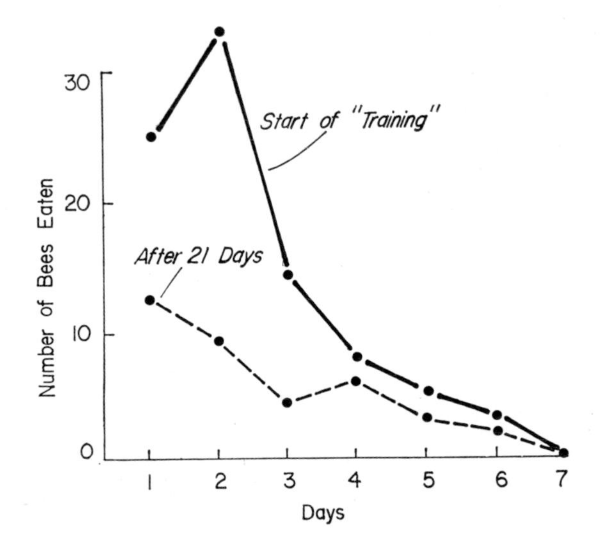

Fig. 239. Number of bees eaten by 19 toads at the start of "training" and after 21 days (from Kott, 1950).

Mimicry includes imitation not only of inedible animals but also of inedible objects. It is widespread among insects and is expressed in similarity of body shape, coloration, and behavior. Examples of flies and butterflies which resemble stinging Hymenoptera are widely known. The form of the body itself may serve as a defense against predators. Among fishes the body form of the bream, skate, and flounder is of this type. Finally, the large, strong animals (ruminants, elephants, whales) are protected from predators by their very size.

Protective coloration, size, and body shape are important as a protection against predators, but they also play a large role in the organism's heat and water exchange. But the surprising variety of form, coloration, design, and security displayed by the hunted animals shows that among many species the effect of biotic conditions of existence has been pivotal. The three basic types of adaptive coloration and body form—camouflage, demonstration, and imitation—are inseparably bound up with behavioral properties. This dependence shows that predators are an important factor in the evolution of the peaceful animals.

The digging way of life, which provides a favorable microclimate inside the burrow and food under the surface, at the same time lessens the danger of being stalked by predators. It is widespread among terrestrial and even aquatic animals. Most of the tunneling species lives in the steppes and deserts, where the plant cover is sparse and covers the animals poorly. There the burrows attain great complexity (Fig. 240). The construction of burrows and the terrestrial adaptations associated with it attest to their importance as methods of protection against predators. On an individual (or family) plot of voles there is a nesting burrow and small defense burrows, the surface exits of which are united by specially cut paths allowing for quick travel. The inhabitants of sparse-grass deserts and steppes usually have more complex burrows; many of them stop up the burrows from within, camouflage the entrance holes from the outside, and throw the dirt out at the entrances.

Like all adaptations, the tunneling way of life gives only relative protection, lowering, but not preventing entirely, the mortality rate from predators and unfavorable physical conditions. Even the "subterranean" animals, which seldom venture onto the surface (moles, shrews, zokor (*Myospalax*), etc.), are often the victims of predators. Nevertheless, the defense proves effective, and the transition to subterranean life is accompanied by a considerable decrease in the fecundity of the digging species.

The above facts enable us to conclude that the food relationships of predators and their victims have served historically as an important factor in the evolution of both groups of species. They explain many of the structural properties and behavioral characteristics of both predators and prey.

SIGNIFICANCE OF PREDATOR-PREY RELATIONSHIPS IN THEIR POPULATION DYNAMICS

The relationship between the number and distribution of predators and the number and distribution of prey causes no doubts. This relationship is the basis for several methods of finding game fish and counting rodents—

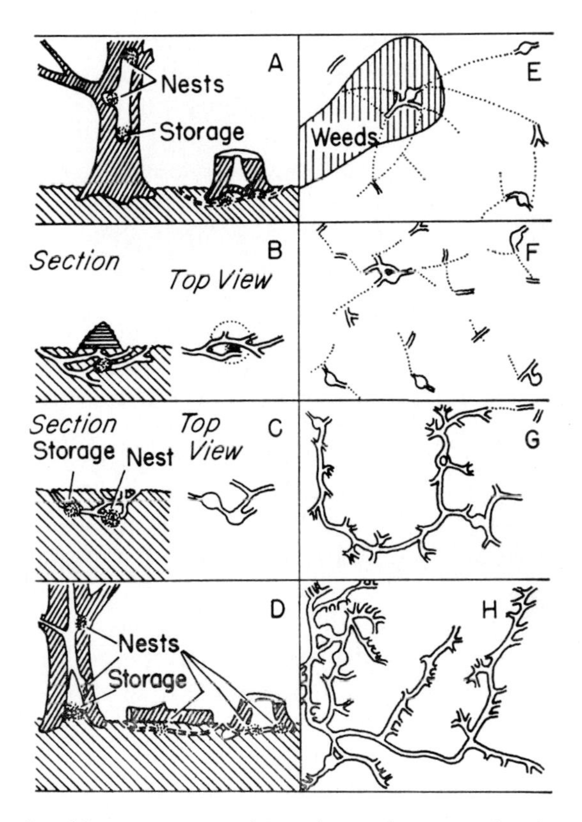

Fig. 240. Properties of burrow construction and nest placement of various species of small rodents (after N. P. Naumov, 1948). A: yellow-throated forest mouse (*Apodemus flavicollis*) of the Tula oak forests; B: mound-building mouse (*Mus musculus hortulanus*) of the southeastern Ukrainian steppes; C: gray hamster (*Cricetulus migratorius*) of the southeastern Ukrainian steppes; D: red vole (*Clethrionomys glareolus*) of the Tula oak forests; E: common vole (*Microtus arvalis*) of the southeastern Ukrainian steppes; F: steppe lemming (*Lagurus lagurus*) of the southeastern Ukrainian steppes; G: society vole (*Microtus socialis*) of the southeastern Ukrainian steppes; H: mole (*Ellobius talpinus*) of Kirghizia and the Issyk-Kul' depression.

by the location of easily noticed fish-eating and predatory birds. No wonder the northern coastal peoples still have the saying "Where there's a seagull, there's a fish" ("Chayka yest', ryba yest' "). A. N. Formozov (1934) successfully used counts of predatory birds made from an automobile and from the window of a railroad coach to make an approximate estimate of the abundance of rodent pests (Fig. 241).

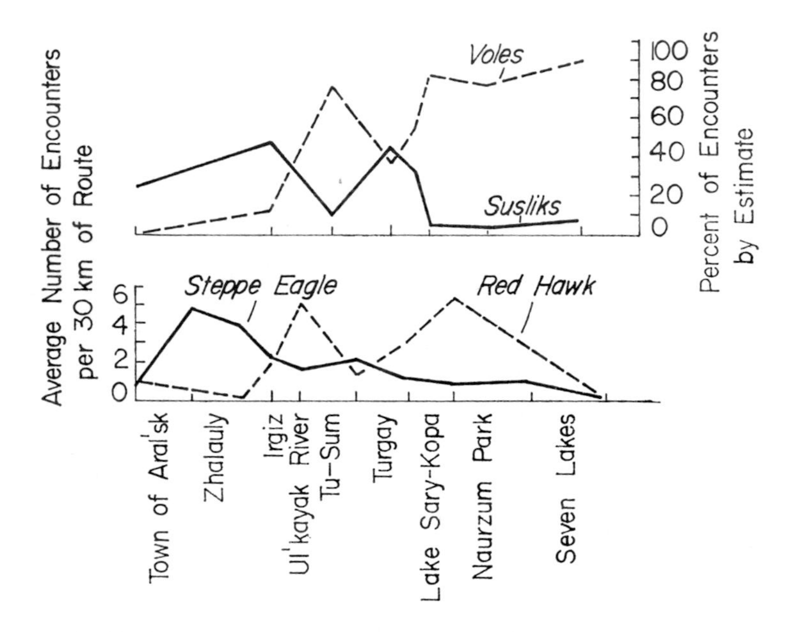

Fig. 241. Distribution of steppe eagles (*Aquila nipalensis*) and red hawks (*Buteo rufinus*) on the route from Aral'sk to Semiozernoye compared with the abundance of voles and susliks (after Osmolovskaya and Formozov, 1952).

The relationship between changes in the number of predators and the abundance of prey is also distinct in time (Fig. 242). An increase in the number of victims is accompanied by an increase in the number of stalkers, but it reaches its greatest height not at the time of maximum food abundance but later. Table 63 shows that foxes preserved their high abundance not only in the year when hares were at a maximum but also for 2 succeeding years, when a fatal epizootic raged among the hares. The epizootic was a basic cause of the drop in rabbit numbers and was augmented by the exterminatory actions of the predators.

An abundance of food intensifies reproduction and improves the survival possibilities of the young predators. The more specialized their feed-

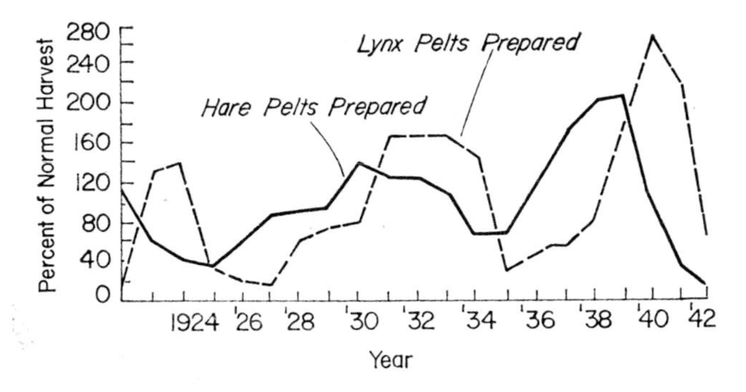

Fig. 242. Changes in abundance of the varying hare *Lepus timidus* and the lynx (*Lynx lynx*) in the Komi ASSR (after S. P. Naumov, 1947).

TABLE 63. *Fluctuations in the abundance of varying hares and red foxes in the Pechoro-Ilych Preserve from 1937 to 1943 (after Teplov, 1947). Regular winter computations on the same routes.*

	WINTER SEASON					
	1937/38	*1938/39*	*1939/40*	*1940/41*	*1941/42*	*1942/43*
Average number of hares encountered per 100 km of route	5.0	2.0	1.5	1.0	0.2	0
Average number of fox tracks per 100 km of route	3.6	5.0	5.1	2.7	2.0	1.0
Frequency of encounter of hares in food of foxes	40%	50%	50%	45%	20%	10%

ing habits and the more difficult it is to replace the basic food (when it is in short supply) with others, the more noticeable will be the effect of the food supply on reproduction, survival, and abundance. A drop in the supply or availability of food causes a drop in the intensity of reproduction, poorer survival, and a higher mortality rate among predators. This is the time when they are often observed to turn to cannibalism: weaker offspring are eaten by the stronger ones, and sometimes the parents eat their young. In especially hungry years the Strigiformes, hawks, eagles, and polar and red foxes stop reproducing entirely, and the increase in mobility during the famine aids in raising the death rate. In such years polar and red foxes change their hunting grounds and carry their little pups from one burrow to another many times.

Predators die off more slowly than their prey (Fig. 243). This is explained by the high mobility and nonspecialized feeding habits of most predators. Predatory birds and mammals easily seek out surviving clusters

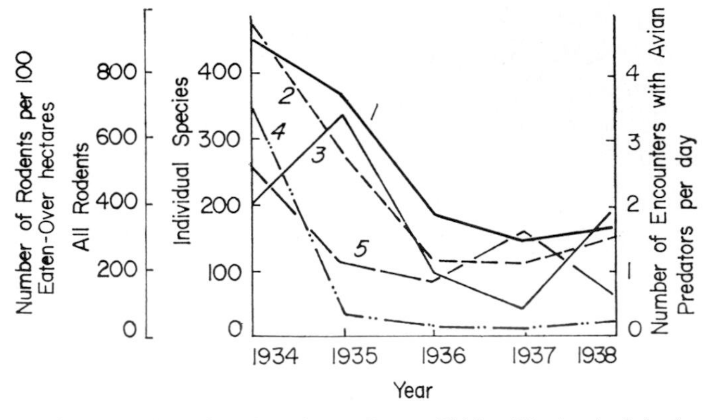

1 Diurnal Feathered Predators (*Buteo,* Chicken Hawks, *Accipiter*)

2 All Rodents

3 *Lagurus lagurus*

4 Common Vole

5 Mound-Building Mouse

Fig. 243. Changes in abundance of rodents and feathered predators in the steppes of the southern Ukraine (after N. P. Naumov, 1948).

of prey and quickly concentrate there. The transition to feeding on new foods, even if it does not give rise to reproduction, still saves the lives of many individuals.

The change in the predator-prey relationship owing to the lag in the predators' death rate permits them to lower the abundance of the peaceful animals even more. On the Izyum Forestry Farm, in a nidus with a high abundance of oak-leaf butterflies, the birds had destroyed 27.6 percent of the pupae by the end of May, while in a nidus with fewer of these pests they destroyed 53 percent (Shilova, 1952). According to observations made in the Kiev oblast' (1946–1948), caterpillars and pupae of *Nygmia phaeorrhaea* were 94.3 percent destroyed over the winter, spring, and summer by insectivorous birds in a forest where there were many predators, 5.3 percent being destroyed by parasitic entomophages. In the fields, where enemies were few, 69.5 percent were destroyed by birds and 11.5 percent by entomophages (Fedotova, 1950).

An omnivorous predator following a species which is not its basic food may actively lower its abundance. In the isolated Kaban region of the Irkutsk oblast', where sables have been on the increase for the past 15 years, the squirrel has practically disappeared. It has been preserved in

noticeable numbers only in areas where sables are not raised (Fig. 244). Extermination of the squirrels proved possible because the sable stalks them successfully, but it feeds basically on small rodents, and a decrease in the number of squirrels does not substantially worsen its feeding situation (Lavov, 1959).

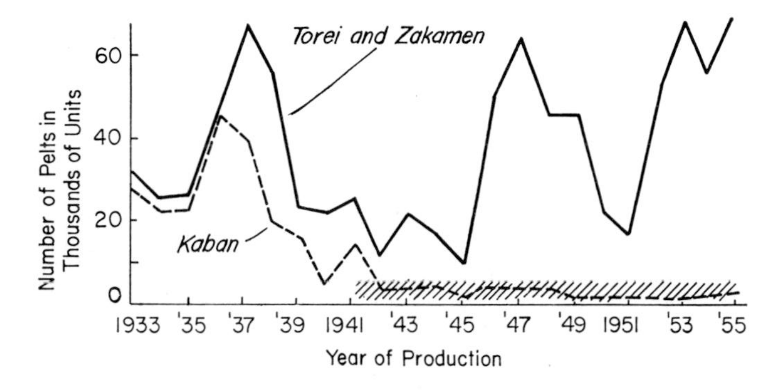

Fig. 244. Changes in abundance (pelts prepared) of squirrels in the Kaban, Torei, and Zakamen regions (after Lavov, 1959). Shading shows period of active effects on sables on abundance of squirrels in the Kaban region.

In various zones the effect of predators on the abundance of the victims is not the same; it depends on the relationship of the populations (Table 64). In the middle belt and in the south, especially in the cultivated agricultural landscapes, the role of predators in regulating the number of victims becomes very large. Because of the abundance of predators there, the protective adaptations of the prey increase, their "watchfulness" is improved, better and more reliable refuges are selected for mating, care of the young is strengthened, and the number of repeated litters rises.

When the number of stalking enemies becomes relatively great, it begins to influence actively the content of the population, the life span, the intensity of reproduction, and the spatial placement of the prey. Under favorable conditions and when enemies are absent, mice and voles live 3 years at the most, and rarely over a year in the wild. Extermination by predators plays a leading role here, and this role becomes greater as the population of prey increases (Fig. 245, Table 65). In the first 2 years not one of the females studied delivered more than two litters, while in the last 2 years from 5 to 22 percent of the females gave birth to three or more.

Predators destroy primarily males or females or concentrate on the adults or the young; when enemies are numerous, this substantially alters the age and sexual bases of the prey populations. Among the common voles caught by *Buteo* hawks before the grain harvest, the dispersing adolescents

TABLE 64. *Ratios of predator-prey abundance in different natural zones (after Teplov, 1957).*

Zone	Place	Number of Years Observed	Length of Route in cm	NUMBER OF TRACKS PER 100 KM OF ROUTE			NUMBER OF TRACKS PER FOX TRACK	
				Fox	*Hare*	*Bearded Vulture*	*Hare*	*Bearded Vulture*
Northern taiga	Pechoro-Ilych Preserve	12	23,600	2.4	18.9	26	7.7	10.8
Southern taiga and mixed forests	Darwin Preserve	7	1,826	11.5	51.4	26	4.5	2.3
Mixed broad-leafed forests	Belovega Forest	5	1,812	41.4	91.3	163	2.2	4.0
Northern forest steppe	Oka Preserve	14	392	100.5	1,141.3	38	1.4	0.4
Southern forest steppe	Voronezh Preserve	2	100	270.0	352.0	–	1.3	–

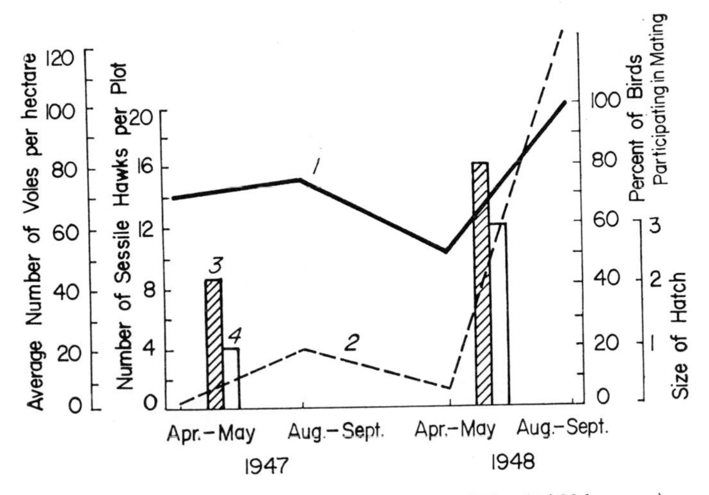

Fig. 245. Changes in abundance and intensity of reproduction of *Buteo buteo* in the southern Moscow oblast' in relation to the abundance of common voles (after Lavrova, original).

TABLE 65. *Proportion of adults among female common voles and* Lagurus lagurus *weighing more than 12 g in the southern Ukraine in relation to abundance of feathered predators (after N. P. Naumov, 1948).*

Year of Observation	*Predators Encountered in 6 Hours*	COMMON VOLE		LAGURUS LAGURUS	
		Females Studied	*Percent Fully Matured*	*Females Studied*	*Percent Fully Matured*
1934	7.9	104	10.5	–	–
1935	3.5	29	20.7	94	19.2
1936	2.1	35	22.9	59	35.5
1937	1.5	23	17.6	9	22.2
1938	1.8	43	32.0	52	53.9

predominated. Among them the majority of the males were beginning spermatogenesis, and the females were in the first stages of pregnancy. The adult and sessile creatures suffered relatively few deaths at this time. But after the harvest they became more available and were exterminated more intensively than the young. Because of this, the sexually immature young predominated in the population by autumn. On the different fields and meadows these changes took place in perfect correlation with the

harvest periods and were one of the reasons for a drop in vole reproduction (Fig. 246). This was responsible for the seasonal peak in vole abundance coming relatively early in years when predators were abundant—in September (1946) and even in August (1948)—while in years when predators were few, the peak came in October (1947) and even in December (1949). In the 2 succeeding years this "juvenilization" of the population by predators was weakly expressed.[1]

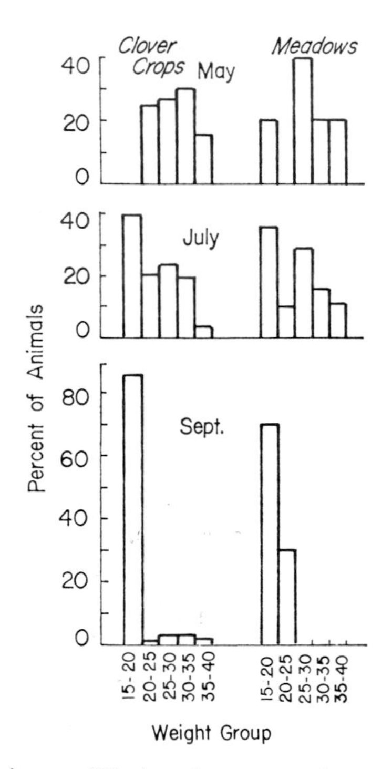

Fig. 246. Seasonal changes in age affiliation of common voles on clover crops and meadows in the southern Moscow oblast', showing percentage of animals of various weights in the population of voles and changes over the months (after Karaseva, original).

Under the influence of increased extermination in populations of small rodents the percentage of adult males decreases in summer, especially in the older groups; this sometimes leads to temporary polygamy or to an increase in the number of unmated females (Fig. 247). Thus selective extermination of individual ages and sexes by predators, shortening the life span of adult individuals, lowers the intensity of mating and sometimes leads to its elimination. This influence is not noticeable in years when

[1] These data are for the southern Moscow oblast' (N. P. Naumov, 1953).

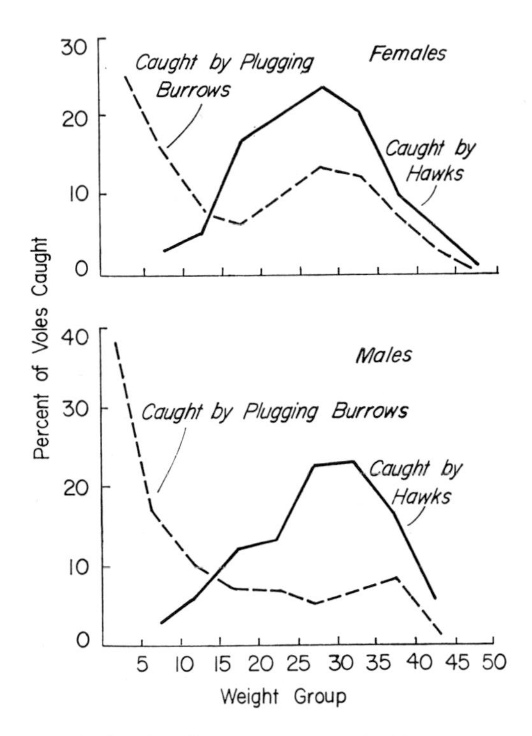

Fig. 247. Selective catch by hawks of common voles of older age groups in the southern Moscow oblast' (after Lavrova and Karaseva, 1956).

prey is abundant and predators are few, but it is substantial when the relationship is reversed.

The role of predators in the population dynamics of different species depends also on the behavioral characteristics of the victims. The very mobile and highly scattered tamarisk aphid *Eulachnus tuberculostemmata* is hardly ever exterminated by predators. Young nymphs of the aphid *Myzus persicae* disperse as soon as they hatch and are practically unavailable to predators, while the less mobile, colonial *Cinara palaestinae, Brevicorine brassicae,* or *Aphis nerii* is often subject to a higher degree of extermination by predators. The significance of the migrating females, which establish new colonies, is especially great for these species (Bodenheimer, 1958).

Predators influence the abundance dynamics of their prey, limiting their ability to use favorable places of habitation, as in rodents attempting to use winter grain crops. The low stand of grass enables the predator to catch very easily any animal that ventures into it. Therefore, the existence of mice and voles on fields of winter grain becomes possible only in years

when predators are relatively few (Fig. 248). P. L. Errington's (1945, 1946) observations show that predators eliminate a small percentage of their victims until the abundance of the latter reaches a level critical for the given biotope, after which the mortality rate due to predators, diseases, and inclement weather rises sharply (Fig. 249). Thus the effects of predators are not limited to the destruction of prey, but they change the structure of the population and the fecundity of the victims as well. It is not surprising that major increases in abundance of plant-eating species are usually preceded by decreases in the number of predators (Fig. 250).

The significance of predation becomes especially convincing when alien species move in. The duck *Dendrocygna autumnalis,* which nests on the ground, was exterminated in a few years when the mongoose *Herpestes griseus* was imported. However, a few were saved in protected areas, and later their numbers began to rise. The acclimatization of the raccoon dog in the European USSR was accompanied in some areas by a sharp drop in the number of heath and black cocks. Applied entomology provides many examples of mass reproduction and substantial harm on the part of imported species, which were brought under control only after importation and acclimatization of their natural enemies (blood aphids, isceria coccids, Comstock coccids, etc.). Much is also known concerning mass destruction of pests by their enemies (aphids by ladybugs (coccinellids), cabbage worms and butterflies by the ichneumon fly, the pupa of the hawthorn

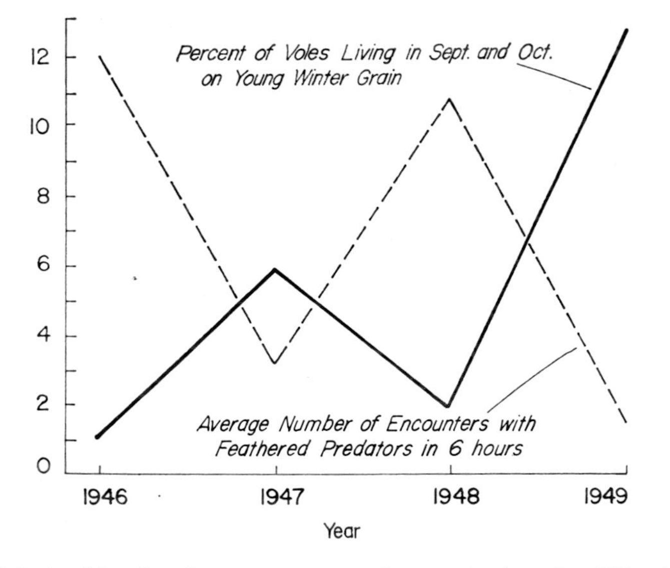

Fig. 248. Relationship of settlement on young winter grains by voles (*Microtus arvalis*) and abundance of feathered predators in the southern Moscow oblast' (original).

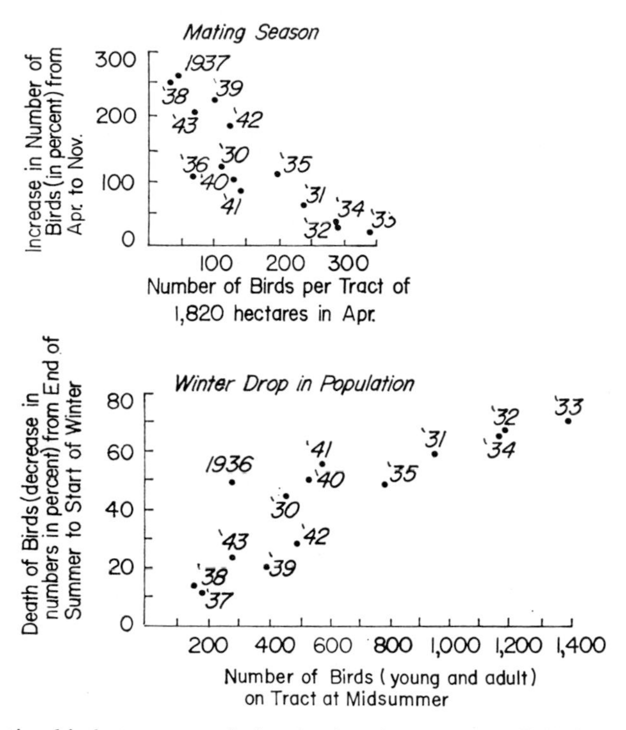

Fig. 249. Relationship between population density of bobwhite quail (*Colinus virginianus*) and amount of weight gain during mating season and during the winter drop in bird population (after Errington, 1945).

butterfly by the ichneumon fly, striped butterfly eggs by the parasite *Trichogramma*, etc.). In these cases the numerical ratio of predators to victims accords, in general, to a theory developed by V. Volterra and A. J. Nicholson, to be discussed below. This theory attempts to express the dynamics of the relationship between food and its consumers.

But the significance of predators in the population dynamics of the victims depends not only on the numerical ratios but also on the conditions of existence. And the latter are associated with the availability of prey, its motility, and the amount of protection offered by its places of habitation, which in turn depend on the development of plant cover, the state of food supplies, the weather, etc. Favorable weather for plant-eating species and good growth and bearing among the plants usually prove to be unfavorable for predators. On cultivated fields the resolving factor is agricultural activity. The survival of important farm pests (such as the larvae of Elateridae) depends on the amount of food available to them and the number of their enemies, the predatory ground beetles (Carabidae). On

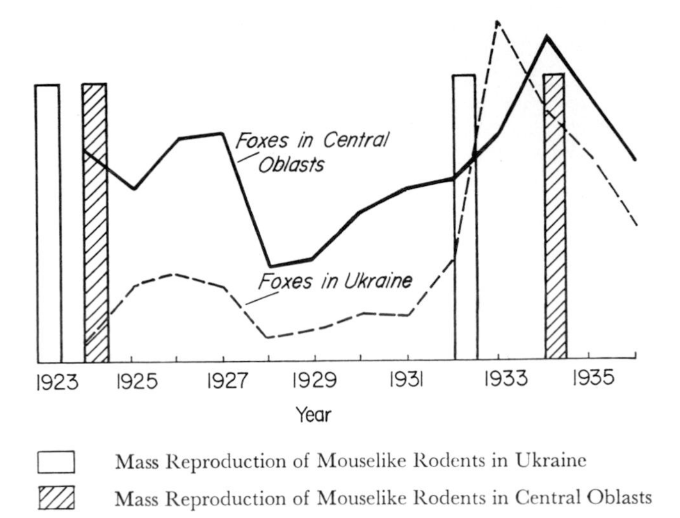

Fig. 250. Changes in abundance of red foxes (from skins prepared) and in mass reproduction of small rodents (after N. P. Naumov, 1948).

black fallow fields, which offer little food but have many ground beetles, only 8 percent of the *Agriotes* larvae are saved; on plots seeded with spring grain crops survival rises to 16.5 percent; on plots with winter grains to 35 percent; and on clovers, where there is a great deal of food and few predators, to 40–57 percent (Grigor'yeva, 1950). Burning off dry plant remains in the steppes, grain stubble on arable land, and reeds in river deltas changes the hunting conditions of predators, which usually concentrate quickly on the burned-out areas (Fig. 251), and thus leads to a sharp drop in the numbers of their prey.[2]

In the middle and high latitudes the snow cover is of great significance. As we have seen, most of the seasonal shifts in feeding habits are associated with it. It is natural that in years with differing amounts of snow the hunting opportunities of the predators will change (Table 66). Increased hunting by predators, and often by man, of chionophobic animals (such as the elk) is responsible for changes in their distribution in winter. Increased snowfall and increased hunting may lead not only to a drop in numbers but also to a curtailment of the habitation area. When this happens, the elks disappear first from the snowier regions (Nasimovich, 1955; Formozov, 1959).

The success of a predator's hunting and its influence on the population

[2] In the case shown in Fig. 251 the drop in the number of small rodents on the burned-over areas was not associated with the shortage of food after the fire; rather, it depended on extermination by predators and on emigration of the creatures.

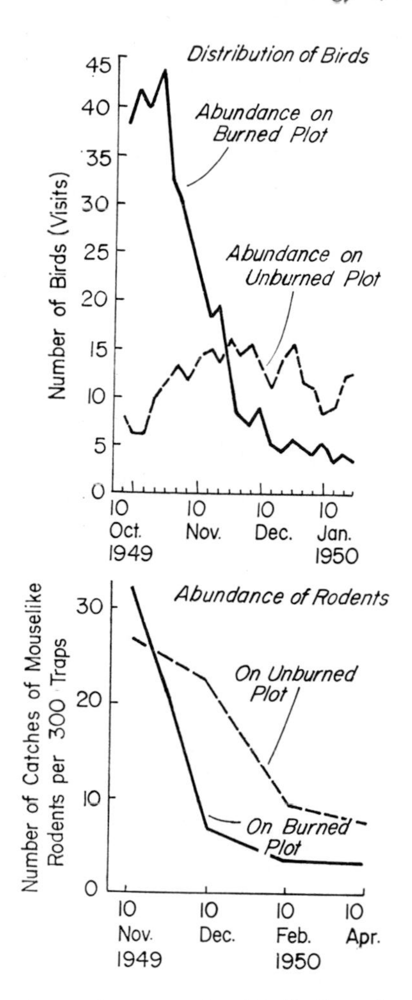

Fig. 251. Effects of burned-over forest on the distribution of predatory birds hunting small rodents and on changes in the latter's abundance in the Ili delta (after Gusev, 1953).

of the prey may be determined both by the width of the victim species' resettlement after multiplication throughout the less favorable habitats and by the rise of unfavorable circumstances in the basic biotopes. In the latter case increase in the significance of predators with regard to the mortality rate of the victims attests to the fact that inclement weather or other external factors did not by themselves reach the critical state and cannot serve as the direct cause of death (Errington, 1946, 1956). The ability of the predators to eliminate an increase in the population of the prey is increased by their catching surplus animals; under these conditions

TABLE 66. *Occurrence of wild hoofed animals in food of wolverines in years of different amounts and conditions of snow cover (after Teplov, 1947, 1955).*

	WINTER SEASON				
	1938/39	*1939/40*	*1940/41*	*1941/42*	*1942/43*
Average depth of snow cover in areas where elk were camping in cm	130	90	120	85	110
Occurrence of elk in food of wolverines in percent	15.9	4.5	39.4	29.3	47.1
Depth in cm and character of snow cover in mountain tundras	?	50–100; solid with ice crust	15–20; lightly packed	20; ice crust	lightly packed
Occurrence of wild northern reindeer in food of wolverines in percent	25.6	82.5	35.5	46.3	38.3

some predators accumulate supplies numbering dozens of dead animals.

The complexity of the relations between predator and prey is increased by the fact that when predators eliminate one species, they influence the interspecies relationships to some extent and thus indirectly favor some of them and aid in suppressing the abundance of others. Thus an experiment in stocking zander in newly created reservoirs proved successful because this species feeds on perch and other competitors of the valuable species and sharply limits their numbers in a body of water. Because of this, food resources are freed for carp and other important game fish, which the zander almost never touches. It feeds on bream only when the latter is undergoing mass reproduction, but this does no harm and even helps regulate the size of the bream's schools. Whereas the annual catch of fish in the Veselov Reservoir (on the Manych) prior to the stocking of zander was 14,548 centners, with carp and bream accounting for 18 percent of the catch and so-called "small change" (secondary and rough fish) only 36 percent, the catch rose to 18,099 centners after acclimatization of the zander, with carp and bream making up 70 percent of the catch, zander 14 percent, and "small change" only 3 percent (Syrovatsky, 1953).

Titmice of the genus *Parus* which were attracted into a nidus of the pine caterpillar *Panolis flammea* scraped away the forest litter and turned up the exposed soil under the tree crowns to destroy not only their pupae, which suffered a threefold drop in abundance, but also the puparia of the tachinid fly, which parasitizes pine caterpillars. At the same time they left *Enicospillis merdarius* untouched in its cocoons, nor did they have anything to do with pine-caterpillar pupae which were infected with fungal diseases. As a result, the titmice not only lowered the number of pine caterpillars but they also changed the quantitative relationships between the

pest and its parasites by increasing the relative abundance of the parasite (Unterberger, 1953).

Thus, not only the numerical ratio of predators to prey but also all the conditions of existence impinging on both of them are governed by their mutual relations. When predators are few or conditions are poor for hunting, the predators cannot retard the growth in numbers of the prey species. Under these conditions abundance increases quickly, and because of the geometric progression of reproduction, the mass species seems to appear very suddenly. When predators are in relative abundance and conditions are poor for the prey, the predators may actively retard an increase in the prey and cause the latter's populations to stay at a low level for some time.

The best proof of this is to observe the rules of agrotechnology, in which good treatment of the soil and proper care of crops deprive pests of cover and food on the fields to the extent that their abundance is always low. A campaign against weeds, definite computed times for harvesting crops and good harvesting practices, fast threshing, autumn plowing, and many other measures have led to a situation in which the fields of the leading state farms and collectives are practically free of pests. Predatory and insectivorous birds and mammals have proven to be man's basic helpers in this campaign; they are aided by predatory insects and parasitic entomophages, which are a permanent harvest-protection service. In forests the ants play a special role by keeping down the numbers of harmful insects. Proper understanding of the activities of all these enemies of the harmful species is necessary for maintaining and developing the results already achieved.

One condition necessary for success is the development of a theory of the interrelations between predator and prey. From this point of view the attempts at mathematical integration contained in the theories of Volterra, Nicholson, and others deserve attention. However, they analyzed simplified cases which rarely occur in nature. Theories describing this interrelationship rest on the Spencerian concept of a "movable balance," which does not actually occur in nature. They are useful, however, insofar as they show how under permanent conditions of existence the interrelations of predator and prey can by themselves cause the numbers of both to fluctuate. This happens because the possibility of a predator's making a catch increases, when the prey population is growing, at a rate greater than that of the population increase. The growing supply of food increases the reproduction of predators, thus progressively increasing the pressure on the victim; at a certain state the abundance level of the victims begins to drop, and soon the number of predators begins to drop also. But the latter condition allows the prey to begin multiplying again, and the cycle repeats itself (Fig. 252). Predator and prey are here regarded as a closed system

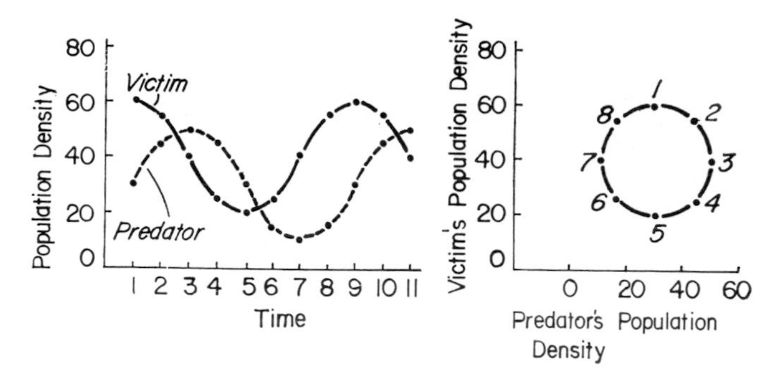

Fig. 252. Relationship between population densities of predator and victim in Lotka's and Volterra's classical fluctuations (after Gause, 1934).

in an unchanged environment with equal possibilities for attack (among predators) and defense (among prey).

G. F. Gause (1934) reproduced these interconnected fluctuations in an experiment by creating in the laboratory an artificial system of two populations of partially settled yeast (*Saccharomyces exiguus*) and the protozoon *Paramecium bursaria,* which is not very motile and which can feed only on yeast cells in suspension (the yeast was partially settled on the walls of a vessel). The walls of the vessel in this case played the role of a refuge, where the yeast was in no danger from the paramecia. In this system a cyclic fluctuation in numbers of predator and prey actually arose, just like that described in the theory of A. J. Lotka and V. Volterra. In Gause's other experiments with the nonpredatory *Paramecium aurelia,* which feeds on bacteria, and the predatory *Didinium nasutum,* the predator either ate up all of the prey and then itself disappeared, or the paramecia found refuges which *Didinium* could not reach, causing the latter to die of starvation, after which the paramecia began to multiply again. The fluctuations demanded by the theory did not exist in this case. They occurred in the experimental *Paramecium-Didinium* populations only if both species were periodically added to the culture from outside, in imitation of animals "immigrating" into the biotope (Fig. 253). This last experiment showed quite convincingly that in simple systems of two populations a stable interrelationship occurs only in rare cases. It is necessary for each of the species to have refuges which cannot be reached by enemies, or for the population to be re-established by newcomers from outside. In both cases the simple system of two interconnected food relationships is no longer closed.

D. Lek (1954) found that in nature the external appearance of the numerical interrelationships between predators and prey is sometimes very

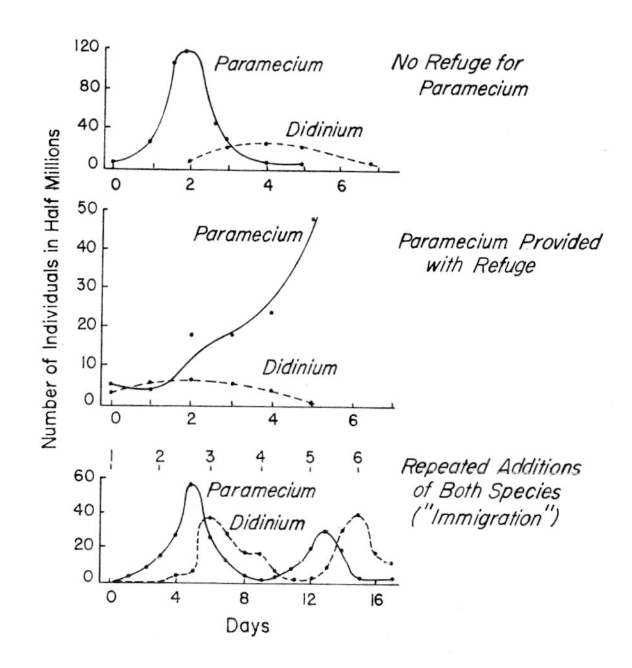

Fig. 253. Changes in abundance of populations of *Paramecium caudatum* (prey) and *Didinium nasutum* (predator) when cultured in a common vessel (after Gause, 1934).

similar to the results one would expect from the mathematical theory (Fig. 254). But this resemblance is only external; at the basis of the more or less correct periodic changes in the abundance of some species lies their correlation with cyclic changes in the weather and the phenomena which accompany them (squirrels and the nut crop, locusts and the climate cycles). Among others, the more or less regular alternation of rises and drops in abundance are due to surplus reproduction when food is in good supply, a periodically created unfavorable situation which ends in death from various causes, especially often from lack of food. These causes may include predators, but they always depend on a compound situation (weather, food supply, and many others). Their effects on the population are very complex. Actually, it is not pairs of species which are acting on one another (predator-prey) but complex, overlapping groups, since predators almost always feed on many species, and each of the latter is hunted by several enemies. These groups, as a rule, include species with various degrees of specialization, with different styles of attack and protection. Finally, the interrelationships unfold into interactions with all the conditions of existence of each of the members of this complex system. The numerical ratio is only one of these conditions and need not be pivotal. Only under

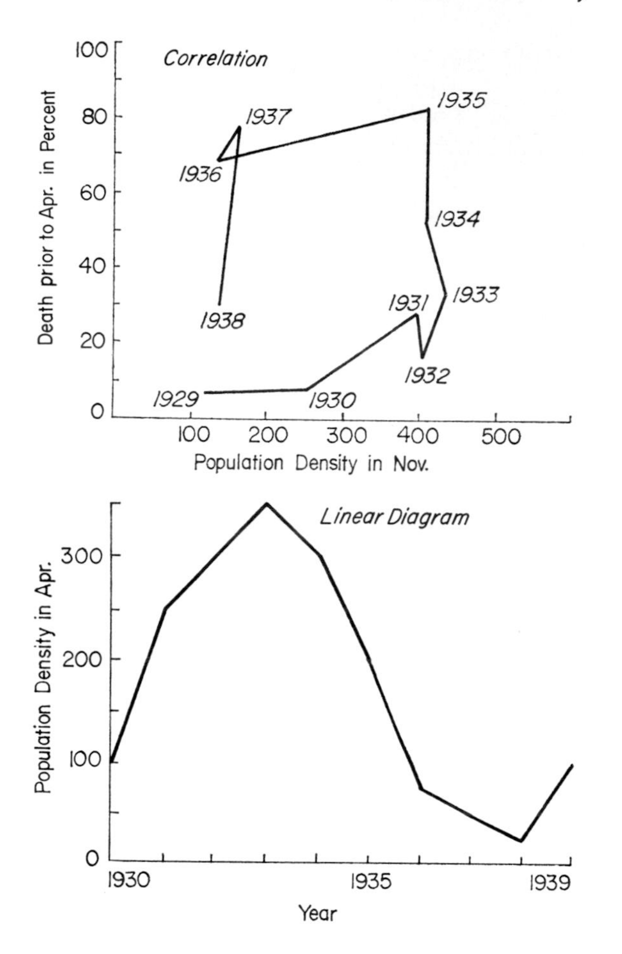

Fig. 254. Winter mortality of bobwhite quail (*Colinus virginianus*) in relation to population density (after Errington, from Lek, 1957).

definite external conditions can predators retard the population increase of the prey species. This situation is most often created after a drop in the number of victims from diseases or unfavorable weather, i.e., after the effects of conditions which did not harm the predators. The cyclic balance between the fluctuations of predators and prey suggested by the mathematical theory of a struggle for existence does not and cannot exist in such a complex system.

The population dynamics of any species is the result of its relationships with all the conditions of existence. Among these the activities of predators can play a leading role only under certain circumstances. The knowledge

and ability to create these circumstances are of great practical importance, since the biological principles of controlling destructive pests are founded upon such knowledge. To deny the role of predators as active exterminators capable of suppressing the population level of the victims, however, would disarm and menace our practice.

The complexity of the relationships between predators and the animals which serve as their food consists in the fact that by attacking their prey, predators often exercise a positive influence on their victims' populations. They catch sick and weakened animals and thus reduce the number of disease carriers and the amount of disease spread. When varying hares suffer from worm infection, wolves, foxes, and dogs begin to catch the infected animals (Fig. 255). Marmots suffering from plague are more often caught by steppe eagles and hawks. In the Trans-Baykal, of a total of 178 remains of rodents caught by predators, three of them, or 1.7 percent, yielded a culture of plague microbes, while not one of 21,929 animals caught by traps in the same place was found to be infected with plague (Bezrukova and Linnik, 1944; P. P. Tarasov, 1944). Nowadays research into predator-prey relationships is practiced widely during searches for infections in the wild.

The primary catches of infected and sick voles is shown in Table 67. Chicken hawks in natural nidi of leptospirosis catch the infected animals

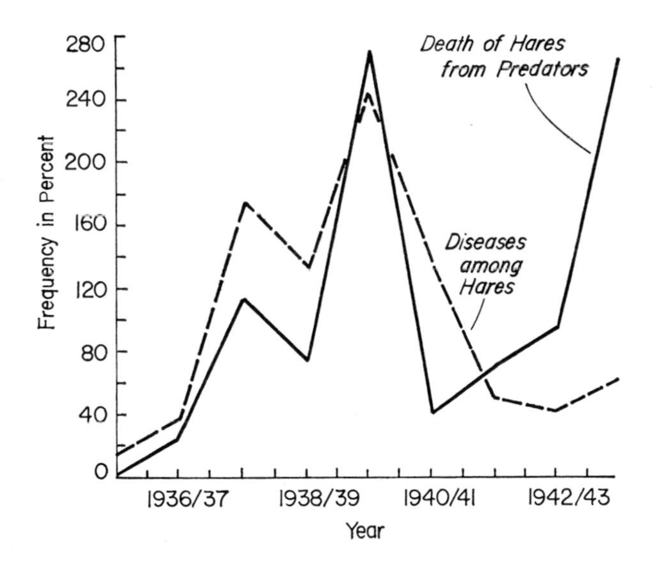

Fig. 255. Frequency of elimination of varying hares (*Lepus timidus*) by predators in connection with hare diseases in the northern and eastern regions of the European USSR (after S. P. Naumov, 1947).

TABLE 67. *Condition of common voles caught in the Belovega Virgin Forest by hawks as compared with the condition of the population as a whole (after Folitarek, 1948).*

	CAUGHT BY HAWKS			NATURAL POPULATION CAUGHT BY TRAPS		
State	*Males*	*Females*	*Average*	*Males*	*Females*	*Average*
Infected with parasitic worms	41.2%	29.4%	34.5%	22.9%	19.2%	17.1%
Internal organs congested	3.4	1.3	2.2	–	–	–
Pathological changes in liver	20.3	21.1	20.7	14.3	2.4	7.9
Pathological changes in spleen	68.5	47.2	56.3	34.3	26.8	30.3

first; the significance of these predatory birds becomes especially great in spring (Karaseva, German, and Korenberg, 1957). The removal or limitation of the sanitary role of predators has led to widespread disease. The shooting of wolves, coyotes, pumas, and other large predators in the United States has led to an increase in worm infections among protected hoofed animals. The increase in the infection rate was accompanied by a rise in the death rate and even a drop in the size of flocks and herds, since proper elimination of sick and infected animals was not carried out.

In order to protect the willow ptarmigan, which is of great importance as a commercial and game bird, the Norwegians adopted a bounty system for controlling its predators as far back as the last century. The drop in the numbers of feathered and four-footed enemies of the ptarmigan was one of the reasons for the spread of coccidiosis and other diseases among these birds, causing mass deaths during the first part of the twentieth century. These epizootics usually hit in years which have a hot, dry summer and a poor harvest of berries (Brinkman, 1926, 1927; Nordhagen, 1928; and others). The same type of kill was observed in England among grouse, heath cocks, and partridge and was also apparently associated with the extermination of feathered and four-footed predators observed by many English zoologists (Middleton, 1935).

However, predators do not catch sick animals during every epizootic. When small rodents are suffering from erysipeloid, the natural nidi of which are associated with small mammals, the latter usually die in their burrows and seldom become the victims of predators (Olsuf'yev, 1954). Sometimes predators aid in spreading disease by carrying the infection within themselves or in the parasites (fleas, ticks, lice, etc.) they pick up from their victims.

Some of the peaceful species derive advantages from living with predators. The house sparrow *Passer domesticus griseogularis* of Central Asia

usually lives in the nests of eagles, *Buteo* hawks, or other large predators and abandons them only when the predators themselves leave. The nests of the peregrine falcon and the rough-legged hawk in the tundra are often situated near the nests of geese, while to the east they are associated with marmot villages. At the end of summer the marmots find food primarily near the nests of predators, where the fertilized soil supports tender grasses. The marmot "toilets" in turn attract coprophagic insects and insectivorous birds, which the predators then eat. Small rodents also settle in marmot colonies. And the cries of danger to which both species react are an important mutual warning (Kapitonov, 1957).

The examples cited underline the contradictory aspects of the effects of predators on the population dynamics of their prey and show the closeness of the evolutionarily developed ties between these two groups of species. Over the course of co-development and mutual adaptation, the one-sided dependence of the predator on its prey has grown into a mutual one. The predator has become an important and sometimes necessary condition for its victim's existence.

CHAPTER 17

Parasites and Hosts

TYPES OF PARASITISM

Parasitism and the forms of cohabitation related to it (mutualism, or mutually useful existence; synoecy, or tenancy; commensalism, or boarding) arose against a background of nutritional and spatial connections between animals. "Parasites are those animals which live at the expense of individuals of another species and are closely associated with them in their life cycle for a greater or lesser amount of time. Parasites feed on the body juices, the tissues or the digested food of their hosts, such parasitic behavior being a specific species-associated feature of the given parasite, which repeatedly (unlike predators) uses its host for feeding. In addition, parasites permanently or temporarily use the body of the host as their territory of habitation" (Pavlovsky, 1946).

Pavlovsky's definition embraces the whole variety of parasitic relationships, enabling us to distinguish between true and so-called "false" or "guest" parasites. In the latter case penetration into the organism's body by another organism and even temporary feeding upon the host are accidental, such as the temporary feeding of larvae of the house fly *Musca domestica,* the gray flesh fly *Sarcophaga cornaria,* and other flies in the human intestine or the maggot of *Lucilia* feeding on the dead tissues in wounds. The mutual relations between parasites and the external environment are established chiefly through the body of the host, which acts as the immediate environment of the parasite (Moshkovsky, 1946; Dogel', 1947).

Mutualism includes cases of close co-residence in which the partners derive mutual benefit from one another, either by mutually supplementing the exchange of matter or by mutually protecting one another against unfavorable influences or enemies. The tightness of these connections in

extreme cases of mutualism reaches such a degree that the existence of one partner without the other is impossible.

Both nutritional and spatial associations may lie at the basis of symbiotic relationships. One example of spatial association is the free cohabitation of *Amphiprion* fishes and large actiniae, hermit crabs, and the hydroid polyp *Hydractinia.* The *Aspidosiphon* worm during its youth hides the end of its body in the small, empty gastropod shell on which the larva of the *Heteropsammia* polyp is resting; as it grows up, the polyp grasps the worm solidly, like a sleeve. The worm provides the polyp with a means of locomotion, while the polyp protects the worm with its stinging cells. All these are examples of ectosymbiosis.

Endosymbiosis is more likely to represent not only spatial but also nutritional associations. In the intestines of the wood-eating termites live flagella-bearing Hypermastiginae capable of digesting woody matter, although the termites themselves do not produce any enzymes capable of breaking down cellular matter. The young nymphs of the termites become infected with the flagellates when their elders give them an excretion from the anal aperture. Similar relationships between infusorians and bacteria and their hosts—the hoofed animals—are known to occur in the rumens of these animals. The yellow zooxanthellae (in marine animals) and green chlorellae (in freshwater species) encountered among many protozoa (radiolarians, infusorians), certain sponges *(Spongilla viridis)* and coelenterates (*Hydra viridis,* medusae, many Anthozoa), certain Turbellaria, rotifers, Bryozoa, and mollusks are examples of symbiosis between animals and algae. If artificially separated, these algae can also live independently.

There is a close symbiosis between fungi and bacteria and the bloodsucking insects. The blind sacs of the esophagus of the mosquito, the stomach disk of the louse, the tracheal organ of the stomach gadfly, the third abdominal segment of the chinch bug, the walls of the middle gut of the tsetse fly, the abdomen of the aphid and the mealy bug, the fat body of the cockroach—all contain special formations of the mycetome consisting of cells inhabited by bacteria or unicellular fungi. Symbionts which secrete enzymes for digesting sugar (in the aphids and mealy bugs) or blood (in the bloodsucking insects) reproduce in them.

Synoecy, or tenancy, is based on spatial associations often supplemented by nutritional relationships. In protozoic cases the tenant settles in the host's living quarters or in the region of its activities. Thus young fish stay among the tentacles of the large medusae and hide under their umbrellas when danger approaches. The nine-legged crab *Arete dorsalis* often lives in the "burrows" of the sea urchin *Heterocentrotus.* In the burrows of rodents and other digging animals, in the nests of birds, and in other instal-

lations built by terrestrial vertebrates live large numbers of tenants, some of which are parasites of the host and some of which only use its refuge. Near Ashkhabad 212 species of tenants, including mammals, birds, reptiles, amphibians, mollusks, insects, ticks, crustaceans, various worms, etc. were found in the burrows of the large gerbil (Vlasov, 1938).

The complex underground passage system of the mole *Talpa europaea* sometimes embraces thousands of hectares of forest (Pavlinin, 1948). In the Raif Forest Preserve near Kazan' 8 species of worms (families Enchytraeidae, Lumbriculidae, and Lumbricidae), 3 species of arachnids, 11 species of spiders (including the typical burrow dweller *Anguilupalpes*), and no less than 6 species of mites (the burrow dwellers *Pergamasus* and *Parasitus, Eugamasus, Poecilochirus, Amblygamasus,* Oribatei) were observed. The same location yielded 4 species of Collembola, 34 species (in 10 families) of beetles, many of which remain more or less permanently in burrows, 10 species of fleas specific for moles, shrews, and rodents, the bumblebees *Bombus lucorum* and *B. hortorum,* and several species of Diptera. Insect larvae also turned up there (20 families of Carabusiidae, Silphidae, Elateridae, skin-eating beetles, cicadas, butterflies, daddy long-legs, etc.). The passages contained 2 species of frogs and 2 species of tritons. Mammals using the passages included all the mice, voles, and shrews; many of them apparently used the passages of the mole as emigration routes (N. P. Voronov, 1957).

The refuges of a given species may be used in different ways by the tenants in various biotopes. In villages of small susliks in the Black Earth region (near the Caspian) 8.5 percent of the catches made did not include the hosts of the burrow but, rather, animals which had taken refuge there. Of these, 67 percent were individuals of the same species and 33 percent of different species. In the Il'men' belt, however, only 2.3 percent of the cases recorded were of individuals fleeing to foreign burrows, of which 32 percent included flights made by individuals of the same species and 68 percent by foreign species. In the villages in the Black Earth belt, contact runs high between susliks and gerbils, which creates a special danger during epizootics of plague. But in the Il'men' the same species of fleas contact greater numbers of rodents than in the Black Earth belt (Pavlov *et al.*, 1959).

In the case of epioekia the tenants settle on the body of the host, using it as a substrate, as in the case of certain barnacles on sharks and whales; these barnacles feed on plankton organisms. Examples of entoekia, or tenancy within the body of the host without dependence on its food supply, include the small fish *Fieresfer,* which stays in the cloaca of *Holothuria tubulosa* and leaves periodically to feed on crustaceans, or certain nema-

todes which live in the intestines of horses and feed on the infusorians they find there. The bitterling *Rodeus sericeus* lays its eggs in the mantle cavity of *Anodonta,* where the larvae of this fish develop safely.

In commensalism, or boarding, the boarder feeds on food remains rejected by the host without causing the latter any noticeable deprivation. One example is the temporary conjunction of mammals and birds. Skylarks, gray and willow ptarmigans, *Parus,* and many others feed in winter in places where domestic and wild ruminants are pastured and stay with the herds on the steppes and in the pastures. Polar foxes follow the polar bears across the ice of the polar basin. Such conjunctions are short-lived and accidental. In other cases they are tighter, usually combined with cohabitation. This is the way *Hyrichtys cyclothoni* settles on the skin of a fish not far from its anus and feeds on its excretions. The plate-gilled mollusk *Montacuta ferruginosa* attaches itself to the anus of the sea urchin *Echinocardium.* Finally, many intestinal protozoa live in the intestines of fish, birds, or mammals but feed on bacteria or fungi and their spores.

A transitional group between tenancy and parasitism includes the food absorbers, which eat materials already prepared by the digestive juices of the host. Such osmotically feeding protozoa are inhabitants of animals' digestive tracts.

Parasites are divided into facultative types, which do not absolutely need their hosts in order to exist, and obligate types, in which the parasite permanently or temporarily, but always of necessity, lives in individuals of another species and feeds at their expense, usually without killing them. Obligate (true) parasites are divided into ectoparasites, which live on the surface of the host's body, and endoparasites, which inhabit its organs, tissues, cells, or body cavities. Examples of ectoparasites include the blood-sucking arthropods (lice, fleas, ticks, mosquitoes, and others) as well as the parasitic sucking chaetopod worms of the subclass Monogenea and the parasitic Polychaeta. Endoparasites include the other parasitic worms, protozoa, crustaceans, and others. Forms with a transitional way of life are also known.

Temporary and stationary parasites are distinguished by the amount of time they spend in or on the body of the host. Temporary parasites enter into contact with the host only when taking food. Their adaptations to a parasitic way of life are slight, since the greater part of their time is spent free, and they attack a wide circle of hosts. Many fleas spend most of their lives on the host ("fur fleas"), while others ("nest fleas") switch to the host only when they are going to suck its blood (Ioff, 1941).

Stationary parasitism is characterized by the long-term or permanent presence of the parasite on the host and is usually accompanied by the presence of many mutual adaptations. It is divided into periodic (the

parasite spends part of the time away from the host) and permanent parasitism (the parasite is connected with the host throughout its life). Periodic parasitism is more likely to be encountered than the permanent type and may take various forms. An alternation of parasitic and free-living generations is characteristic of the nematode *Rhabdias,* which lives in the lungs of frogs. The hermaphroditic form lives in the host, and the bisexual generation lives outside it.

When individual phases of development parasitize the host, larval or imaginal parasitism may occur. The former is characteristic of nematodes of the family Mermithidae, the larvae of which live in mollusks and insects and the adult stages of which live free. Larval parasitism is common among the Pantopoda and is widespread among the parasitic insects (ichneumon flies, the Tachinidae, oestrid gadflies, and many others). Parasitization by the adult stages is widespread among nematodes of the families Trichostrongylidae, Ancylostomidae, the crustacean Copepoda, the cestodes, and the monogenetic flukes.

A third type of periodic parasitism is repetitive parasitization at different stages of development. Among larval, nymph, and adult ixodid ticks the relatively short period of bloodsucking, each time on a new host, alternates with a long period of existence away from the host. This change of hosts is important in the transmission by ticks of bacterial and viral diseases. The permanent parasite lives all its life on or within the body of the host and cannot exist in the external environment. Such are sucking and biting lice, mange mites, trichinae, and certain protozoa. Many permanent parasites change their hosts during the course of metamorphosis; an example is the malarial *Plasmodium.*

COADAPTATIONS OF PARASITES AND HOSTS

Adaptations to the parasitic way of life are numerous and include structure, physiological properties, and way of life. Simultaneously, residence of the parasite on the host or in its body is accompanied by response reactions, and therefore it is more correct to speak of mutual adaptations, or coadaptations, of both organisms. Changes in the structure of the parasite as compared with its free-living ancestors involve the disappearance of organs not needed in the new surroundings and the appearance or re-formation of organs associated with the features of life on the surface of or inside the host's body.

In the case of ectoparasites the body becomes flatter and shorter and its segmentation diminishes. Flattening is more likely to be dorsoventral and

aids the parasite in clinging to the host's body surface (lice, bedbugs, parasitic nemertines, leeches, ticks, etc.). Among fleas the body is flattened laterally, which is associated with their habitat on hairs or among feathers and with their special mode of locomotion (jumping). Intestinal parasites are distinguished by their lengthened bodies; in many species the body is segmented (flatworms, acanthocephalids, pentastomids, and others). Among the alimentary parasites the body is spherical (young cestodes as compared with the older forms); they acquire sometimes a flattened and sometimes a branchlike form, which increases the surface for sucking up food.

For holding onto the surface or interior of the host there are various organs of attachment (hooks, suckers, sticky filaments, and others). In the digestive tract of those bloodsucking ectoparasites which feed periodically, the dissolving power of the tract increases (bedbugs, mosquitoes, bloodsucking flies) or else the tract develops an ability to expand, thus increasing its capacity (leeches, isopods, ticks). Among ixodid ticks the bloodsucking period is accompanied by marked growth in the intestine and cuticle, with genuine strain setting in only during the last 12–24 hours. This combination of uninterrupted growth and rapid stretching permits ticks to swallow extraordinarily large amounts of blood (Balashov, 1957). Parasites have the ability to go hungry, sometimes for 3 or more years (ixodid ticks). The glands of bloodsuckers secrete an anticoagulant. Among the endoparasites the digestive system is simplified and sometimes atrophies, as do the nervous system and the sense organs.

Better feeding conditions for parasites have aided them in developing more rapidly and have increased their fecundity. The increase in fecundity is biologically necessary, since the necessity of penetrating the host diminishes the chances of survival. Parasitic nematodes lay up to 10,000 eggs, while free-living species produce hundreds or only a few dozen eggs. In 5 or 6 months of sexual maturity the nematode *Ascaris lumbricoides* produces up to 50 or 60 million eggs weighing 1,700 times as much as the original worm itself. *Taenia saginata* produces 600 million eggs a year and in its relatively long life (up to 18 years) produces about 10 billion eggs (Shapiro, 1937).

The eggs and larvae of parasites, if left in the external environment for long periods of time, are resistant to unfavorable conditions, can survive long periods of hunger, or can exist in an anabiotic, encysted state. When dehydrated, the larvae of trichostrongylids from the intestines of hoofed animals remain alive for more than 2 months. They resist the action of saturated copper sulfate solution for several hours with no ill effects. Corrosive sublimate (1 percent) and formalin (4 percent), as well as other

substances, likewise have no effect on them. They can live for 6 months at a temperature as low as —10° C and can stand short increases in temperature up to 80° C. Their stability recalls that of the cysts of protozoa.

The host organism responds to the penetration of parasites with protective changes, creating immunity. Immunity, or the ability to resist disease organisms, is an evolved property of a species, the result of its mutual relationships with parasites. Special protective mechanisms lie at the basis of immunity. A species' first protection is its external covering, which prevents the parasite from entering. Mucus is a means of mechanically removing pathogenic organisms, and it contains bactericidal substances. Phagocytosis by formed elements of the "white blood" and cells of the reticulo-endothelial system is a second line of defense. The third line consists of the protective reactions of the blood. When parasites enter the bloodstream, substances are produced which are capable of counteracting the effects of the poisons secreted by them (antitoxins), coagulating the foreign proteins (agglutinins), or precipitating them out of the bloodstream (precipitins).

Immunity includes congenital forms (total, effective against many infections and parasites, and specific, effective against a definite species of parasite) and acquired forms (the growing ability of the host organism to resist when the parasite enters). Acquired immunity may be sterile or nonsterile; in the latter the parasite is preserved in the host, but its reproduction is retarded by the defense reactions of the host. Congenital and acquired immunities (nonsusceptibility) against various infections are more common if the host has been in contact with the parasite over a long period of time. Immunity grows stronger as the host re-encounters the infection.

Four groups of animals are distinguished in regard to their relationship to infectious diseases. The first group includes the susceptible, highly sensitive, and acutely infectable species which have practically no protective adaptations and which die if even the smallest amount of pathogenic material enters their bodies—even one microbe cell. The second group includes species which are susceptible but poorly sensitive; the lethal dose of parasites is hundreds or thousands of times larger than for the first group. When infected with small doses, they suffer a chronic infection and, after recovery and the acquisition of a nonsterile immunity, remain carriers of the infection and excrete it into the external environment. The third group includes animals which are relatively insusceptible to the disease and practically insensitive to infection. They cannot be infected even with enormous doses of the agent. The fourth group is composed of those animals which have such a high degree of adaptation between parasite and host that both species are actually symbiotic to one another.

The existence in nature of the rodent tularemia parasite, a microbe, is possible only in the first and second groups of animals (Olsuf'yev *et al.*, 1950, 1952). But the relationships between this microbe and its basic host, the ixodid ticks *Dermacentor pictus* and *D. marginatus*, are distinguished by adaptations which enable the microbes to reproduce inside the ticks without causing them any substantial harm (Olsuf'yev and Rudnev, 1960).

Many parasites have conditionally pathogenic significance for their hosts and do not cause disease in healthy animals with sufficiently strong powers of resistance (immunity). But in hosts weakened by hunger or other influences, such parasites may multiply intensely and cause death (the necrobacilli among northern reindeer, pasteurellae among voles and marmots, *Erysipelothrix* among many rodents, etc.). The same thing happens when the parasites are unusually virulent. The defense reactions of the population are strengthened or weakened depending on the state of the animals and their conditions of existence. At the start of a plague epizootic among marmots, acute, rapidly fatal cases predominate. But at the end of the epizootic many of the sick animals recover, and their organs betray many traces of the fight against the disease. The causes of the disease's weakening are associated both with a decrease in the pathogenicity of the agent, which is infected with a unique hyperparasite (the plague bacteriophage), and with an increase in immunity in the marmot population (Tinker, 1940).

The virulence of the parasite and the immunity resisting it in the host are species-associated properties. The sensitivity of a host varies with regard to various infections, while the pathogenicity of a parasite varies with regard to different hosts. Both vary widely depending on environmental factors. Variability of virulence and immunity is governed primarily by their close dependence on one another. The external environment changes the state of the host, strengthening or weakening its ability to resist the parasite. Through the body of the host, and less often directly, the same effects are experienced by the parasite, the virulence of which appears against a living and variable substrate, the body of the host (Pavlovsky, 1951).

Species immunity has a special biological importance as a means of struggle and competition with other species. During tularemia epizootics the more resistant field mice survive better than other species and after the epizootic often become the most numerous. Their wide geographic distribution is due, at least partially, to their low susceptibility to many diseases. The resistance of South African antelopes to trypanosomes transmitted by the tsetse fly permits them to live in regions with natural nidi of these diseases, regions into which highly susceptible hoofed animals cannot penetrate.

RISE AND EVOLUTION OF SYMBIOTIC AND ANTIBIOTIC RELATIONSHIPS

Symbiotic relationships arose and developed evolutionarily on a basis of combined habitation and ever-closer nutritional and spatial associations. Mutual adaptations necessary for the close co-residence characteristic of interspecies relationships were formed out of these associations. Antibiosis, a phenomenon every bit as natural as symbiosis, is the result of competition for food and space leading to greater and greater spatial disjunction or ecological isolation of various species.

Symbiosis in all of its forms could have risen in at least three ways (Fig. 256). In the first case simple tenancy, followed by a move to the animal's body and then into it, create conditions for supplementing spatial relationships with nutritional ones. The latter may begin as boarding, causing the host no harm. After settlement in the body boarding is easily replaced by a form of living off the body and its juices, which *does* harm the host; i.e., parasitism has appeared. If this does not happen, a mutual coordination of matter exchange arises, with mutual use of metabolites, i.e., mutualism.

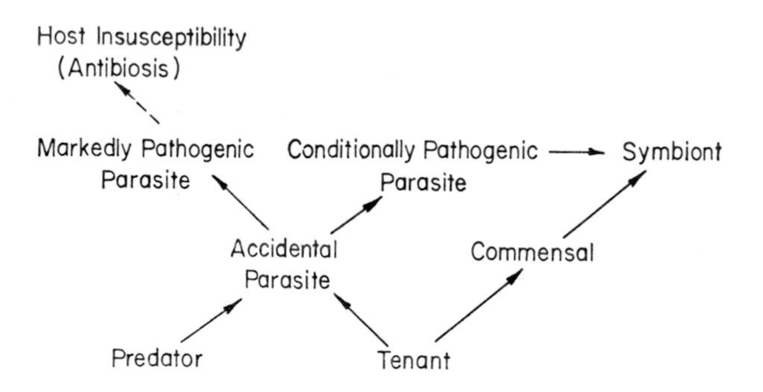

Fig. 256. Development of symbiotic (parasitic) relationships (original).

A second path leading to the parasitic way of life is predation. In nature traces of such transitions still exist. The European leech *Glossosiphonia complanata* is a free-living parasite infecting small invertebrates. The horse leech *Haemopis* eats small animals completely and sucks the blood of larger ones. The Ceylonese leech *Haemadipsa ceylonica* lives in bushes, attacking its prey for blood from time to time and never using any other method of feeding. But the fish leech *Piscicola heometra* spends its whole life on the body of the fish, leaving it only during the mating season to lay

cocoons of eggs on the bottom of the body of water. Finally, the leech *Iohansonia pantopodum,* which occurs on the pantopod *Collosendeis,* even lays its eggs on the body of the host. Among bugs of the genus *Reduvius* there are true free-living species which live on various insects; however, *R. personatus* may turn from predation and on occasion will suck human blood. The bedbug *Cimex lectularius* has turned completely to sucking the blood of man, domestic birds, and other animals (Dogel', 1947).

The third path along which parasitism arose is associated with the accidental penetration of a parasite-to-be into the body of an animal in which it may be able to exist at least temporarily. The nematode *Aloionema appendiculatum,* which penetrates into the mucosa of *Arion ater,* lives and reaches its greatest size inside the "host," but it cannot mature there and must emerge onto the surface in order to reproduce. One can see how deterioration of external circumstances may force free-living individuals to disappear and facultative or false parasites to turn into true ones.

Having arisen from predation, accidental parasitism, and tenancy, parasitic relationships have developed in several directions. Mutual adaptations which diminished the pathogenicity of the host toward the parasite were subsequently capable of leading to conditional harm, one-sided benefit, and, finally, to mutual benefit. This path has apparently been used by many parasites. The virus of tick encephalitis is not pathogenic for ixodid ticks, its chief carriers; the malarial *Plasmodium* is not pathogenic for mosquitoes; *Leishmania* is weakly pathogenic for reptiles and rodents (gerbils). This type has been referred to as "parasitism for survival," or symbiosis in the narrower sense of the word.

In other cases the relationships between parasites and hosts have become sharper and sharper. Parasitism leads to the death of the host, since the parasite acquires a greater and greater pathogenicity toward it. Usually this takes place in the presence of a complex developmental cycle of the parasite which requires a change of hosts. In these cases the death of the intermediate or ultimate (definitive) host forces the parasite to find a new host. Thus the mass reproduction of plague or tularemia bacteria in a sick rodent best guarantees[1] the infection of the fleas, lice, ticks, and other bloodsucking parasites which feed on it and pass the infection on to other rodents. Strains of these bacteria, which cause acute septic diseases in the rodents and which usually bring death, do not penetrate into the bloodsucking vectors and have no chance to spread. Because of this, there is in nature a permanent selection of the most pathogenic lines (strains) of the

[1] By means of a generalized illness and the development of total sepsis, in which there is a mass of microbes in the blood and organs.

agents of such infections; the less virulent strains which do not cause sepsis are not transmitted and die off.

When such long-term connections exist between the host and the pathogenic parasite, the former may develop an immunity that leaves the animal relatively insensitive to infection. The given species then ceases to be the basic host of such an infection, and its former role passes either to those species which retain a high susceptibility or to newly appeared species. Such "migrations" of parasites from one host to another are apparently common in the evolution of parasitism.

The development of acutely antagonistic relationships, or "parasitism for slaughter," is also encountered among the helminths. The larvae of *Ligula intestinalis* enter fishes along with *Diaptomus* crustaceans, the primary intermediate hosts. Ligulosis acutely weakens the fishes and simultaneously causes them to puff up; as a result, the sick fish rises to the upper layers, where it easily becomes the prey of fish-eating birds (especially seagulls), the final hosts of the cestodes. The sexually mature stage of the parasite causes no substantial damage in the birds ("parasitism for survival"). In rodents (hares) and hoofed animals the liver fluke is a pathogenic parasite which weakens the host, but it does not harm predators.

Thus the variety of properties now observed to exist among parasites and hosts, and the variety of relationships shared by them, express the variety of ways in which they developed.

Parasitism is often encountered among the lower and primarily smaller animals. In the 17 classes of deuterostomate animals there are almost no parasitic species, with the exception of the hagfish, the lampreys to some extent, the fish *Stegophilus*, and the leaf-nosed bat *Desmodes*. Among the 37 classes of protostomate animals more than half (20) have parasitic representatives, and 6 of them are completely parasitic (trematodes, cestodes, Acanthocephala, Nematomorpha, Myzostomida, Pentastomida). Of the total number of animal species, now thought to be around 1,500,000, approximately 60,000–65,000 (4–5 percent) are parasitic. The broad distribution and variety of forms associated with parasitism testify to the proposition that parasites, as a life form with a special type of relationship to the environment (through the body of the host), arose independently in the different taxonomic groups.

PARASITOCENOSES AND THEIR DYNAMICS

The parasitic population of the host, or its parasitocenose, takes shape under the influence of the morpho-physiological properties of the animal

(the environment of the first order for the parasites) and the external conditions (the environment of the second order for the parasites) (Pavlovsky, 1951).

Great importance is attached to the character, place, and methods of feeding of the host. These factors explain the similarity between the parasitofaunas of the intestinal tracts of hoofed animals and rodents, which belong to different orders of the Mammalia. Among forest mice, voles, lemmings, and South American rodents we find the same genera of infusorians of the families Ophryoscolecidae and Cycloposthiidae which are widespread among ruminants and single-hoofed animals. Infusorians of the family Pycnotrichidae are found in the intestines of ruminants, rodents, and hyraxes, which feed similarly but which belong to three different orders of Mammalia. Flagellates of the subfamily Hypermastigina, which break down cellular matter in the intestines of insects, are found in the wood-eating termites and the cockroaches *Blatta* and *Cryptocercus*. If the food is not identical, the endoparasite fauna is probably different. The intestinal parasites of the large gerbil and the hedgehog *Hemiechinus auritus*, which live in the same places, are different, while their ectoparasites are the same. The plant-eating rust-colored vole has 7 species of intestinal helminths, the seed-eating forest mouse has 5, while the insect-eating common shrew has 19. V. A. Dogel' (1936) observed the same correlation in birds (Table 68).

TABLE 68. *Endoparasitic infections in birds in relation to the character of their feeding (after Dogel', 1947, with changes).*

Bird	*Number of Birds Dissected*	*Number of Parasite Species*	*Average Number of Parasites per Bird*	*Type of Feeding*
Finch	45	5	7.3	mixed
Swallow	27	11	50.5	insectivorous
Martin	34	13	70.0	insectivorous

Among frogs the relationship between infection and place of residence, character of food, and methods of catching it is easily observed (Table 69). The tree-dwelling squirrel and dormouse and the burrowing mole are species of mammals not much infected with parasites. Because of its nest parasitism, the cuckoo *Cuculus canorus* transmits its specific Mallophaga not to the young birds in the nest but at the moment of copulation (Dogel', 1936).

The relationships arising between the various parasites are of great importance in the formation of the parasitocenose. The penetration of species which exercise a mutually favorable (synergistic) effect increases the reproduction and growth of each, while antibiotic relationships pre-

TABLE 69. *Fluke infections in frogs in the Volga delta (after Dubinina, from Dogel', 1947, with changes).*

Species	*Place of Habitation*	*Percent of Infection*	*Number of Parasite Species*
Lake frog	bodies of water	90.5	14
Green frog	bodies of water	96.2	12
Swamp frog	swamps and damp areas	72.8	8
Grass frog	relatively dry spots in shade	40.0	5
Gray toad	relatively dry spots in shade	30.7	4
Green toad	drier, often steppe, areas	20.0	3
Red-bellied frog	drier, often steppe, areas	16.6	6
Tree frog	tree crowns	15.6	1

clude or limit the possibility of coexistence. These relationships are little studied, but they deserve careful attention, since they are of great practical importance. The development of gram-negative bacteria in animal intestines is accompanied by an increase in the number of parasitic amoebae *Entamoeba muris* and *Chilomastix bettencourti.* On the other hand, the growth of proteolytic anaerobes leads to a decrease in the number of gram-positive and gram-negative bacteria and protozoa (Pavlovsky, 1951). Regulation of the content and relationships of species in the parasitocenose with the aid of antibiotics and other tools opens up tremendous possibilities in controlling human diseases and in raising the productivity of domesticated animals.

The parasitocenose undergoes substantial changes as the host ages. Usually the intensity (number of parasites per individual) of the infection increases, its extent (number of animals infected) rises, and the number of species of parasites involved increases. The last factor grows because of the addition of species with complex cycles of development or of those which are typical only of the adult host. Young society voles, forest mice, and house mice in the Trans-Caucasus are infected with worms which develop primarily without intermediate hosts. The adult rodents have species with complex developmental cycles, and the total number of parasites increases (Kirshenblat, 1938). The parasitocenoses of helminths change with age among frogs (Table 70).

The seasonal dynamics of the parasitofauna depends on changes in the external environment, on the state and way of life of the host, and on the parasites' cycles of development, intermediate hosts, and vectors. Most of the parasitic protozoa, roundworms, and certain other parasites having a short developmental cycle (without a change of hosts) are encountered in the same numbers at all times of the year. When bats go into hibernation, they are not purged of their endoparasites, but parasites do suspend their development while the host is in torpor.

TABLE 70. *Helminth infections at various age levels of the lake frog* Rana ridibunda *in the Volga delta (after Dubinina, from Dogel', 1947).*

	NUMBER OF PARASITE SPECIES			
Age of Frog	*Flukes*	*Cestodes*	*Nematodes*	*Total*
Tadpole	3	–	1	4
Under 1 year	11	–	4	15
1 year	11	–	4	15
2 years	15	–	5	20
3 years	11	–	5	16
4 years	17	1	7	25
5 years	20	1	7	28

Parasites which have seasonal species of insects as intermediate hosts are themselves strictly seasonal. A compressed period of development is significant in that it succeeds in getting the eggs laid at the place where the intermediate hosts are found and when they are abundant. The fluke *Phaneropsolus micrococcus,* which is encountered only from May through August, parasitizes the intestine of the barn swallow in this way. It appears only when mass swarms of insects are flying. During this period the parasites finish their development in the swallow and lay their eggs, which then infect the insects. Other species of flukes spend the same short amount of time in amphibians (salamanders) during the middle of summer. The seasonal drop in abundance and variety of the intestinal parasitofauna may be connected with the eating of specific foods which apparently have a therapeutic effect. The disappearance of coccidia from the intestines of hazel hens during the autumn is apparently connected with the ingestion of berries (Oliger, 1940). It is probable that susliks and hoofed animals derive a worming effect from eating wormwood and that hares derive the same benefits from eating bark and shoots.

The dynamics of ectoparasites which are connected with their hosts only periodically is distinctly seasonal. Fleas are divided into "summer," "autumn," "winter," and "spring" species. In the steppes of the Cis-Caucasus, lower Volga, and western Kazakhstan two basic species of fleas parasitize susliks: the "summer" *Ceratophyllus tesquorum* and the "autumn-winter" *Neopsylla setosa* (Fig. 257). The time when the former species appears is marked by the plague epizootics which hit the suslik populations as the young are emigrating. The end of the epizootics corresponds in time with the beginning of hibernation and the appearance of *Neopsylla,* the body of which contains the agents of plague and preserves them through the winter until the animals awaken in spring. The seasonal alternation of the two species of fleas has created a stable preservation of the infection (Tinker, 1940; Fenyuk, 1948).

The seasonal appearance of bloodsucking flies corresponding to changes

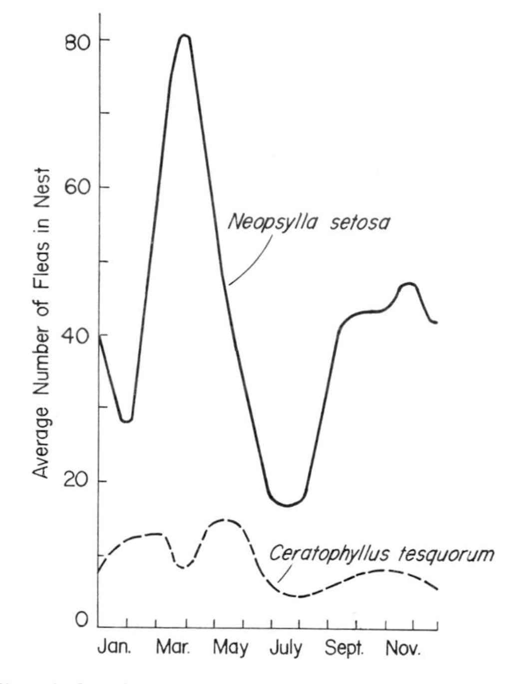

Fig. 257. Seasonality of abundance of fleas of the small suslik in its nests in the south-eastern steppes of the RSFSR (after Ioff, 1941).

in temperature and humidity is well known (Fig. 258). Infection of man and animals by *Leishmania* corresponds to the season when the vectors of *Leishmania* (sandflies) are flying,[2] while malaria corresponds in time to the flight of the malarial mosquitoes. Autumnal or "mosquito" encephalitis infection is possible only at the end of summer, when the mosquitoes *Culex* and *Aëdes* are swarming. Summertime-transmissible outbreaks of tularemia are associated with the development of tularemia epizootics among water voles and transmission of the infection from sick animals to man by horseflies, mosquitoes, and stable flies, which are abundant during haying in the meadows. The rise of epidemics of tick (spring-summer) encephalitis, spring epizootics of tularemia among small rodents, and tick-borne rickettsial fevers is associated with the mass appearance of adult ixodid ticks in summer and spring (Pavlovsky, 1946; Olsuf'yev, 1960).

Substantial changes in the parasitofauna take place when the hosts change their place of residence (migrations). The salmon *Salmo salar* lives

[2] Leishmanioses are cutaneous (Pendinsk sore) or generalized (*kala-azar*) diseases caused by the protozoa *Leishmania donovani* and *L. tropica* and are widespread in the south (here in Central Asia and the Trans-Caucasus).

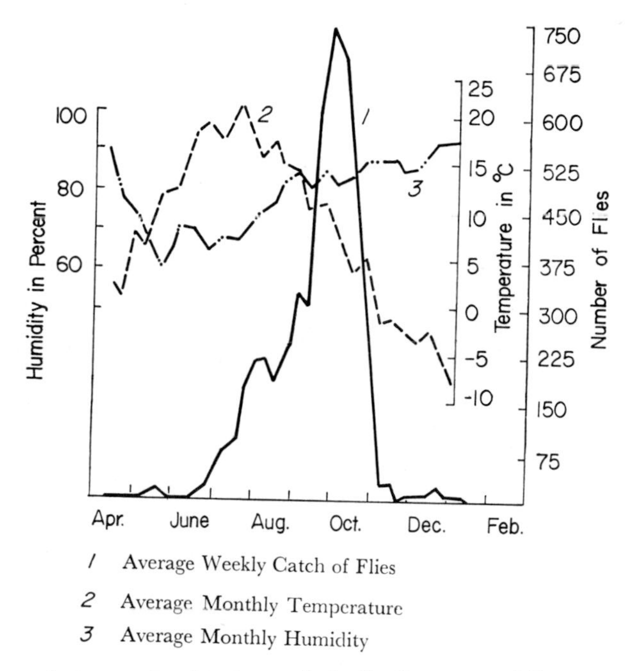

Fig. 258. Seasonal changes in abundance of the fly *Stomoxys calcitrans* near Moscow in 1947 (after Kuzina, 1950).

in the upper courses of rivers until 3 or 4 years of age, then swims to the sea, where fattening and maturity take place. At the age of 5–7 they return to the river to spawn for the first time; during this period they do not feed, but after spawning they return again to the sea for fattening. In 2 or 3 years they again go upstream to spawn. The migrations are accompanied by changes in parasitofauna. When the young fish go out to sea, their freshwater parasitofauna gradually disappears, being replaced by a marine parasitofauna. Every time the salmon return to the freshwater environment, the saltwater parasitofauna becomes impoverished and is replaced by a freshwater variety which dies out when the fish go back to the sea. The Novaya Zemlya *chav Salvelinus alpinus* spends most of its life in rivers, spending only 2 or 3 months of each year feeding in the sea. Therefore, both freshwater and saltwater parasites are usually found in it, but the numbers of the former increase when the fish is in the river, and the latter are abundant when it is in the ocean (Fig. 259).

The migrations of birds are also accompanied by a change in parasitofauna. V. A. Dogel' (1947) distinguished ubiquitous species, which are encountered in the host or upon it through the year (in both north and south); southern forms, which infect the birds during the winter stay; and

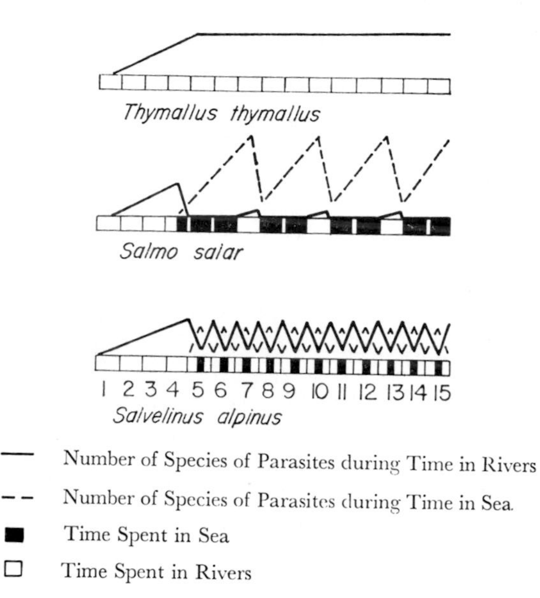

Fig. 259. Changes in the parasitofauna of freshwater and migratory fishes in relation to their migration (from Dogel', 1947).

northern ones, encountered only at the nesting ground. The ubiquitous species include ectoparasites, which are closely connected with the body of the bird (Mallophaga and feather mites), and endoparasites, which infect the bird in both north and south because they are able to pass their complete cycle within the bird's body. The southern and northern forms include species of parasites with short cycles of development which are capable of developing only under certain circumstances. Fig. 260 shows that the starling, which migrates and winters nearby in southern Europe and northern Africa, has three groups of parasites, while the barn swallow, which winters in southern Africa, lacks ubiquitous parasites. A change in the parasite population has also been observed in birds which migrate vertically.

SIGNIFICANCE OF PARASITES IN THE DYNAMICS OF THEIR HOSTS

The influence of parasites depends not only on the character of their relationships with the host but also on the external environment, which alters the states of both. The presence of weakly pathogenic parasites usually

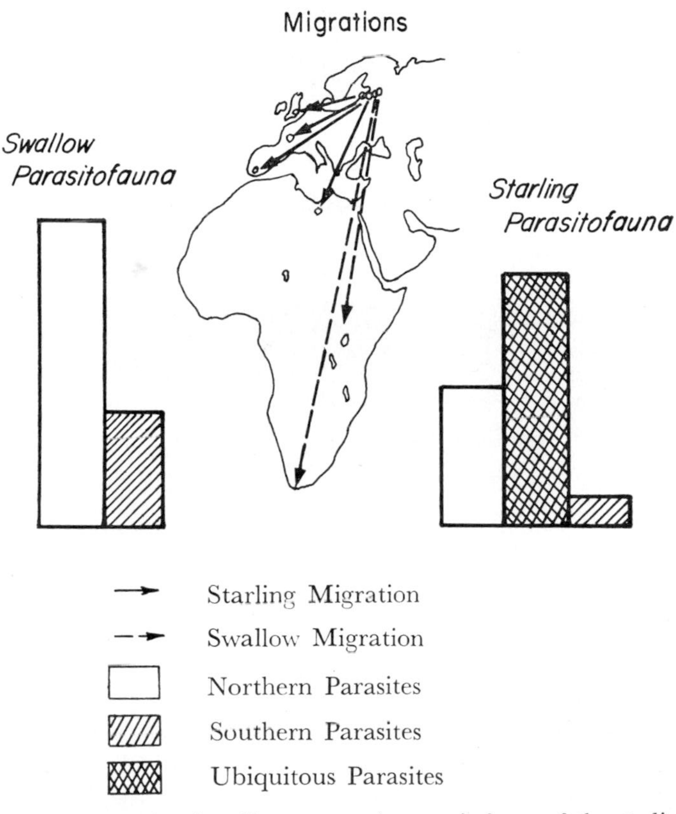

Fig. 260. Influence of migration distances on the parasitofauna of the starling *Sturnus vulgaris* and the barn swallow *Hirundo rustica* (after Dogel', 1947).

causes the host no substantial harm. But if the host is weakened, both the numbers and then the destructiveness of the parasites may rise markedly. In years which are favorable for Brandt's voles and tarbagans, *Pasteurella* infections are few and deaths rare. If the winter is poor and spring comes late, the emaciated rodents die en masse of pasteurellosis, which is widespread among them during such years. The disease usually dies down after the appearance of fresh green feed and the return of normal feeding habits among the creatures. Necrobacillosis spreads among herds of northern reindeer in years with a hot summer, when the reindeer suffer from overheating and from large swarms of mosquitoes and gnats to the extent that they cannot feed normally. The wild northern reindeer, however, hardly ever suffer from necrobacillosis because they pasture in cool places where swarms are few.

The reproduction of parasites depends on the abundance of their hosts.

According to the so-called Farr's law, the development of epizootics among animals is more probable as the animal population increases. This general quantitative principle rests on those same quantitative relationships between food and its consumers which lie at the basis of the dependence of predator abundance on the number of victims. The result of this dependence, as in those cases, should be periodic fluctuations in the numbers of the interacting species. This principle is the biological basis for prognoses of infectious diseases of domestic and wild animals. However, the relationship between the distribution of parasites and the population density of the hosts is very complex.

The possibility of infecting a susceptible animal and spreading the infection in the population is, in the simplest case, determined by the frequency of contact between one animal and another and between animals and the source of the infection. This frequency depends on the distribution of the animals and on their behavior, especially their motility. Both change under the influence of changes in the environment. The matter is complicated by the fact that many pathogenic parasites are transmitted to the definitive hosts through vectors (intermediate hosts). This is how the more or less complicated parasitic systems arise, systems which are parts of food chains in which the individual links (species) may react differently to the same changes in the external environment.

It is well known that epizootics among the mass species of insects, rodents, and other animals usually break out when the numbers of these animals are high, but they are also associated with a deterioration in the conditions of existence following several years of conditions favorable to growth. This growth aids in promoting directly or indirectly (through vectors, pastures, refuges, etc.) the contact between hosts and parasites and the wide spread of the latter. At the same time the presence of parasites in the weakened body of the host often increases their pathogenicity. Since such a situation favors the simultaneous development of many parasites, complex epizootics often arise which are due to several agents.

One example of the complex relationships between parasites and hosts is the development of fox mange, the agent of which is the mange mite *Acarus siro vulpes*. Healthy red foxes become infested with them in the burrows of sick animals, especially in the damper areas where the mites survive better. Young and weak foxes are more easily infested. The distribution of mites in the fox population is accompanied not only by a decrease in the number of animals, sometimes a very steep one, but also by a rise in resistance among the remaining foxes. Both lead to a decrease in the strength of the epizootic. During the period when fox abundance is depressed the mite population is maintained through infections of the young, but in some places it disappears. The number of mite nidi falls to

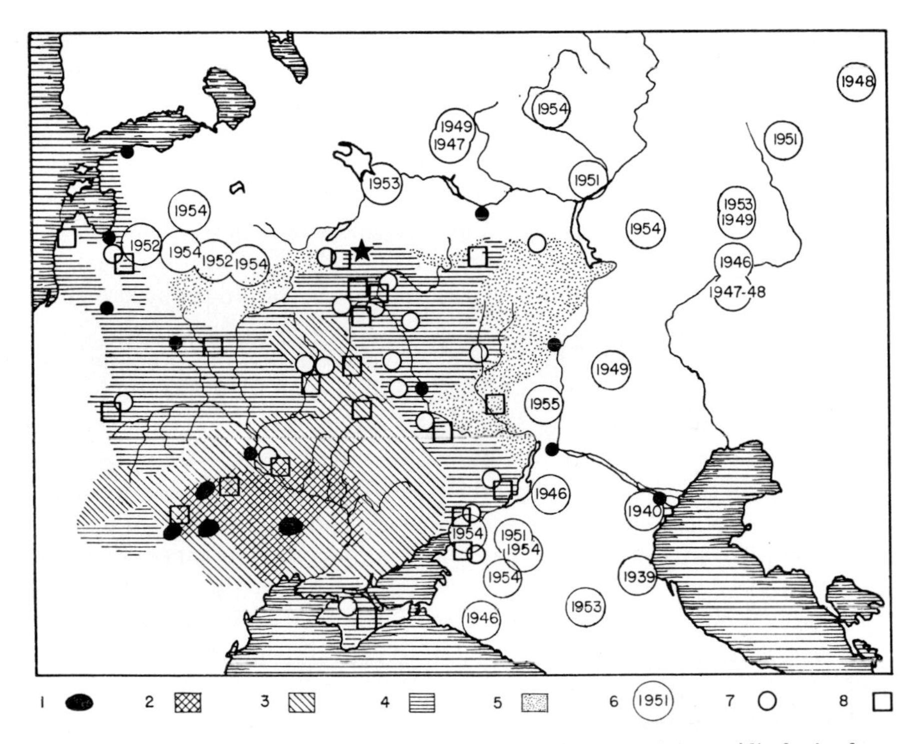

Fig. 261. Distribution of the fox mange mite (after Chirkova, 1957). 1: nidi of mite from 1919 to 1922; 2: distribution of disease from 1923 to 1933; 3: the same from 1934 to 1940; 4: the same from 1945 to 1949; 5: the same after 1950; 6: individual illness (in circle appears the year of discovery); 7: finding of mange spots on pelts; 8: unfavorable region according to data from regional tanneries in 1948 1949; 9: southern border of taiga;

a very low level. A new wave of mange arises only when the mite population rises again (Becker and Wharton, 1955; Dubinin, 1955). The concrete embodiment of this scheme depends on the weather and the food supply. In the USSR fox mange embraces large territories. Its most stable nidi are situated in the southwest, from which it has fanned out to the north and east (Fig. 261). The map, however, creates a false impression of the mange's spread from several centers; a careful inspection will show that in many regions the mange has spread from local nidi (Ling, 1956; Chirkova, 1957). In the forest zone it has developed approximately twice as slowly as in the steppes; this phenomenon is associated with the different speeds at which fox populations resurge (Fig. 262). When a high population density is reached, as shown in Fig. 262, the mange may be an important factor in decreasing the abundance of the foxes (Dubinin, 1955).

In the opinion of most contemporary entomologists, parasitic insects

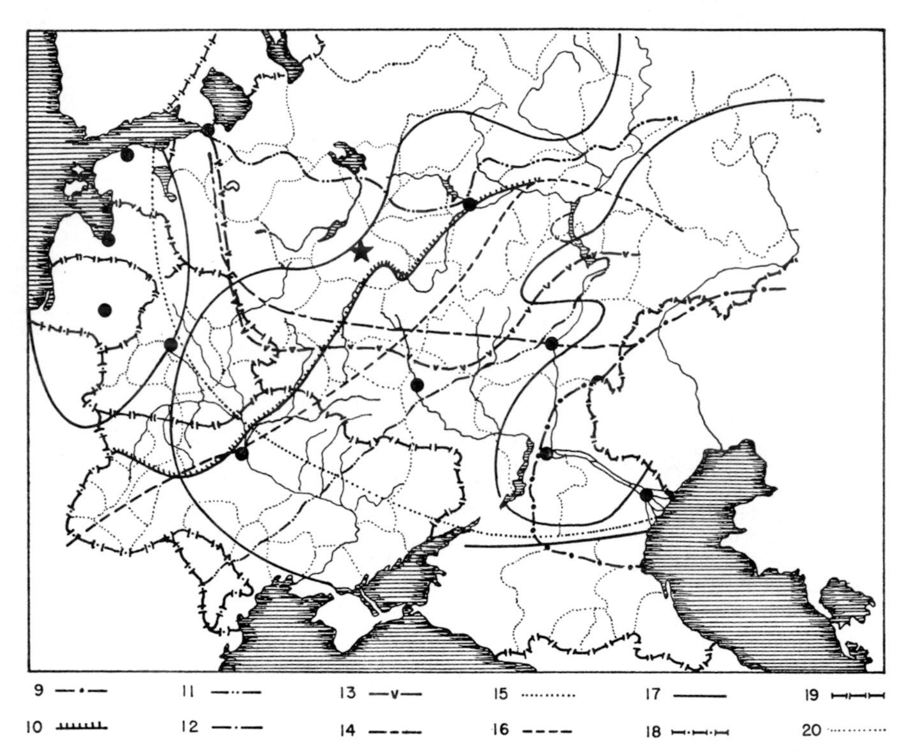

10: southern border of mixed forest; 11: southern border of forest steppe; 12: boundary of steppes and semidesert; 13: depth of snow cover (40 cm); 14: duration of snow cover (140 days); 15: January isotherm (–6° C); 16: July isotherm (20° C); 17: area with absolute humidity in July 1911 of 12–19 mm (after Borisov, 1948); 18: border of USSR; 19: border of autonomous or union republic; 20: border of oblast'.

play an important role in the abundance dynamics of their hosts (Rubtsov, 1944, 1953). Most important are the parasitic Diptera and Hymenoptera, especially the ichneumon flies (suborder Heterophaga). Their influence on host abundance depends on external conditions which aid the parasites in attacking and weakening the resistance of their hosts. Temperature, humidity, and the amount of snow cover are tremendously important in determining the abundance of silver Y moths, but a large role also belongs to predators, parasites, and diseases. Usually their effects are not displayed simultaneously and depend on the state of the physical factors. The greatest level of parasitic infection occurs during dry, hot weather, while the diseases themselves are most manifest in moist weather. Therefore, first-generation caterpillars are destroyed chiefly by parasitic insects, while second-generation caterpillars die mainly from fungal and bacterial diseases (Yermolayev, 1950).

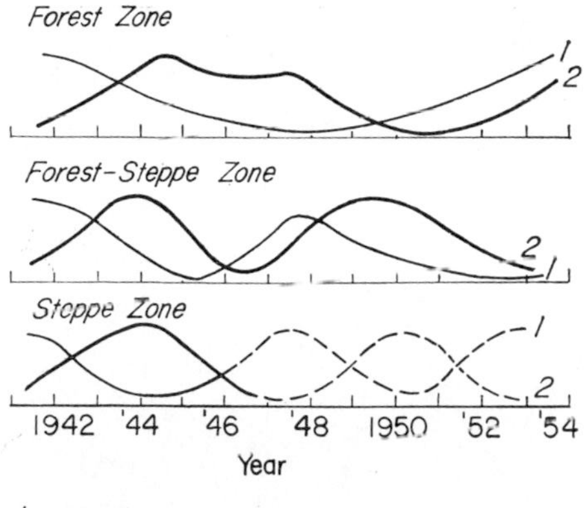

Fig. 262. Change in the number of prepared pelts (in thousands) and of foxes caught suffering from mange during the years 1942–1954 in the forest, forest-steppe, and steppe zones of the European RSFSR (after Dubinin, 1955).

A study of the wood pests *Eulecanium prunastri, E. corni,* and others has shown that parasites and predators are the chief regulators of their abundance. The entomophages (predatory beetles, parasites) which limit the numbers of the pests are different in different regions, and they often replace one another in the same region at different times. Under favorable conditions they may completely destroy the pests; in this case the predators themselves die off. Apparently this explains the existence of the well-known blank spots in the distribution of many of the specialized entomophages *(Rodomia, Aphelinus)*; they probably disappeared from these areas after they killed off their food. The reverse case is also known. I. A. Rubtsov (1950) described a case of uninterrupted mass reproduction of blackflies (Simuliidae) along the upper courses of rivers as well as in certain parts of the middle and lower courses. There blackfly larvae are free of parasites (Microsporidia and mermithids) because such spots contain quiet pools with no blackfly larvae and no parasites capable of serving as a source of infection for populations farther downstream.

Most of the coccids in this country are destroyed by parasites and predators. Successful regulation of Coccoidea and scale insects has been attained with the aid of their enemies. Significant increases in abundance, lasting 4–8 years and repeated every 10–25 years, are characteristic of the karakurt spider *Latrodectus tridecemguttatus.* These rises come in warm years with abundant food (Orthoptera and meal beetles) and few ichneumon flies. The latter is especially important, since when the parasites multiply,

the spiders are preserved in those few places where there are no ichneumons (Marikovsky, 1950). Apparently the significance of such biotic conditions is greater in the south than in the north (Rubtsov, 1953).

Protozoa and helminths even more often prove to be the direct cause of animal plagues. In the 1930s there was a marked decrease in the number of black cocks, especially the gray and Scotland grouse, in England. The basic cause was the one-host parasite *Trichostrongylus tenuis*. The birds became infected by scratching up its eggs and larvae while trying to obtain young shoots of grass on infested pastures. The intensity of the infection was extremely high, and 15,000–20,000 parasites could be found in one bird. The extensive spread of the disease was aided by a series of moist years, which increased the survival rate of the worm, and by increased extermination of predators, which markedly limited their sanitary role (Middleton, 1932, 1934).

In Norway the basic cause of the die-off of willow ptarmigan was the coccidium *Eimeria avium,* which is widespread in years when the berry harvest is poor; the parasites infested feeding grounds and watercourses. A long-drawn-out spring, with a delay in the emergence of preserved berries from under the snow, and a hot summer favor the spread of coccidiosis because berries are the basic food of these ptarmigan in spring and autumn. In such years as many as 70 percent of the young and 50 percent of the adult birds fell ill, leading to a serious disease problem. The development of coccidiosis was aided by the extermination of predatory birds which ordinarily would have played a sanitary role (Brinkman, 1927; Closter, 1928; Nordhagen, 1929; and others). Along with the ptarmigan kills there were plagues among capercailzies, hazel hens, and tundra ptarmigan.

Parasitic infections are one of the most important causes of depressions in the number of varying hares throughout almost the entire range of this species (S. P. Naumov, 1947, 1960). They are caused by the flukes *Fasciola* and *Dicrocoelium,* the tapeworms *Mosgovoyia, Cysticercus,* and *Coenurus,* and the nematodes *Synthetocaulus, Trichostrongylus, Nematodirus,* and *Protostrongylus.* The survival of eggs and larvae of these parasites and the abundance of their intermediate hosts are favored by habitats with a high humidity. Therefore, the helminthoses often repeat themselves and prove especially fatal in regions wth abundant precipitation, hard soils, and close groundwaters (Fig. 263). On well-drained ground with deep groundwaters the hares are less likely to die. As a result, the death rate is not solid, even when the helminths are widespread, and the animals survive on the dry islands (survival stations), from which they later resettle. In years of increased moisture, infection and death rates rise among the hares (Fig. 264). Extensive spread of helminthoses demands not only surroundings favorable to the survival of the parasites (humidity) but also a large num-

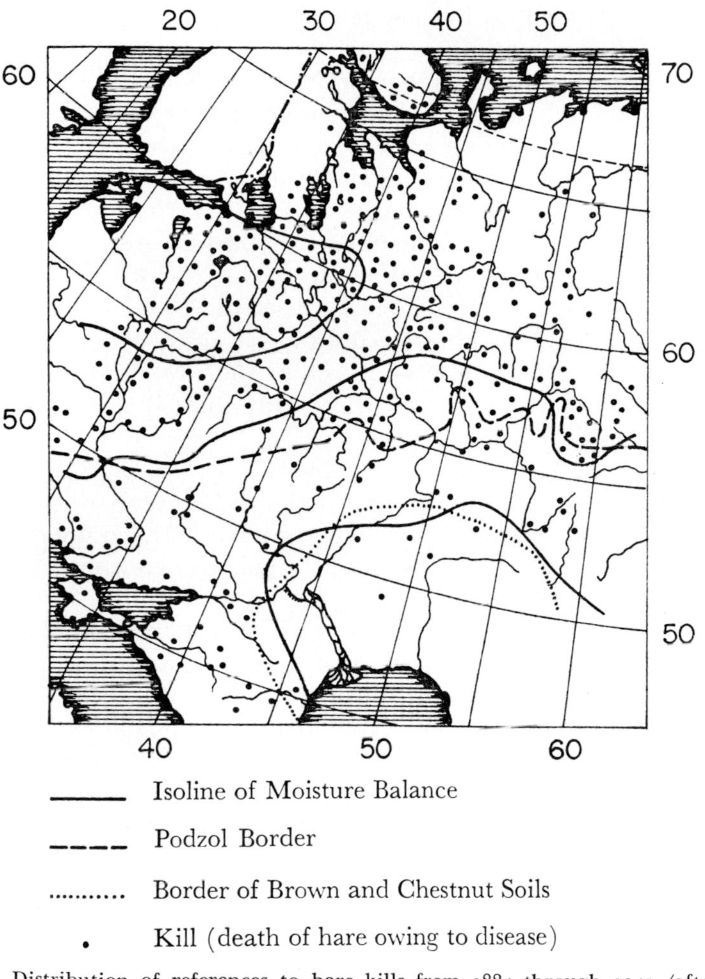

Fig. 263. Distribution of references to hare kills from 1884 through 1941 (after S. P. Naumov, 1947).

ber of hosts to infest the habitation area with worm eggs and larvae. However, in addition to helminthoses, bacterial infections play a large role in changes of abundance of hares; the distribution of these infections during years of high population among hares actually exceeds the distribution of helminthoses. In the northeastern USSR lack of food, which weakens the creatures, is of great importance, while after a drop in rabbit populations predators become important (S. P. Naumov, 1960; M. V. Popov, 1960).

When the nematodes *Skrjabingylus nasicola* or *S. petrovi* parasitize the frontal sinuses of ermines, least weasels, martens, Siberian weasels, pole-

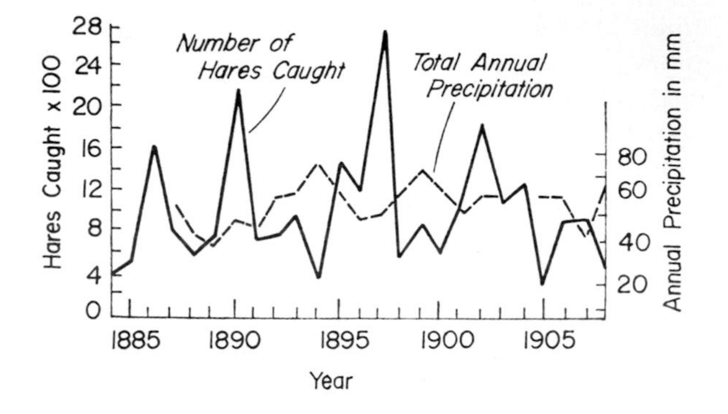

Fig. 264. Changes in abundance (catches) of varying hares at Gatchina as compared with amount of precipitation per year (after S. P. Naumov, 1947).

cats, and minks, they cause changes which may lead to the death of the animals (Fig. 265). Intense helminthoses are known to occur in small forest rodents. Infections with Anguillulata nematode larvae in beetles and locusts are also accompanied by mass death. Here, too, the helminthoses precede or accompany bacterial, fungal, and viral infections. Parasitic worm infection is not only the direct or indirect cause of death but also can influence the fecundity of animals, sometimes causing the "parasitic castration" which is well known among insects.

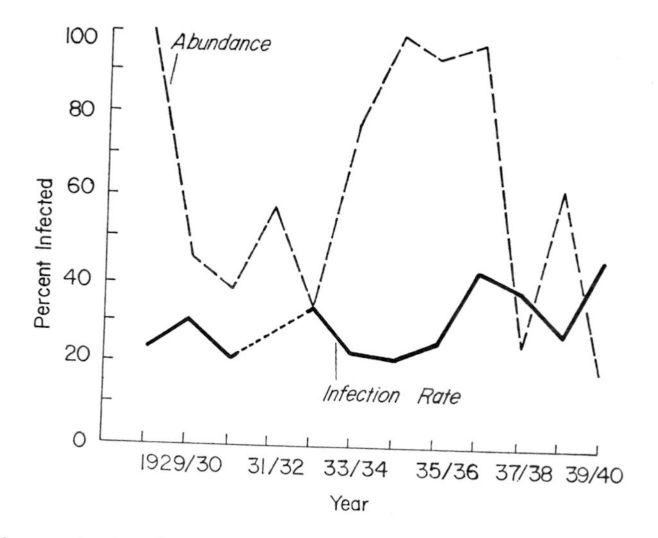

Fig. 265. Changes in abundance (pelts prepared) of the ermine and in infection rate due to skrjabingylosis in the Tatar ASSR (after V. A. Popov, 1947).

The examples cited show the extent to which the simple interactions of the two-member parasite-host system can be complicated by the influence of the external environment, which changes individually the abundance, intensity of reproduction, mobility, and other properties of both partners' ways of life. One need merely imagine clearly the extent to which the complexity of the phenomena is complicated if just one host can be infected by several pathogenic parasites, while each of the latter has several basic or intermediate hosts which live independently and react in various ways to changes in the environment.

In such a complex parasitic system some of the species entering into it not only may be associated by relationships between food and consumer but may be competitors as well. It is for this reason that nature seldom displays the precise numerical relationship between the parasite population level and the abundance of the host species specified in Farr's law and the Lotka-Nicholson biological theories, which always simplify and schematicize the most complex natural phenomena. In nature, however, apparently paradoxical phenomena take place. Thus the caterpillars of the lepidopteran *Etiella zinckenella,* when infected with the parasite *Phanerotoma rjabovi* (Hymenoptera, Braconidae), survive in larger percentages (45.7–22.4 percent) than when not infected (34.0–9.0 percent) if they spend the winter at a temperature of —10° C. The ratios are reversed if the winter is passed at a temperature around 0° C (the infected ones survive in the range 86.4–58.1 percent, noninfected ones in the range 90.9–86.5 percent). The important thing is that wintering at different temperatures changes the numerical ratios because of the variation in cold resistance manifested between parasite and host (Viktorov, 1956).

In many cases the abundance of parasites in the basic (definitive) host depends on the abundance and state of the intermediate hosts. Thus black cocks in the Yaroslavl oblast' are heavily infected with cestodes (primarily *Raillietina urogalli*), which sharply increase the mortality rate of the young birds (Table 71). Distribution of the cestode in fledglings depends on the abundance of the ants which are intermediate hosts (vectors) of *Raillietina*; the ants are more abundant in the heath cock's nesting grounds than they are in unfavorable areas. The disease not only retards growth and development but also doubles the death rate of the young birds (Rykovsky, 1960).

The effectiveness of a parasite's influence may be lowered by hyperparasites, which diminish its abundance and destructiveness with regard to the host population. In Yugoslavia 16 percent of the lepidopteran *Iponomeuta palinella* are infected with parasites, but the basic species of the latter (*Angitta armillata*) is itself 37 percent infected with hyperparasites (Vukassovic, 1933). One important limitation factor of many harm-

TABLE 71. *Survival rate of young black cocks in various biotopes (after Rykovsky, 1960).*

Place of Habitation	*Number of Anthills per Hectare (Myrmicinae)*	*Average Number of Chicks in Brood*	*Average Weight of Young Summer Bird in g*	*Average Number of Cestodes per Bird*	*Percent of Chicks Surviving to End of Summer*
Basically favorable for black cocks	183	8.3	603.7	59.5	38.6
Others less favorable	90	7.1	869.3	14.3	80.2

ful insects, the ichneumon fly *Apanteles melanoscelus,* has 35 species of secondary parasites, so that in the first generation only 25–30 percent of the cocoons develop, and only 1 percent of the second, or winter, generation matures successfully (Bodenheimer, 1958). The complex interrelationships of the beetle *Callosobruchus chinensis* and its parasites are explained by their effects on one another (Fig. 266). Increases in abundance of the basic parasite, the ichneumon fly *Neocatolaccus mamezophagus,* follow increases in abundance of the host. But increases in the abundance of the secondary parasite, *Heterospilus prosopidis,* come only during decreases in the abundance of its competitor, *Neocatolaccus* (Utida, 1955).

Many destroyers and parasites of plants and animals are forerunners which prepare the ground for penetration by new parasites. The Siberian silkworm lives in the larch forests of the mountainous forest steppes of Mongolia. A consequence of the acute weakening of the trees which have lost their leaves is an intense infestation by trunk pests, especially the

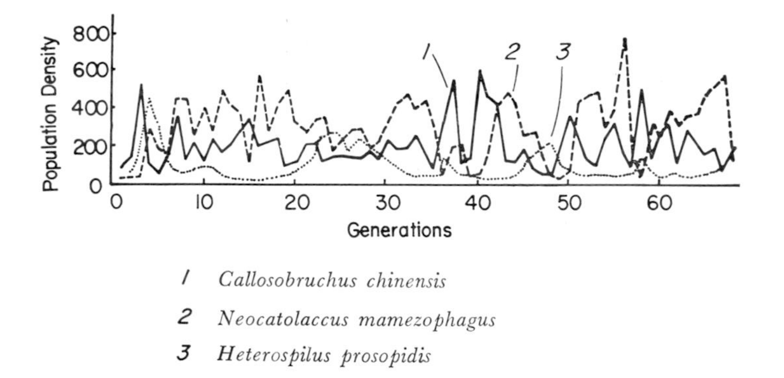

Fig. 266. Fluctuations in population abundance of a host and two of its parasites (after Utida, 1955). Host is the beetle *Callosobruchus chinensis*; parasites are the ichneumon flies *Neocatolaccus mamezophagus* and *Heterospilus prosopidis.*

large larch-bark beetle *Ips subelongatus,* causing mass death to the trees (Grechkin, 1960).

Animals are often disseminators of diseases not only of animals but of plants as well. The onion fly and the onion bulb fly (imago, larvae, and eggs) were infected with bacteria in 50–55 percent of the cases. Among the strains isolated, about one-fourth were pathogenic for plants (dry rot) (Gorlenko, Voronkevich, and Maksimova, 1956). Predatory birds and wheatears, which nest in rodent burrows, as well as many other terrestrial vertebrates, may spread rodent ticks and fleas, the vectors of many infections.

The complexity of parasitic systems, whose individual links comprise not one but several species, aid the stability of parasite abundance. Thus, when the population level of their basic host (small forest rodents) is depressed, the larvae and nymphs of the tick *Ixodes persulcatus* (the vector of tick encephalitis) feed on secondary hosts—birds, shrews, moles, and even lizards (Fig. 267).

The studies of Soviet physicians and biologists have shown that the origin, development, and mechanism of distribution of epizootics of various types depend on the state of the organism and on the abundance and contact possibilities of parasites, vectors, and hosts. All these conditions are associated with the external environment. The frequency of direct

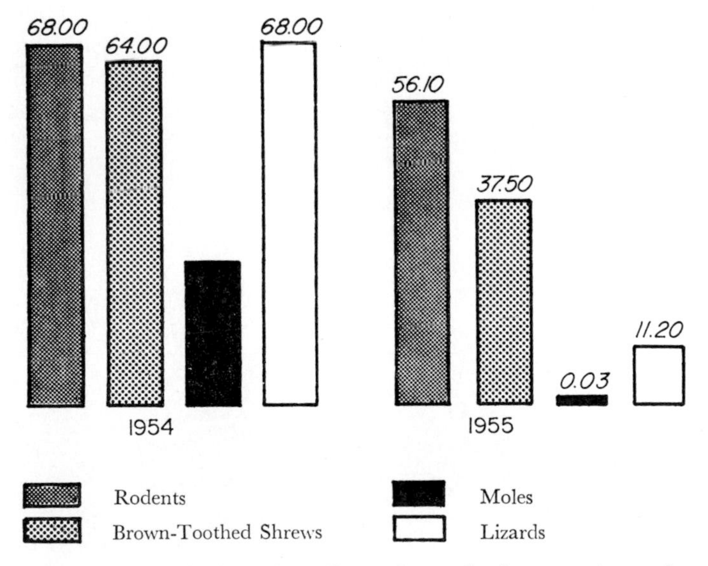

Fig. 267. Percentage of infection of small vertebrates by larvae and nymphs of *Ixodes persulcatus* in the Perm oblast' during years of low (1954) and high (1955) rodent abundance (after Shilova, Mal'kov, Chabovsky, and Meshcheryakova, 1956).

or indirect contact between sick and healthy animals necessary for broad spread of a disease is determined by the magnitude and character of their mobility and their susceptibility to infection. These in turn depend on feeding conditions, weather, and other environmental factors. So in some cases epizootics do not occur when host abundance is high, and in others an epizootic may occur when population density is low; when this happens, the epizootic is usually accompanied by unfavorable conditions of some sort.

Thus parasitic relationships develop in the presence of interaction between all the conditions of existence of hosts, parasites, and vectors, sometimes forming complex parasitic systems of many species. These relationships are only one and not necessarily the main condition of existence of the species entering into the system. This explains the differing significance of parasitic relationships not only in the lives of different species but also among animals of a single species in different regions or at different times.

The indirect role of parasitic relationships is also of great importance. It depends on the different reactions of parasites and hosts to the same factors in the abiotic environment. This changes their numerical relationships and sometimes causes an unexpected result. Examples of such phenomena have been given above. Thus the influence of abiotic factors on the abundance of species is mediated by their connections as members of a parasitic system. As we have seen, the same thing is true for the relationships between predators and prey. This forces us to make a sharp distinction between the concepts of physiological and synecological optima, which in most cases do not correspond (Rubtsov, 1937, 1938).

NATURAL NIDALITY OF DISEASES

One valuable and practical application of the study of parasitism is the theory of natural nidality of diseases of man and domestic animals worked out by Academician Ye. N. Pavlovsky (1946). The diseases of man include anthroponoses, the agents of which circulate only within the boundaries of human society (measles, chicken pox, scarlatina, influenza, and many others), and the zoonoses, which infect both man and certain species of animals (brucellosis, tularemia, rabies, seasonal encephalitides, rickettsioses, leishmanioses, and many others). Malaria and murine typhus represent anthroponoses transmitted by arthropod vectors.

The zoonoses are associated with man to various degrees. Some of them have almost become anthroponoses, having lost most of their connections

with animals. Thus visceral leishmaniosis, a serious disease caused by the protozoon *Leishmania donovani,* is encountered almost exclusively in human settlements. Its sister disease, cutaneous leishmaniosis or Pendinsk sore, is a typical zoonosis. Desert rodents (gerbils and the thin-fingered suslik) are its reservoir, and *Phlebotomus* is the vector. However, nidi of cutaneous leishmaniosis have also been observed in several cities; there the agent is transmitted by sandflies directly from one sick human being to another, rodents taking no part in the process (Latyshev, 1948).

Zoonoses exist only where the hosts of the agent and its vectors live and where uninterrupted contact between the parasitic agent and the susceptible host (through a vector, if necessary) maintains constant reproduction of the parasite. Pavlovsky called such areas natural nidi of disease. The natural nidus is an evolutionarily developed complex of host, vector, and agent species in a given area acting as a part of the local association. As a result of development on the same spot, all of these species are adapted to one another.

The existence of a natural nidus requires not only cohabitation of the members of such a complex but also a sufficient, stable abundance level among the reservoirs and vectors of the disease. The disease must be able to be passed on uninterruptedly from one susceptible host to another; i.e., it must be reproduced in nature as part of a system of close nutritional associations. The latter also lie at the basis of the natural nidus. "Analysis of all these factors in the natural nidality of transmissible diseases is necessary for an understanding of their genesis and the conditions of existence bringing about their outbreak. In the same way this analysis is necessary for the construction of radical systems of measures aimed at rendering harmless or completely liquidating such nidi" (Pavlovsky, 1946). A high density of population and stability of numbers are characteristic of such basic carriers of plague as the susliks, marmots, and gerbils. The significance of the less common species (jerboas, hamsters) and of animals with variable abundance is lower. In nidi of tularemia, rickettsioses, and relapsing fever the abundance of warm-blooded hosts (small rodents) is variable, while populations of ticks, the basic reservoirs of these agents in nature, are stable.

In large, uninterrupted colonies of infection carriers the disease may exist in the form of acute epizootics which move sequentially throughout the territory and die down as the abundance level of the hosts drops off in a given area, as the immunity level rises, and as other obstacles to the progress of the disease are erected. Its progress is possible only as it can spread into new territories as yet unattacked by the disease. As the disease dies out in one area, where its penetration is no longer possible, the epizootic flares up in new spots. In the area where it died out the animal

population gradually re-establishes itself, and after a certain interval of time the disease can again develop. Such natural nidi with "wandering" infections have been described for the enormous, uniform steppe and desert valleys (Rogozin, Fedorov, and Fenyuk, 1955).

When animals are distributed unevenly in insular or ribbon-type colonies, the stable nidi of infection, in which the agent is maintained permanently, tend to be located in those areas with a high and stable population of reservoirs and vectors. These spots have been given the name "elementary nidi of infection" (N. P. Naumov, 1955). Tularemia is preserved in the southern steppe zone and in the forest steppes on meadow plots with a constantly high level of pasture ticks (Olsuf'yev and Rudnev, 1960). Nidi of tick encephalitis are found in the deeper and moister parts of the broad-leafed forests, where the abundance level of *Ixodes persulcatus* ticks and the populations of their hosts (hoofed animals, small rodents, and insectivores) are high. Finally, the elementary nidi of plague in the deserts of Kazakhstan either are sandy plots in the clayey desert (where the large gerbils which carry plague survive even under the most unfavorable conditions and live in very complex burrows with a microclimate favorable for the fleas which are its vectors) or are the places where these same gerbils gather on their permanent wandering routes in the ribbon-type settlements (Naumov *et al.*, 1960; Dubyansky, 1961).

Reproduction of the agent and continuity of its transmission from one host to another are in many cases guaranteed by the "polyhostality" and "polyvectoricity" of the elementary nidus. Under these terms is subsumed participation in the agent's circulation in the nidus by several species of hosts and several species of vectors (Fig. 268). Multiplicity of hosts and vectors plays an important role in the stability of tularemia and plague nidi. Polyhostality plays an important role in the existence of natural nidi of transmissible zoonoses transmitted from a sick animal to a healthy one directly or by infestation of the surrounding environment. Such are the leptospiroses, listerioses, and certain other infections dangerous to man and domestic animals. But even in polyhostal nidi we usually distinguish the basic species which support the nidus, for it is in such populations that the agent of the disease is preserved continuously and later transmitted to populations of the other species. The participation of the latter strengthens the nidus and makes it more stable.

Ye. N. Pavlovsky's concept of the natural nidality of diseases put the campaign against them on a scientific basis. Our country's successes in this area have been great. Malaria has been eliminated as an epidemic disease in the USSR, and the task of eradicating tularemia has been established. Major successes in eliminating these diseases have been attained by as-

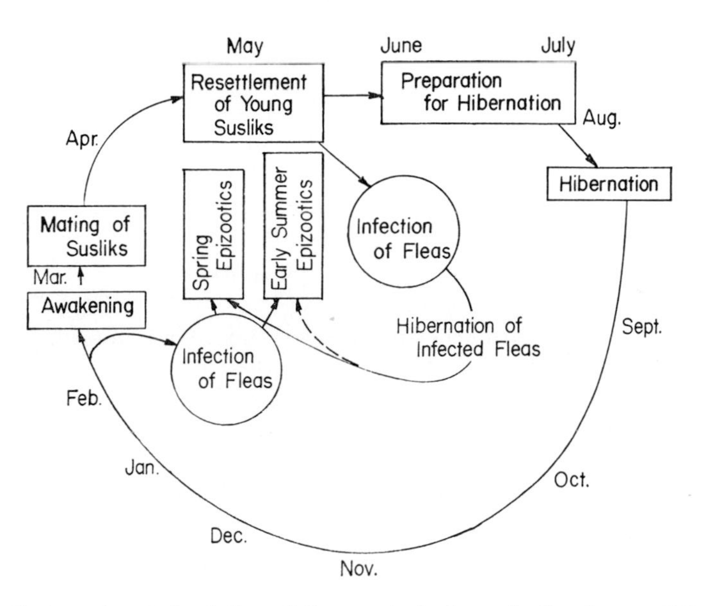

Fig. 268. Basic paths of circulation of the agent of plague in the steppe natural nidus (after Fenyuk, from Pavlovsky, 1946).

siduous planned extermination of their carriers and vectors. The success of the fight was due primarily to a proper understanding of the biological bases of nidality and its evolutionary roots. In addition to extermination of the disease carriers, active reconstruction of the landscape for agricultural and other purposes was used as an important control measure and has led to interruption of the contact between parasites and hosts on the one hand and between parasites and man on the other. In some cases this has brought about self-eradication of natural nidi of disease. But sometimes the changes which are useful and necessary for agriculture or animal husbandry aid in the rise or entrenchment of natural nidi. In the protective forest strips of the steppe belt survival areas of mouselike rodents and ticks have arisen, and the appearance of bodies of water has increased the numbers of mosquitoes, etc. This demands a step-up in the campaign against carriers and vectors of disease and in the development of new measures designed for the new conditions. The campaign against carriers of disease and man-made changes in the landscape, along with special prophylactic measures (vaccination of the populace and domestic animals against diseases), provide the country with sanitary conditions. At the same time our literature is beginning to take up the question of active campaigns against diseases of the useful wild species.

BIOLOGICAL METHOD OF CONTROLLING FARM AND FOREST PESTS

The second practical application of the study of parasitism is the biological method of controlling farm and forest pests. It represents the use of such enemies and parasites as carnivorous vertebrates, predatory invertebrates, parasitic insects, and infectious fungi, bacteria, and viruses to keep down the abundance of pests. The use of insectivorous mammals, birds, and insects (especially ants) to retard the reproduction of harmful insects is widely known. Many countries have experienced cases in which reproduction became more frequent and massive and losses grew precipitously when the abundance and number of species of natural enemies were reduced. We have many reports on the ichneumon flies (19 families of the order Hymenoptera) and the parasitic flies of the family Tachinidae. Of the ichneumon flies, the genera *Trichogramma* and *Telenomus* are especially important. Among the predatory insects the beetles, especially the families Coccinellidae and Carabidae, are very important.

The first step in controlling pests biologically is to acclimatize their enemies or parasitic entomophages. This method was used with great effect in Hawaii and the Fiji Islands, with their unusually regular and favorable climates. However, even here complete extermination of parasites was not achieved. The parasite *Scolia manilae,* which was imported into Hawaii to control the sugarcane beetle *Anomala orientalis,* effectively controlled its numbers. But the pest is still present in many areas to this day, though in lower numbers. I. D. Tothill (1930) reduced the damage due to the coconut moth *Levunana viridescens* by importing its parasites into the Fiji Islands, but the pest remains common there. Lantana still flourishes in all moist areas in the Hawaiian Islands, although several animals which destroy the plant have been imported; it has disappeared completely only on the dry parts of the islands.

Thus acclimatization of enemies and parasites does not, as a rule, lead to the complete destruction of the pests. Sometimes not even a small effect can be attained. This has made it necessary to support acclimatization in the wild by breeding entomophages in special laboratories and releasing them at the necessary time in areas threatened by pests. Bacterial campaigns have the same goal. The bacterial struggle consists in releasing batches of insects infected with fungi, bacteria, or viruses wherever the pests have multiplied, so as to spread diseases among them. The effectiveness of parasites and predators in controlling pests depends on their specific (species-related) properties (Fig. 269), state, abundance, and activities in nature.

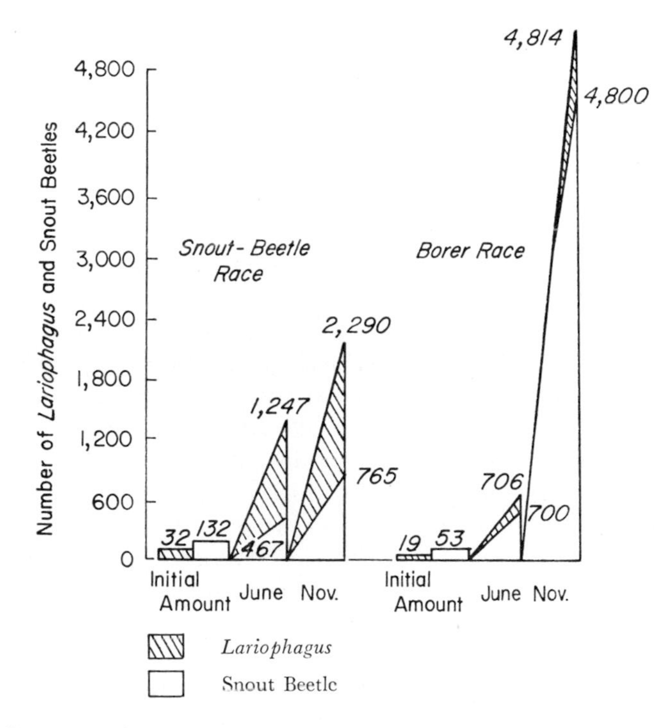

Fig. 269. Reconstruction of the dynamics of reproduction of the Ambar snout beetle under the influence of various biological races of its parasite, *Lariophagus* (after Telenga, 1953).

The effectiveness of entomophages may be increased considerably by selection and hybridization of the natural forms. This is how the modern race of aphid ichneumon flies was obtained—through selection and crossing of three forms from distant areas; it is now widespread throughout the range of aphids. *Encarsia formosa,* which is used effectively against the whitewing *Trialeurodes vaporariorum,* was obtained from one narrowly local form of this parasite first observed in 1926 in Hertfordshire. *Prospaltella perniciosi,* which is very useful in controlling the California scale insect, was obtained in the same way. As a result of the spread of such effective, but at first narrowly local, forms throughout the whole range of the pest, a very satisfactory effect is achieved in most cases (Rubtsov, 1960).

Ignoring the complex interaction of factors determining the influence of the parasite on the host has led to exaggeration of the significance of numerical ratios. This has found expression in the so-called "flood theory," according to which a large number of parasites is released on plots infested by pests, regardless of the weather and other conditions. Practice

has shown the error of this method. For *Trichogramma*, for example, the release of 100,000–300,000 individuals per hectare proved excessive. Good results were obtained not by "flooding" fields with parasites but by selecting races, raising them properly, and estimating correctly their activity in the wild. Thus the effectiveness of infecting the pest's eggs with *Trichogramma* depends on the degree to which races can adapt to local conditions and the conditions of laboratory breeding. In collective farm laboratories the parasite was raised under strictly standardized conditions (25–26° C and 75–80 percent relative humidity), but in the wild it encountered variable temperatures and humidities to which it was not adapted. As a result, its resistance dropped, and so did its activity and its effect on the pests. These failures were eliminated by breeding *Trichogramma* in the laboratory under comparable conditions (Kovaleva, 1951).

CHAPTER 18

Biocenoses: Dynamics and Productivity

GENERAL CHARACTERIZATION

The animal, plant, and microbe population of the earth lives as a unified whole. Its individual parts (populations of various species) are closely and directly associated with one another if they live on one territory and enter into one biocenose, or they are indirectly, distantly, and weakly associated if they are territorially dispersed. The character and tightness of the connections depend on the characteristics of the species and the spatial relationships of their populations, which occupy the same territory, neighboring plots, or far-distant portions of the surface.

The living population of the earth can be broken down naturally into a series of different-sized territorial groupings (topical groups), or biocenoses, representing associations of several (and more often, many) species. Thus the small associations are parts of larger ones. Each biocenose in turn has its hierarchical structure (stratification) and consists of synoecous or ecologically similar groupings of species occupying a given part of the whole association's area of habitation.

The smallest associations, which are not attached to territories, include complexes of parasites in the same host or in one of its individual organs (parasitocenoses). The same feature is characteristic of aggregations of microorganisms and animals on one plant. The cohabitants of one burrow (burrow biocenoses), the inhabitants of caves and tree holes, groups of sessile plants and animals (periphyton) infesting marine installations, and other groupings occupying similar microbiotopes are the next link. Associations of individual tiers, or synoeces, are combinations of species in a tighter relationship than species in the other tiers. Such are the plants and

animals inhabiting the soil (edaphon), the tier of mosses, the tier of grassy plants and shrubbery, and the tier of woody vegetation (dendrophon). In bodies of water these hierarchical unions include the population of the ground (infauna) and the inhabitants of its surface (epifauna), which together make up the bottom fauna, or benthos, as well as the inhabitants of the deep waters (nekton, plankton).

Structure of the Living Cover of Earth and the Hierarchy of Territorial Associations (Biocenoses)

Territorial Association	*Structure*		
	Tier associations (synusia, plankton, nekton, benthos, dendrophon, etc.)	Microbiotopic associations (caves, burrows, hollow trees, "overgrowths")	Parasitocenoses (populations of parasites in one host organism)
Geomerid—the population of the biosphere			
Associations (populations) of continents, oceans, seas			
Zonal associations—populations of natural zones and vertical belts			
Associations of landscape-geographic regions (landscapes of order I, according to Berg)			
Associations of biotopes (after Hesse) or of landscapes of order II (after Berg)			
Associations of facies—the population of elements of the mesorelief (after Hesse)			

Acting as fairly dependent component parts, they form an integral and more autonomous territorial association in the biotope (according to Hesse's terminology). Examples might be fir groves, pine groves, oak woods, couch-grass meadows, pitted sands, dunes, etc. They make up even larger units, the landscape associations of order I (pine forests, leafy forests, river floodplains, clayey desert, sandy desert, etc.). The highly mobile animal groups unite the landscape associations into even larger ones, the zonal associations, which are aggregations of associations characterizing one natural (geographic) zone or another. The latter may be combined into associations of individual continents, oceans, and seas and, finally, into the population of the biosphere, the geomerid.

All of these associations are not accidental combinations of species shifted by external conditions (the "sieve" of environment) but are evolutionarily compounded systems of populations which have been brought together by nutritional and spatial relationships. They arose on the basis of a single turnover of matter which gave rise to the mutual adaptations of the species. The result of their connections is a definite associational structure, the greater or lesser stability of which is maintained by mutual

regulation of the abundance and relationships of the species. Practical measures for controlling agricultural pests and carriers of diseases of man and farm animals, for game management, and for acclimatization of useful species are all founded on these dependencies. The biogenic turnover of matter unites the different associations into an integral living population of the earth.

The existence of mutual regulation makes it possible for biocenoses to be regarded as "organisms of a higher order," or "superorganisms," and their regular changes (birth, development, and death) as similar to those undergone by any living creature. Actually, associations, like organisms, are systems capable of internal regulation. This constitutes their similarity, but it is only external, since these systems are actually completely different in essence.

The organism is a highly integrated system of parts—cells, tissues, organs—the action of which is subordinated to a complex and multifaceted coordination with the aid of humoral and nervous factors. Such a system not only is an integral unity of parts which cannot exist outside the whole but also is a unity which can reproduce and regenerate itself if its integrity is disturbed. A species, as we have seen, also lives as a whole, and the intraspecies connections furnish this unity primarily with the aid of intraspecies organization and so-called congruence—the adaptations which members of one species display toward another. However, the tightness of the connections within individual populations or biological groups into which the species is divided is immeasurably less, while autonomy is greater, than the connection and autonomy of the individual parts of an organism. The species is a system which is incomparably more "friable" and mobile. This is true to an even greater degree for a system of mutually connected populations of various species, the association or biocenose.

At the basis of associations lie interspecies relationships which differ markedly from intraspecies relationships and even more so from the relationships between the parts of a single organism. This is true because populations of different species are autonomous, and only in rare cases (the monophages, for example) does their existence depend unconditionally on a partner in the association. In the overwhelming majority of other examples, however, interspecies connections, unlike the connections within a species (and especially within an organism), have no obligatory significance. Species may even enter into new relationships and interrupt or change the character of their relationships for short periods of time. All this makes the association a formation of a different type, with short-term, quickly changing, and generally relatively unstable connections among the individual species. The organism is usually distinctly limited in space; the borders of individual biocenoses, however, are difficult or impossible

to establish. The population of any forest or meadow plot (strictly forest or strictly meadow species) is different, while many other species may be common to both biotopes. Daily and seasonal migrations of animals complicate even further our attempts to delineate the boundaries of associations. In some biocenoses the wanderers predominate over the sessile species (as in the tundra and high mountains), and their reproduction, survival, and abundance depend not only on conditions at the mating grounds (summer habitation areas) but also on conditions at the winter stations. Mass emigrations of animals (locusts, butterflies and dragonflies, birds, mammals, etc.) sometimes go far beyond the boundaries of individual biocenoses. This kind of exchange among the populations of associations has no analog in the structure and life of organisms.

Although the basis of the association is generally formed by species of one floristic and faunistic complex, there is always an admixture of species of other origins which enter the structure of the association from time to time. This mixture causes the association's heterogeneity and heterochronicity, which likewise have no parallel in the structure of an organism. In the forest associations of the middle belt of the European USSR, the upper tree tier is sometimes formed from fir and pine, with which the taiga insects, birds (woodpeckers, crossbills, *Parus, Tetrao urogallus,* and several others), and mammals (flying squirrels, red, red-gray, and economist voles, etc.) are associated. But the second wood and shrub tier may be dominated by filberts, limes, and honeysuckles, with which insects, birds, and mammals of the oak groves are associated.

Finally, in the associations there is no analog with the individual organism's development or ontogenesis, which begins with birth and ends with death. The specific item which distinguishes the association from the individual is the extraordinary adaptability of the system, its ability to take in new species easily and allow others to disappear without dying off, all without destroying the integrity of the biocenose. This is determined only by the presence or absence of free territory, food, and natural physicochemical conditions.

"Saturated" and "nonsaturated" biocenoses (with free ecological niches) are distinguished according to the degree to which their existing vital resources are being used. This terminology is imprecise and conditional. An ecological niche is primarily a definite position of a species in the food chain, on which its place in the biocenose depends and on which the character and tension of its association with its partners in the association depend. It is also characterized by the presence of free resources, i.e., supplies of food, water, gases, and acceptable microclimates not being used by other species. The niche does not exist independently of the species occupying it or independently of that species's cohabitants. Hence the num-

ber of niches in the association depends not on the initial variety of nature in a given area but on the content of the association. The depth and extent to which the biocenose uses the resources of inorganic nature are associated with the latter. Therefore, the biological capacity of the biotope is not only of variable magnitude, but the figure also depends on its living cover. It changes at various seasons and over periods of several years and takes on a different character whenever the affiliates of the association are replaced for whatever reason.

The entrance of a new species into the association and its naturalization therein depend on the extent to which the species already in the association are using the vital resources which it will need. "Nonsaturated" biotopes are what we call associations which are not complex in their structure and in which the few kinds of resources in the food chain are poorly exploited or remain free. The success with which a species penetrates a complex, "saturated" association usually depends on its ability to compete with the aborigines which occupy the ecological niches closest to the newcomer. In this case one species is replaced by another, or else the niche is broken down into two niches, further complicating the association. A change in the species affiliates of the association does not, as a rule, lead to its destruction and does not necessarily lead even to a change in its basic structure. The inclusion of a new species in the food chain or the replacement of a species which has disappeared is not, in many cases, expressed substantially through the turnover of matter.

A natural association or biocenose is an aggregation of autonomous, though in many ways interrelated, species. The autonomy is maintained by the independence of the distribution of individual species which enter into the same associations but which do not have corresponding ranges. Associated with this are (a) a lack of distinct borders between the individual biocenoses, the "mutual penetration" of biocenoses into one another, or the gradual transformation of one into another in a complex mosaic landscape; (b) the existence of populations which enter simultaneously into several biocenoses; (c) seasonal and daily aspects of the biocenoses, which are sometimes poorly connected with one another; (d) heterogeneity and heterochronicity of associations, as expressed in the fact that the individual species or their complexes arose and developed in different places and entered into the makeup of the association at different times.

None of these properties has any parallel with the organism, so we are forced to regard living associations as unique and very mobile systems. They are evolutionarily compounded forms into which the earth's living population is organized and which arose in the course of development of the biological turnover of matter. Territorial associations or biocenoses are aggregations of species whose interrelations are mechanisms for estab-

lishing this turnover under concrete, natural conditions. It is for this very reason that different-sized associations are subordinated to one another (associations at the continental level, natural-zone associations, landscape (biotope) associations, microlandscape associations, individual-tier associations, associations within one host organism), while those of equal size are interconnected and overlap. All these constitute the various territorial variants or individual links in the turnover of matter; the streams, rivers, and tributaries between them, now blending, now diverging, make up the integral flow of matter and energy in the biosphere.

BASIC RELATIONSHIPS BETWEEN SPECIES IN AN ASSOCIATION

Nutritional and spatial connections are the basic relationships of an association. The former lie at the basis of the food chain, the individual links of which comprise the different stages in the biogenic turnover of matter. One association generally has several parallel food chains; each of them consists of a small number of food-associated relationships among a number of species which seldom exceeds ten. Some species take part in different food chains and connect them in so-called "food cycles" (Fig. 270). The possibility of lengthening the chain is limited by two conditions; these are the so-called "rule of individual size" and the "rule of pyramids of numbers." By the rule of individual size we understand the sequential growth of consumers as compared with the size of those on which they feed. Thus in the sea phytoplankton consists of organisms of very small dimensions. The plankton animals which live on them are larger, while the peaceful fishes which feed on them are larger than their prey but smaller than their enemies, the predatory fishes.

Growth in the number of links in the food chain is hindered by the motor possibilities of the organism, which diminish with growth in size and depend on the solidity of its skeleton and the strength of its muscular system. The necessity for expanding the feeding grounds of the individual, also limited by the drop in motor qualities, is another important factor. But the dimensions of the animal do not depend only on food relationships: the ratio of body mass to body surface plays an important role in water and heat exchange (Rubner's rule), and the size of the body may serve as protection against enemies. Thus there are many exceptions to the rule of individual size. Parasites are always smaller than their hosts; the same is true of "semi-parasitic feeding" in the case of the plant-eating insects, hoofed animals, certain rodents, etc. Finally, those predators which

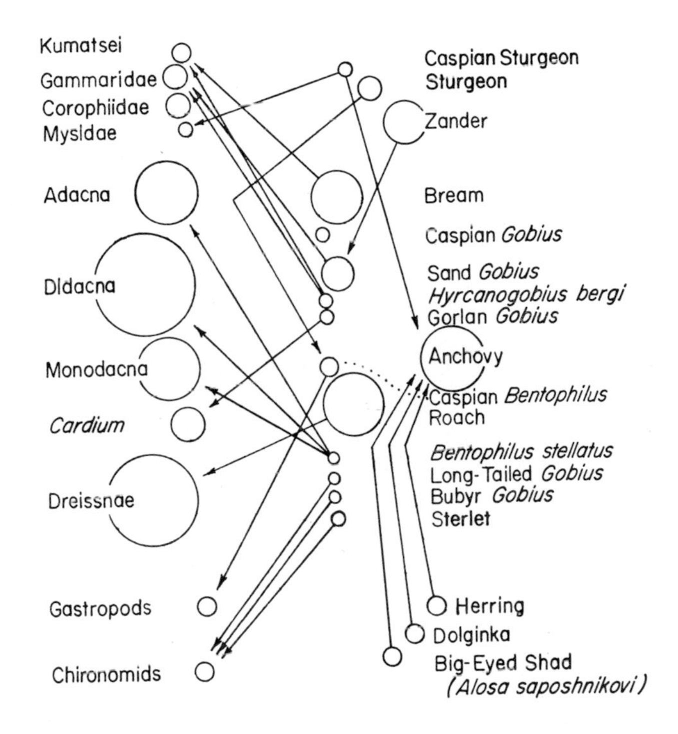

Fig. 270. Important food relationships among fishes in the northern Caspian (after Shorygin, 1952). Arrows show direction from predator to victim; only groups accounting for at least 25 percent of a species' food are shown. ["Dolginka" and "herring" are different subspecies of *Caspialosa brashnikovi*, according to Naumov (*in lit.*).—ED.]

hunt in groups or which wait and spring unexpectedly may cope with even very large prey.

By the rule of pyramids of numbers we mean the diminishing biomass of the consumer, usually accompanied by a drop in the number of its individuals, as compared with the mass and numbers of the individuals on which it feeds. K. G. Semper's (1881) assertion that the biomass of each succeeding link in the food chain is approximately one-tenth of that preceding it is now being tested. In some cases the ratios are actually rather close (Fig. 271), but exceptions are frequent. The sequential diminution in the biomass of the consumer as compared with that of the consumed is associated with the fact that the food goes mainly to supply energy demands, with only a small portion remaining for growth and development. The ratio of "constructive" exchange to "energetic" exchange is different for poikilothermic and homoiothermic animals. Small mammals expend 10–30 times more matter and energy on vital activities than they use for

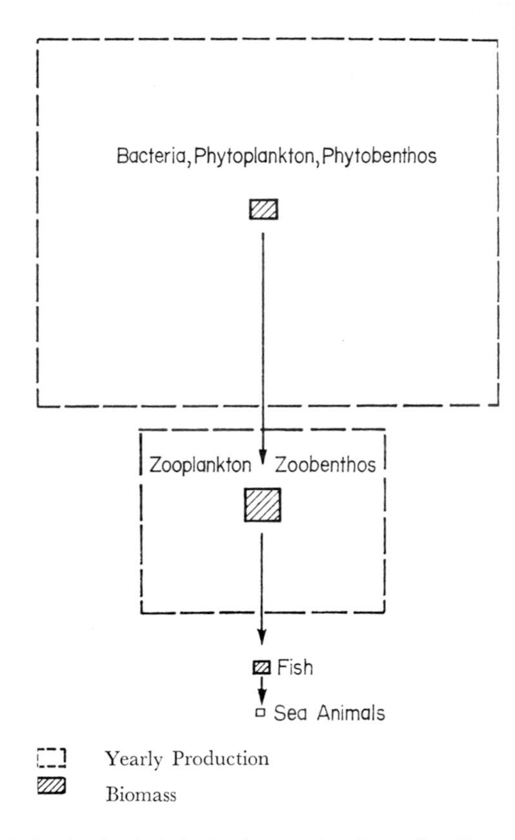

Fig. 271. Basic links in the food chain in the Caspian Sea (after Bruyevich, 1939). Arrows show route from food to consumer.

growth. The energy expenditures of fish are three to four times greater, while that of chironomid larvae is seven to eight times higher, than that needed for growth. Any changes depend on the state and age of the organism and especially on the surrounding environment. Diminution of the biomass of each succeeding link in the food chain is also limited by the number of links.

In addition to direct food connections, indirect ones (competition, boarding) also exist, and these play an important, though different, role in the association. Similar species of one faunistic complex usually compete less with one another, since in the course of evolution they have split into different food niches or different areas. In the former case they may live together; in the latter they live in different associations. Thus destroyers of pine trunks attack the tree and take it over in a definite sequence. The larvae of the bark-eating *Ips typographus* appear first and settle in the

lower part of the trunk; the snout beetle *Pissodes harcinae* and the bark-eating *Polygraphus polygraphus* attack the middle part. As the tree weakens, new pests appear, each of which occupies a specific zone and rarely shares it with others. Insects living on the dandelion do not compete with one another, since they feed on different parts of the plant (Gilyarov, 1940). In the flocks of titmice which wander about our forests in winter, each species hunts in its own food area (Fig. 272). In the Amu Darya River species of one faunistic complex of fishes (the Turkestan pseudoscaphirinx and *Aspiolucius*) compete for food weakly, while between them and representatives of the Ponto-Caspian complex—the bream, the sichel, and the rudd—competition is considerable (Fig. 273).

Permanent food connections exist between the basic food and its consumers. The connections between consumers and seasonal and accidental food are transient and temporary. The former group makes up the nucleus of the biocenose and takes the name of builders or dominants. Spatial connections and relationships are very important in associations. They

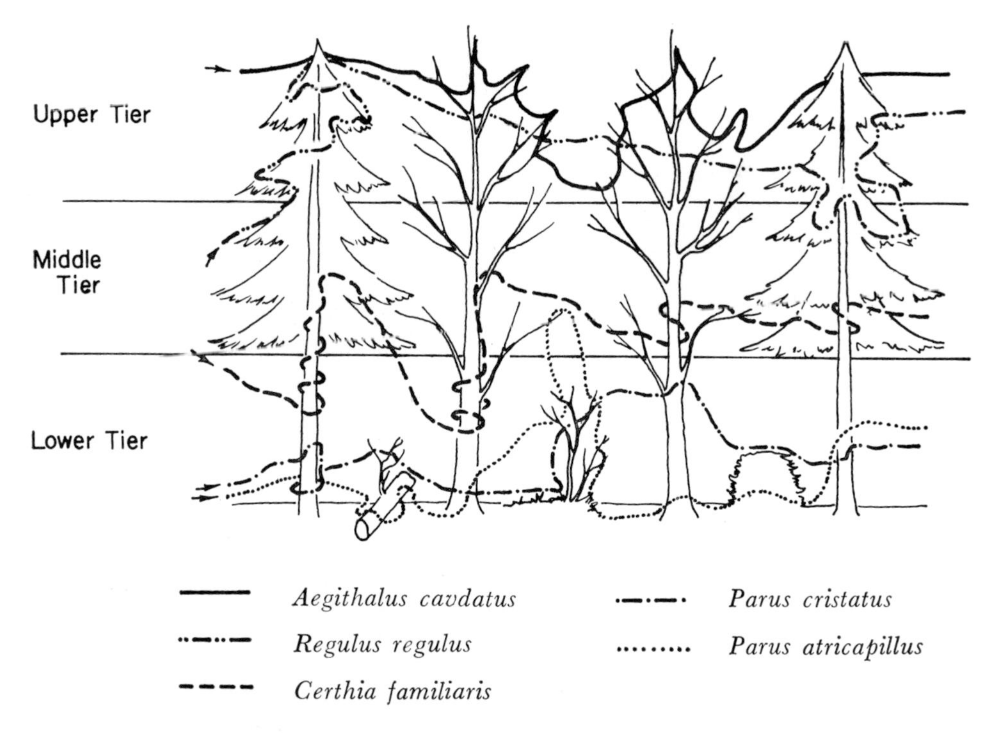

Fig. 272. Tier distribution of different species of birds in a mixed flock in the forest of the Zvenigorod Biostation of Moscow State University on 15 November 1957 (after Chang-Ho Sheng, 1960).

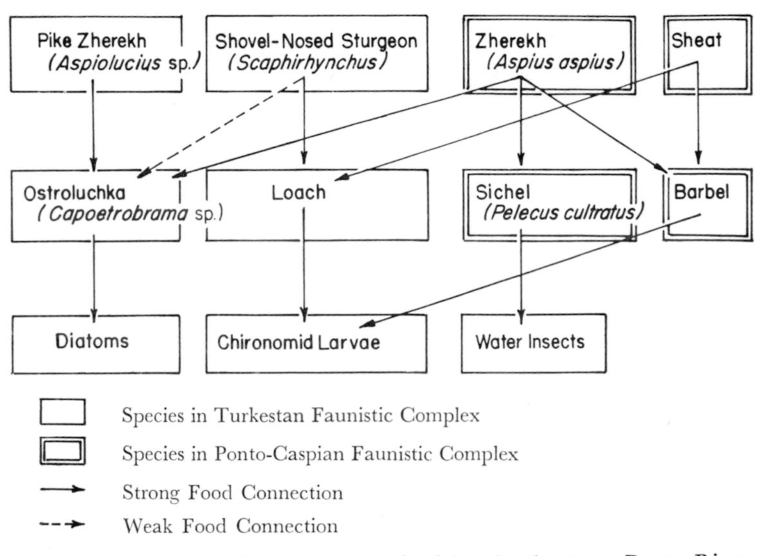

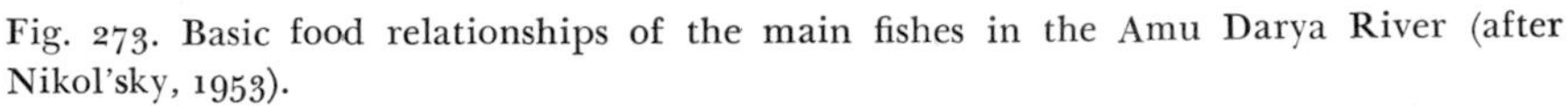

Fig. 273. Basic food relationships of the main fishes in the Amu Darya River (after Nikol'sky, 1953).

are expressed in different forms of synoecy and arise when territories and the refuges and shelters thereon are occupied by two species. When this happens, a definite degree of independent hierarchical union, involving animals, plants, and microorganisms, takes place, thus determining the stratification of the association.

Stratification is most distinct in the forest, but it also exists in tundra, steppe, desert, and aquatic associations. The tier of the association usually consists of several groups of species with similar relationships to the physico-chemical conditions of existence (hydrothermic regime, illumination, etc.). The basis of the tier is the plant synoecy, by which term we mean "groups of species growing up together, belonging to the same or closely related species of life forms, having identical forms of development, and creating a definite vital environment" (Alekhin, 1944). Synoecous combinations" of animals are associated with them. These groups of ecologically close species may form a food chain, but this is not obligatory. Some of the plants and animals occupy several tiers at the same time. Among animals this is brought about by their daily migrations and their use of different tiers during different stages of development or at different seasons.

In terrestrial associations the lower tier is the edaphon, or the association of soil inhabitants, which consists of species living permanently in the soil —the geobionts—plus those species living there temporarily. Part of the

animals feed in the soil and have a great effect on its chemistry, while others spend only their rest periods there (Gilyarov, 1949). The number of species and individuals associated with the soil exceeds the number of those living on the surface. The soil population is numerous in places where the soil is rich in organic matter and has a great effect on soil-building processes. The animals dig up the soil, changing its structure, porosity, aeration, and hydrothermic regime, mix up the different layers, and bring in various organic substances which aid in the enrichment of the soil microflora and speed up microbial decomposition. Terrestrial animals also affect soil formation by changing the plant cover and the upper soil layers. Animals play an especially great role in the initial phases of soil-building. The primary soils on rocks are settled by animals just as densely as the developed soils. Motile, nondigging forms predominate here; they are resistant to desiccation and feed primarily on detritus. They process plants, mix up their remains with masses of dung and mineral particles, and create water-absorbing aggregations (Stebayev, 1958).

The significance of animals is preserved in the development of soils. Earthworms, ants, termites, and deerflies, by piercing the ground with their passageways, aid in distributing humus, raise the level of porosity and aeration, and change the hydrothermic regime of the soil. All this increases soil-building and aids in maturing the soil. At the same time certain animals, by carrying off salt-bearing ground, aid in salting the surface layers of the soil. The digging activities of rodents are similar and just as important in soil-building.

Above the soil tier in the dry-land associations there are generally several terrestrial tiers. The dead litter, with its rich population of invertebrates and small vertebrates; the tier of moss and lichen cover in the forests and sometimes the meadows; the grassy tier; and the shrub and bush tiers in the steppes and deserts—all are of this type. In the desert one might add the shrubbery and one or two of the tree-height tiers to this one. In each of them live specific species, but the highly motile animals (such as the birds) may use several tiers, nesting on the ground but feeding in the trees and bushes (chiffchaffs) or vice versa (thrushes, *Erithacus*).

The same hierarchical system of associations with more or less distinct differences in population at the different levels may be observed in bodies of water. Here, too, the abundance of animals and plants in each tier depends on the amount of food available and favorability of the microclimate. Although individual tiers in the terrestrial and aquatic associations each have their own populations, the tiers are closely connected to one another, and changes in one of them must of necessity affect the whole association.

Within most associations one usually observes a certain degree of mosaic structure, in which the small individual plots differ in the character of

their soils and microclimates. They tend to be associated with individual elements of the mesorelief, are occupied by different vegetations, and are settled by different animals. Such a mosaic of microbiotopes is characteristic of complex steppes and semideserts, sandy deserts, many types of tundra, and, on a larger scale, of the forest-steppe, forest-tundra, and other transitional zones. By increasing the variety of conditions, mosaics have a favorable influence, and the animal populations of such areas are always richer than on monotonous and uniform stretches. For the same reason the animal world is richer on the borders of two biotopes (water and dry land, forest and meadow, etc.) settled by different associations. In such areas species appear which live primarily on only the border between the two biotopes. Examples include such species of birds as the common bunting, the forest *Anthus*, certain warblers, etc.

Along the slopes of elevations—in gorges, ravines, and river valleys—the complexity may take on the character of microbelts. Here a definite altitude, with its characteristic soil and ground conditions, microclimate, and vegetation, is host to typical animals living in the microbelt. The conditions of existence are different in the various parts of the intertidal region, where individual zones are easy to delineate. The usual scheme is as follows (after G. L. Clarke, 1954):

Littorina zone	Maximal tide
Balanus zone	Tidal zone
Laminaria zone	Ebb tide

The same belted arrangement may be observed in those parts of streams and rivers where different species of benthos, plankton, and nekton live. Such narrowly distributed species are delicately adapted to local conditions. Thus for *Planaria alpina,* an inhabitant of cold upper waters of streams and rivers which lives in the 0–10° C temperature range, conditions are optimal between 3 and 7° C. Raising the environmental temperature to 10° C increases feeding less than oxygen consumption, and the animal starves. For *P. gonocephala,* which lives along the lower courses of rivers and streams at a temperature of 12–20° C, conditions are optimal at 15° C. If the temperature is lowered, food consumption drops off faster than oxidation, which also leads to starvation.

Even smaller groupings are found in the biocenoses of caves, hollow trees, and burrows, which sometimes include hundreds of species of invertebrates alone. A varied and abundant population of crustaceans, millipedes, insects, arachnids, worms, and others lives in the burrows of many desert rodents, where there is a rather high humidity and a stable tem-

perature. There, too, we find the refuges of the green toads, which are sometimes found 10 km from the nearest permanent body of water, ambystomids (in Mexico), many species of lizards and snakes, turtles, small mammals, birds, and in Texas even the lungless salamanders of the superfamily Plathodontinae, which breathe the permanently moist air through their skin surface. The total number of species of such tenants and codwellers approaches 300 in some cases (Vlasov, 1937). In the beaver burrows and dams near Voronezh, where even in winter the temperature gets no lower than 0–5° C, 17 species of vertebrate and 32 of invertebrate animals associated by commensalism, mutualism, parasitism, and synoecy have been observed (Barabash-Nikiforov, 1956; Dezhkin, 1959) (Fig. 274). Finally, we have already mentioned the parasitocenoses of individual plants and animals (Pavlovsky, 1951).

These small complexes really exist, enter into the structure of larger territorial associations, and at the same time are connected to one another by a multitude of threads. The smaller the association, the less is its

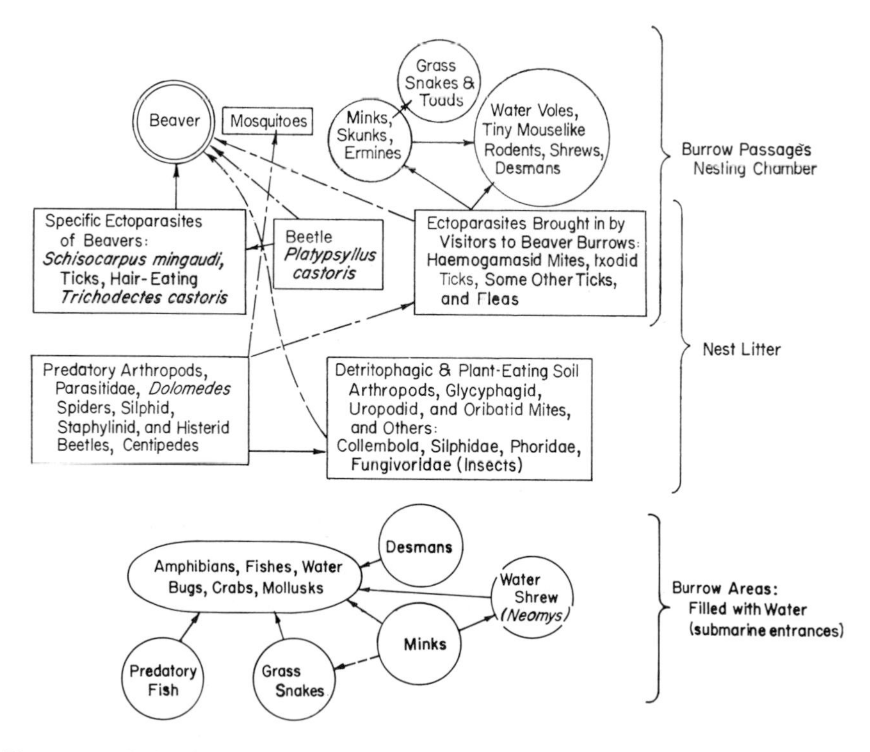

Fig. 274. Relationships in a beaver-dam association in the Voronezh Preserve (after Barabash-Nikiforov, 1959).

autonomy and individuality. But even the larger zonal associations are connected with one another by transzonally distributed species. As one goes from zone to zone, the stations they occupy succeed one another in a regular fashion (see Fig. 281).

ASSOCIATIONS OF THE NATURAL ZONES

The associations of the natural zones were formed evolutionarily under the influence of their climatic and geomorphological properties. The biocenoses of the tropical rain forests are the richest and most complex. Tropical forests, such as are found in the Congo, Cameroons, India, the islands of the Malay Archipelago up to New Guinea, northern Queensland, the basins of the Orinoco and the Amazon, and Guiana, occupy regions with a stable, warm, moist climate. Annual fluctuations in the mean monthly temperature do not exceed 6° C, and the fluctuation over 24 hours is even less. The associations of the tropical forests are distinguished for the richness of their species and the complexity of their structure (many tiers); plants alone number some 45,000 species here (Vul'f, 1944). The species list of insects (Table 72) and other animals is great.

TABLE 72. *Species variety of fauna of certain vertebrates (after Hesse, Allee, and Schmidt, 1951).*

	NUMBER OF SPECIES	
Family	*Tropics and Subtropics*	*Nontropical Zones*
Locustidae	2,726	1,120
Acrididae	2,811	1,842
Dragonflies	2,096	921
Pentastomidae	3,675	1,565
Total	11,308	5,448

The abundance of species is accompanied by a comparatively low number of individuals of each species. In the image suggested by A. Wallace, it is easier to capture representatives of 100 species here than it is to catch 100 individuals of one species. This attests to the relative "narrowness" of the ecological niches of the individual species and to the high level of specialization characteristic of most of the animals inhabiting the tropical forest. Stenobionts tend to predominate. The stability of the climate is linked with the relatively low motility of the animals, the almost complete absence of seasonal migrations, and the low amplitude of seasonal changes

in abundance and fluctuations in different years. An important role in the population dynamics of tropical animals is played by biological connections (competition for food, the influence of enemies and parasites). The role of climatic factors can scarcely be noticed.

The daily aspects of the association are distinctly expressed in tropical forests. Two considerably different, and in many ways independent, associations may live in the same place (diurnal and nocturnal animals). On the other hand, seasonal aspects, owing to the absence of migration and seasonal changes in abundance, are weakly expressed and usually consist chiefly of migratory birds.

This design of the biocenoses in the tropical forest was formed under the direct influence of the stable climate. Its benignity has supplied the richness and strength of the living population of the zone and has also led to the development of stenobiontism. An important factor is the length of time that these rich and relatively stable associations have existed. The forests of the tropical belt have been preserved with very few changes since Tertiary times, and they have many features in common with the psylophytic, lepidodendronous, and sigilliary forests of the Mesozoic and even the Paleozoic. Great age also distinguishes the laterite soils of the equatorial region, which have changed insignificantly since the end of the Cretaceous period (Polynov, 1944). The stability of this single, archaic, natural complex corresponds also to the relative geological stability of the climate of the lower latitudes (Markov, 1951). The long development in comparatively unchanging and favorable conditions aided in increasing the variety of living forms and their mutual adaptation. Hence there was an increase in the role of biological relationships in the lives of the tropical animals.

Associations of the tropical forest steppes (savannas) are rather closely related to those of the tropical forests in the character of the climate and general features. But the species affiliations here are not as numerous, and the tier structure is simpler. The savanna has distinct seasonal migrations (hoofed animals and birds) and distinct seasonal changes in their numbers.

Subtropical forests exist in the USSR on the Black Sea coast of the Caucasus and on the southern coast of the Crimea, occupying regions where the average temperature of the coldest month of the year is no lower than 2° C. The plant species include evergreens (yew, box) and deciduous trees (oak, beech, hornbeam, etc.). The tier scheme of these forests is still very complex. The local fauna includes many stenobiotic forms, although they do not predominate here. Animals which demand moisture and heat are richly represented. Thus in Kolkhida there are 13 species of amphibians, while there are 10 in the central oblasts of the USSR and 3 in the European north (taiga). Seasonal migrants are char-

acteristic of this area to only a small degree, but seasonal guests from the north are common in the tropical forests, since they often winter here. Thus the seasonal aspects of the association are distinct but not definite here. Variability of abundance among the local species is greater here than in the tropical forests, but great bursts of reproduction are rare here too.

North of the tropical and subtropical forests in the west and far east of Eurasia and in the east of the American continent is a belt of leafy and coniferous forests in the temperate zone. In the American west and over large stretches of Eurasia the tropical and temperate forest zones are separated by a strip of steppes and deserts. The temperate types of deciduous forests in Europe and the Soviet Far East are made up of broad-leafed species (oaks, lime, maples, ashes, hornbeam, beech) and in Siberia of narrow-leafed birch forests. They are distributed in the zone of "oak climates." The winters here are not severe but are snowy. The average temperature of the 4 summer months exceeds 10° C; seasonal variations are significant.

The animals of the association of leafy forests of the temperate zone are still richer in species and are comparatively more complex, but they are much poorer and simpler than those of the tropical forests and the subtropics. The number of stenobiotic forms is usually exceeded here by the number of types capable of existing in a variable environment. Their ecological niches are wider than those of the tropical-forest species. The leafy-forest species are typified by distinct seasonal changes in abundance. Many of them make distinct migrations. Winter hibernation is one widespread method of adapting to seasonal fluctuations in the conditions of existence. Food is stored up by species which are active all year. The number of diurnal species begins to predominate quite strongly over that of nocturnal ones. The list of species in this association includes those with highly variable abundance; some of them break into mass reproduction during certain years. These are the small rodents and many insects (the oak fortrix moths, the prominents (Notodontidae), the silkworms, and others). Small rodents (mice, voles) may vary as much as 100 times here. In certain periods, distinguished by the prevalence of various types of weather, species of different ecological design (warmth-loving and cold-resistant types, hygrophiles, and xerophiles) find various advantages and multiply. Thus arise the so-called climatic aspects of fauna, which change the picture substantially from year to year.

The taiga is a territory to the north where pine forests predominate. Its characteristic "larch and fir climate" is distinguished by a rather warm summer (average July temperature no less than 10° C) and a severe winter with much snow in the west and little in the east on the Eurasian con-

tinent. The forests in this zone are distinguished by their monotony. Their tier structure is relatively simple: in the Iceland-moss forests there is a tree tier, sometimes a sparse forest litter, lichens, or dead cover, and the soil. The species supply is poor, primarily of cold-loving eurybionts. Population density falls as one goes north (Fig. 275). In the wintertime many of the animals sleep and hibernate or collect food supplies. Reproduction here is timed strictly to the spring-summer period, with rare exceptions (crossbills multiply when pine seeds can be harvested in winter). A high percentage of the mammal and bird species makes distant seasonal migrations. A bad harvest of tree seeds often occurs on wide stretches of territory, causing more intense wandering among seed eaters (crossbills, nutcrackers, squirrels, etc.); they wander far from the poor harvest zone in search of food. Deep, friable snow limits the existence possibilities of chionophobes and causes hoofed animals to wander.

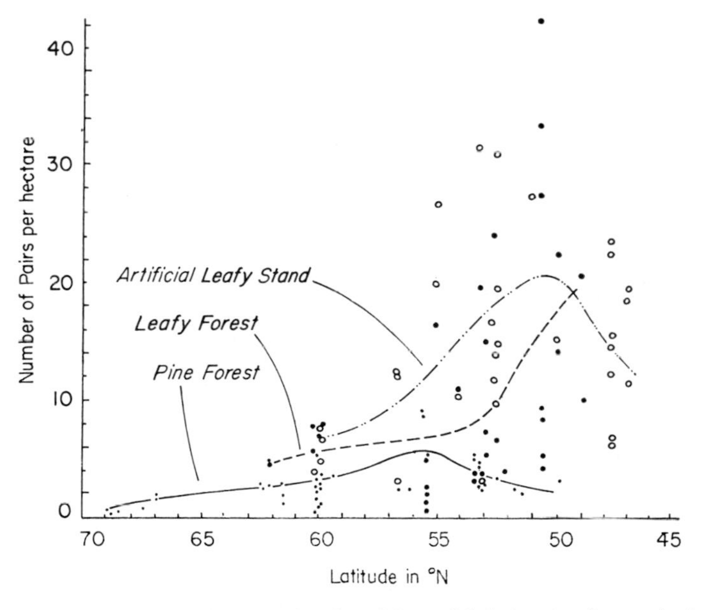

Fig. 275. Zonal changes in population density of forest birds in pine forests, leafy forests, and artificial leafy stands (after Novikov, 1960).

The pattern of taiga associations changes markedly at different seasons. The active part of the population in winter is only one-tenth that of summer. The daily aspects, on the other hand, are expressed weakly, and the number of typically nocturnal animals is not large. The abundance index of taiga animals is subject to large fluctuations, and years of abundance

alternate with periods of great depression. Weather conditions which now favor, now hinder the reproduction and survival of the animals, and which increase the death rate and lower fecundity, play a large direct and indirect role in these fluctuations. The possibilities for mass reproduction of individual species in favorable years are increased by low numbers of competitors, enemies, and parasites. Consequently, the ecological niches of the local species are broad, while the biotic relationships in their lives play a smaller role than in the southern forms. The eurybiotic qualities of the animals confirm this.

The tundras are distinguished by their especially severe climatic conditions: a long winter and a short, cold summer. The average temperature of the warmest month is less than 10° C. Frosts can occur in any of the summer months, and strong winds increase the severity of the climate by many times. Therefore, the tundra associations have few species. In the northern part of the tundra zone about 70 percent of the area is unfit for animal habitation, so the few suitable spots are all the more heavily populated. Thus in the *Eriophorum*-sedge-moss tundras there is an average of 16–24 birds' nests per sq. km (Sdobnikov, 1959). Animals which leave the tundra for the winter predominate. Those which remain are invertebrates which go into torpor during the winter and the lemmings and voles, which winter under the snow. Hares, ermines, and weasels migrate to the river valleys, where they find a heavier, more powdery snow cover, while polar foxes sometimes move to the south or to the ocean coasts.

One general property of tundra animals is their high metabolism in summer, which provides for rapid growth, completion of reproduction in a short time, and rapid accumulation of reserve matter. Among homoiothermic animals the expenditure of energy for physical thermoregulation drops off, which is especially useful in the high latitudes, where food is scarce during the winter. Animals which do not have these adaptations leave the tundra for the winter (Sdobnikov, 1957). The animal population of tundra soils is poorer than that of the taiga, but the population density of the active horizon (0–6 cm) is no lower. Enchitreids and Diptera (tipulid larvae) predominate. The swamp soils are characterized by coccids, onychiurids, and *Sericeus brunneus,* while the brushy tundra is typically inhabited by tipulid larvae; both groups are found in the mossy and lichenous tundra. Heated slopes and dry elevations are most densely populated, while the populations are most impoverished near stands of forest (Stebayev, 1959).

Owing to the unique light regime, the daily aspects of tundra biocenoses are practically absent, but seasonal changes are expressed here as in no other zone. The abundance levels of most tundra animals change sharply in different years. Because of this, the tundra association is a highly motile

system with flexible food connections between species; they are quickly established when abundance rises and quickly terminated when it drops. Under these conditions specialized feeding is almost impossible, and euryphagy is common in the tundra, even among species which are highly specialized in other zones. Even the falcon takes lemmings if they are abundant when willow ptarmigan and shorebirds are in short supply (Osmolovskaya, 1948). The features of the tundra associations, so different from those of the tropics, may be explained by the relative youth of the tundra landscape (Pleistocene). Even poorer and less stable are the Arctic and Antarctic associations, where major colonies of birds' nests (bird bazaars) are found only along the seacoasts.

Steppes and deserts develop in regions distant from the seacoasts where arid conditions are created. As a zonal phenomenon they exist wherever the summer is hot and dry; average temperature of the summer months is 20° C or more, while up to 450 mm of precipitation falls per year.

In the steppe, which is distinguished by the predominance of turfy cereals—feather grass, fescue, and hair grass—the fauna is usually poorer than in the forest region. Steppe associations have a simpler structure composed of one or several tiers of grassy vegetation and the soil. The seasonal temperature variation and the already-mentioned moisture deficit leave their characteristic impress on steppe animals. The latter are well known as storers of reserve matter and foodstuffs, and many go into a winter or summer hibernation. The motility of steppe animals is high, and many of them make long or short migrations in connection with changes in general or local conditions of existence. The seasonal aspects of the association are distinct; at the same time there are marked daily aspects which can be observed especially well in summer. The abundance of many of the steppe species fluctuates sharply from year to year, in connection with which the pattern of the biocenoses changes even over short intervals of time. For these reasons, and by virtue of the mobility of the animals, steppe biocenoses are similar to those of the tundra, with which they are also united by the simplicity of their tier structure. This is explained not only by the noticeable similarity between conditions of existence but also by the similarity between steppe and tundra flora and fauna.

The deserts of the temperate belt are areas of xerophytic brush vegetation, isolated trees (*Haloxylon*), salty shrub bushes and wormwood, ephemerals, and ephemeroids. Less than 300 mm of precipitation fall in a year, mainly in spring and fall. There is no snow cover in winter, or else it is not deep and is of short duration. The winter is relatively severe, and the coldest month in the year has an average temperature of less than 2° C. The summer is hot and dry. The fauna of the desert is specific. In the sandy deserts the psammophiles predominate, while in the clayey and

stony deserts we find species associated with hard ground. The general distinguishing feature of desert animals is their very economical water exchange, brought about by conditions of acute moisture deficit, and their low, seasonally fluctuating metabolism, which is a result of poor feeding conditions (Rustamov, 1955; Kalabukhov, Nurgel'dyyev, and Skvortsov, 1958).

In the desert associations the percentage of animals which hibernate in winter is high, and often there are many animals which hibernate in summer as well or which store up food supplies. Many make seasonal migrations. The percentage of nocturnal species is high. Seasonality of reproduction and seasonal aspects of the associations are well expressed, as are periodic perennial fluctuations in numbers of animals. Mass reproduction usually comes during a period of high moisture. Therefore, the design of desert biocenoses changes in different years, but less acutely, since the typical desert animals (jerboas, gerbils, many birds) are distinguished by what is, on the average, a more stable abundance than we find among the steppe species.

Mountainous regions are distinguished by a unique climate and landscape. In Central Asia low temperatures and great aridity combine with strong winds and lack of snow. Because of this, life here is not only pressed to the earth but to a considerable degree goes under the earth. Associations are poor, and their pattern is variable. The abundance of most of the species fluctuates markedly, primarily under the influence of unfavorable weather. Among the remaining active animals winter storage of food is common, even among species which do not normally accumulate foodstuffs, such as Brandt's vole. The absence of snow favors the existence of a rich group of hoofed animals.

Even a cursory outline of the properties and dynamics of the terrestrial associations of individual geographic zones shows the sequential changes connected with zonal climatic changes. A benign climate is accompanied by a variety and abundance of life forms, complicated societies, specialized species, and compression of ecological niches, reflected in stable numbers and a high level of sessility among the species. A rise in climatic severity and variability in the temperate and high latitudes is accompanied by a drop in species variety and biomass and by a simplification of the structure of the association. This corresponds to the acquisition by most of the species of eurybiotic traits, the appearance and intensification of the seasonal aspects of the association, and a diminution in the daily aspect, which completely disappears in the higher latitudes. Animal mobility increases and is transformed into more and more distant migrations. At the same time the instability of each species' abundance increases, and so does the variability of the pattern of the biotope as a whole. In connection with

such variability, the force, and sometimes the character, of the connections between different species change. Stable interspecies connections in the tropical biocenoses are transformed in the temperate and high latitudes into mobile and temporary ones. Food changes arise, then collapse, and the entire association becomes variable in its content and structure. The extreme degrees of these properties are attained in the open landscapes with simple, often single-tiered associations (tundras, steppes, deserts). In connection with this the relative role of biotic conditions of existence (enemies, parasites, competitors) drops off, and first place is occupied by abiotic factors which directly or indirectly determine the outcome of the interrelationships of the species. These changes are diagrammed in Fig. 276.

The features of the physical environment (climate, geological substrate, relief) and their variability are the prime cause of the different zonal associations. But they are apparently also associated with the geological growth of each. The length of time over which species have existed together

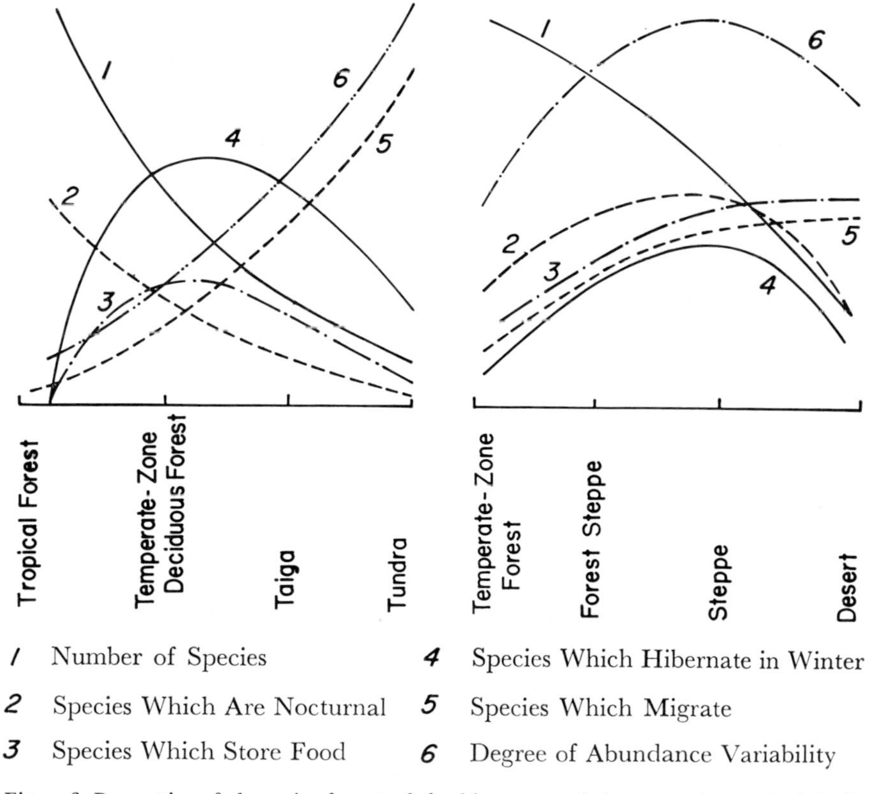

Fig. 276. Properties of the animal part of the biocenoses of the natural zone (original).

is an important factor in their mutual adaptation. "Young" landscapes are settled by a few species, often of different origins. Their mutual adaptations are indistinct, while their connections with one another have not solidified, which contributes to the instability and variability of the associations in time.

Zonal properties are also characteristic of aquatic biocenoses. In the high latitudes of the ocean the vital phenomena have very low indices because of the ice regime, the long polar night, the low temperatures, and the poor vertical circulation of the waters. In the middle latitudes of the ocean there are regions where food is abundant, biological productivity is at its highest level, and the greatest fish catches are made. These regions occur in areas with the greatest vertical circulation of waters, whereby the nutritive salts from the lower layers are carried to the surface (the polar front zone). Farther south is a broad belt of ocean less favorable for the development of abundant pelagic life because of the limited vertical circulation of the waters. At the same time the climate favors an abundance of species, the number of which in the tropics is about 100 times greater than in the high Arctic. In the temperate latitudes, where species are 10–20 times fewer, the biomass of plankton is 5–20 times greater than the biomass of plankton in the tropics.

In addition to latitudinal symmetry, the oceans have a regular change in the living population along meridianal lines from the middle of the deeper and saltier parts toward the coastal areas. In this case, too, the drop in the species content of the body of water does not correspond with and even runs against the quantitative abundance which governs favorability of feeding conditions. Because of the abundance of food brought in by continental currents, plankton and benthos develop abundantly in the coastal areas of the oceans and seas; a great number of fishes and other animals live on these forms of life (Zenkevich, 1951).

SEASONAL AND DAILY CHANGES IN ASSOCIATIONS

It has been mentioned that regular changes in content and relationships between different species at various seasons or hours of the day are a property of many associations. As a result of replacements or changes in activity, a given species may be excluded from participation in the life of the biocenose for a certain length of time. The daily periodicity of biological processes and the existence of "diurnal" and "nocturnal" species has led to the rise of two different and, in some ways, independent and

alternating daily aspects of associations. This path has led not only to a complication of the associations but also to a more complete use of the environment by the organisms living in a given area.

Seasonal changes in associations depend on some species falling into a summer or winter sleep and on others migrating when winter or drought conditions set in (Fig. 277). The sessile species comprise the basic nucleus of the biocenoses, while the seasonal ones determine its pattern at the different periods. Their appearance changes not only the pattern of the association but also the connections which unite its members. The extent to which the terrestrial tundra association is complicated by their arrival in summer is well shown in Fig. 279.

Seasonal biological changes in the aquatic environment are closely associated with the hydrological seasons. They do not correspond in the different zones and have both a quantitative (fluctuation of mass) and a qualitative (changes in species) character. In marine waters the biological spring is usually short and is characterized by an abundance of phyto-

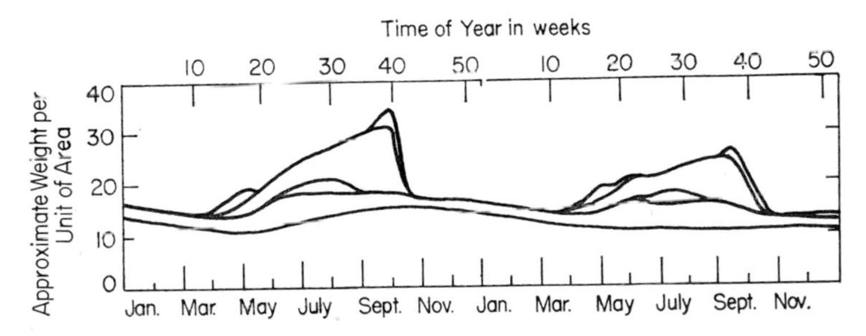

Fig. 277. Seasonal changes in abundance (biomass) of basic groups of animals in the leafy forest of western Manitoba (after Bird and Shelford, from Kashkarov, 1945).

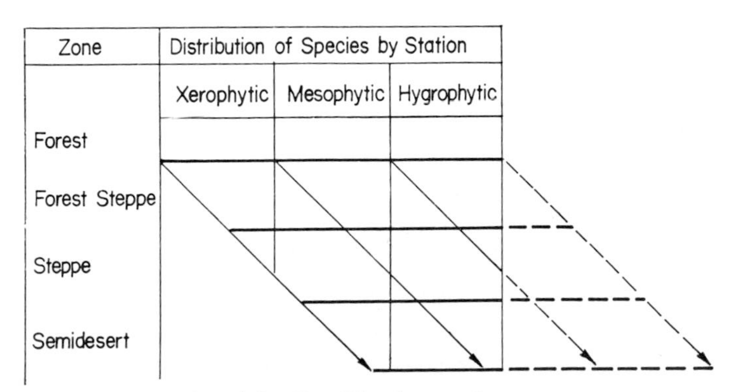

Fig. 278. Zonal shift in stations (after Bey-Biyenko, 1959).

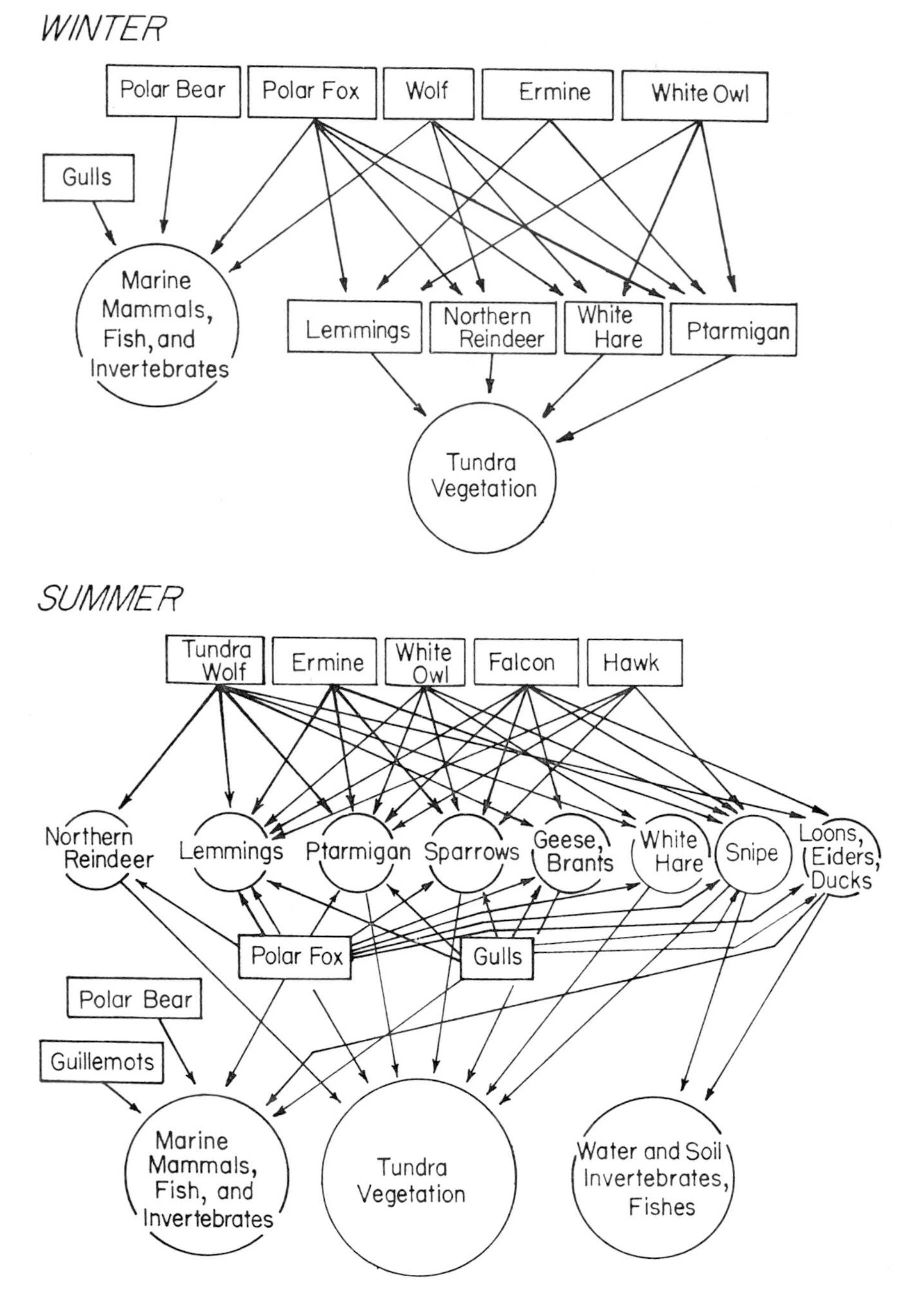

Fig. 279. Food connections in the Arctic tundra biocenose (after Sdobnikov, 1958).

plankton, especially diatoms (the sea's flowering season), as well as eggs and larvae of bottom animals; most species of zooplankton multiply intensely in this period. The biological summer is distinguished by something of a diminution in the number of plankton; although its animal part reaches a maximum, the plant part dwindles acutely. The biological autumn in the Arctic seas is characterized by a generalized marked drop in the plankton population, especially the plant portion. In the seas of the temperate zone there is a second outbreak of development among phytoplankton (the autumn flowering). Finally, the biological winter is characterized everywhere by a continued decrease in and, ultimately, a minimum of plankton. Many animal and plant species exist at this time in a quiescent winter stage.

The lengths of the biological seasons differ at the various latitudes, and the seasons occur in different months. The growing season in the tropics stretches over 12 months, while in the Arctic it lasts only 2 or 3, sometimes as little as 1. In the tropics the plankton spring occurs in winter, while in the Arctic Ocean it is in August. Biological winter is practically absent from the tropics, while in the Arctic it lasts for 10 months (the time of the polar night). The seasonal dynamics of the plankton mass in the Arctic and the tropical seas has one maximum, while in the seas of the temperate belt it has two (Fig. 280). The difference between warm and cold seas consists in the number of generations. In the Arctic the planktonic copepods produce one or rarely two generations a year, in the temperate latitudes there are four to six, and in the tropics there are many generations a year (Zenkevich, 1951). Similar situations prevail among marine benthos and freshwater populations.

DYNAMICS OF ASSOCIATIONS

The replacement of one type of association by another is called succession. Successions occur as the result of changes in the environment, which may be caused by external or internal factors. If the effect of factors leading to a profound change in the character of the environment, and hence to a change in the association, is temporary, it is possible that the original type of association can be re-established. If the effects of the conditions continue and lead to a change in the association, the latter may exist for a long time in its changed state and may never return to its initial condition. Major alterations in associations are made possible by small, usually quantitative, shifts.

Changes of flora and fauna embracing whole geological periods were

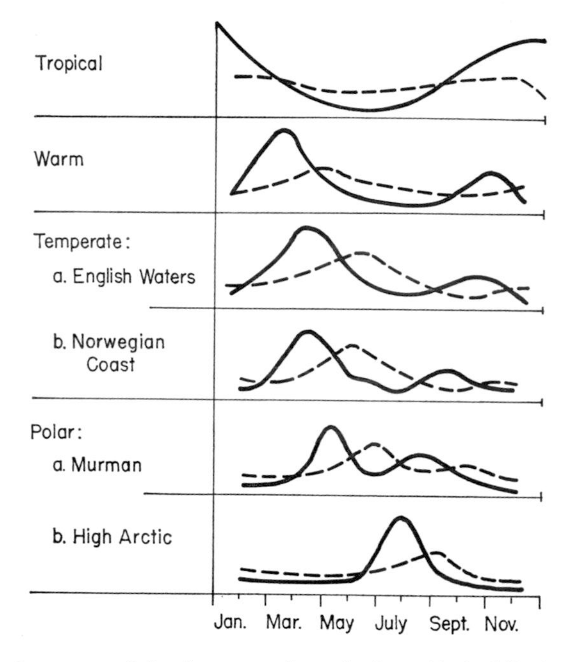

Fig. 280. Seasonal progress of development of zooplankton (dashed line) and phytoplankton (solid line) in the ocean at different latitudes (after Borogov, from Zenkevich, 1951).

brought into being as shifts in types of associations. They occurred in connection with climatic and geological replacements and were associated with the evolution of species. The evolution of associations and the evolution of species are closely related to each other. Replacements in the environment must of necessity change the relationships of the species in the association and speed up or cause the evolution of species, while the appearance of new species reconstructs the association and thus changes the environment again. However, both processes, though they accompany one another, are at the same time independent. Changes may occur in the associations without evolution of species being involved, while the appearance of a new species does not necessarily change the type of association. Evolution of both species and associations takes place only when the accumulated quantitative properties reach a magnitude at which qualitative changes in the structure and character of relationships with the environment can take place.

In surveying the dynamics of the territorial biocenose, it is necessary to remember that the association is a mechanism for setting up a biological turnover of matter. Therefore, its stability depends on the metabolic balance of the species populations making up the association. As they

consume food, oxygen, and water and contaminate the environment with their excretions, the individual species could not exist if their partners in the association failed to create a suitable environment by using their metabolic products and thus neutralizing their one-sided effects. The stability of the association is greater as the balancing action of the species in it becomes more efficient. This is attained in the so-called "climax," or "perfected association," distinguished by its high degree of adaptations between one species and another and between all species and the abiotic environment. In such a biocenose, under the existing soil-climatic or hydrological conditions, the results and products of the vital activities of some species are used most completely by others, and the turnover of matter is very harmonious and hence intensive.

After the destruction or considerable disturbance of a climax association by fire, flood, landslide, or mudslide (in the mountains), ground-breaking (in the steppes), lumbering (in the forests), etc., the association usually re-establishes itself after a certain time. As the area progresses toward its climax state, a succession of life forms replaces one another until the final, stable form of biocenose has been reached—the one most closely resembling the original. This makes it possible for us to view the series of transitional biocenoses as stages of development of the climax association and to treat the series as we would the ontogenesis of an organism. In actuality, the course of succession is characterized by the replacement of one temporary association with another as the result of the immigration of newer and newer species; these impinge on and limit each other's abundance in sequence until the most stable combination has been reached. When this happens, each temporary association is not an accidental collection of species but a true association which changes the environment in a definite way. But none of these associations is stable because the changes which they cause are unfavorable and aid in bringing in and establishing new, more competitive species. This kind of succession may be artificially halted at any stage, and if the conditions characteristic of that stage are preserved, the temporary association will continue to exist for an indeterminate length of time. This also happens in cases of overpasturing, in which a pasture which has been overgrazed is retarded for the whole time that the cattle stay on it.

The establishment of a coniferous forest serves as a good example of succession and is a case which has been well studied. After lumbering or a forest fire the forest area is covered in the first (1–3) years by brushy shoots and such grassy plants as thistle, rose bay, reeds, etc. They find favorable conditions (abundant humus) but later are cramped by shoots of young aspen and birch, while fir and pine are developing under their cover. The young of these species quickly cover the site of the fire or lumbering

operation, shutting off light from the grassy weed vegetation, which is replaced by mosses and forest multigrass. At the same time the young birches and aspens retard the development of the young pine trees. The pine forest has been replaced by a birch or aspen plantation, sometimes with a small admixture of pines. After the birches and aspens attain an age of 10–20 years and light once again reaches the forest floor, conditions are favorable for the growth of pine trees. Their growth makes it impossible for light-loving leafy trees to become established. So, as a result of the long-term pressure on the "temporary" birch-aspen plantation, the pine forest begins to flourish.

The animal population of the area changes similarly. It also passes through a "weedy-meadow" phase, a "brushy" phase, "a leafy-glade" phase, and a mature-forest phase. In the leafy-glade phase the number of species is at its lowest level (Table 73). Such is the usual scheme of succession in pine forests. Pasturing cattle along the cut-down portions, which destroys bushes and saplings, turns the previous forest into a dry meadow where the establishment of a forest encounters serious obstacles in the meadow vegetation even after pasturing has ceased. In many places birch-aspen forests replace the original one and become permanent, even without special forestry management techniques. Eventually, the rise of a pine or broad-leafed forest is impossible in such a locality.

A plowed-up virgin steppe first become overgrown with annual and perennial weed species that have flying seeds and grow quickly. They form an unstable association of short and tall weeds. Later the rooty cereals appear, such as wheat grass, forcing out the previous weeds and forming a more or less solid cover. However, they cannot hinder the penetration of subsequent growth of turfy feather grass, *Festuca sulcata,* and hair grass. Thus arises the typical steppe vegetation association and its characteristic fauna. It is the most stable in the steppe soil-climate complex. Here, too, the tempo of re-establishment depends on the surrounding conditions. Reproduction of small rodents (voles) retards the weedy stage. Heavy pasturing with cattle halts the succession at the couch-grass stage. Pasturing of cattle on the virgin steppe is accompanied by the elimination of feather grass and later by the elimination of all turfy cereals and the development of low-valued vegetation consisting of short and tall weeds. The new associations which then arise are stable as long as pasturing is continued. Thus any of the successional phases may become the "climax formation" under new conditions.

The balance in the relationships of species and the mutual neutralization of one-sided use of the inorganic environment in the climax association are only relative. Therefore, the stability of the climax association is also relative, mainly because the turnover of matter is never a complete

TABLE 73. *Change in species assortment of birds on Segrowa clearings in the northern Urals (after N. N. Danilov, 1958).*

	PINE FOREST				PINE AND FIR FOREST					FIR GROVE			
	1–10 Years	*10–20 Years*	*20–30 Years*	*Maturity*	*1–10 Years*	*10–20 Years*	*20–30 Years*	*30–70 Years*	*Maturity*	*1–10 Years*	*10–20 Years*	*30–70 Years*	*Maturity*
Number of species of birds nesting	13	13	13	30	5	12	8	13	27	10	7	13	13
Number of species common to mature stands	3	5	10	30	2	3	6	12	27	1	2	6	13
Number of pairs per hectare	2.1	2.3	1.6	2.6	2.0	1.7	1.0	1.0	3.3	1.3	1.1	0.9	1.2
	1.2	1.0	1.2	1.8	–	–	–	–	1.7	–	–	–	–

Note: Bottom figures show only pairs per hectare.

repetition of the same biochemical cycles. As the association interacts with inorganic nature, some of the substances are attracted back into the turnover, while others are eliminated from it in the form of inert caustobiolites (silts, sapropels, petroleum, coal, organogenic limestones, flints, salts, sulfur, etc.). Neither process comes anywhere near balancing the other, and changes accumulating in sequence in the surrounding environment are usually unfavorable for the association. Until a known degree of magnitude is reached, they cause the biocenose to undergo reconstruction. Under some forest plots a layer of orstein impermeable to water is set up, which limits the productivity of the underlying soil and raises the groundwater level to the point where swamps are formed. As a result, the pine forest turns into a marsh, if the climate, relief, and ground favor it. The growth of a mossy layer on the top of a sphagnum swamp isolates the living sphagnum from the soil solutions and may lead to its extinction, subsequent degradation of the turf, and establishment of forest trees on the former turf.

The accumulation of such changes accelerates, decelerates, or changes its character and direction under the influence of changes in climate, relief, and other geological factors as well as under the influence of changes in species or the content of associations. Therefore, the division of such changes in associations into endogenous ones (caused only by internal biotic factors, i.e., changes in interspecies relationships and the content of the associations) and exogenous ones (external changes caused by a change in climate or the chemistry of the environment), although admissible, is still conditional. At the basis of every reconstruction of associations there always lie changes in their interactions with the nonorganic environment, but this may occur when the numerical ratio of the species, the content of the associations, or the climate changes. The dimensions and direction of such changes are summed up in the form of changes in the turnover of matter on a given plot of ground on the earth's surface which includes both the biocenose and its environment.

An example of this effect is the influence of animals on the plant cover. Pasturing of hoofed animals leads to trampling of plants, destruction of the dead litter, hardening of the upper soil horizons, and destruction of its structure. This is accompanied by changes in aeration, moisture turnover, and temperature conditions in the soil, a drop in moisture absorption, and an increase in the surface flow. As a result, even in the north rising currents of soil solution begin to prevail, and in dry periods evaporation is intensified, leading to salinization of the soil, to "steppification," or even to "desertization." On such plots a vegetation appears which is characteristic of drier zones (wormwood, bulbous meadow grass, narrow-leafed cereals), and it is followed by a characteristic fauna (Formozov, 1929). Susliks,

jerboas, and other steppe and desert animals settle in the north along such pastures, which are formed in the immediate proximity of villages and along roads. Such changes in associations take a greater amount of time and are associated with larger changes in the mutual relations of association and environment.

A large role in periodic and nonperiodic changes in the design of steppes, semideserts, and deserts is played by the depth of the snow cover. On it depend the development and mass of the mesophilic plants which are so important as food. Periodic drying up and filling of lakes in Kazakhstan, West Siberia, and many other regions depends first of all on the amount of snow falling in the previous winter. Changes in the level and area of the lakes are accompanied by changes in salinity, chemistry, and clearness of the waters and changes in the benthos and plankton, followed by changes in the number and abundance of the species of animals which feed on these foods. Drying or filling up of bodies of water is a long process, occupying 10–20 years or more.

When shallow, well-heated steppe lakes are filled up, the number of plankton in them increases quickly, as do benthos and fishes, and cane and reeds flourish abundantly in the small tributaries and shallows around them. These are valuable as food, as are the underwater meadows of sunken plants, pond weeds, etc. At this time the population of the lake reaches its greatest variety and richness of species. When the lake begins to dry up, the increasing salinity of its water leads to impoverishment of the plankton and benthos and thus to a decrease in the number of fish. When the lake is completely dried up, only the tributaries and springs have any water left. The spoonbill, great cormorant, widgeon, and whistling teal stop nesting; the number of Fuligulinae decreases markedly; and the common pochard becomes numerous as compared with the tufted duck, which is dominant in the years when there is moisture. Only the shoveler and the pintail become abundant.

When the water level drops further and salinity increases, the inhabitants of the lake remain specifically saltwater species, such as the *Tadorna* or ruddy sheldrake. At the same time the numbers of gulls, *Chilodias leucoptera, C. nigra,* warblers, and other birds decrease, or the species disappear completely. Predatory birds and mammals associated with bodies of water are forced to search for new sources of food. Some of them shift to catching terrestrial prey (small rodents, susliks). In connection with the periodic drying up and refilling of steppe lakes, the northern border of nesting grounds for the black-headed gull, great cormorant, spoonbill, and several other birds moves from north to south and back again over a stretch of dozens and even hundreds of kilometers (Formozov, 1936).

Climatic periodicity is expressed in the changes of other bodies of water

too (see Fig. 180). In Lake Baykal the content and biomass of plankton are not the same in years of different meteorological conditions. When the previous autumn has been warm and of long duration and the first half of summer brings a great deal of sudden heat to the water, the diatom alga *Melosira* and the crustacean *Cyclops baicalensis* undergo mass development. With an early, cold autumn the previous year and slow heating of the water next spring, *Melosira* is almost absent, being replaced by the diatom *Cyclotella* and other algae whose abundance is associated with the mass appearance of the crustacean *Epischura*. In such years the total harvest of plankton is usually smaller (Kozhov, 1955).

PRODUCTIVITY OF ASSOCIATIONS

The practically important question about the biological productivity of associations is closely connected with the study of the structure and life of biocenoses, especially with their dynamics. By "productivity of associations" we mean the yearly production of supplies of wild animals and plants which may be used by man without destroying the biocenose or disturbing any of the species which enter into it. The biological productivity of wild associations is similar to the concept of soil fertility in agriculture. In this connection "ideas" have penetrated into ecology on the unavoidable exhaustion of the natural associations under the influence of cultivation; these "ideas" are the original theories of diminishing soil fertility. Such are the considerations as to degradation of game hunting or a decrease in the productivity of fishing as an unavoidable consequence of the development of industry and cultivation of the landscape.

The problem of total productivity is being worked out intensively in hydrobiology. The task has been set to find a balance of organic matter at different stages in the turnover of matter in a body of water. The ability of the body of water to produce and accumulate organic matter is evaluated according to the amount of mineral salts in the water, since their abundance determines the possibilities of developing water plants, the producers of organic matter. On this basis lakes have been classified into rich (eutrophic) and poor (oligotrophic). There is no doubt that an abundance of minerals and solar radiation influences the development and abundance of living creatures in a body of water. But the degree and effectiveness of their use depend on many circumstances, and in practice they rarely assimilate great amounts of salts or solar energy.

The degree to which inorganic resources are used is associated with the species makeup of the association. There may be periodic and nonperiodic

fluctuations in abundance and biomass of each species and association as a whole, as a result both of their interactions with their inorganic surroundings and of the relationships between the species. Therefore, the concept of association productivity as a whole, and especially the working out of means for raising this productivity, are impossible without a profound study of the ecology of basic, economically important species which enter into a given association. The latter is actually the center of attention of Soviet ecologists occupied with the study of the productivity of game and fish farms and of the means for raising their productivity.

Productivity also depends on the means of using the game species. The direct and indirect actions of man on game populations, which change their population density, food supply, and refuges, interrelations with populations of other species—all influence their reproduction, productivity, and survival, i.e., their abundance and dynamics. More than anything else the productivity of the association depends on the form and means of economic use of game animals. When predatory use is made of game species or when the association as a whole is not managed properly, a decrease in productivity is unavoidable ("diminishing fertility"). When proper conservation of natural supplies of valuable species is maintained, and especially when breeding is introduced on a large scale, productivity increases.

Thus the productivity of the association can be used by man to produce economically valuable game animals under the existing system of exploitation. Productivity depends both on supplies of primary resources (mineral substances, food plants) and on the form of economic use made of the association by man. Man influences the content and numerical ratios of the species in the association and the population density, intensity of reproduction, and survival of the different species. The goal of studying the factors of productivity is to raise it, and this may be accomplished by changing inorganic nature, the species content and ratios, and the means of using and managing the populations. The last way is simple; in practice it is the most applicable. It demands detailed study of ecology and especially of the principles of the population dynamics of the economically important species. This is the path which Soviet ecology is following.

CHAPTER 19

Human Activity and the Animal World

GENERAL SIGNIFICANCE FOR NATURE

With the appearance of man on the earth, the animal world began to experience the ever-growing influence of human activity. For a long time it was unsystematic. Certain species of wild birds, mammals, and insects were domesticated and ceased, or nearly ceased, to exist in their wild forms. Their domestication and breeding diminished and complicated the possibility of existence for ecologically similar species. The extermination of valuable game animals continued: elephants, hoofed animals, and, later, several fur-bearing animals for which there was an increased demand. These changes and the even greater changes in the landscape caused by lumbering, plowing, and cultivation of agricultural plants proved favorable to other species. Numerous pests of cultivated plants, forests, meadows, and pastures appeared, as did the carriers and vectors of diseases of man and domestic animals. The lack of system in the changes in living nature was accompanied by its decimation: ". . . a crop, if it develops naturally, and is not directed consciously . . . leaves a desert behind. . . ."[1] All of this constitutes a strong argument for conducting a struggle against harmful species and an effort to protect useful ones.

In our country the conquest of nature and the rational use of her resources must be combined harmoniously with the intense growth of agriculture and industry. The growth of these two main branches of the national economy serves as the basis for building the material base of communism. Such a combination is furnished by planning all branches of production and culture and by an intensive study of nature, which per-

[1] Letter to Engels, 25 Mar. 1868, in K. Marx and F. Engels, *Works* (1931), 24: 35.

mits us to "understand correctly its laws and comprehend both the nearest, as well as the most distant, consequences of our active interference in its natural course."[2]

The influence of man on the animal world may be direct or indirect. Both forms of influence overlap and are closely blended with both biotic and abiotic factors in the environment. Direct influence is expressed in the immediate effect of man on the life of wild populations. Such are campaigns against plant pests and the disease vectors dangerous to man, catching of game species (furbearers, sport animals, fish), active conservation activities, and breeding of supplies of useful species. Indirect influences (the more widespread and powerful type) consist of man's efforts to change the surrounding natural environment. Agriculture and forestry, industry, transport, and other forms of human activity, even without acting directly on animals, do have an immediate effect on the conditions of their existence by changing the landscape.

ANIMAL HUSBANDRY AND ITS SIGNIFICANCE FOR WILD SPECIES

Domestication of wild animals is a special form of the use of natural resources in which man conquers them most effectively and assures their expanding production. He changes profoundly the conditions of existence of domesticated species, furnishing their food, which is usually different from what they obtained in the wild, creating a favorable set of circumstances for water and heat exchange, and protecting them against predators and diseases. Man has almost unlimited possibilities for making changes in a direction which answers his needs; as a result of this, many breeds of domestic animals appear.

At the present time 15 species of mammals, 10 species of birds, several species of fishes (mainly decorative types), and 2 species of insects have been domesticated. The number of breeds of these species is difficult to compute, but it is undoubtedly very large. There are about 500 breeds of pigeons alone, about 200 of dogs, 100 of rabbits, and as many as 2,000 of the mulberry silkworm. Their total number exceeds many thousands and emphasizes the powerful effects of artificial selection.

The degree of domestication is determined by the amount and type of care which man devotes to the animal. Man exercises his greatest degree of domestication when he keeps animals in stalls or cages, takes care of them, and feeds them food whose contents and amounts he himself has deter-

[2] F. Engels, *The dialectic of nature* (1957), p. 57.

mined. However, keeping animals in stalls sometimes reduces their fecundity and results in barrenness (among horses, for example), especially if each animal is isolated from the others. This is a result of the low mobility allowed to animals which formerly traveled in herds. Selectional genetic work on a large scale is necessary to eliminate these dangers, and the stall confinement of farm animals must be perfected, along with the methods used to feed them. This demands liberal expenditures of labor and is a more refined form of agriculture.

The opposite state is found in so-called nomadic cattle raising, in which the animals live year round on whatever feed they find underfoot. Even temporary supplemental feeds are absent. Until recent times northern reindeer and nomadic steppe animals were bred in this manner. Success or failure depended to a large degree on the weather, which could be avoided only by driving the livestock to a protected area with a good food supply. In many cases this method proved insufficient, and the hot summer's abundance of bloodsucking disease vectors, the winter's ice crusts, jute, etc. were often accompanied by mass death among the animals. Because of the way animals in herds cluster, the death rate was even higher than it would have been among wild animals.

The appearance of domestic animals feeding on the same foods as the wild species markedly changes the living conditions of the latter. Pasturing of domestic livestock changes the plant cover and the animal population of the pastures. In the steppes uncontrolled pasturing is accompanied by "desertization" of the plant cover; in the forest zone it retards second growth and causes the appearance of the so-called waterless valley meadows in clearings where timber has been cut. When pastures are overgrazed, they lose their food value and turn into barren areas. As a result, the wild hoofed animals, which are closely related to the domestic livestock, disappear or suffer sharp declines in their numbers.

Domestic animals aid in the transmission and distribution of diseases among wild species. The most resistant nidi of pastoral ticks are acclimatized to the pastures, migration routes, and water holes used by domestic cattle. Infection of pastures during epizootics among domestic animals plays a large role in the spread of contact diseases and helminthoses. Attacks of anthrax among wild ruminants are usually associated with its spread among domestic livestock or with infection of wild animals at places where the carcasses of anthrax victims have been buried. Hog cholera of domestic swine has been the cause of death among wild boars in the Caucasus State Preserve (Teplov, 1938). Transmission of disease from domestic animals to wild ones has also been verified for reindeer, chamois, beaver, ptarmigan, pheasant, and certain other species.

The solidification of upper soil horizons caused by grazing, the destruc-

tion of its structure, and the deterioration of its aeration and hydrothermic regimen lead to impoverishment of the soil fauna and a decrease in the abundance of the soil population. Trampling of the standing grass cover and underlying litter increases the death rate among birds nesting on the surface and among all other animals associated with the surface of the soil, especially with litter.

Cutting of the grass and the appearance of xerophytes in it are accompanied by an increase in the number of xerophilic animals—susliks, jerboas, steppe species of voles, steppe and desert birds, and insects (geophilic locusts)—and by expansion of the area occupied by them. In the north they tend strongly toward places which have been changed by grazing (Fig. 281). The well-documented dispersal of small susliks in the northwestern Cis-Caucasus and Cis-Caspian, which continues even into the present period, was apparently made possible by the concentration there of large flocks of sheep and herds of horses and cattle during the end of the last and the beginning of the present century. This concentration was the result of the development of wheat cultivation in the Ukraine, the commercialization of which rose sharply after the development of the railroad network and the construction of the Black Sea ports. Intensification of livestock breeding in the Cis-Caucasus was accompanied by the appearance of vast stretches of wormwood and meadow grass beaten down until they were favorable for the existence of small susliks. This also created conditions for their spread (Formozov, 1959).

Groups of dung-feeding coprophages and domestic animal parasites flourish on pastures. Their spread and abundance depend completely on

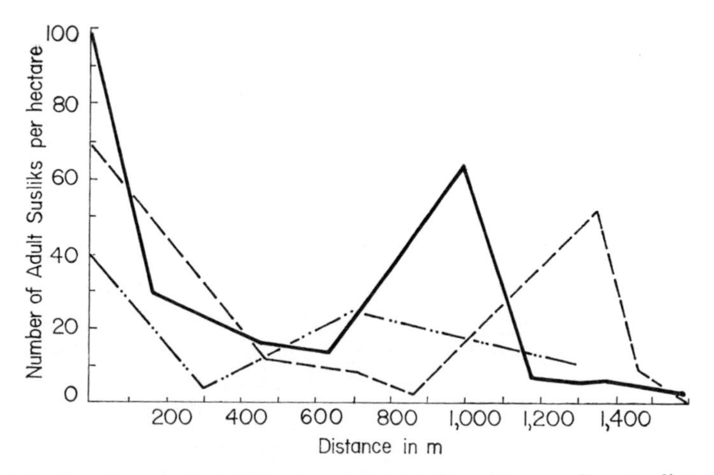

Fig. 281. Changes in population density of the small suslik according to distance from human dwellings (after Mamontov, 1948). Various changes.

the new mass hosts. Carrion vultures and certain predators (wolves) have always been closely associated with man and are most numerous in regions of primitive "nomadic" animal raising, where many types of domestic animals have clustered. As far back as 1908 P. P. Sushkin noticed that in central and northern Kazakhstan the vultures nested only in years when there were many deaths among livestock. Near the herds there were many birds using domestic livestock as "beaters" to frighten their catch from the grass (swallows, starlings, rooks, magpies). In winter the pasturing herds would scrape away the snow and expose food for skylarks, ptarmigan, and other animals.

GROUND-BREAKING

Ground-breaking has tremendous direct and indirect effects upon the wild animal population, since it changes the landscape to a greater degree than most other types of human activity. The antiquity of plowing, which arose in our country at the end of the Neolithic and the beginning of the Bronze Age,[3] and its economic significance explain the enormous influence it has had on nature. Ground-breaking is associated with the clearing of forests and the plowing and cleaning of virgin steppe lands; i.e., it is accompanied by destruction of the initial vegetation and its replacement by cultivated crops. When the area of the forest is cut down, *Tetrao urogallus,* hazel grouse, forest martens, sables, elks, and other forest species are reduced in number or disappear altogether. Destruction of the virgin steppe is accompanied by the disappearance of the little bustard and the marmots.

For species which formerly occupied a subordinate position, changes in the landscape mean an increase in abundance and a wider range of distribution. Many of them become agricultural pests, while some become the carriers and vectors of natural-nidal diseases of man and animals.

Before the advent of ground-breaking and forest-clearing in the middle belt of the European USSR, the common voles took shelter in the meadows, floodplains, burned patches of forest, and well-illuminated woods, occupying one of the lowest places on the scale of small mammals. The development of ground-breaking permitted the vole to disperse into the fields, where it became a major pest. The appearance of valuable food crops and the practice of leaving the gleanings outside during the winter allowed them to multiply year round and raised their fecundity. Harvesting and breaking up of fields forced the voles to seek out suitable areas; this was accompanied by a rise in their death rate. As a result, a variable abundance

[3] Approximately the third millennium B.C.

was established, and the vole was turned into a species capable of mass reproduction during favorable years.

In areas of poorly developed plowing (certain regions of Central Asia) certain insects live on wild vegetation and usually do not attain a high abundance. The appearance of cereal crops raises the food supply by an extraordinary degree, changes the bug's way of living, and sharply limits intraspecies competition, the number of enemies, and the effect of parasites. This turns a species which previously had occupied a modest place into an important cereal-grain pest. The beet snout beetle, Hessian and Swedish flies, *Anisoplia anstriaca*, noctuid moths, and many others lived on different species, no single one of which was a mass food source, before the introduction of cultivated beets, grains, and other crops. The granary pests (mites and insects) previously lived in the burrows and nests of rodents, birds, and other animals, feeding on their supplies of food or on the scraps they rejected. With the appearance in prehistoric times of the first human storage areas for bread products, some of these species moved in and became true pests, while others continued to live in the burrows and nests of animals. The establishment of the potential granary pests near man might have occurred along with the movement of their vertebrate hosts, which were becoming cohabitants of man (synanthropes). The grain trade of the Middle Ages, which flourished among many countries, aided in spreading the granary pests, and many of them have become cosmopolitan in the present era. As grain-storage methods have become perfected, some of the mites and insects have been squeezed out of the granaries, but others still retain their destructive capabilities today.

The destroyers of rubber plants, such as the koksagyz fly, green root aphid, *Nisius* bug, and other insects, previously lived on wild composites. All of them were oligophagic insects and were capable of flight in the imago stage. They became pests because the koksagyz (the Russian dandelion) was similar to their natural food plants. Stands of this plant created favorable microclimatic conditions for the insects, and the phenology of the koksagyz corresponded in time to the growing and bearing seasons of their natural foods. The species which became pests of the koksagyz had short (no longer than 1 year) developmental cycles and were capable of comparatively distant migrations. This made it possible for them to tolerate changes in the plant cover caused by crop turnover (Gilyarov, 1950).

All these examples show that the establishment of pests takes place not as a result of the simple expansion of their food possibilities but as a result of the rise of new relationships between the species and the environment. These relationships are involved with the more simple associations of the cultivated landscape and are almost always accompanied by one

change or another in the way of life of the species and its connection with its co-members of a biocenose. When this happens, the sessile forms might become mobile or vice versa. Thus, after the appearance of field-protecting forest belts in the steppes, *Opatrum sabulosum,* a destructive beetle which previously led a sessile life, began to move into the forest en masse to spend the winter. On the other hand, the alfalfa snout beetle, which previously wandered far afield in search of food plants, began to feed on yellow acacia and lead a sessile life (Likventov, 1950). A species which lives on cultivated plants does not become a dangerous pest in all regions, only in those where it can attain high abundance at least temporarily.

The appearance of uniform stands of plants or the wide use of single-species forest plantings (so-called monocultures) has created advantages for certain species by expanding their ecological niches and at the same time suppressing their competitors and, frequently, their enemies and parasites. This has created favorable conditions for high abundance or repeated reproduction. Changes in the microclimate may also be of great importance.

The change in the landscape caused by plowing and its penetration to the north is the cause of many species becoming established in the forest (hamster, common vole, steppe lemming, field and miniature mice, European hedgehog, black and bright polecats, European hares, susliks, gray ptarmigan, quail, landrail, rook, skylark, yellow bunting, finches, and many others). They were accompanied by many insects, pasture ticks, etc. Transport of seed material and products of agriculture often caused emigration and dispersal of harmful insects, ticks, and even rodents. The formation of sessile agricultural settlements in the forest and steppe zones and cultivated oases in the desert enriched the fauna and expanded the ranges of many species. Thus many species of bats, forest and forest-steppe rodents, insectivores, predators, etc. came into the steppes and deserts along with the construction and use of wooden villages.

The influence of plowing on freshwater fauna is great and multifaceted. Changes in climate arising from the cutting of forests and the plowing of virgin steppes are of special importance. Faster melting of snows and more intense runoff of soil layers change the character of freshets, increase the muddiness of the water, and accelerate the filling of the waters with mineral precipitates. A change in the profile of the body of water is expressed in its population profile (the relationships between limnophilic and rheophilic species). Coincidentally, the waters are enriched with mineral and organic substances, and this increases the plant and animal populations. An increase in wind strength increases the number of mineral particles, organic detritus, seeds, and animals (insects) carried into the water.

FORESTRY

Forestry changes the design of forests and consequently the affiliates and abundance of the animals which inhabit them. The changes are especially great when large tracts are all cut down simultaneously. The temporary associations which arise on the clearings usually form a unified fauna. The relative paucity of the animal population is associated with natural monotypicity and simplification of the tier structure which occurs in such places. However, hares, elks, black cocks, and certain other species associated with clearings and fire-devastated areas find important advantages in such a situation. For a long time we have known of fluctuations in the number of elks and pulsations in the borders of their ranges, apparently in connection with changes in the areas of tracts devastated by fires and timbering operations (D. N. Danilov, 1948).

The appearance of fields inside solid forest massifs lengthens the forest margins and thus has a favorable effect on the abundance of squirrels, since trees on the forest margins bear more frequently and abundantly than those in the depths of the woods. This partially explains the relative abundance of squirrels in the Moscow forests as compared with the Siberian taiga game regions.[4] Forestry measures—sanitary clearings, cuttings for health purposes, collection of dead wood and wind-fallen branches, and planting of trees—change the content and structure of a forest tract. These measures aid in reducing the abundance of many pests and raise the profitability of lumbering operations. But destruction of hollow trees diminishes the number of woodpeckers, owls, and other hollow-nesting birds as well as the number of martens, flying squirrels, etc. Most of these species are important exterminators of forest pests or are themselves valuable commercial animals.

Forest fires and the deliberate burning of prairie vegetation, cane, and reeds along river valleys are especially hard on animals. Forest fires have existed since the rise of modern forests, but they became widespread only after the appearance of man and his mastery of fire. The number of forest fires rose sharply in the eighteenth, nineteenth, and twentieth centuries. Judging from the larch obtained from the Cis-Amur region, there were 2 great fires from 1701 through 1750, 30 from 1751 through 1800, 79 from 1801 through 1850, and 154 from 1851 through 1900 (Strogy, 1920).

Fires are especially destructive in the huge stretches of uniform taiga. In the basin of the very lightly populated Lower Tunguska River there is no spot, however small, on which traces of some ancient or recent fire cannot

[4] Records of squirrel pelts in an average year total 1.2 units per sq. km in the Moscow oblast', 1.3 in the Archangel oblast', 0.5 in the Soviet Far East, and 0.4 in Yakutia.

be found. The wide use of aviation in our country has reduced the danger of fires spreading even in these regions, but previously they covered huge territories. In the drought of 1915 some 1,600,000 sq. km of forest were destroyed by fire in Siberia, while the smoke covered 6,000,000 sq. km, an area as big as Europe (Fig. 282). Solar radiation reaching the ground under this pall was diminished to 65 percent of its usual intensity in August. The fire lasted some 50 days and set back the ripening of grain crops 10–15 days. Numerous emigrations of squirrels, bears, elks, and other animals were recorded during the summer of that year.

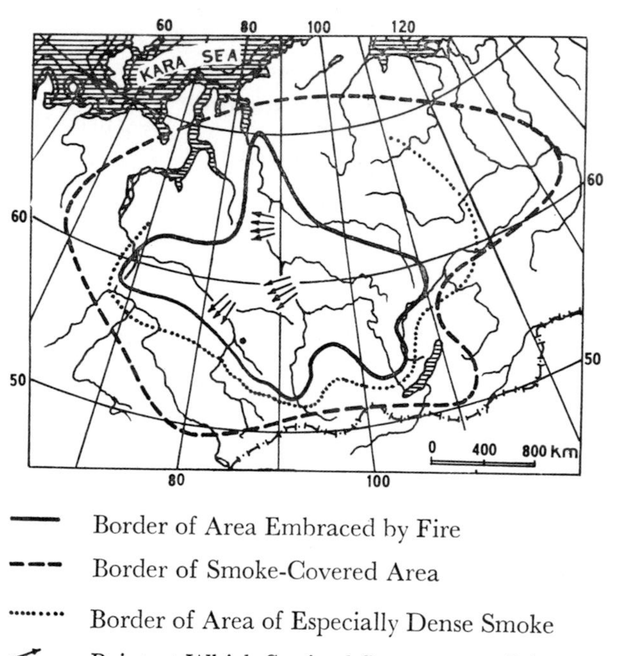

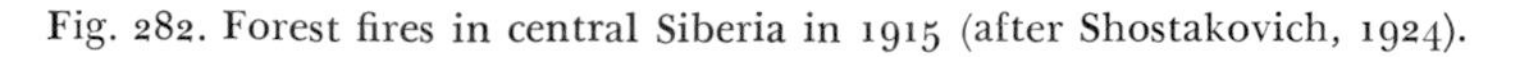
Fig. 282. Forest fires in central Siberia in 1915 (after Shostakovich, 1924).

By disrupting previously existing associations, fires give rise to successions similar to those caused by forest-clearing. But complete destruction of the grass cover, mosses, and lichens is accompanied by a greater unification of the animal population, especially in the first stages of recovery. Fires are a great threat to the northern-reindeer industry because they destroy pastures of Iceland moss, and the re-establishment of the moss cover in the north takes about 20–40 years. Re-establishment of reproducing timber takes 50–80 years, and during this time squirrels and other

seed eaters cannot live in the area. Springtime firing of steppe vegetation is unfavorable for most animals and sometimes leads to mass death, but for hoofed animals and some rodents the improved plant growth which follows a prairie fire is favorable. Burning off underbrush in the valleys and river deltas kills many of the inhabitants of these areas and causes others to migrate.

HUNTING OF ANIMALS

Trapping of fur-bearing animals, game, and fish changes not only their abundance but also their ability to exist at all as significant populations. The degree of effect depends on the intensity of hunting and its form (times and methods of catching animals).

With heavy trapping or hunting, the number of sexually mature animals is reduced, and those of older age drop completely out of the biocenose (Fig. 283). Among fishes this is expressed as a general decrease in average size (Fig. 284). These phenomena attest to a reduction in length of life and a decrease in productivity of the population. Continued overexploitation leads to the animals' disappearance, beginning in the less

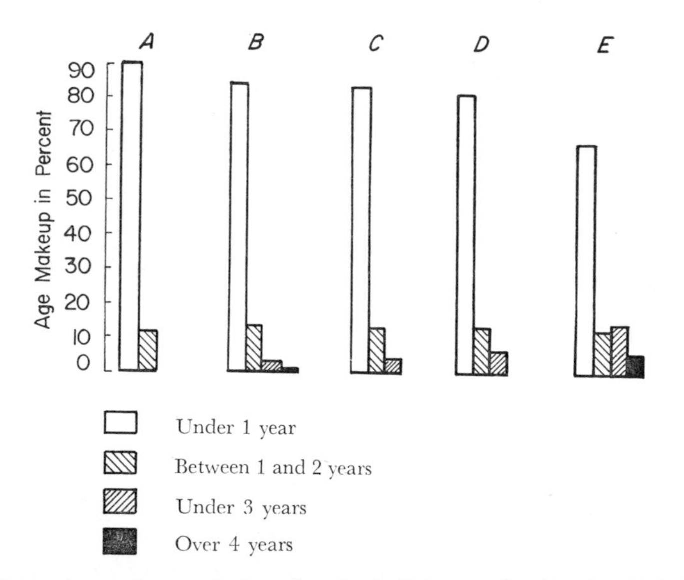

Fig. 283. Age makeup of a population of squirrels *(Sciurus vulgaris)* (after N. P. Naumov, 1934). A: squirrels from ribbon-shaped tracts of pine in West Siberia *(teleutka* squirrels); B: from the Lower Tunguska; C: from the middle course of the Ob' (Surgut region); D: from the upper Vychegod region of the Komi ASSR; E: from a preserve near Moscow.

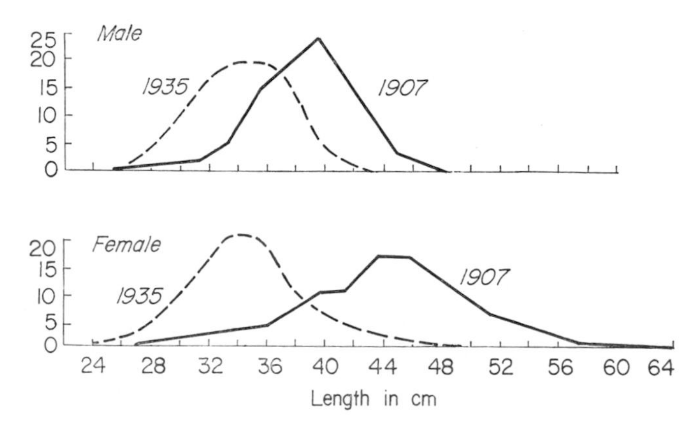

Fig. 284. Changes in size of the marine flounder *Platessa platessa* in the Barents Sea as a result of intensified fishing (after Milinsky, 1944).

favorable regions. This is the position in which the red squirrel found itself in the ribbon-shaped pine forests of Kazakhstan at the end of the 1920s and the beginning of the 1930s, when the material shown in Fig. 283 was being collected. It was threatened with extinction, which was finally prevented by protective measures. The amount of hunting is normal in the northern regions of Siberia and the European USSR, and the squirrel population shows no signs of progressive reduction in numbers. In the preserves, where no hunting is permitted, squirrels even live to old age; they have been found there with completely worn-down tooth enamel and incapable of reproduction. A pair in such a region will produce fewer young, on the average, than will a pair in an area where hunting is permitted. Thus both overhunting and insufficient hunting can cause a decrease in animal productivity.

The history of managed hunting knows many examples of complete extermination or great reduction in the numbers of valuable game species. Even before the October Revolution intense trapping had led to the almost complete extinction of the beaver and the desman, to a drop in the number of sables and a collapse of their range of distribution to a series of isolated tracts, and to a sharp reduction and in many cases complete extermination of many of the wild hoofed animals (the antelope *Saiga tartarica*, wild ass, reindeer, wild boar, and many others). In the last thousand years alone the Caucasus has lost nine species of large mammals: the lion in the tenth century, the aurochs in the twelfth, the wild ass and the cheetah in the thirteenth, the beaver and the elk in the beginning of the nineteenth, Przewalski's horse in the 1880s, and the bison and the tiger in the twentieth (Vereshchagin, 1952).

I. G. Pidoplichko (1951, 1952) felt that even the disappearance of the mammoths and the northern rhinoceroses occurred with the direct participation of man the hunter. Many birds were also rendered extinct: wingless auks in the North Atlantic, wingless doves and dodos in the Mascarene Islands, and many others. Among the exterminated species or those which suffered a marked decrease in abundance were the relatively infertile colonial or flocking forms, for which a curtailment in flock and colony size leads to a general rise in the death rate. Meat- and fur-bearing animals, as well as inhabitants of the steppes and forest steppes, which are easier to catch than the forest and mountain species, suffered more intensely from extermination.

Under conditions of proper game management a species' abundance is not reduced, and its fecundity may even rise. Expanded productivity of game animal supplies is also achieved by improving the conditions of their existence. This demands the removal or weakening of the effects of limiting factors, i.e., an increase in the capacity of the game preserve district. The usual limiting factors are a lack of refuges, places for mating, or food (seasonal or permanent insufficiency, qualitative or quantitative shortages of food, water holes, etc.), unfavorable microclimate, an abundance of predators, favorable conditions for the spread of diseases, etc. The construction of artificial nesting grounds, spawning areas, refuges, and water holes, supplementary feeding, control of disease, and an increase in protection of the habitat all diminish the death rate and raise the animal's fertility. Such methods for managing supplies of wild and game animals are acquiring greater and greater significance.

The catching or trapping of animals is a matter of "selectively harvesting" the game species. The time and technique of catching them determine the quality and quantity of the catch and the ultimate productivity of the game flock or herd. Trapping or hunting animals sharply changes the character of their seasonal dynamics and the conditions of existence of the remaining individuals. A sudden decrease in abundance as a result of hunting aids in providing the remaining animals with refuges and food and cuts down on the losses caused by predators, since the latter partly die off and partly migrate from the region of sparse prey. The reduced population remaining after a hunting or trapping operation is less dense and thus reduces the possibility of disease being spread. Selective catching, which is possible through many means of trapping animals, changes the population content, often by raising the proportion of females, increases the survival rate of the remaining individuals, increases the tempo of growth and development, and thus intensifies mating. The productivity of the population thus grows, making for an increased total catch.

If the extent of hunting does not exceed the norm, the hunted popula-

tions do not show any signs of a decrease in abundance. But the absence of hunting and trapping activities can have an unfavorable result. Thus the autumn hunting season for the California quail *Lophortyx californica* increased their survival on the hunting grounds, but in a preserve district with the same type of natural features a smaller number of quail was left in spring than could be found at the hunting grounds, and they were in a poorer state (Errington, 1949). Similar data have been obtained for the muskrat (Table 74). The highest percentage of muskrats surviving the winter was observed in a body of water which was not so well supplied with food but which was subjected to intensive trapping.

TABLE 74. *Effects of commercial trapping on survival of the muskrat in the West Siberian muskrat grounds (after Korsakov, 1950).*

Body of Water	*Food Supply of Water (percent of overgrowth)*	*Muskrats Caught as Percent of Pre-Trap Levels*	*Muskrats per Hectare Remaining in Nesting Areas*	*Those Surviving until Spring per Hectare of Suitable Nesting Areas*	*Percent of Those Surviving until Spring*
Barsuch'ye	36	71	1.8	1.05	59
Vikolovo	57	59	2.7	0.70	27
Cherepn'	78	15	4.0	0.85	26
Kolomenskoye	65	0	4.1	1.00	24

The fertility of a population of normally hunted animals increases as a result of change in the content and character of its dynamics, in which biotic factors play a comparatively smaller role and factors of feeding and heat exchange play a greater one. As a result of reduced predator activity, competition, and disease distribution, the population makes better use of each favorable change in the external environment. Because of this, increases in the abundance of many game species (squirrels, foxes, polar foxes, ermines) distinctly coincide with changes in the weather or the food supply in nature. Only among the varying hares do worm infections and epizootics act as the chief cause of fluctuations in abundance. In most of the game regions this species is poorly hunted and is caught primarily with nooses and traps in the latter half of the winter. Consequently, during a large part of the postmating period, i.e., the time of high abundance, hunting has little effect on the conditions of existence of this species. The other hare—the European hare—is caught primarily in autumn and the beginning of winter, during its yearly maximum of abundance, and its population changes in close correspondence with the intensity of reproduction and survival rate of the early harvests, which depend on the spring weather. The active effect on abundance, reproduction, and state of the

population of game species caused by trapping (hunting) and by simple protective measures opens up wide possibilities for increasing the productivity of game fauna.

INDUSTRY AND TRANSPORT

Industry and transport exercise their effects for the most part in an indirect manner—through changes in the landscape. Individual types of industrial activity have a direct negative or positive effect on various species of animals. Thus the gaseous and fluid wastes from many factories are fatal to many species of plants and animals (air pollution, poisoning of waterways, soils, etc.).

Water, rail, air, and other forms of transport have aided in the distribution of several animals. The penetration of the black rat *Rattus rattus* into the European USSR occurred from south to north and was associated with the ancient river trade routes "from the Varangians to the Greeks." Its contemporary distribution is associated with the basins of the Dnieper, the Don, and partially the Volga. The gray rat *Rattus norvegicus,* which is more able as a competitor, used various means of transport and became more widely distributed (Fig. 285). Railroad cars carry not only rats and house mice but also other small rodents (hamsters, voles, etc.) over long distances. Steamships, trains, and airplanes transport insects and ticks. A case has been described in which malarial mosquitoes were carried from the railroad station at Chiili (middle course of the Syr Darya) to Syrzan' and even greater distances. The saltwater crab *Eriocheir sinensis* has been carried in the freshwater tanks of steamships from northern China to Europe. In its native land this animal lives in the basin of the Yangtse and other rivers. It was first observed in Europe in the Weser River in 1912 and afterward quickly settled the European rivers from the Baltic to the Seine (C. Elton, 1960). Agricultural pests also move in this manner. Among the 602 species estimated to occur in the United States up to 1897, 111 were of foreign origin. In many cases they were distributed very rapidly.

The development of air transport aided in the unusually rapid import of dangerous disease vectors into new areas. In some cases this led to the rise of new nidi of disease. Thus the vector of malaria, the mosquito *Anopheles gambiae,* probably arrived in Port Natal, Brazil, from Dakar by aircraft in 1929. In 1938 and 1939 severe epidemics of malaria with a high mortality rate broke out in Brazil. *Aëdes aegypti* (the vector of yellow fever), mosquitoes of the genera *Culex* and *Mansonia,* the flies *Stomoxys* and *Glossina* (the vector of sleeping sickness), cockroaches, bedbugs,

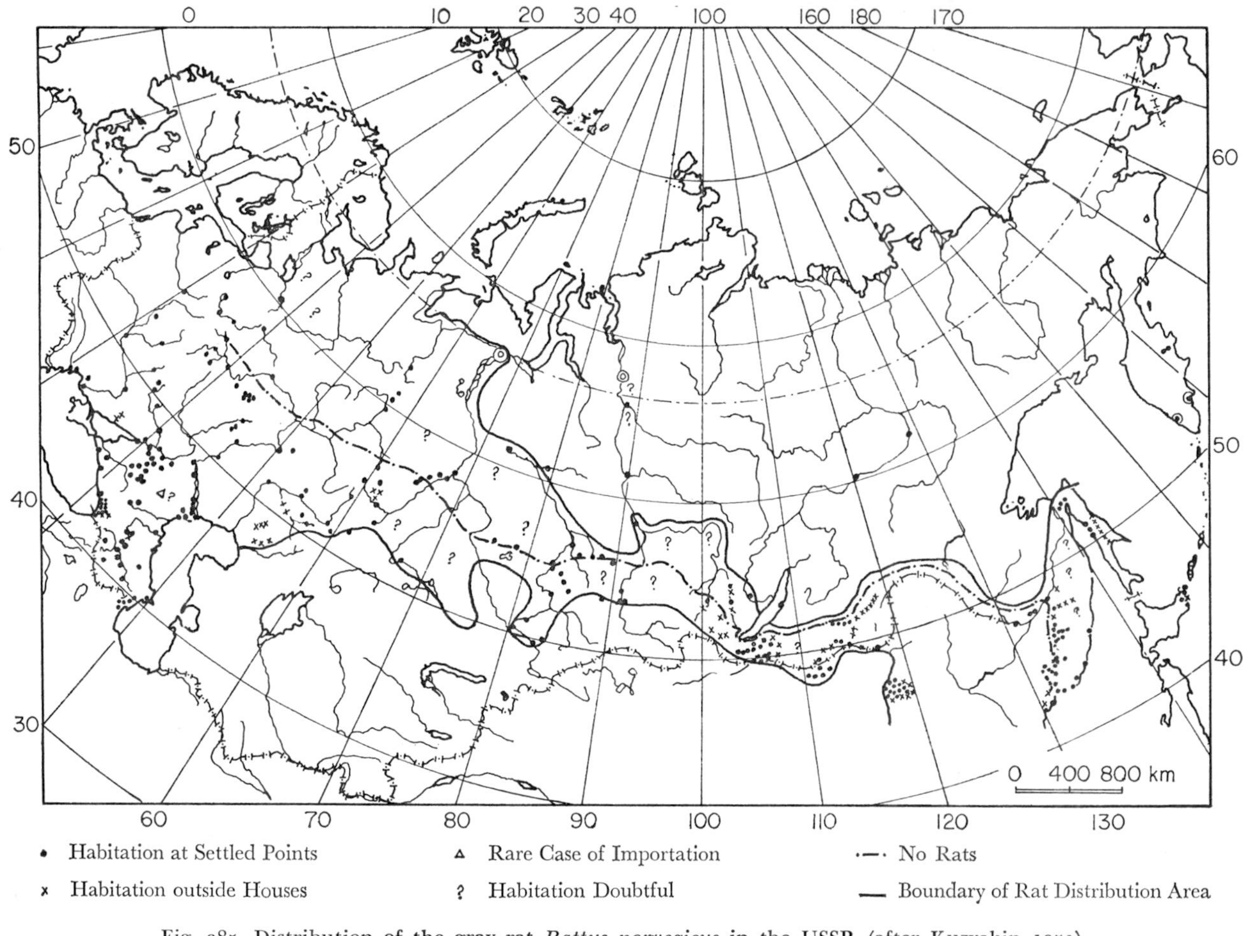

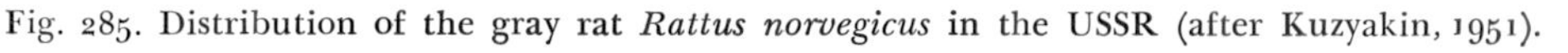

Fig. 285. Distribution of the gray rat *Rattus norvegicus* in the USSR (after Kuzyakin, 1951).

tabanids, and certain ticks—more than 30 species of arthropods—have been observed during inspections of aircraft. It has been established that mosquitoes can live in an aircraft up to 6.5 days and have covered distances of 70,000 km (Reykhardt, 1941).

Night-flying migratory birds die at night on lighthouses, telegraph wires, high-voltage transmission lines, etc. Their death rate is so high that special measures have to be taken. Diurnal predators use telephone poles as observation points while watching for prey, and their numbers along the main lines are always higher than in the neighboring areas. Cuckoos use these lines during their migrations. Some animals, primarily invertebrates (beetles, ants, snails, and many others), die on highways and other roads under the wheels of vehicles. Calculations have shown that in 24 hours many thousands of individuals may die on 1 km of road. Some birds (hobbies, swallows) make use of moving persons while hunting; they catch animals which have been disturbed by the movement.

Petroleum and lubricating oils which enter the water from steamships, tankers, etc. are dangerous to waterfowl, since they gum up their feathers and dissolve the fatty oils which protect their plumage. As a result, the birds die of overcooling. In the region of Baku petroleum has contaminated the bottom in a strip 8–10 m deep and 3–4 km wide near the shore. This has killed off the eelgrass, which forms underwater pastures, and has stopped the multiplication of mollusks and many other animals.

The dense network of suburban roads and highways which covers the country aids in the spread of many species and acts as an ecological channel or trough along which steppe and desert fauna can penetrate into the forest zone. Gallinaceous birds collect the pebbles and gravel so necessary for digestion along the village roads. Birds from the open areas (gray ptarmigan) stop near roads during their flights; they feed there in spring on the fresh grasses, which appear earlier than in other places, and so do hares and other animals. In the steppes and meadows the roads are agents for the movements of animals avoiding dense stands of grass; foxes and wolves use them to move around in winter. Polygenaceae, which stay green and nutritious there after the grass has dried up in other places, are fed upon by susliks. Many grain-eating birds (buntings) collect seeds which have fallen by the roadside or seek oats in stables (Formozov, 1937).

BUILT- UP AREAS AND SYNANTHROPIC ANIMALS

The construction of settlements changes the local conditions of existence very much, and therefore their fauna is distinguished by its great variety.

Its most specific element consists of the so-called synanthropes, which find an especially favorable environment near man and are closely associated with his activities. The settlements are also penetrated by species which live outside the city or village. Thus among the 141 species of birds recorded in Moscow we find even such infrequent ones as the pelican, the roller, and several others. Moscow's parks and gardens contain blue titmice, nuthatches, redstarts, gray and spotted flycatchers, finches, and many other wild birds. Nightingales have been recorded on more than one occasion in the bushes of Sverdlov Square. A sessile population of Cracow ducks lives freely on the ponds of the zoo.

Synanthropes are associated with man by their habits of feeding on the products or scraps of his own activity (many granary and house pests and boarders), by parasitizing man or domestic animals, and by using refuges created by man. Consequently, synanthropism is parasitism, boarding, or tenancy in human settlements. It may be temporary or permanent, partial or complete. Permanent and complete synanthropes live only in human settlements. They include many of the specific endo- and ectoparasites of man, the pests, and the boarders (cockroaches, clothes moths, gray rats, house mice, sparrows, and others). But the human flea is often found on wild carnivores, marmots, and other animals, while house mice and gray rats live year round outside human settlements in the south. In the middle latitudes a large proportion of such synanthropes returns to the house in autumn and leaves again in spring.

Two groups of synanthropes are distinguished according to the character of their synanthropism. The first group includes tenants, which use only refuges near man (swallows, except for the cliff swallow, martlets, starlings, rooks). They can live not only in dwellings but also in the ruins of cities and settlements, where they find the same favorable conditions for nesting. The second group also feeds in the human dwelling and is thus associated more closely with man (parasites and such pests and boarders as rats, mice, and sparrows).

Partial and temporary synanthropes include the tenant animals, of which only a small part lives inside human dwellings. Rodents, insectivores, small carnivores, birds, reptiles, amphibians, and many invertebrates (especially insects) willingly settle in buildings, which they use as a nesting or hibernation area, although the majority live outside the settlements. In the steppe and the desert even temporary yurts attract gerbils, hamsters, shrews, and other small mammals. The small mounds of earth around peasant houses are used by small jerboas and other creatures for building their burrows. The Caucasian stone houses (*saklias*) are used as nesting areas by wild birds (*Fringilla montifringilla*). Mountain and snow voles (*Chionomys, Alticola*) are common guests in high mountain cabins, while the chough

Pyrrhocorax is found in larger structures. In the desert the adobe structures attract stone sparrows of the genus *Petronia, Oenanthe hispanica, pleschanka,* and many others.

The degree of synanthropism within a single species often reveals itself to different extents in various areas of its range and is usually greater on the periphery of its range. The house mouse is most closely attached to man in the north and in the clayey deserts. The gray hamster becomes a synanthrope in the desert regions of Central Asia. The yellow-throated mouse almost never occurs in homes in the broad-leafed forest zone, but it can pass the winter successfully only in human dwellings in the northern forests (Leningrad, Pskov, Velikiye Luki, and several other oblasts) and in the Caucasus Mountains. The same is true of most other synanthropes. The circumstances under which synanthropism arose have been poorly studied. It is presumed that it arose out of such ecologo-physiological properties as plastic feeding habits and the possibility of using various foods, labile heat exchange and daily activity, and the high demand for refuges associated with these phenomena. Anatomic properties are also important, especially the relatively small size of the animals.

RECONSTRUCTION OF NATURE IN OUR COUNTRY

The foregoing shows that even elemental, unplanned changes in nature caused by man are not necessarily accompanied by its impoverishment. The entire history of man is the conquest and development of natural productive powers. But the pursuit of profits and the chemical exploitation of natural riches under capitalism lead to their exhaustion. This danger is eliminated by planned use and development of natural resources—the proper transformation of nature—which are possible only under socialism and communism. The resolutions on the development of industry and the sharp increase in agriculture adopted by the 20th and 22nd sessions of the Communist Party of the Soviet Union and printed in the program of the CPSU also include the conquest of virgin and fallow lands, construction of energy and hydrotechnical installations, cultivation of steppe-land forests, increasing the fertility of non-chernozem lands, and planned use of the forest riches of the northern and eastern parts of the country.

Breaking up the virgin lands, especially in the southeast, returning basic forestry projects to the timber-rich northeastern regions, and intensifying forestry activities in the middle, south, and west are profoundly changing the natural surroundings in the greater part of the Union. A program has

been laid out for further development along these lines. This general path of improvement and melioration of climate in many ways aids in solving the main problem: a significant rise in agricultural production. At the same time this creates better forest growth, higher field production, etc. This path had already been marked out by the basic formulators of the concept of landscape, V. V. Dokuchayev, A. I. Voyeykov, P. A. Kostychev, and others.

A change in the plant cover caused by plowing up virgin lands and by forest management has a special effect on soil moisture turnover. The yearly increase in soil moisture as a result of forest cultivation and snow retention is 45 mm in the southern Ukrainian steppe zone, 64 mm in the northern steppe zone, and 89 mm in the forest-steppe zone. The index having the greatest agricultural importance (the ratio of moisture intake to expenditure) changes as shown in Table 75. Plowing, sowing, and sub-

TABLE 75. *Changes in moisture index (after V. G. Popov, 1950).*

	MOISTURE INDEX	
Zone	*Present*	*Future*
Southern steppe	less than 1	1.5
Northern steppe	1.2	2.0
Forest steppe	1.7	2.8

sequent agrotechnical measures, especially stubble plowing, cultivation, and fall plowing, greatly diminish surface runoff and evaporation and aid the soil in accumulating moisture.

As a result, the borders of the landscape zones change. In the Ukraine the forest zone is moving southward to a line from Kirovograd to Poltava to Khar'kov. The forest steppe occupies all the space between this line (the southern border of the forest zone) and the shores of the Black and Azov seas. The steppe has practically ceased to exist as a landscape within the Ukraine. Great changes are taking place along the lower Volga, in the steppe and semidesert Cis-Caucasus, in the Volga-Ural interfluve, and in North Kazakhstan, where virgin and fallow lands are being mastered at top speed and hydrotechnical installations are under construction. Next in line is the reconstruction of the living associations of the deserts of Kazakhstan, Central Asia, West and East Siberia, and other regions.

An integral part of the re-formation of nature is the active, directed reconstruction of the fauna. It is brought about by limiting or squeezing out the destructive species, by breeding and distributing the useful ones, and by introducing new and valuable forms. The Soviet Union's successes in controlling animal pests have been great. Planned establishment of con-

trol measures has practically paralyzed locusts, moths, and other species. The losses caused by other mass pests have been sharply reduced. Successful solution of these tasks has been accomplished by the technical perfection of extermination methods which are possible only in a large, mechanized, socialist agriculture. In most cases extermination is mechanized. Specialized aviation is used widely to treat vast tracts, not only of cultivated crops but also of deserts, which are hard to reach and poorly settled.

The protection and restoration of useful animals is provided first by planned, regulated trapping of game animals and sport fish and then by construction of artificial breeding grounds and hatcheries, organization of supplementary feeding in famine years, and many other protective measures. There is a wide network of state preserves in the Soviet Union—laboratories in nature where wild species of plants and animals are not only protected but studied, and where experiments are conducted to see if they can be domesticated or breed freely or semifreely. Large projects have been carried out aimed at domesticating the elk, which is a valuable meat animal and a transport animal in the taiga regions, and at breeding beaver, nutria, and several other valuable game species. The preserves occupy an important place in the projects which are attempting to reshape nature in our country.

Protection of nearly extinct species has yielded positive results. In many areas the sable has been re-established, and the numbers of elk have risen sharply. The beaver has been distributed into many regions from the few places where it existed before the October Revolution (Fig. 286). The first stage in the re-establishment of the sable and elk has now been passed, and broad-scale exploitation of these animals has begun; this occurs in the form of regular, licensed trapping. Observations have shown that lack of trapping or poorly conducted weeding out of animals can even have a negative effect by retarding the growth of the sable population.

Fur and game resources have been enriched by the successful acclimatization of certain alien species. Acclimatization is the term applied to the introduction into the association of a new species (or several species), accompanied by its adaptation to the local physico-chemical conditions (climate), the establishment of connections (especially food connections) with other species, reproduction, and resettlement. Except for a few cases in which the new species encounters circumstances identical to those of its native area, acclimatization involves a change in its morpho-physiological properties and way of life. Conditions for successful acclimatization include the presence of food available to the acclimatized species, general correspondence between the local physico-chemical, especially climatic, conditions and the species' demands, and absence of enemies and parasites

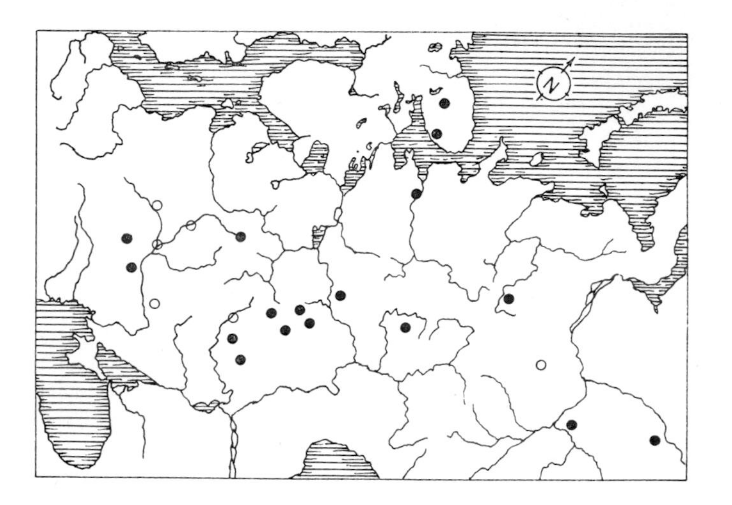

○ Where Beavers Preserved before 1917

● Where Beavers Released after 1917

Fig. 286. Results of reacclimatization of the beaver (*Castor fiber*) in the USSR (original, compiled after Lavrov, 1946).

which might destroy the species to be acclimatized. In other words, the association must have some kind of ecological niche, however small, which responds to the basic demands of the new species. Only under conditions of successful reproduction and survival can a new species enter into the roster of an association and then adapt to the local biotic and abiotic conditions which it meets.

Success in further acclimatization depends on the degree of correspondence between local conditions and the demands of the new species, on the possibility of adapting to these conditions (mobility of organization), and on the pressure of competition with aboriginal species. Experience has shown that the initial abundance of the species to be acclimatized is of great importance. Its introduction into the association takes place more rapidly and reliably as the number of individuals in the released batch increases. Attempts at acclimatization with small groups (consisting of a few individuals) are more likely to prove unsuccessful. When large batches are released, competition is more successful against ecologically similar species living in the area, and the necessary intraspecies environment or population structure becomes established. This is especially important for herd and colonial animals.

The establishment of exterminated animals in their old habitats is

known as reacclimatization. It is difficult to draw a distinction between this concept and that of acclimatization, since reacclimatization is almost always accompanied by acclimatization. The animals to be settled in the previously inhabited region usually encounter new conditions, since human activity quickly changes the landscape.

The acclimatization projects in the USSR are very broad in scope. In the 20 years from 1925 through 1944 attempts were made to acclimatize 26 species of fur-bearing animals. They included 2 species of insectivores (mole and desman), 14 species of carnivores (common and Canadian red foxes, corsac, continental and Komandorsky polar foxes, American raccoon, raccoon dog, sable, stone marten, skunk, American mink, kolinsky, steppe polecat, and sea otter), and 10 species of rodents (gray and bobac marmots, yellow suslik, squirrel, muskrat, European and American beavers, nutria, European hare and rabbit). Into the new regions 76,500 individuals were poured during this period (almost 4,000 desmans, 3,200 raccoons, more than 3,000 American minks, almost 500 sables, and more than 48 muskrats).

As might be expected, the results were not the same for all species. The rabbits, Canadian foxes, and polar foxes did not survive in all areas. The skunks, American raccoons, stone martens, and desmans did not prove wholly successful. But the remaining species became acclimatized well in most areas. The muskrat spread out over vast areas; at present it has entered solidly into the roster of our commercial fauna and each year yields a large number of valuable pelts (Fig. 287). The nutria has become successfully acclimatized in the Trans-Caucasus, while the American mink has been successful in the northern forest and southern mountain regions. The latter has adjusted especially well in the mountainous regions of West and East Siberia and in the central oblasts of the European USSR. Attempts to acclimatize the dark races of sables in the west and the red squirrels in the Crimea have proven unsuccessful, since under these new conditions the dark coloring of the sables has grown brighter, the fur of the red squirrel has lost its protective properties, and the squirrels have practically eliminated the beech nuts. These results show to what extent the prospective acclimatization must be calculated.

In recent years projects have been developing aimed at acclimatizing aquatic animals. Valuable species of fish (sturgeons) have been imported into the Aral Sea. Many forms are being acclimatized in lakes, rivers, and reservoirs. Collective fish farming is being practiced in ponds. After successful acclimatization of the valuable food species *Nereis* in the Caspian Sea, the next task seems to be the question of how to acclimatize the food sources of the fish (Zenkevich, 1952).

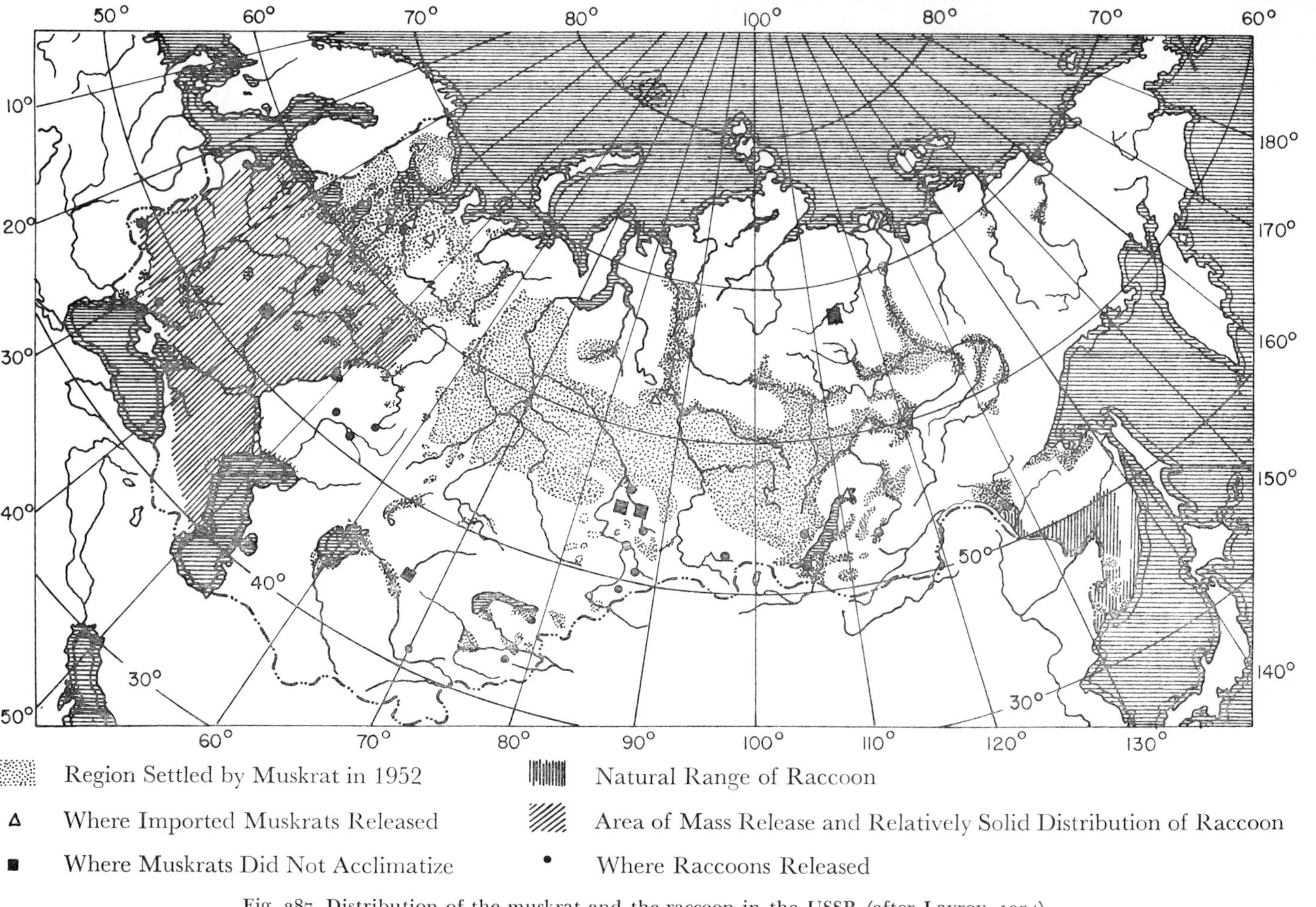

Fig. 287. Distribution of the muskrat and the raccoon in the USSR (after Lavrov, 1954).

REFERENCES

[The names of Russian authors are rendered in the transliteration suggested by the U.S. Board on Geographic Names, with the following exception: the Russian suffix "СКИИ" is transliterated as "sky" rather than "skiy." This has been done for two reasons: (a) the ending "sky" is now a common one in the United States and is the accepted spelling for Slavic family names of this type; (b) most of the Russian authors who write in English sign their names this way. Soviet authors who publish in Russian but who bear foreign names will have their surnames transliterated in the same manner, even though the results may sometimes appear strange. Thus the German "Schwartz," when borne by a Russian, is transliterated as "Shvarts," the English "Williams" as "Vil'yams," etc. The surnames of non-Russian authors publishing in their native languages appear in Latin script. M = Moscow, L = Leningrad, SPB = St. Petersburg. —Tr.]

Adol'f, T. K. 1951. Toward the question as to the influence of spring flooding on small mammals in the floodplain of the Mologi River. *Uch. Zap. Mosk. Gorod. Ped. in-ta Potemkina* 18 (1). M.

Agafonov, A. V., *et al.* 1957. Toward an ecology of the steppe eagle. *Byull. MOIP* 62 (2).

Alekhin, V. V. 1944. *Geography of plants.* M.

Aleksandrova, V. D. 1940. The nutritional characteristics of the plants of the far north. *Tr. in-ta Polyarn. Zemledeliya, Zhivotnovodchi. Promyslov. Kh-Va. Ser. Olenevodstvo* 2. M. and L.

Alikina, Ye. V. 1959. The influence of a water regime of feeding on ovogenesis and spermatogenesis in common and society voles. *Zool. Zhurn.* 28 (4).

Alisov, B. P., O. A. Drozdov, and Ye. S. Rubinshteyn. 1952. *A course in climatology,* chs. I and II. M.

Allee, W. C. 1931. *Animal aggregation.* Chicago.

———. 1934. Recent studies in mass physiology. *Biol. Rev.* 9.

———. 1938. *The social life of animals.* New York.

———. 1951. *Cooperation among animals.* New York.

Allee, W. C., *et al.* 1949. *Principles of animal ecology.* Philadelphia and London.

Allen, E. J. 1909. Mackerel and sunshine. *J. Mar. Biol. Ass. U.K.* 8.

Anderson, J. W. 1954. The production of ultrasonic sounds by laboratory rats and other animals. *Science* 119.

Andrewartha, H. G. 1952. Diapause in relation to the ecology of insects. *Biol. Rev.* 27.

Andrewartha, H. G., and L. C. Birch. 1948. Measurement of "environmental resistance" in the Australian plague grasshopper. *Nature* 101.

———. 1954. *The distribution and abundance of animals.* Chicago.

Andreyev, L. A. 1936. Characteristics of the aural analyzer of the dog on the basis of experimental data obtained according to the method of conditioned reflexes. *Zhurn. Tekh. Fiziki* 6 (12).

Andriyashev, A. P. 1939. *An outline of the zoogeography and origin of the fish fauna of the Bering Sea and allied waters.* Izd. LGU, L.

Andrushko, A. M. 1939. *The activity of rodents on dry pastures in Central Asia.* Izd. LGU, L.

Arnol'di, K. V. 1957. On the theory of the range in connection with the ecology and origin of species populations. *Zool. Zhurn.* 36 (11).

Babayan, A. S. 1950. A study of the principles of development in the hollyhock moth *Gelechius malvella* Hub. In Second Ecological Conference on the problem "Mass reproduction of animals and their prognosis," *Tezisy dokladov,* ch. 1. Kiev.

Babenyshev, V. P. 1938. The significance of fur cover and body size in certain rodents for heat elimination and their resistance to the effects of external factors. *Zool. Zhurn.* 17 (3).

Bachmetjev, P. *Experimentelle entomologische Studien Temperaturverhältnisse bei Insekten.* Vol. 1, Leipzig, 1901; vol. 2, Leipzig, 1907.

Backer, G. R. 1938. The evolution of breeding seasons. In *Evolution: Essays presented to E. S. Goodrich.* Oxford.

Bacot, A. W., and C. J. Martin. 1924. The respective influences of temperature and moisture upon the survival of the rat flea (*Xenopsylla cheopis*) away from its host. *I. Hyg.* 23.

Bakeyev, N. N., O. N. Skalon, and Yu. D. Chugunov. 1957. New data on the distribution and ecology of the seagull *Larus genei* Br. *Tr. Tsent. Byuro Kol'tsevaniya* 9. M.

Bakhmet'yev, P. I. 1898. The temperature of insects. *Zhurn. Nauchnoye Obozreniye* 5.

———. 1899. The temperature of bees in particular and insects in general. *Russk. Pchel. Listok* 14 (3–4).

Balashov, Yu. S. 1955. Mixing of temperature boundaries of cold and heat torpor in the housefly under the influence of changes in the temperature of the surrounding environment. *Zool. Zhurn.* 34 (2).

———. 1957. Some adaptations to the acquisition of large amounts of blood among the ixodid ticks. *Zool. Zhurn.* 36 (6).

Baldwin, S., and C. Kendeigh. 1932. Physiology of temperature of birds. *Sci. Publ. Cleveland Mus. Nat. Hist.* 3.

Balogh, J. A. 1953. *Zoocönologia ALAPJAI (Grundzüge der Zoocönologie).* Budapest.

Barabash-Nikiforov, I. I. 1956. New data on the intensive emigration of certain species of beasts over the last decade. *Zool. Zhurn.* 35 (2).

———. 1959. Symbiotic connections in the population of a beaver dam. *Zool. Zhurn.* 38 (5).

Baranov, F. I. 1918. Toward the question of the biological bases of the fishing industry. *Izv. Otd. Rybovodstva i Priklad. Ikhtiologii GIOA* 1.

———. 1928. On the predicted catch of the Caspian roach. On the Astrakhan failures. *Byull. Rybn. Khoz.* 1, 8.

———. 1947. Index of productivity of the fish school. *Rybn. Khozyaystvo* 2.

Baranov, G. I. 1957. On the applicability of Allen's law to the mouselike rodents. *Vestn. LGU* 15.

Baranov, P. A. 1955. Plants. *BSE* 36.

Baranovskaya, T. N. 1957. Moving of rodents by various types of transport. *Zool. Zhurn.* 36 (5).

Barbashova, Z. I. 1941. *Materials toward the problem of acclimatization to low partial pressure of oxygen.* Izd. AN SSSR.

Barbashova, Z. I., and G. A. Ginetsinsky. 1942. Properties of adaptation to altitude in Hissarian sheep. *Izv. AN SSSR, Ser. Biol.*

Barkroft, Zh. 1937. *The basic features of the architecture of physiological functions.* Biomedgiz, M. and L.

Bashenina, N. V. 1951. Materials on the abundance dynamics of small rodents in the forest zone. *Byull. MOIP* 2.

———. 1958. On the critical point in small voles. *Zool. Zhurn.* 37 (12).

Beklemishev, V. N. 1931. The basic concepts of biocenology as applied to the animal components of terrestrial associations. *Tr. po Zashch. Rasteniy* 1, 2.

———. 1934. Daily migrations of invertebrates in a complex of terrestrial biocenoses. *Tr. Permskogo In-Ta* 6 (3–4).

———. 1944. *Ecology of the malarial mosquito.* Medgiz, M.

———. 1949. *Medical entomology,* ch. I. Medgiz, M.

———. 1954. Parasitism of arthropods on terrestrial vertebrates. Ch. II. The basic directions of its development. *Med. Parazitologiya i Parazit. Bolezni* 1.

———. 1959. Populations and micropopulations of parasites and nidicola. *Zool. Zhurn.* 38 (8).

———. 1960. Spatial and functional structure of populations. *Byull. MOIP, Otd. Biol.* 65 (2).

Beklemishev, V. N., and I. Chetyrkina. 1935. Toward a biology of floods. *Vopr. Ekol. Biotsenol.* 2.

Belanovsky, I. D. 1956. The uniqueness of parasitism among insects and its economic significance. In *Problemy parazitologii, Tr. II Nauch. Konf. Parazitol. USSR.* Kiev.

Belopol'sky, L. O. 1957. *Ecology of colonial sea birds of the Barents Sea.* Izd. AN SSSR.

Belyayev, D. K. 1950. The role of light in the governing of biological rhythms of mammals. *Zhurn. Obshch. Biologii* 11 (1).

Belyayev, D. K., and L. G. Utkin. 1949. The influence of a reduced period of daylight on the maturing period for fur among foxes. *Karakulevodstvo Zverovodstvo* 2.

Belyayev, D. M. 1951. Osmotic pressure of the cavitial fluid of aquatic invertebrates in waters of differing salinity. *Tr. Vs. Gidrob. Obshch.* 3.

Belyayev, M. M. 1927. Ein Experiment über die Bedeutung der Schützfarbung. *Biol. Zentralbl.* 47.

———. 1947. *Coloration of animals and natural selection.* M.

Berezina, A. N. 1957. Midge attacks under natural conditions along the middle course of the Angara River. *Parazitol. Sbornik* 17; Izd. AN SSSR.

Berg, L. S. 1908. The Aral Sea. *Izv. Turkest. Otd. Geogr. Ob-Va* 5.

———. 1922. *Nomogenesis, or evolution on the basis of principles.* Pg. Gos. Izd.

———. 1935. Materials on the biology of the salmon. *Izv. VNIORKh* 20.

———. 1938. *Fundamentals of climatology,* vol. 2. L.

———. 1947. *Climate and life.* Geografgiz, M.

———. 1953. Spring and summer races of migratory fishes. *Ocherki po Obshch. Voprosam Ikhtiologii.* M.

Bergmann, S. 1847. Über die Verhältnisse der Wärmeökonomie der Tiere zu ihrer Grösse. *Abgedruckt aus den Göttinger Studien.*

Beskrovny, M. A. 1953. *A practical work on the ecology of animals,* ch. I. Izd. Khar'k. In-Ta, Khar'kov.

Bey-Biyenko, G. Ya. 1959. The principle of exchange of stations and the problem of initial divergence of species. *Zhurn. Obshch. Biologii* 20 (5).

———. 1961. On certain principles in the change of invertebrate fauna during conquest of the virgin steppe. *Entomol. Obozrenie* 40 (4).

Bezrukova, M. I., and T. G. Linnik. 1944. The nests of predatory birds as a place of concentration of the plague virus in nature. *Izv. Irkutsk. Protivochumn. in-ta Sibiri i DV* 5. Irkutsk.

Bibikov, D. I., L. V. Zhirnov, and V. P. Kulikova. 1956. Seasonal changes in terrestrial activity and intrapopulational contact among gray marmots in the Tyan'-Shan'. *Tr. Sred.-Aziat. N.-I. Protivochumn. In-Ta* 3. Alma-Ata.

Birch, L. C. 1944. Two strains of *Calandra oryzae* L. (Coleoptera). *Aust. J. Exp. Biol. Med. Sci.* 22 (4).

———. 1953. Experimental background to the study of the distribution and abundance of insects. II. The relation between innate capacity for increase in numbers and the abundance of three grain beetles in experimental populations. *Ecology* 34.

Bissonette, T. H. 1936. Sexual photoperiodicity. *Quart. Rev. Biol.* 11.

Bissonette, T. H., and A. P. R. Wadlund. 1932. Duration of testis activity of *Sturnus vulgaris* in relation to the type of illumination. *J. Exp. Biol.* 9.

Bityukov, E. P. 1960. On the ecology of *Limnocalanus grimaldii* (Guerne) of the Gulf of Finland. *Zool. Zhurn.* 39 (12).

Blair, W. 1940. Home ranges and populations of the meadow vole in southern Michigan. *J. Wildl. Mgmt.* 4.

———. 1943. Populations of the deer mouse and associated small mammals in the mesquite association of southern New Mexico. *Contr. Lab. Vertebr. Biol. Univ. Mich.* 20.

———. 1951. Population structure, social behavior, and environmental relations in a natural population of the beech mouse (*Peromyscus polionotus leucocephalus*). *Contr. Lab. Vertebr. Biol. Univ. Mich.* 48.

Blinov, L. K. Toward the question as to the origin of the salt content of sea water. *Sb. Meteorol. Gidrolog.* 4, 1946; 4, 1947.

Bobrinsky, N. A., L. A. Zenkevich, and Ya. A. Birshteyn. 1946. *Geography of animals.* Sov. Nauka, M.

Bodenheimer, F. S. 1934. Über die Temperaturabhängigkeiten der Insekten. *Zool. Jahrb.* 66 (1–2).

———. 1938. *Problems of animal ecology.* Oxford.

———. 1958. *Ecology today.* The Hague.

Bodrova, N. V., and B. V. Krayukhin. 1958. On the reaction of fishes to the action of an electric current. *Tr. Soveshch. Fiziol. Ryb.*; Izd. AN SSSR.

Bogorov, V. G. 1938. Biological seasons of the polar sea. *DAN SSSR* 19 (8).

———. 1938. Daily vertical distribution of plankton under polar conditions (in the southeastern part of the Barents Sea). *Tr. PINRO* 2 (3).

———. 1939. Properties of seasonal phenomena in plankton of the polar seas and their significance for ice predictions. *Zool. Zhurn.* 18.

Bokova, Ye. N. 1939. Consumption and conquest of food by the Caspian roach. *Tr. VNIRO* 11.

Boldaruyev, V. P. 1955. Fecundity and nutritional specialization of the Siberian silkworm (*Dendrolimus sibiricus* Tshtv.). *Zool. Zhurn.* 34 (4).

Bondarenko, N. V. 1950. The influence of a shortened day on the yearly cycle of development of the common spiderweb tick. *DAN SSSR* 70 (6).

———. 1958. Properties of diapause in the spiderweb tick *Tetranychus urtice* Koch. *Zool. Zhurn.* 37 (7).

Bongard, M. M., and M. S. Smirnov. 1959. Color vision in man and animals. *Priroda* 5.

Borodin, L. P. 1951. The role of the springtime high-water period in the ecology of small mammals on floodplain biotopes. *Zool. Zhurn.* 30 (6).

Borodulina, T. O. 1951. On the latent period in the development of the embryo of the Altay mole. *DAN SSSR* 80 (4).

Bozhko, G. V. 1956. Prognosis of changes in the abundance of deerflies (Tabanidae) in the valley of the lower Dnieper in connection with the formation of the Kakhov Reservoir. In *Problemy parazitologii, Tr. II Nauch. Konf. Parazitol. USSR.* Kiev.

Braarud, T. 1951. Salinity as an ecological factor in marine phytoplankton. *Physiol. Plant.* 4.

Brand, T. 1951. *Anaerobiosis in invertebrates.* Izd. IL, M.

Brinkman, A. 1926. Coccidiosen hosliripen. *Bergens Mus. Aurbon Naturvid. Raenk.* 9.

Broekhuysen, G. J. 1941. A preliminary investigation of the importance of desiccation, temperature, and salinity as factors controlling the vertical distribution of certain intertribal marine gastropods in False Bay, South Africa. *Tr. Roy. Soc. S. Afr.* 28.

Brues, C. T. 1946. *Insect dietary*. Cambridge, Mass.

Bruyevich, S. V. 1939. Distribution and dynamics of the living matter in the Caspian Sea. *DAN SSSR* 25.

Buch, L. von. 1825. Physikalische Beschreibung der Canarischen Inseln. *Kaiserl. Akad. Wiss.* Berlin.

Buddenbrock, W. von. 1928. *Grundriss der vergleichenden Physiologie.*

———. 1952. *Vom Farbensinn der Tiere.* Stuttgart.

Bukhman, E. G., and V. Ya. Andreyevsky. 1940. A change in sperm production and thermoregulation in sheep during the summertime. *Doklad. VASKhNIL* 9.

Bullock, T. H., and F. P. J. Diecke. 1956. Properties of an infrared receptor. *J. Physiol.* 134 (1).

Bullough, W. S. 1951. *Vertebrate sexual cycles.* London and New York.

Burdelov, A. S. 1956. Toward the question as to the causes of perennial changes in abundance of the large gerbil. *Tr. Sred.-Aziat. N.-I. Protivochumn. In-Ta* 3.

Burkholder, P. R. 1952. Cooperation and conflict among primitive organisms. *Amer. Scient.* 40.

Busnel, M.-C., B. Dumortier, and R.-G. Busnel. 1960. Recherches sur la phonocinese de certains insects. *Bull. Soc. Zool. Fr.* 8 (4).

Buturlin, S. A. 1913. Observations of mammals made during the Kolyma expedition of 1905. *Dnevn. Zool. Otd. O-Va Lyubit. Yestestv.*, Nov. Ser. 1 (5). M.

Buxton, P. A. 1923. *Animal life in deserts.* London.

Bykov, K. M., and A. D. Slonim. 1949. An experimental study of vertical regulation under natural life circumstances. *Sb. Opyt Izucheniya Regulyatsii Fiziol. Funktsiy.* M. and L.

Byunning, E. 1961. *Rhythms of physiological processes.* Izd. IL, M.

Carriker, M. R. 1951. Ecological observations on the distribution of oyster larvae in New Jersey estuaries. *Ecol. Monogr.* 21 (1).

Cesnola, A. P. 1904. Preliminary note on the protective value of colour in *Manthis religiosa. Biometrika* 3.

Chagin, K. P. 1948. The level of activity of mosquito attacks against man and its daily rhythm under natural conditions in thc Primorsky kray. *Entom. Obozr.* 30 (1–2).

Chang, Ho Chen. 1960. Winter observations of flocks of small birds in the forests of the Zvenigorod Biological Station of the Moscow State University. *Zool. Zhurn.* 39 (10).

Chapman, R. N. 1928. The quantitative analysis of environmental factors. *Ecology* 9.

———. 1931. *Animal ecology with special reference to insects.* New York.

Chernyavskaya, S. I. 1956. Seasonal displacement and wanderings of wild hoofed

animals and the bear in the region of the Caucasus Preserve in connection with the distribution of fruit- and nut-tree harvests. *Byull. MOIP* 61 (4).

Chernyshev, V. B. 1959. Reactions of certain species of insects to different parts of the spectrum. *Zool. Zhurn.* 38 (5).

Chirkova, A. F. 1947. Materials on the ecology of the red fox. *Tr. VNIO* 7.

———. 1947. Calculation of foxes by burrows and prediction of their numbers. *Nauch.-Metod. Zap. Gl. Upr. po Zap.* 9.

———. 1948. Materials on the ecology of the fox. II. The food regime and endoparasitic infection of the red foxes in the Stavropol' kray. *Tr. VNIO* 8.

———. 1951. Preparatory methodology for prognosis of changes in the abundance of polar foxes. *Vopr. Biol. Pushn. Zverey Tekh. Okhot. Promysl. Tr. VNIO* 11.

———. 1953. Materials on the dynamics of abundance of the red foxes of the Voronezh oblast' in connection with prognoses of their "harvests." *Tr. VNIO* 13.

———. 1957. Distribution of fox mange among the foxes of the USSR in connection with geographic factors. *Zool. Zhurn.* 36 (5).

Chitty, D. 1954. Tuberculosis among wild voles, with a discussion of other pathological conditions among certain mammals and birds. *Ecology* 35.

———. 1955. Allgemeine Gedankengänge über die Dichteschwankungen bei der Erdmaus (*Microtus agrestis*). *Z. Säugetierk.* 20.

———. 1955. Adverse effects of population density upon the viability of later generations. In *The numbers of man and animals,* published by Oliver and Boyd Ltd. for the Institute of Biology. Edinburgh.

———. 1957. Self-regulation of numbers through changes in viability. *Cold Spr. Harb. Symp. Quant. Biol.* 22.

Christian, J. J. 1955. Reserpine suppression of density-dependent adrenal hypertrophy and reproductive hypoendorinism in populations of male mice. *Amer. J. Physiol.* 187 (2).

Chugunov, Yu. D., N. I. Kudryashova, and N. D. Chugunov. Properties of thermoregulation in steppe and mountain varieties of small susliks in the North Caucasus. *Tr. N.-I. Protivochumn. in-ta Kavkaza Zakavkaz'ya* 1. Stavropol'.

Clark, L. R. 1953. The ecology of *Chrysomella gamellata* Rossi and *C. hyperici* and their effect on St. John's wort in Bright district, Victoria. *Aust. J. Zool.* 1.

Clarke, G. L. 1954. *Elements of ecology.* New York and London.

Clarke, J. R. 1955. Influence of numbers of reproduction and survival in two experimental vole populations. *Proc. Roy. Soc. B* 144.

Clausen, C. P. 1958. Biological control of insect pests. *Ann. Rev. Entom.* 3. Palo Alto.

Clements, F. E., and V. K. Shelford. 1939. *Bio-ecology.* New York.

Coats, C. W. 1954. Electric fishes. *Anim. Kingd.* 57 (6).

Cowan, I. McT. 1938. Geographic distribution of color phases of red fox and bear in the Pacific region of North America. *J. Mammal.* 19.

Crowcroft, P., and F. P. Row. 1957. The growth of confined colonies of the wild house mouse (*Mus musculus* L.). *Proc. Zool. Soc. Lond.* 129 (3).

———. 1958. The growth of confined colonies of the wild house mouse (*Mus musculus* L.). The effect of dispersal on animal fecundity. *Proc. Zool. Soc. Lond.* 131 (3).

Cuénot, L. 1925. *L'adaptation.*

———. 1936. *L'espece.*

Danilevsky, A. S. 1957. Seasonal rhythm and intraspecies geographic differentiation among insects. *Vestn. LGU* 21.

Danilevsky, A. S., and Ye. I. Glinyanaya. 1950. On the influence of rhythm of illumination and temperature on the rise of diapause in insects. *DAN SSSR* 71 (5).

Danilov, D. N. 1941. Methodology for evaluating harvests of foods for squirrels when predicting the latter's abundance. *Tr. Tsentr. Lab. Biol. Okhot. Promysla* 5.

———. 1944. Food resources of the fir forest and use of them by the squirrel. *Tr. Tsentr. Lab. Biol. Okhot. Prom. Tovaroved.* 6. M.

———. 1947. *Preparation of muskrat areas.* Zagotizdat.

———. 1950. Geographic displacement of pine seeds in connection with environmental circumstances. *Probl. Piz. Geog.* 15.

———. 1951. Expansion of the range of the elk in the south of the European USSR. *Okhrana Prirody* 13.

———. 1952. *Periodicity of fruit-bearing and geographic displacement of harvests of pine seeds.* Goslesbumizdat, M. and L.

Danilov, N. N. 1958. Changes in the ornithofauna of regrowing clearings in the northern Urals. *Zool. Zhurn.* 37 (12).

Dantsig, Ye. M. 1939. On the biological forms of the comma-shaped worms (*Lepidosaphes ulmi* L.) Homoptera, Coccoidea. *Zool. Zhurn.* 38 (6).

Darlington, P. J., Jr. 1938. The origin of the fauna of the Greater Antilles with discussion of dispersal of animals over water and through the air. *Quart. Rev. Biol.* 13.

Darwin, C. 1939. *The origin of species by means of natural selection.* 3rd ed. Izd. AN SSSR, M. and L.

DeBach, P. 1949. Population studies of the long-tailed mealybug and its natural enemies on citrus trees in southern California, 1946. *Ecology* 30.

Deevey, E. S. 1947. Life tables for natural populations of animals. *Quart. Rev. Biol.* 22.

Dement'yev, G. P. 1935. *Hunting with catching birds.* KOIZ, M.

———. 1935. Toward the question of Bergmann's rule. *Byull. MOIP* 44 (7–8).

Dement'yeva, T. F. 1957. Principles of fluctuation of abundance of the basic commercial fishes and methods of commercial prognosis. *Tr. Vs. Konf. Vopr. Rybn. Khoz-Va.*; Izd. AN SSSR.

———. 1958. Dynamics of marine fish populations. *Tr. VNIRO* 34.

———. 1961. The significance of the resolving factor in the light of yearly and perennial fluctuations in the abundance of populations. *Tr. Soveshch. Dinamike Chislennosti Ryb.*; Izd. AN SSSR.

Dergunov, N. I. 1923. A method of mass calculation of the number of birds and

an experimental application of it in the case of woodpeckers. *Tr. Vseros. S'yezda Zool. Anat. Gistol.*

Deryugin, K. M. 1915. The fauna of the Kola Gulf and the conditions of its existence. *Zap. AN SSSR, Fiz.-Mat. Otd.* 2 (34).

———. 1928. The fauna of the White Sea and the conditions of its existence. *Issledovaniye Morey SSSR* 7–8.

Derzhavin, A. N. 1922. The sevruga (*Acipenser stellatus*) (a biological outline). *Izv. Bakin, Ikhtiol. Labor.* 1.

———. 1947. Reproductive supplies of the sturgeons. *AN Azerb. SSR.* Baku.

Dethier, V. G. 1937. Gustation and olfaction in lepidopterous larvae. *Biol. Bull.* 72.

Dezhkin, V. V. 1959. Results of measurements of temperature in dams of the river beaver. *Zool. Zhurn.* 38 (1).

Dice, L. R. 1940. Speciation in *Peromyscus. Amer. Nat.* 74.

———. 1945. Minimum intensities of illumination under which owls can find dead prey by sight. *Amer. Nat.* 79.

———. 1947. Effectiveness of selection by owls of deer mice (*Peromyscus maniculatus*) which contrast in color with their background. *Contr. Lab. Vertebr. Biol. Univ. Mich.* 34.

———. 1952. *Natural communities.* Ann Arbor, Mich.

Diesselhorst, G. 1935. Hörversuche an Fischen ohne Weberschen Apparat. *Z. Vergl. Physiol.* 25.

Dijkgraff, S. 1952. Über die Schallwahrnemung bei Meeresfischen. *Z. Vergl. Physiol.* 34.

Dimo, N. A. 1945. Wood lice and their role in the formation of soils in the deserts. *Pochvovedeniye* 2.

Dinesman, L. G. 1948. Adaptation of amphibians to various circumstances of moist air. *Zool. Zhurn.* 27 (3).

———. 1948. Toward the question as to the ecological differentiation of species among amphibians. *Byull. MOIP, Otd. Biol.*, Nov. Ser. 53 (6).

———. 1949. On the distribution and ecology of reptiles in connection with zones of solar radiation. *Probl. Fiz. Geog.* 14.

———. 1960. *A change in the nature of the northwest Cis-Caspian lowlands.* Izd. AN SSSR, M.

———. 1961. *The influence of wild mammals on the formation of stands of timber.* Izd. AN SSSR, M.

Dinesman, L. G., and M. L. Kaletskaya. 1952. Methods for quantitative calculation of amphibians and reptiles. *Sb. Metody Ucheta Chislennosti i Geograf. Raspredeleniya Nazemn. Pozvonochnykh.*; Izd. AN SSSR.

Dizer, Yu. B. 1955. On the physiological role of the elytra and subelytral cavity among steppe and desert Tenebrionidae. *Zool. Zhurn.* 34 (2).

Dmitriyev, V. V. 1938. Hoofed animals of the Altay Preserve and adjacent areas. *Tr. Altaysk. Gos. Sap.* 1.

Dogel', V. A. 1936. Properties of mallophagan infestation among cuckoos. *Priroda.*

———. 1947. *A course in general parasitology.* Uchpedgiz, L.

Dokuchayev, V. V. 1883. The Russian black earth. Imp. Bol'n. Ekon. Obshch. SPB.

———. 1886. Great moments in the history of land evaluation in European Russia since the classification of Russian soils. Materials on the evaluation of the lands of Nizhegorod guberniya, 1, SPB.

———. 1954. *Toward the question of the opening in Russian universities of chairs of soil science and studies on soil microorganisms.* 1896. Izbr. Soch. Sel'khozgiz.

———. 1954. *Toward the study of the zones of nature.* Izbr. Soch. Sel'khozgiz.

Donaurov, S. S., and V. P. Teplov. 1938. The wild boar in the Caucasus Preserve. *Tr. Kavkazsk. Gos. Zap.* 1. M.

Donaurov, S. S., V. P. Teplov, and P. A. Shikina. 1938. The feeding habits of the forest marten under the conditions in the Caucasus Preserve. *Tr. Kavkazsk. Gos. Zap.* 1. M.

Dryagin, P. A. 1949. Sexual cycles and fish spawning. *Izv. Vs. N.-I. in-ta Ozern. Rechn. Khoz.-Va.* 28.

Dubinin, V. P. 1948. The significance of migration of animals in the distribution of illnesses. *Izv. AN Kazakh. SSR, Ser. Parazitol.* 5.

———. 1953. Feather ticks (Analgesoidea). *Fauna SSSR, Paukoobraznyye* 6 (6).

———. 1955. Mange ticks (Acariformes, Sarcoptoidea) and mange illnesses of wild mammals. *Zool. Zhurn.* 34 (6).

———. 1956. Orders of ticks and their position in the system Chelicerata. *Tr. II Nauch. Konf. Parazitol. USSR.* Kiev.

———. 1959. Tasks and principles of mapping populations of parasitic animals in connection with the feeding habits of their final and intermediate hosts. *Sb. Geog. Naseleniya Nazem. Zhiv. i Metody Ego Izucheniya.* M.

Dzhaya, I., and S. Dzhelineo. 1935. The tension of atmospheric oxygen and resistance to means of cooling. *Glas. Srpsk. Kral'. Akad.* 1 (82B).

Dzhelineo, S. 1959. The development of homoiothermy in mammals. *Usp. Sovr. Biol.* 47 (1).

Ecological Conference on the problem "Mass reproduction of animals" (15–20 Nov. 1940). *Tezisy dokladov,* chs. 1 and 2. Izd. AN SSSR, Kiev, 1940.

Second Ecological Conference on the problem "Mass reproduction of animals and their prognosis." *Tezisy dokladov,* chs. 1 and 2, 1950; ch. 3 and Resolutions, 1951. Izd. Kiyevsk. In-Ta, Kiev.

Third Ecological Conference (3–10 Dec. 1954). *Tezisy dokladov,* chs. 1–3, ch. 4 (Resolutions). Izd. Kiyevsk. In-Ta, Kiev, 1954.

Edney, E. B. 1951. The evaporation of water from wood lice and the millipede *Glomeris. J. Exp. Biol.* 28.

Edwards, G. A., and L. Irving. 1943. The influence of temperature and season upon the oxygen consumption of the sand crab *Emerita talpoida* Say. *J. Cell. Comp. Physiol.* 21.

Eisentraut, M. 1933. Winterstarre, Winterschlaf, und Winterruhe. *Mitt. Zool. Mus. Berl.* 19.

———. 1937. Die deutschen Fledermäuse, eine biologische Studie. *Verb. Schöps Leipzig* 13 (4).

Elton, C. 1934. *Animal ecology*. Biomedgiz.
———. 1960. *The ecology of invasions of animals and plants*. Izd. IL, M.
Elton, C. S. 1924. Periodic fluctuations in the numbers of animals: Their causes and effects. *Brit. J. Exp. Biol.* 2.
———. 1925. Plague and the regulation of numbers in wild animals. *I. Hyg.* 24.
———. 1927, 1939. *Animal ecology*. New York.
———. 1938. Animal numbers and adaptation. In *Evolution: Essays on aspects of evolutionary biology*. Oxford.
———. 1942. *Voles, mice and lemmings (problems in population dynamics)*. Oxford.
Émme, A. M. 1944. Diapause in insects. *Usp. Sovr. Biol.* 18 (1).
Erman, L. A. 1956. On the quantitative aspect of the feeding habits of rotifers. *Zool. Zhurn.* 25 (7).
Errington, P. L. 1942. On the analysis of productivity in populations of higher vertebrates. *Wildl. Mgmt.* 6 (2).
———. 1943. On analysis of mink predation upon muskrats in the north-central United States. *Iowa St. Coll. Agric. Exp. Sta. Bull.* 320. Ames.
———. 1945. Some contributions of a fifteen-year local study of the northern bobwhite to a knowledge of population phenomena. *Ecol. Monogr.* 15.
———. 1946. Predation on vertebrate population. *Quart. Rev. Biol.* 21.
———. 1951. Concerning fluctuations in populations of the prolific and widely distributed muskrat. *Amer. Nat.* 85.
———. 1956. Factors limiting higher vertebrate populations. *Science* 124.
Evans, F. C. 1938. Physiological relationships between insects and their host plants. I. The effect of the chemical composition of the plant on reproduction and production in *Brevicoryne brassicae* L. (Aphididae). *Ann. Appl. Biol.* 25.
Eygenson, M. S. 1948. Sun, weather, and climate. *Tr. II Vs. Geogr. S'yezda* 1.
Faleyeva, Z. N. 1959. The relationship between the onset of hibernation and the level of fat reserves and the state of the nervous system in the speckled suslik. *Zool. Zhurn.* 38 (2).
Farkas, G. 1936. Zur Kenntnis des Horvermögens und des Gehörorgans der Fische. *Acta Oto-Laryng.* 23.
Farr, W. 1843. Causes of mortality in town districts. *V. Ann. Rep. Reg. Gen. Births, Deaths and Marriages in England.*
Fedorov, V. N., I. I. Rogozin, and B. K. Fenyuk. 1955. *Plague prophylaxis*. Medgiz, M.
Fedotov, D. M. 1944. Observations of the internal state of *Eurygaster integriceps*. *DAN SSSR* 42 (9).
———. 1949. Methods of prognosing the abundance of *E. integriceps* according to the internal state. *Dokl. VASKhNIL* 9.
Fedotova, K. M. 1950. The role of biological factors in mass reproductions of the *zlatoguzka* and *boyaryshnika*. In Second Ecological Conference on the problem "Mass reproduction of animals and their prognosis," *Tezisy dokladov*, ch. 1 Kiev.
Fenyuk, B. K. 1936. Resettlement of steppe rodents. *Priroda* 10.

———. 1937. The influence of ground-breaking on the abundance of mouselike rodents and the biological bases of combating them. *Vestn. Mikrob. Epidem. Parazitol.* 16 (3–4).

———. 1941. The home instinct in rodents. *Priroda* 5.

———. Ecological factors of the nidality of plague among rodents. *Vestn. Mikrob. Epidem. Parazitol.* 23, 1944; *Tr. Yubil. Nauch. Konf. "Mikrob" Saratov*, 1948.

Fenyuk, B. K., and M. P. Demyashev. 1936. A study of the migrations of gerbils (Mammalia, Glires) by the banding method. Life span in nature of gerbils (*Pallasiomys meridianus*, Mammalia). *Vestn. Mikrob. Epidem. Parazitol.* 15 (1, 3–4).

Ferhat-Akat, S. 1939. Über den Gehörsinn der Amphibien. *Z. Vergl. Physiol.* 26.

Fesenkov, V. G. 1949. *Contemporary concepts of the universe.*

Filatova, L. G. 1944. Ecologo-physiological research among the mammal inhabitants of the semideserts of Kirghizia. *Avtoref. Diss.*, Frunze.

———. 1946. Properties of lung respiration and physiological hypoxemia in the hedgehog. *Byull. Exsp. Biol. Med.* 20 (5).

———. 1949. Thermoregulation of mammal inhabitants of the semideserts. *Sb. Opyt Izucheniya Regulyatsii Fiziol. Funktsiy*; Izd. AN SSSR.

Filinger, G. A. 1931. The effect of temperature on feeding and development of the leaf-tier *Phlyctaenia ferruginalis. J. Econ. Ent.* 24.

Flerov, B. V. 1958. Methodological references concerning the working out of prognoses of the reproduction of the Siberian silkworm by computing its nidi and observing them. *M-vo S/kh RSFSR.* L.

Flint, V. Ye. 1955. On an experimental study of the activity of setting eider ducks. *Zool. Zhurn.* 33 (1).

———. 1958. Toward the question as to the structure of the range and types of settlements of certain mouselike rodents in North Kazakhstan. *Byull. MOIP, Otd. Biol.* 63 (6).

Florken, M. 1947. *Biochemical evolution.* Izd. IL, M.

Folitarek, S. S. 1947. Materials on the commercial biology of the Bargruza sable. *Tr. VNIO* 7.

———. 1948. Predatory birds as a factor in natural selection in natural populations of small rodents. *Zhurn. Obshch. Biol.* 9 (1).

Forbes, S. A. 1882. The regulative action of birds upon insect oscillations. *Bull. Ill. Lab. Nat. Hist.* 1 (6).

———. 1887. The lake as a microcosm. *Bull. Sci. Acad.* Peoria.

———. 1907. On the local distribution of certain Illinois fishes (an essay in statistical biology). *Bull. Ill. Lab. Nat. Hist.* 7.

Ford, E. B. 1945. Polymorphism. *Biol. Rev.* 20.

Formozov, A. N. 1929. Livestock slaughtering: Its significance for steppe fauna and the struggle against pests. *Priroda* 11.

———. 1933. The cedar-nut harvest, the flights of the Siberian nutcracker into Europe, and fluctuations in the abundance of the squirrel. *Byull. NIIZ MGU* 1.

———. 1934. Basic questions of the ecology of the squirrel (*Sciurus vulgaris* L.)

and a program of projects in this area. In *The ecology of the squirrel.* KOIZ, M.

———. 1934. The lake forest steppe and the West Siberian steppe as an area of mass habitation of waterfowl. *Byull. MOIP, Otd. Biol.* 48 (1).

———. 1934. Predatory birds and rodents. *Zool. Zhurn.* 13 (4).

———. 1934. Materials toward a biology of the hazel grouse according to observations in the north of the Gor'ky kray. *Byull. MOIP, Otd. Biol.* 10 (3).

———. 1934. Toward the question of interspecies competition. *DAN SSSR* 111 (3).

———. 1935. *Fluctuations in abundance of commercial animals.* KOIZ, M.

———. 1936. Migrations of the common squirrel (*Sciurus vulgaris* L.) in the USSR. *Tr. ZIN AN SSSR* 3.

———. 1937. On the question as to the conquest of the terrestrial vertebrate fauna and questions of reconstructing it. *Zool. Zhurn.* 16 (3).

———. 1937. A program and methodology of projects of observation points aimed at calculating mouselike rodents so as to predict their mass appearances. *Uch. Zap. MGU* 11.

———. 1937. Materials toward the ecology of waterfowl according to observations on the lakes of the Naurzum State Preserve (North Kazakhstan). *Sb. Pamyati Akad. Mensbira.* M. and L.

———. 1939. The role of epizootics in the abundance dynamics of commercial mammals and birds. *Soveshch. Parazitol. Probl. AN SSSR, Tezisy Dokl.*

———. 1942. A study of the fluctuations in abundance of commercial animals and organization of "harvest prognoses" on a hunting farm of the USSR during the period 1917–1942. *Zool. Zhurn.* 21 (6).

———. 1946. *The snow cover in the life of mammals and birds of the USSR.* Izd. MOIP, M.

———. 1947. An outline of the ecology of the mouselike rodents carrying tularemia. *Mater. Pozn. Fauny Flory SSSR, Otd. Zool.*, Nov. Ser. 7 (22); *Mater. po Gryzunam* 1.

———. 1948. Small rodents and insectivores of the Shar'ya region of the Kostroma oblast' during the period 1930–1940. *Mater. Pozn. Fauny Flory SSSR, Otd. Zool.*, Nov. Ser. 17; *Mater. po Gryzunam* 3.

———. 1950. On certain features of the biology of birds in connection with questions of protection against forest and tree-farm pests. In *Birds and forest pests.* Izd. MOIP, M.

———. 1951. The quantitative method in the zoogeography of terrestrial vertebrates and tasks in the rebuilding of nature in the USSR. *Izv. AN SSSR, Ser. Geog.* 2.

———. 1952. *The pathfinder's companion.* Izd. MOIP, M.

———. 1952. *Methods of computing the abundance and geographic distribution of terrestrial vertebrates.* Izd. AN SSSR, M.

———. 1953. Materials on the biogeography of the USSR. I. Zoogeography and ecology of the terrestrial fauna of Kazakhstan. *Tr. in-ta Geog. AN SSSR* 54. M. and L.

———. 1959. On the movements and fluctuations of the boundaries of distribution of mammals and birds. In *Handbook: Geography of land animal population and methods for its study*. Izd. AN SSSR, M.

Formozov, A. N., and A. G. Voronov. 1939. The activity of rodents on pastures and haying fields of West Kazakhstan and its agricultural significance. *Uch. Zap. MGU, Zool.* 12.

Formozov, A. N., K. S. Khodashova, and B. A. Golov. 1954. The influence of rodents on pasture and hay-field vegetation of the clayey semideserts of the Volga-Ural interfluve. *Vop. Uluchsh. Korm. Bazy Stepi Polupust. Pustyni. Zonakh SSSR*. M. and L.

Formozov, A. N., N. P. Naumov, and I. D. Kiris. 1934. *Ecology of the common squirrel*. KOIZ, M.

Formozov, A. N., V. I. Osmolovskaya, and K. N. Blagosklonov. 1950. *Birds and forest pests*. Izd. MOIP, M.

Fraenkel, G., and M. Blewett. 1946. The dietetics of the caterpillars of three *Ephestia* species, *E. kuehniella, E. elutella,* and *E. cautella,* and of a closely related species, *Plodia interpunctella. J. Exp. Biol.* 22.

Fraenkel, G., and D. L. Gunn. 1940. *The orientation of animals: Kineses, taxes and compass reactions*. Oxford.

Frank, F. 1953. Untersuchungen über den Zussamenbruch von Feldmausplagen. *Zool. Jb. (Syst.)* 82.

———. 1954. Die Kausalität der Nagetierzyklen im Lichte Neuer populationsdynamischer Untersuchungen an deutschen Microtinen. *Z. Morph. o Okol. Tiere.* 43.

———. 1957. The causality of microtine cycles in Germany. *J. Wildl. Mgmt.* 21.

Frideriks, K. 1932. *The ecological fundamentals of applied zoology and entomology*. Sel'khozgiz, M. and L.

Frings, H., and M. Frings. 1958. Uses of sound by insects. *Ann. Rev. Ent.* 3.

Frisch, K. 1923. Das Problem des tierischen Farbensinnes. *Naturwiss.* 13.

Frisch, K. 1933. Sind die Fische farbenblind? *Zool. Jb. Abt. Allg. Zool.* 33.

———. 1946. Die Tanze der Bienen. *Öster. Zool. Z.* 1.

———. 1950. Die Sonne als Kompas im Leben der Bienen. *Experimenta* 6.

Frisch, K. 1955. *The bees: Their sense of vision, smell, taste, and language.* Izd. IL, M.

Frisch, K., and H. Stetter. 1932. Untersuchungen über den Siz des Gehörsinnes bei der Elritze. *Z. Vergl. Physiol.* 17.

Froloff, J. 1925–1928. Bedingte Reflexe bei Fischen. *Pflüg. Arch. Ges. Physiol.* 1–2.

Gagina, T. N. 1958. Toward recognition of mass nonperiodic migrations of wild gallinaceous birds in the Siberian taiga. *Izv. Irkutskogo. S.kh. in-ta Irkutsk* 8.

Galakhov, I. N. 1937. Autumn flight of cranes and geese as an indicator of cold fronts. *Priroda* 2.

Galambos, R., and D. R. Griffin. 1942. Obstacle avoidance by flying bats. *J. Exp. Zool.* 89.

Galuzo, I. G., and V. I. L'vova. 1945. Materials on the ecology of *Hyalomma detritum* P. Sch. *Tr. Tadzhik. FAN SSSR* 14; Zool. Parazitol., M. and L.

Garret-Jones, C. A. 1950. Dispersion of mosquitoes by wind. *Nature* 165.

Gause, G. F. 1934. Experimental studies of the struggle for existence between *Paramecium caudatum, P. aurelia,* and *Stylonichia mytilus. Zool. Zhurn.* 13 (1).

———. 1934. *The struggle for existence.* Baltimore.

Gayevskaya, N. S. 1948. Trophological direction in hydrobiology: Its object, some basic problems and tasks. *Sb. Pamyati Akad. Zernova*; Izd. AN SSSR.

———. The feeding habits and nutritional associations of the animals inhabiting the bottom vegetation and the shoreline outcroppings of the Black Sea. *Byull. MOIP* 61 (5), 1956; *Zool. Zhurn.* 37 (11), 1958.

Gaysky, N. A. 1944. Infections and immunity among animals which go into winter hibernation. *Izv. Irkutsk. Gos. Protivochum. in-ta Sib. DV* 5.

Gein, D. L. 1944. Temperature of the body in poikilothermic animals. *Usp. Sovr. Biol.* 17 (1).

Gekker, P. F. 1957. *An introduction to paleoecology.* Gosgeotekhizdat, M.

Geoffroy Saint-Hilaire, Isidore. 1859. *Histoire naturelle générale des regnes organiques.*

Geptner, V. G. 1936. *General zoogeography.* Biomedgiz, M.

———. 1956. On the number of species of fauna in the USSR and its relationship to world fauna. *Zool. Zhurn.* 35 (12).

Gerasimenko, G. G. 1950. The effects of light on certain seasonal changes in the organism of the speckled suslik. *DAN SSSR* 71 (3).

Gerbil'sky, N. L. 1957. Intraspecies biological differentiation and its significance for the species in the fish world. *Vestn. LGU* 21.

Gerstell, R. 1939. Physiological variations in wild turkeys and their significance in management. *Rev. Bull. Penn. Game Com.* 2.

Getsova, A. B., and L. K. Lozina-Lozinsky. 1955. The role of insect behavior in the process of their adaptation to vegetable food matter. *Zool. Zhurn.* 34 (5).

Geyger, R. 1931. *The climate of the air near the surface.* Gos. Izd. S.-kh. Kolkh. Koop. 1-ry, M.

Geyspits, K. F. 1957. On the mechanism of perception of light stimuli in the photoperiodic reaction of caterpillars of Lepidoptera. *Zool. Zhurn.* 36 (4).

Gilyarov, M. S. 1942. Toward an entomological evaluation of seed turnover in the kok-saghyz [*Taraxacum kok-saghyz*—Ed.]. *Dokl. VASKhNIL* 3.

———. 1944. The relationship between size and abundance of soil invertebrates. *Dokl. AN SSSR* 43 (6).

———. 1945. Fundamental properties of harmful insects which adapt to field sowing of seed. *Dokl. AN SSSR* 47 (3).

———. 1949. *Properties of the soil as media of habitation and their significance in the evolution of insects.* Izd. AN SSSR, M. and L.

———. 1954. The species as a population and the biocenose. *Zool. Zhurn.* 33 (4).

Gladkov, N. A. 1958. Some questions on the zoogeography of the cultivated landscape (the fauna of birds, for example). *Uch. Zap. MGU, Ornitol.* 197.

Golikov, A. N. 1959. The influence of factors of the external environment on the intraspecies variability of *Neptunea arthritica* and *Littorina squalida. Zool. Zhurn.* 38 (9).

Gorizontova, M. N., L. A. Krasnaya, and T. S. Perel'. 1957. Observations of the distribution and abundance of rain worms in the soil over a 1-year span. *Uch. Zap. Mosk. Gorod. Ped. in-ta Potemkina* 65.

Gorlenko, M. V., I. V. Voronkevich, and T. Ye. Maksimova. 1956. The interrelationships of the onion fly and the onion bulb fly with the bacterial agents of dry rot in plants. *Zool. Zhurn.* 1.

Goryachev, P. P. 1958. The influence of the flood level of a river on the process of development of the agent of opisthorchosis. *Zool. Zhurn.* 37 (12).

Grayevsky. 1946. Thermoreferendum and temperature optimum of freshwater mollusks and arthropods. *Zhurn. Obshch. Biol.* 7 (6).

———. 1948. The glassy state of protoplasm under conditions of deep cooling. *Usp. Sovr. Biol.* 25 (2).

Grechkin, V. P. 1960. The Siberian silkworm (*Dendrolimus sibiricus* Tschetw.)—the pest of the Mongolian forests. *Zool. Zhurn.* 39 (1).

Griffin, D. R. 1955. Hearing and acoustic orientation in marine mammals. *Deep Sea Res.* 3 (suppl.).

———. 1961. *The echo in the lives of humans and animals.* GIZ Fizmat. Lit., M.

Grigor'yev, N. G., and V. P. Teplov. 1939. Results of a study of feeding habits of fur-bearing animals in the Volga-Kama region. Volzh.-Kam. Okhot.-Prom. Biostantsiya, Kazan'.

Grigor'yeva, T. 1950. Ways of using agronomy measures in the struggle against *provolochniki.* In Second Ecological Conference on the problem "Mass reproduction of animals and their prognosis," *Tezisy dokladov,* ch. 1. Kiev.

Grobov, A. G. 1956. Changes in the mosquito fauna of the Crimea in the last 10 years in connection with the conducting of antimosquito measures. In *Problemy parazitologii, Tr. II Nauch. Konf. Parazitol. USSR.* Kiev.

Groebbels, F. *Der Vogel.* Vol. 1, Berlin, 1932; vol. 2, Berlin, 1937.

Gunter, R. 1954. The discrimination between lights of different wavelengths in the cat. *J. Comp. Physiol. Psychol.* 47 (2).

Gusev, V. M. 1953. Materials on the feeding habits of the *solongoy* (*Kolonocus altaicus* Pall.) in the delta of the Ili River. *Zool. Zhurn.* 32 (3).

Haeckel, E. 1866. *Generelle Morphologie.*

———. 1869. Über Entwicklungsgang und Aufgabe der Zoologie. *Jena. Z.* 5.

Hall, F. G. 1922. The vital limits of exsiccation of certain animals. *Biol. Bull.* 42.

Heape, W. 1931. *Emigration, migration and nomadism.* Cambridge.

Helland-Nansen, B., and F. Nansen. 1909. The Norwegian Sea. Die jährlichen Schwankungen der Wassermassen im norwegischen Nordmeere in ihrer Beziehung zu den Schwankungen der meteorologischen Verhältnisse. *Rep. Norweg. Fish. Invest.* 11 (1).

Hendrikson, G. O., and C. Swan. 1938. Winter notes on the short-eared owl. *Ecology* 19 (14).

Herter, K. 1934. Eine verbesserte Temperaturorgel und ihre Anwendung zu Insekten un Saugetieren. *Biol. Zentralbl.* 54 (9–10).

———. 1940. Über das Weesen, der Vorzugstemperatur bei Echsen und Nagern. *Z. Vergl. Physiol.* 28 (3).

———. 1957. Das Verhalten der Insektivoren. *Handb. Zool.* 8, 9.

Hesse, R., W. C. Allee, and K. P. Schmidt. 1937. *Ecological animal geography.* New York.

Howard, E. 1920. *Territory in bird life.* London.

Howard, L. O., and W. F. Fiske. 1911. The importation into the United States of the parasites of the gypsy moth and the brown-tail moth. *U.S. Dept. Agr. Bur. Ent. Bull.* 9.

Howes, N. H., and G. P. Wells. 1934. The water relations of snails and slugs. *J. Exp. Biol.* 11.

Humbach, J. 1960. Geruch und Geschmack bei den augenlosen Höhlenfische *Anoptichys jordani* Hubbs und Innes und *A. hubsi. Alvarez Naturwiss.* 47 (23).

Hustich, I., ed. 1952. *The recent climatic fluctuation in Finland and its consequences (a symposium).* Helsinki.

Huus, J. 1928. Darmparasiten des norwegischen Moorschneehuhns. *Bergens Mus. Aarb.* 2.

Huxley, J. S. 1939. Clines: An auxiliary method in taxonomy. *Bijdr. Dierk.* 27.

———, ed. 1940. *The new systematics.* Oxford.

Ikoyev, F. I. 1957. A study of the aural analyzer of the horse in a natural experiment. *Sb. Vopr. Fiziol. S.-kh. Zhivotn.*; Izd. AN SSSR, M. and L.

Il'yenko, A. I. 1958. Factors determining the beginning of reproduction in a population of house sparrows in Moscow. *Zool. Zhurn.* 37 (12).

Ioff, I. G. 1941. Questions of the ecology of fleas in connection with their epidemiological significance. *Pyatigorsk.*

———. 1949. Aphaniptera of Kirghizia. *Sb. Ektoparazity* 1; Izd. AN SSSR.

Ioff, I. G., *et al.* 1951. A high-mountain nidus of plague in Kirghizia. *Sb. Prirod. Och. Transmissiv. Bolez. Kazakh.* 1; Izd. AN Kazakh. SSR.

Ioganzen, V. G. 1959. *Fundamentals of ecology.* Tomsk. Gos. Un-T., Tomsk.

Isakov, Yu. A. 1940. The ecology of wintering among the waterfowl of the southern Caspian. *Tr. Vs. Ornitol. Zap. Gassan-Kuli* 1.

———. 1948. Elementary populations among birds. *Tr. Tsentr. Byuro Kol'tsev.* 7. M.

———. 1949. Toward the question as to the elementary populations of birds. *Izv. AN SSSR, Ser. Biol.* 1.

———. 1952. An experimental study of the distribution of a species within its range. *Byull. MOIP, Otd. Biol.*, Nov. Ser. 57 (6).

Isakov, Yu. A., M. P. Raspopov, and M. K. Krumina. 1947. Materials on the ecology of the common gull. *Sb. Ocherki Prirody Podmosk. Mosk. Obl.*

Ivanov, A. V. 1960. Pogonophora. *Fauna SSSR.*

Ivanov, V. V. 1950. The least suslik–desalinator of soils. *Izv. Vs. Geogr. Ob-va* 82 (5).

Ivlev, V. S. 1955. On the influence of winter conditions on the blood of certain freshwater fishes. *Byull. MOIP* 60 (4).

———. 1958. Some questions of theoretical ecology. *Byull. MOIP, Otd. Biol.* 63 (1).

———. 1960. An analysis of the mechanism of distribution of fish under conditions of a temperature gradient. *Zool. Zhurn.* 39 (4).

Ivleva, I. V. 1953. Growth and reproduction of *Enchytraeus albidus* Heule. *Zool. Zhurn.* 32 (3).

———. 1960. Respiration in *Enchytraeus albidus* Heule. *Zool. Zhurn.* 39 (2).

Jackson, C. M. 1926. Storage of water in various parts of the earthworm at different stages of exsiccation. *Proc. Soc. Exp. Biol. Med.* 23.

Johnson, C. G. 1934. On the stages of *Notostira erratica* L. (Hemiptera, Capsidae). I. Observations on the structure of the egg and subopercular yolk-plug, swelling of the egg and hatching. *Trans. Soc. Brit. Ent.* 1.

———. 1937. Absorption of water and the associated volume changes in eggs of *Notostira erratica* L. (Hemiptera, Capsidae) during embryonic development under experimental conditions. *J. Exp. Biol.* 14.

Kabak, Ya. M., and S. I. Tereza. 1939. The role of temperature and light in the regulation of seasonal changes in the sexual system (an experiment on sparrows). *Tr. Dinam. Razvit.* 11.

Kadochnikov, N. P. 1960. Experiments on the displacement of nests of the large titmouse *Parus major* L. and the redstart *Phoenicurus phoenicurus. Zool. Zhurn.* 39 (11).

Kaftanovsky, Yu. M. 1951. *Chistik*-type birds of the eastern Atlantic. *Mater. Pozn. Fauny Flory SSSR* 28; Izd. MOIP.

Kalabukhov, N. I. 1935. Physiological properties of mountain and prairie subspecies of the forest mouse (*Apodemus sylvaticus ciscaucasicus* Ogn. and *A. s. mosquensis* Ogn.). *DAN SSSR* 2 (1).

———. 1937. Properties of reaction of certain species of prairie rodents to a drop in atmospheric pressure. *Zool. Zhurn.* 16 (3).

———. 1938. Some ecological properties of closely related species of rodents. I. Properties of the reaction of forest mice (*Apodemus sylvaticus* and *A. flavicollis*) and the susliks (*Citellus pygmaeus* and *C. suslica*) to intensity of illumination. *Zool. Zhurn.* 17 (3).

———. 1938. Some data on the effects of temperature of the environment on the growth of the mouse (*Mus musculus* L.). *Byull. MOIP, Otd. Biol.* 47 (3).

———. 1940. Variability and mass reproduction. In First Ecological Conference on the problem "Mass reproduction of animals," *Tezisy dokladov.* Kiev.

———. 1940. The daily activity cycle of animals. *Usp. Sovr. Biol.* 12 (1).

———. 1946. Preservation of the energy balance of the organism as a basis for the process of adaptation. *Zhurn. Obshch. Biol.* 7 (6).

———. 1950. *The ecologo-physiological properties of animals and the conditions in the environment.* Ch. 1. Divergence of certain ecologo-physiological features of closely related forms of animals. KhGU, Khar'kov.

———. 1951. *Methodology of experimental investigation of the ecology of terrestrial vertebrates.* Sov. Nauka, M.

———. 1954. Ecologo-physiological properties of the "geographic forms of existence of a species" and closely related species of animals. *Byull. MOIP, Otd. Biol.* 59 (1).

———. 1955. Ecologo-physiological properties of life forms of the rodents of the forest steppe and the left bank of the Ukraine and European RSFSR. *Zool. Zhurn.* 34 (3).
———. 1956. *Hibernation of animals.* 3rd ed., Khar'kov.
———. 1957. The preferred temperature of animals and its connection with other properties of their thermoregulation. *Sb. Gryzuny Bor'ba Nimi* 5.
———. 1958. Properties of thermoregulation in rodents as one of the factors of their sensitivity to plague infection. *ZhMEI* 9.
———. 1958. Cooling and supercooling of the organism of vertebrate animals. *Usp. Sovr. Biol.* 46 (2–5).
———. 1960. The rise of ecologo-physiological properties of closely related forms of animals as the initial stage of divergence. *Sb. Evolyutsiya Fiziol. Funktsiy*; Izd. AN SSSR, M. and L.
Kalabukhov, N. I., and N. M. Ladygina. 1953. The rise of ecologo-physiological properties in mammals under the influence of the external environment. *Zool. Zhurn.* 32 (2).
Kalabukhov, N. I., and V. V. Rayevsky. 1935. A study of the movements of susliks in the steppe regions of the North Caucasus by the banding method. *Vopr. Ekol. Biotsenol.* 2.
Kalabukhov, N. I., and V. M. Rodionov. 1936. Hemoglobin content and the number of erythrocytes in the blood of prairie and mountain forest mice (*Apodemus sylvaticus ciscaucasicus* Ogn.) and the change in these indices when the habitat altitude is changed. *Byull. MOIP* 45 (1).
Kalabukhov, N. I., O. N. Nurgel'dyyev, and G. N. Skvortsov. 1958. Life forms of rodents of the sandy and clayey deserts of Turkmenia. *Zool. Zhurn.* 37 (3).
Kalela, O. 1934. Einige Hauptzüge der regionalen Verteilung der Brotvögel fauna in dem Seengebiet von Kokemäenjoki. *Ornis Fenn.* 11.
———. 1944. Über den 10-jährigen Massenwechsel bei pflanzenfressen den Vögeln und Säugetieren nebst einigen Gesichspunkten zu einer Erklärung. *Ornis Fenn.* 21.
———. 1950. Zur säkülaren Rhytmik der Areal veränderungen europäischer Vögel und Säugetiere, mit besonderer Berücksichtigung der Überwinterungsverhältnisse als Kausalfaktor. *Ornis Fenn.* 27.
———. 1951. Einige Konsequenzen aus der regionalen Intensitätsvariation im Massenwechsel der Säugetieren und Vögel. *Ann. Zool. Soc. Zool. Botan. Fenn. Vanamo* 14 (5).
———. 1954. Populationsökologische Geschichtspunkte zur Entstehung des Vogelzuges. *Ann. Zool. Soc. Vanamo* 16 (4).
———. 1957. Regulation of reproduction rate in subarctic populations of the vole *Clethrionomys rufocanus* (Sund). *Suom. Tied. Toimit. Ser. R, Biol.* 34.
Kaletskaya, M. L. 1957. The role of the regime of the Rybinsk Reservoir in the life of the mammals of the Darwin Preserve. *Tr. Darvinsk. Gos. Sap. Vologda* 4.
Kapitonov, V. L. 1957. On the relationships between certain predatory birds and the marmots in the tundras of eastern Yakutia. *Zool. Zhurn.* 36 (8).
Karaseva, Ye. V. 1956. Certain properties of development of epizootics of lepto-

spirosis among economist voles as studied by the creature-marking method. *Zool. Zhurn.* 35 (9).

———. 1957. Some properties of the biology of the economist vole as studied by the creature-marking method. *Vopr. Ekol.* 2.

Karaseva, Ye. V., and V. V. Kucheruk. 1954. A study of the mobility of the common voles with the aid of banding. In Third Ecological Conference, *Tezisy dokladov.* Kiev.

Karaseva, Ye. V., A. L. German, and E. I. Korenberg. 1957. Feeding habits of the field hen-harrier and their role in the course of an epizootic of non-icteric leptospirosis in a population of economist voles. *Byull. MOIP* 62 (1).

Karpov, V. V., and V. Ya. Parovshchikov. 1941. The spotted flycatcher. *Priroda Sots. Khoz-vo* 8 (2).

Kartashev, N. N., and V. P. Teplov. 1958. Toward the question as to the role of spring floods in the ecology of waterfowl. *Tr. Oksk. Gos. Zap.* 2; *Raboty Oksk. Ornitol. Stantsii* 1. M.

Karzinkin, G. S. 1952. *The bases of the biological productivity of bodies of water.* Pishchepromizdat, M.

Kashkarov, D. N. 1933. *The environment and the association.* Gosmedizdat.

———. 1937. Soviet ecology: Its state, successes of 20 years, and perspectives for development. *Priroda* 10.

———. *Fundamentals of animal ecology.* 1st ed., L., 1938; 2nd ed., M. and L., 1945.

Kashkarov, D. N., and Ye. P. Korovin. 1936. *The life of the desert.* Biomedgiz, M. and L.

Kashkarov, D. N., and L. Leyn. 1927. Ecological observations on the yellow suslik. *Sred.-Aziat. St. Zashch. Rasteniy*; Uzstayara.

Katasukü, J., and H. Davis. 1954. Electrophysiological studies of the ear of the kangaroo rat (*Dipodomys*). *J. Neurophysiol.* 17 (3).

Kaverznev, A. A. 1952. On the rebirth of animals. *Iabr. Proizvedeniya Russkikh Myslitelcy Vtoroy Poloviny* 18 (1). M.

Kayser, C. 1961. *The physiology of natural hibernation.* Long Island City, N.Y.

Keller, B. A. 1938. The plant and the environment: Ecological types and life forms. *Rastit. SSSR* 1.

Kellog, W. N., R. Kohla, and H. N. Morris. 1953. Porpoise sounds as sonar signals. *Science* 117.

Kerzina, M. N. 1952. Replacement of a population of terrestrial vertebrates on clearings and burned-over areas. *Byull. MOIP, Otd. Biol.* 57 (1).

Kheysin, Ye. M. 1953. The behavior of adult *Ixodes persulcatus* P. Sch. in relation to the temperature and humidity of the surrounding environment. *Zool. Zhurn.* 32 (1).

Khodashova, K. S., and L. G. Dinesman. 1961. The role of least susliks in the formation of a complex soil cover in the clayey semidesert of the Trans-Volga area. *Pochvovedeniye* 1.

Kholodkovsky, N. A. 1908. *Hermes, pest of pine forests.* Izd. Dep. Zemled., SPB.

———. 1910. On the biological species. *Izv. Imp. Akad. Nauk.*

Kipalev, N. M. 1937. Feeding habits of the Gobiidae in the northern Caspian. *Zool. Zhurn.* 16 (4).

Kirikov, S. V. 1940. On the connections between crossbills and certain pine trees. *Izv. AN SSSR, Otd. Biol.* 3.

———. 1946. On the periodic death of animals in the southern extremities of the Urals during severe, nonsnowy winters. *Zool. Zhurn.* 25 (6).

———. 1952. Birds and mammals under conditions of the landscape in the southern extremities of the Urals. *Izv. AN SSSR.*

———. Historical changes in the animal world of our country in the thirteenth–nineteenth centuries. *Izv. AN SSSR, Ser. Geogr.* 6, 1952; *Izv. AN SSSR, Ser. Geogr.* 1, 1955.

———. 1956. Variability of organisms and geographical properties of their life environment. *Izv. AN SSSR, Ser. Geogr.* 5.

———. 1960. *Historical changes in the animal world in the natural zones of the USSR.* Ch. II. The forest zone and the forest tundra. Izd. AN SSSR, M.

Kiris, I. D. 1944. Feeding habits of the squirrel and their significance for its abundance and trade. *Tr. Tsentr. Labor. Biol. Okhotn. Prom. Tovar.* 6.

———. 1947. Principles and causes of changes in the abundance of the squirrel in the USSR. *Tr. VNIO* 7.

———. 1948. Protozoic and worm invasions and their role in the abundance dynamics of the squirrel. *Tr. VNIO* 8.

———. *Migrations of the squirrel in the USSR.* Ch. I, *Tr. VNIO* 16, 1956; ch. II, *Tr. VNIO* 17, 1958.

Kirshenblat, Ya. 1938. *Principles of the dynamics of parasitofauna of the mouse-like rodents.* Izd. LGU, L.

———. 1956. Changes in the sex glands of the host under the influence of parasites. In *Problemy parazitologii, Tr. II Nauch. Konf. Parazitol. USSR.* Kiev.

———. 1958. Telegrons and their biological significance. *Usp. Sovr. Biol.* 46, 3 (6).

Kiselev, F. A. 1948. Winter supplies of the mound-building mouse in the Crimea. *Priroda* 8.

Kleerecoper, H., and E. Chagnon. 1954. Hearing in fish with special reference to *Semotilus atromaculatus. J. Fish. Res. Bd. Can.* 11 (2).

Klyuchareva, O. A. 1956. On certain questions of the intraspecies relationships among fish. *Zool. Zhurn.* 35 (2).

Knipovich, N. M. 1923. *The Caspian Sea and its trade.* Gos. Izd. RSFSR, Berlin.

———. 1938. *Hydrology of the seas and brackish waters as applied to commercial affairs.* Pishchepromizdat, M. and L.

Knoll, F. 1921–1926. Insekten und Blumen. *Abh. Zool.-Bot. Ges. Wien.*

———. 1926. *Die Arum-Blüten stände und ihre Besucher.*

Koblentz, W. W. 1939. Physical aspects of ultraviolet radiation in vitamin-D therapy. In *The vitamins.* Chicago.

Kolosov, A. M., and N. N. Bakeyev. 1947. Biology of the rusak hare. *Mater. Pozn. Fauny Flory SSSR, Otd. Zool.,* Nov. Ser. 9.

Komarov, L. 1940. *The study of species among plants.* Izd. AN SSSR, L.

Kondrashkin, G. A. 1955. On the delta type of tularemia nidus. In *Natural nidality of diseases of man and regional epidemiology*. Medgiz, M.

———. 1957. Principles of reproduction and abundance dynamics of water voles in the Volga delta. *Vopr. Ekol. Kiev.*

———. 1959. On the winter reproduction of water voles. In *Manual: Rodents and the campaign against them*, vol. 6. Saratov.

Korsakov, G. K. 1950. Trapping and its effect on herds of muskrats in the Kurgan forest steppe. *Tr. VNIO* 9.

Korzhuyev, P. A., G. V. Kruglova, and A. N. Sviridova. 1957. Certain ecologo-physiological properties of reptiles. *Zool. Zhurn.* 36 (2).

Koshtoyants, Kh. S. 1950. *Fundamentals of comparative physiology*, vol. 1. Izd. AN SSSR, M.

Kosmachevsky, A. S. 1948. The influence of total illumination on the behavior of certain animals. *Zool. Zhurn.* 27 (5).

Kott, Kh. 1950. *Adaptive coloration in animals.* Izd. IL, M.

Kovaleva, M. F. 1951. Ways of raising the effectiveness of *Trichogramma* in the campaign against agricultural crops and perspectives for further inculcation of this method of control in production. In Second Ecological Conference on the problem "Mass reproduction of animals and their prognosis," *Tezisy dokladov*, ch. 3. Kiev.

Koval'sky, V. V. 1957. New directions and tasks of biological chemistry of farm animals in connection with the study of the biochemical provinces. *Mater. Sesii, Psovyashchennoy 40-Letiyu Vel. Okt. Sots. Revolyutsii.* M.

———. 1958. Geochemical ecology. *Priroda* 9.

Kozhanchikov, I. V. 1935. Experimental investigations on the influence of temperature on the development of the meadow moth. *Zashch. Rasteniy* 7.

———. 1937. *Experimental ecological methods of research in entomology.* Izd. VASKhNIL.

———. 1948. Biological forms of the willow beetle (*Lochmaea caprea*). *Tr. ZIN AN SSSR* 8 (1).

———. 1948. Features of adaptation of insect respiration to conditions in the environment. *Zhurn. Obshch. Biol.* 7 (1).

———. 1948. Intraspecies biological forms of animals. *Priroda* 3.

———. 1948. Seasonal differences in the reaction of gas exchange in the sharp-faced frog (*Rana terrestria* Andra) to thermal influence. *DAN SSSR* 12.

———. 1949. Features of the influence of negative temperature on the embryonic development of insects. *Zhurn. Obshch. Biol.* 10 (1).

———. 1951. Feeding specialization and its significance in the life of insects. *Entomol. Oboz.* 31 (3–4).

———. 1956. Toward recognition of the biological forms and biological species among insects. *Zool. Zhurn.* 35 (5).

———. 1958. Biological properties of European species of the genus *Galerucella* and conditions of formation of form in *Galerucella lineola* F. *Tr. ZIN AN SSSR* 24.

Kozhin, N. I. 1941. Fluctuations in the abundance of commercial fish of the Caspian and raising production of their supplies. *Tr. VNIRO* 16.

Kozhov, M. N. 1955. New data on the life of the deep waters of Lake Baykal. *Zool. Zhurn.* 34 (1).

Kozlov, V. V. 1952. Methods of making quantitative evaluations of the wolf. *Sb. Metody Ucheta Chisl. Geogr. Raspr. Nazem. Pozv.*; Izd. AN SSSR.

Kratinov, A. G. 1947. Hibernation in mammals and seasonal dynamics of the functional state of the vegetative nervous system. VII Vs. Syezd Fiziol. Biokhim. Farmakolog., *Tezisy*.

Kratinov, A. G., and A. T. Shkirina. 1947. On the seasonal dynamics of the function of the thyroid gland in the least suslik (*Citellus pygmaeus*). *Izv. AN SSSR, Otd. Biol.* 2.

Kratinov, A. G., *et al.* 1947. Seasonal dynamics of the ascorbic acid content of the organs of the least suslik (*Citellus pygmaeus*). *Izv. AN SSSR, Otd. Biol.* 2.

Krause, G. 1938. Die Ausbildung der Körpergrundgestalt im Ei der Gewächshausschrecke *Tachycines asymamorus. Z. Morphol. Ökol. Tiere* 34.

Krishtafovich, A. N. 1957. *Paleobotany*. 4th ed. Gostoptekhizdat, L.

Krogh, A. 1914. On the rate of development and CO_2 production of chrysalids of *Tenebrio molitor* at different temperatures. *Z. Allg. Physiol.* 16.

Krotov, A. I., and N. I. Vinogradov. 1940. The Black Sea *Scomber. Tr. Ukr. Ryb. Khoz. Stantsii.* 8.

Krotov, A. V. 1949. *The life of the Black Sea.* Odes. Obl. Izd., Odessa.

Kryshtal', A. F. 1955. Toward a study of the dynamics of the entomofauna of the soils and litter during flooding of valleys along the middle course of the Dnieper. *Zool. Zhurn.* 34 (1).

Kryzhanovsky, S. G. 1949. Ecologo-morphological principles of development of the carps, loaches, and sheats. *Tr. in-ta Morfol. Zhivot.* 1.

Kryzhanovsky, S. G., N. N. Disler, and Ye. N. Smirnova. 1953. Ecologo-morphological principles of development of the perches (Percoidei). *Tr. in-ta Morfol. Zhivot.* 10.

Kucheruk, V. V., and T. N. Dunayeva. 1948. Materials on the abundance dynamics of Brandt's vole. *Mater. po Gryzunam* 3; Izd. MOIP, M.

Kucheruk, V. V., and M. A. Rubina. 1953. The causes determining the species content and abundance of rodents in haystacks, lofts, and ricks in the southern regions of the Moscow oblast'. *Zool. Zhurn.* 37 (3).

Kulik, I. L. 1955. Some properties of the mobility of large gerbils in connection with a study of natural nidi of infection. In *Manual: Natural nidality of diseases of man and regional epidemiology*. Medgiz.

Kumari, E. 1954. The dynamics of the ornithofauna of the Cis-Baltic in the last 100 years and possible directions of its further changes. In Third Ecological Conference, *Tezisy dokladov*, ch. 3. Kiev.

Künkel, K. 1916. *Zur Biologie der Lungenschneckel.* Heidelberg.

Kuroda, B. 1933. Studies of audition in reptiles. *J. Comp. Physiol.* 3.

Kuznetsov, N. Ya. 1948. *Fundamentals of insect physiology*. Izd. AN SSSR, M. and L.

———. 1951. *Handbook of zoology*. Ch. III. Insects. M.
Kuznetsov, V. V. 1956. Some principles of intraspecies morphological and biological differentiation of qualities using the marine invertebrates as an example. *Zool. Zhurn.* 35 (8).
Kuzyakin, A. P. 1936. Conditions of habitation of animals in hollow trees. *Vopr. Ekol. Biotsenol.* 3.
———. 1950. *Bats*. Sov. Nauka, M.
———. 1951. History of emigration, contemporary distribution, and places of habitation of the Norway rat in the USSR. *Mater. po Gryzunam* 4; Izd. MOIP, M.
Kydyrbayev, Kh. 1959. Properties of reproduction of the yellow suslik on the eastern boundary of its range. *Tr. ZIN AN Kazakh. SSR Alma-Ata* 10.
Lamarck, J. B. P. A. *The philosophy of zoology*. Vol. 1, Biomedgiz, M. and L., 1935; vol. 2, Biomedgiz, M. and L., 1937.
Larina, N. I. 1952. On the role of the olfactory and visual receptors in rodents. *Zool. Zhurn.* 31 (3).
———. 1958. On the evolutionary significance of geographic changes and interspecies hybridization among rodents. *Nauch. Dokl. Vyssh. Shkoly, Ser. Biol.* 4.
———. 1958. Toward the question as to diagnosing species of forest and yellow-throated mice. *Zool. Zhurn.* 37 (11).
Larionov, V. F. 1945. Changes of cover and their connection with reproduction in birds. *Uch. Zap. MGU, Biol.* 88.
Lavov, M. A. 1959. The effect of the sable on a squirrel population in the Trans-Baykal. *Zool. Zhurn.* 38 (2).
Lavrenko, Ye. M. 1952. Microcomplexity and mosaicity of plant cover in the steppe as a result of the activities of animals and plants. *Tr. Bot. in-ta Komarova*, Ser. 3, 8.
Lavrov, N. P. 1941. The role of worm invasions and infectious illnesses in the abundance dynamics of the ermine. *Tr. Tsentr. Lab. Biol. Okhot. Prom. Tovaroved.* 6.
———. 1946. *Acclimatization and reacclimatization of fur-bearing creatures in the USSR*. Zagotizdat, M.
———. 1957. *Acclimatization of the muskrat in the USSR*. Izd. Tsentrosoyuza, M.
Lavrova, M. Ya., and Ye. V. Karaseva. 1956. The activities of predatory birds and the common vole population on farmlands in the southern Moscow oblast'. *Byull. MOIP, Otd. Biol.* 61 (3).
Leb, Zh. 1926. *The organism as a whole from the physico-chemical point of view*. Gosizdat, M. and L.
Lebedev, N. V. 1946. Elementary populations of fishes. *Zool. Zhurn.* 25 (2).
Lebedev, V. D. 1959. On the causes of the rise of migrations of the European river eel in the waters of the Atlantic. *Nauch. Dokl. Vyssh. Shkoly, Ser. Biol.* 3.
———. 1960. *Freshwater Quaternary ichthyofauna of the European part of the USSR*. Izd. MGU, M.
Lebedeva, M. O. 1956. On the life span of the birds. *Okhota i Okhotnich'ye Khozyaystvo.* 9.

Lees, A. B. 1943. On the behavior of wireworms of the genus *Agriotes* Esch. (Coleoptera, Elateridae). *J. Exp. Biol.* 20.

———. 1946. The water balance in *Ixodes ricinus* L. and certain other species of ticks. *Parasitology* 37.

Lek, D. 1957. *Animal abundance and its regulation in nature.* Izd. IL, M.

Levin, N. A. 1953. The relationship between anatomical structure of the bone labyrinth of birds and their way of life. *Zool. Zhurn.* 34 (3).

Liebig, J. 1840. *Chemistry in its applications to agriculture and physiology.* London.

Likhachev, G. N. 1951. Reproduction and the feeding habits of the raven in the Tula abatis. *Byull. MOIP, Otd. Biol.* 56 (5).

———. 1952. Winter supply of food of the sparrow owl. *Priroda* 11.

———. 1953. Observations on the reproduction of the blue titmouse under artificial nesting conditions. *Zool. Zhurn.* 32 (1).

———. 1955. Biology of the black kite in the Tula abatis. *Byull. MOIP, Otd. Biol.* 60 (5).

———. 1958. A survey of fox abundance in the region of the Tula abatis. *Byull. MOIP, Otd. Biol.* 63 (1).

Likventov, A. V. 1950. The significance of preferred temperature in the concept of insect behavior. In Second Ecological Conference on the problem "Mass reproduction of animals and their prognosis," *Tezisy dokladov.* Kiev.

———. 1960. The use of preferred temperature in the study of insect behavior. *Zool. Zhurn.* 39 (1).

Lindemann, W. 1951. Psychologie des Igels. *Z. Tierpsychol.* 8.

Ling, Kh. I. 1956. Mange in foxes in the Estonian SSR. *Zool. Zhurn.* 35 (7).

Lobashev, W. Ye., and V. B. Savvateyev. 1959. *The physiology of the daily regime of animals.* Izd. AN SSSR, M. and L.

Lotka, A. J. 1925. *Elements of physical biology.* Baltimore.

———. 1934. Théorie analytique des associations biologiques. *Act. Sci. Indus.* 187.

Lozina-Lozinsky, L. K. 1952. Viability and anabiosis under low-temperature conditions in animals. *Izv. Yest. Nauch. in-ta Lesgrafta* 25. M.

———. 1956. *The organism and conditions of life.* Uchpedgiz, M.

Lozinsky, V. A. 1960. The influence of floods in forests on the formation of nidi of lepidopterous pests. *Zool. Zhurn.* 39 (10).

Lutta, A. S., and R. Ye. Shul'man. 1958. The influence of microclimatic conditions of meadow and forest on the survival rate and activity of the tick *Ixodes ricinus* L. *Zool. Zhurn.* 37 (12).

Lysenko, T. D. 1952. *Agrobiology.* Sel'khozgiz, M.

MacArthur, J. W., and W. H. T. Bailey. 1929. Metabolic activity and duration of life. *J. Exp. Zool.* 53.

MacLulich, D. A. 1937. Fluctuations in the numbers of varying hares (*Lepus americanus*). *Univ. Toronto Stud. Biol.* 43.

Maksimov, A. A. 1956. On the landscape-geographic properties in the abundance dynamics of the water rat in West Siberia. *Tr. Tomsk. Gos. In-Ta* 142.

———. 1958. The hydrobiological regime of bodies of water and prognoses of transmissible outbreaks of tularemia. *Izv. Novosib. Otd. Geogr. O-va* 2.

Mal'chevsky, A. S. 1958. Local songs and geographic variability of the songs of birds. *Vestn. LGU* 9.

———. 1958. On the biological races of the common cuckoo (*Cuculus canorus* L.) in the territory of the European part of the USSR. *Zool. Zhurn.* 37 (1).

———. 1959. *Nest life of the songbirds.* Izd. LGU, L.

Malthus, T. R. 1798. *An essay on the principles of population.* London.

Mamontov, I. M. 1946. The influence of certain types of agrotechnology on the abundance of the least suslik. In *Manual: Rodents and their control,* vol. 2. Saratov.

———. 1948. Distribution in space and abundance dynamics of the least suslik (*Citellus pygmaeus* Pall.). *Tr. Nauch. Konf. Posvyash. 25-Let. Yubil. in-ta Mikrob.* Saratov.

Manteyfel', B. P. 1959. The adaptive significance of periodic migrations of water organisms. *Vopr. Ikhtiol.* 13.

———. *Vertical migrations of marine organisms.* Ch. I, *Tr. in-ta Morfol. Zhivot. AN SSSR* 13, 1959; ch. II, *Tr. in-ta Morfol. Zhivot. AN SSSR* 39, 1961.

Marinovsky, P. I. 1950. Mass reproductions of the poisonous spider of the Karakurt, *Latrodectus tridecemguttatus* Rossi. In Second Ecological Conference on the problem "Mass reproduction of animals and their prognosis," *Tezisy dokladov,* ch. 1. Kiev.

Markov, K. K. 1951. *Paleogeography.* Geografgiz, M.

Markova, A. S. 1948. Properties of the reaction of closely related species of rodents to a drop in atmospheric pressure. *DAN SSSR* 62 (3).

Mashkovtsev, A. A. 1949. The significance for biology of I. P. Pavlov's study of higher nervous activity. *Usp. Sovr. Biol.* 28, 1 (4).

Matthews, H. D. 1942. On the stridulation of insects. *Science* 95.

Matveyev, B. S. 1953. On the biological stages in the postembryonic development of the sturgeons. *Zool. Zhurn.* 32 (2).

Mayr, E. 1947. *Systematics and the origin of species.* Izd. IL, M.

Mayr, E., E. Linsley, and R. Usinger. 1956. *Methods and principles of ecological systematics.* Izd. IL, M.

Mazokhin-Porshnyakov, G. A. 1955. Mass attraction of insects to ultraviolet light. *DAN SSSR* 102 (4).

———. 1956. Nocturnal catching of insects by the light of a mercury lamp and prospects for using it in applied entomology. *Zool. Zhurn.* 35 (2).

———. 1959. The vision of the arthropods and polarized light. *Zool. Zhurn.* 38 (7).

———. 1960. Substantiation of the existence of color vision in the wasps (Vespidae). *Zool. Zhurn.* 39 (4).

Mel'nichenko, A. N. 1936. Principles of mass migrations of the meadow moth and the problem of constructing prognoses of its flights. *Tr. Zashch. Rasteniy* 1 (17).

———. 1949. *The field-protecting belts and the reproduction of animals useful and harmful to agriculture.* Izd. MOIP, M.

Merkova, M. A. 1955. Some data on the ecology of rust-colored voles and the yellow-throated mouse in the south of the Moscow oblast' and the Tellerman grove. *Byull. MOIP, Otd. Biol.* 60 (1).

Meyer, D. R., R. S. Miles, and Ph. Ratoosh. 1954. Absence of color vision in the cat. *J. Neurophysiol.* 17 (3).

Michal, K. 1931. Die Beziehung der Populationsdichte zum Lebensoptimum und Einfluss des Lebensoptimums auf das Zahlenverhältnis der Geschlechter bei Mehlwurm und Stubenfliege. *Biol. Gen.* 7.

Middendorf, A. F. 1869. *A journey to the north and east of Siberia.* SPB.

Middleton, A. D. 1935. Factors controlling the population of the partridge (*Perdrix perdrix*) in Great Britain. *Proc. Zool. Soc. Lond.*

Mikheyev, A. V. 1948. New data on the seasonal displacement and migrations of ducks. *Tr. Tsentr. Byur. Kol'ts. M.* 7.

Miklukho-Maklay, N. N. 1883. Temperature of the body of *Echidna hystrix*. *Proc. Linn. Soc. N.S.W.* 8.

———. 1884. The temperature of the body of *Ornithorynchus paradoxus*. *Proc. Linn. Soc. N.S.W.* 9.

Mileykovsky, S. 1960. Daily abundance dynamics of pelagic larvae of bottom invertebrates and a series of holoplankterons in the surface of the waters of Great Salma Sound at the beginning of the "biological summer." *Zool. Zhurn.* 39 (3).

Milinsky, G. I. 1944. Biology and commercial catch of the marine flounder (*Pleuronectes platessa*) of the Barents Sea. *Tr. PINRO* 2.

Mill, J. S. 1848. *A system of logic.* London.

Miller, A. 1940. Embryonic membranes, yolk cells, and morphogenesis of the stonefly *Pteronarcys proteus* Newman (Plecoptera: Pteronarcidae). *Ann. Ent. Soc. Amer.* 33.

Miller, R. B. 1954. Movements of cutthroat trout after different periods of retention upstream and downstream from their homes. *J. Fish. Res. Bd. Can.* 11 (5).

Minnich, D. E. 1919. The photic reactions of the honey bee *Apis melifera* L. *J. Exp. Zool.* 29.

Mironov, N. P., P. I. Whiranovich, and F. A. Pushnitsa. 1949. Contact connections between rodents under the conditions of the sandy semidesert of the northwestern Cis-Caspian. *Ref. Rabot Rostov. Protiv. In-Ta* 8.

Mitscherlich, E. A. 1921. Das Wirkungsgesetz der Wachstumfaktoren. *Landw. Jb.* 56.

Möbius, K. 1877. *Die Auster und die Austernwirtschaft.* Berlin.

Modestov, V. M. 1939. Feeding habits of the gulls of the eastern Murman and their role in the formation and life of bird bazaars. *Sb. Stud. Rabot MGU, Zool.* 9.

Mokriyevich, N. A. 1957. Seasonal changes in certain ecologo-physiological properties of the southern (*Meriones meridianus* Pall.) and crested (*M. tamariscinus* Pall.) gerbils in the Volga-Ural sands. In *Manual: Rodents and their control,* vol. 5. Saratov.

Mokrousov, N. Ya. 1957. Periodics of life activities and reproduction of the thick-tailed, three-toed jerboa (*Scirtopoda telum* Licht.) in the northwestern Cis-Caspian. In *Manual: Rodents and their control,* vol. 5. Saratov.

Monastyrsky, G. N. 1949. On the types of spawning populations among fishes. *Zool. Zhurn.* 38 (6).

———. 1952. Abundance dynamics of commercial fish. *Tr. VNIRO* 21.

Monchadsky, A. S. 1949. On the types of reaction of insects to changes in the temperature of the surrounding environment. *Izv. AN SSSR, Ser. Biol.* 2.

———. 1956. Attacks by mosquitoes against man in the maritime part of the Volga delta. *Parazitol. Sbornik* 16; Izd. AN SSSR, M.

———. 1958. On the classification of factors in the surrounding environment. *Zool. Zhurn.* 37 (5).

Monchadsky, A. S., and A. N. Berezina. 1959. Intraspecies relationships among live larvae of mosquitoes of the subfamily Chaoborinae (Diptera, Culicidae). *Zool. Zhurn.* 38 (10).

Mordvilko, A. K. 1926. Evolution of cycles and origin of migrations among aphids. *Zashch. Rasteniy* 6, 7.

———. 1936. The black-bean or beet aphid *Aphis fabae* Scop. *Zashch. Rasteniy* 10.

———. 1939. Aphids: Cycles of generations and their evolution. *Priroda* 24 (11).

Moshkovsky, Sh. D. 1946. Functional parasitology (outlines 1–3). *Med. Parazitol. i Parazit. Bolezni* 4–6.

Moskacheva, Ye. A. 1960. Survivability of oribatid mites on the annually flooded river floodplains of Byelorussia. *Zool. Zhurn.* 39 (3).

Movchan, V. A. 1948. *Ecological bases of intensified growth in the carp.* Izd. AN SSSR, M.

Murie, A. 1944. *The wolves of Mount McKinley.* Washington.

Nadezhdin, V. M. 1959. The influence of hydrological and meteorological conditions on the concentration of Kandalaksha and Onega herring. *Zool. Zhurn.* 38 (2).

Nasimovich, A. A. 1948. Ecology of the fox in the Lapland Preserve. Ecology of the forest marten. *Tr. Laplandsk. Gos. Zap.* 3. M.

———. 1955. *The role of the snow cover regime in the life of hoofed animals in the territory of the USSR.* Izd. AN SSSR, M.

Nasimovich, A. A., G. A. Novikov, and O. I. Semënov-Tyan-Shansky. 1948. The Norwegian lemming. *Mater. Pozn. Fauny Flory SSSR, Otd. Zool.*, Nov. Ser. 17; *Mater. po Gryzunam* 3.

Naumov, N. P. 1930. Materials toward the recognition of squirrel "harvests." *Tr. Lesnomu Opyt. Delu TsLOS* 7.

———. 1934. Periodicity in fluctuations in the abundance of the common squirrel in the USSR. Biology of reproduction of the common squirrel. In *Manual: Ecology of the squirrel.* KOIZ, M.

———. 1934. Determination of the age of the squirrel. *Uch. Zap. MGU, Zool. Ser.* 2.

———. 1936. On the properties of stationary determination of the distribution of mouselike rodents in the southern Ukraine. *Zool. Zhurn.* 15 (4).

———. 1939. Ecological properties of steppe mice and voles. *Zool. Zhurn.* 18.

———. 1940. Ecology of the mound-building mouse *(Mus musculus hortulanus* Nordm.). *Tr. in-ta Morf. Zhivot. AN SSSR* 2.

———. 1945. Geographic variability of abundance dynamics and evolution. *Zhurn. Obshch. Biol.* 6 (1).

———. 1948. *Sketches of the comparative ecology of the mouselike rodents.* Izd. AN SSSR, M.

———. 1951. A new method of studying the ecology of small forest rodents. In *Manual: Fauna and ecology of rodents. Mater. Pozn. Fauny Flory SSSR, Otd. Zool.,* Nov. Ser. 22; *Mater. po Gryzunam* 4.

———. 1953. Abundance dynamics of the common vole and methods for predicting them in the middle belt of the USSR. *Zool. Zhurn.* 32 (2).

———. 1954. Types of settlements of rodents and their ecological significance. *Zool. Zhurn.* 33 (2).

———. 1955. A study of the motility and abundance of small mammals, using trapping ditches. *Vopr. Krayev. Obshch. Eksper. Parazit. Medits. Zool.* 9.

———. 1955. *Elementary nidality in natural nidi of disease.* ZhMEI.

———. 1956. Marking of animals and study of their intraspecies associations. *Zool. Zhurn.* 35 (1).

———. 1956. Interspecies and intraspecies relationships among animals (primarily vertebrates). *Usp. Sovr. Biol.* 41 (1).

———. 1958. Some basic questions on the dynamics of animal settlement. *Zool. Zhurn.* 37 (5).

Naumov, N. P., and S. S. Folitarek. 1945. Geographic properties of the abundance dynamics of the mouselike rodents. *Zhurn. Obshch. Biol.* 6 (5).

Naumov, N. P., S. A. Shilova, and V. I. Chabovsky. 1957. The role of wild vertebrates in the natural nidi of the tick encephalites. *Zool. Zhurn.* 36 (3).

Naumov, N. P., *et al.* 1960. The conditions of existence and the most important epizootiological properties of the Cis-Aral portion of the Central Asian plague nidus. In *Handbook: Natural nidality and epidemiology of the especially dangerous infectious diseases.* Saratov.

Naumov, R. V. 1959. Toward the question as to the causes of the reduction in outbreaks of mass reproduction of *zlatoguzki. Zool. Zhurn.* 38 (1).

Naumov, S. P. 1939. Fluctuations in the abundance of hares. *Vopr. Ekol. Biotsenol.* 5–6.

———. 1941. General questions on the fluctuations in numbers of creatures and organization of research. Methodology of compiling prognoses of changes in the abundance of the belyak here. *Tr. Tsentr. Labor. Biol. Okhot. Prom. Tovaroved. V/o Zagotzhivsyr'ye* 5. M.

———. 1947. Ecology of the belyak hare. *Mater. Pozn. Fauny Flory SSSR, Otd. Zool.,* Nov. Ser. 10 (25); Izd. MOIP, M.

———. 1956. General properties of the abundance dynamics of the belyak hare in Yakutia. *Uch. Zap. Mosk. Gos. Ped. in-ta Lenina.* 96.

———, ed. 1960. *Research into the causes and principles of the abundance dynamics of the belyak hare in Yakutia.* Izd. AN SSSR, M.

Nekipelov, N. V. 1952. Seasonal motility and contact among the Trans-Baykal rodents. *Izv. Irkutsk. Gos. Protivochumn. in-ta Sib. DV* 10.
Nice, M. M. 1943. Studies in the life history of the song sparrow. II. The behavior of the song sparrow and other passerines. *Amer. Midl. Nat.* 6.
———. 1944. The role of territory in bird life. *Amer. Midl. Nat.* 7.
———. 1956. Four generations of a song sparrow family. *Jack Pine Warbler* 34 (2).
Nicholson, A. J. 1933. The balance of animal populations. *J. Anim. Ecol.* 2.
———. 1941. The homes and social habits of the wood mouse (*Peromyscus leucopus noveboracensis*) in southern Michigan. *Amer. Midl. Nat.* 25 (1).
———. 1947. Fluctuation of animal populations (presidential address). *Aust. N.Z. Adv. Sci. Sect. D.*
———. 1958. Dynamics of insect populations. *Ann. Rev. Ent.* 3.
Nikanorov, G. I. 1959. Differences in size, coloration, and stickiness of fish eggs of *Coregonus albula* L. in different schools in the Latvian SSR. *Zool. Zhurn.* 38 (6).
Nikitina, N. A. 1958. Properties of use of territory by field mice (*Apodemus agrarius* Pall.). *Zool. Zhurn.* 37 (9).
Nikol'sky, G. V. 1944. *The biology of fishes.* Sov. Nauka.
———. 1947. On the biological specifics of faunistic complexes and the significance of its analysis for zoogeography. *Zool. Zhurn.* 26 (3).
———. 1949. On the principles of intraspecies nutritional relationships among the freshwater fishes. *Byull. MOIP* 1.
———. 1950. *Special ichthyology.* Sov. Nauka.
———. 1950. On the biological basis of the catch contingent and means of directing the abundance of a school of fish. On the abundance dynamics of a school of fish and on the so-called problem of the biological productivity of bodies of water. *Zool. Zhurn.* 29 (1).
———. 1952. On the type of school dynamics and character of spawning of the humpbacked and chum salmon in the Amur. *DAN SSSR* 86 (4).
———. 1953. On the theoretical bases of projects covering the abundance dynamics of fish. *Tr. Vs. Konf. Probl. Rybn. Kho-Va.*; Izd. AN SSSR.
———. 1953. On certain principles of the dynamics of fecundity of fishes. *Ocherki Obshch. Voprosam Ikhtiol.*; Izd. AN SSSR.
———. 1955. On the contents, theoretical bases, and basic tasks of animal ecology. *Zool. Zhurn.* 34 (1).
———. 1956. Information on animal ecology in the Indian epos of *Mahabharata* and *Ramayana. Vopr. Istorii Yestesvozn. o Tekhniki* 2; Izd. AN SSSR, M.
———. 1958. On the influence of catching on the structure of a population of commercial fish. *Zool. Zhurn.* 37 (1).
———. 1958. On the biological bases of fishing regulation. *Vopr. Ikhtiol.* 11.
———. 1961. *The ecology of fishes.* Izd. Vysshaya Shkola, M.
———. 1961. On the causes of fluctuations in the abundance of fishes. *Vopr. Ikhtiol.* 4 (21).
Nordhagen, R. 1928. Rypeaar og Baeraar. *Bergens Mus. Aarb.* 2.

Novikov, G. A. 1939. *The European mink.* L.

———. 1944. Properties of the nesting life of birds on the Kola Peninsula. *Priroda* 1.

———. 1949. Daily life of forest birds in the Sub-Arctic. *Zool. Zhurn.* 28 (5).

———. *Field research on the ecology of terrestrial vertebrates.* 1st ed., 1949; 2nd ed., 1953.

———. 1957. Toward a history of the domestic ecology of terrestrial vertebrate animals. *Tr. in-ta Istorii Yestestvozn. o Tekhniki AN SSSR* 16 (3).

———. 1958. From the history of evolutionary teaching. *Istoriya Biol. Nauk.* 5; *Tr. in-ta Istorii Yestestvozn. o Tekhniki* 24.

———. 1959. *Ecology of the animals and birds of the forest-steppe oak groves.* Izd. LGU, L.

———. 1960. Geographic variability of the population density of forest birds in the European part of the USSR and adjacent countries. *Zool. Zhurn.* 39 (3).

Ognev, S. I. 1926. *Mammals of northeastern Siberia.* Vladivostok.

———. 1940. *Beasts of the USSR and contiguous countries,* vol. 4. Izd. AN SSSR, M.

———. 1951. *Outline of the ecology of mammals.* Izd. MOIP, M.

Oliger, I. M. 1940. Parasitofauna of the hazel grouse in the northern Gor'ky oblast'. *Uch. Zap. LGU* 59.

———. 1940. Parasitic protozoans and their role in abundance fluctuations of the hazel grouse in the northern Gor'ky oblast'. *DAN SSSR* 28 (5).

———. 1957. Parasitofauna of the black grouse of the forest zone of the European part of the RSFSR. *Zool. Zhurn.* 36 (4).

Ol'nyanskaya, R. P., and A. D. Slonim. 1947. On the adaptations of organisms to very low environmental temperatures. *Izv. AN SSSR, Ser. Biol.* 2.

Olsuf'yev, N. G. 1954. Natural nidality of swine erysipeloid and listerelliosis. In *Natural nidality of diseases in Kazakhstan,* vol. 2.

Olsuf'yev, N. G., and G. P. Rudnev. 1960. *Tularemia.* Medgiz.

Olsuf'yev, N. G., *et al.* 1950. A study of the characteristics of *Bacterium tularense* and its biological interrelationships with carrier animals and with vector ticks. *Vestn. AMN* 3.

Oparin, A. I. 1951. The problem of the origin of life in contemporary natural studies. In *Philosophical questions of contemporary biology.* Izd. AN SSSR, M.

———. 1954. The problem of the origin of life on earth in the light of the achievements of contemporary nature studies. *Izv. AN SSSR, Ser. Biol.* 2.

———. 1957. *The rise of life on earth.* Izd. AN SSSR, M.

Orgel and Smith. 1954. Test of the magnetic theory of homing. *Science* 120 (3126).

Orlov, O. Yu. 1957. Materials on the biology of jerboas in the western Kyzyl-Kum. *Sb. Stud. Nauch. Rabot MGU, Biol.*

Osmolovskaya, V. I. 1948. The ecology of predatory birds of the Yamal Peninsula. *Tr. in-ta Geog. AN SSSR* 41.

———. 1949. The ecology of steppe predatory birds of North Kazakhstan. *Tr. Naurzumsk. Gos. Zap.* 2.

Osmolovskaya, V. I., and A. N. Formozov. 1950. Notes on the ecology of certain useful birds of the forest. In *Manual: Birds and forest pests.* Izd. MOIP, M.

———. 1952. Methods of computing the abundance and geographic distribution of diurnal and nocturnal predatory birds. In *Manual: Methods for computing the abundance and geographic distribution of terrestrial vertebrates.* Izd. AN SSSR, M.

Ostroumov, N. A. 1957. Forest-float and fish farming in several of the northern rivers of the European part of the USSR. *Izv. Yest. Nauchn. in-ta Permsk. Gos. un-ta Gor'kova* 14 (1).

Pachossky, I. K. *A description of the vegetation in the Kherson guberniya.* Vol. 1, Kherson, 1915; vol. 2, Kherson, 1917; vol. 3, Kherson, 1927.

Paly, V. F. 1960. The beet-root aphid (*Apis fabae* Scop.) in the beet-raising regions of the central-chernozem oblasts of the RSFSR and causes of fluctuations in its abundance. *Zool. Zhurn.* 39 (4).

Panteleyev, P. A. 1959. Types of settlements of the water rat (*Arvicola terrestris* L.) in the northern part of the Volga-Aktyubinsk floodplain. *Zool. Zhurn.* 38 (9).

Pantyukhov, G. A. 1956. Geographic variability of cold resistance among several insects. *Zool. Zhurn.* 35 (9).

———. 1958. Cold resistance of the larvae of *Scolytus multistriatus* Marsh. *Zool. Zhurn.* 37 (9).

Parker, I. R. 1930. Some effects of temperature and moisture upon *Melanplus mexicanus* Souss. and *Camnula pelliccida* Scud. (Orthoptera). *Bull. Univ. Mont. Agr. Exp. Sta.* 223.

Pavlinin, V. N. 1948. Materials on banding of the mole (*Talpa europaea*) in the Urals. *Zool. Zhurn.* 28 (6).

Pavlov, A. N., *et al.* 1959. On the intraspecies and interspecies contacts among gerbils in the Black Earth and Il'men regions of the northwestern Cis-Caspian. *Zool. Zhurn.* 38 (7).

Pavlov, I. F. 1958. Survival of larvae and the number of generations of the Hessen fly. *Zool. Zhurn.* 37 (12).

Pavlov, I. P. 1926. *Lectures on the work of the great hemispheres of the brain.* OGIZ.

———. 1938. *A 20-year experiment in the objective study of the higher nervous activity (behavior) of animals.* Biomedgiz, M. and L.

Pavlovsky, Ye. N. 1934. The organism as a medium of habitation. *Priroda* 1.

———. 1937. A study of biocenoses as applied to certain parasitological problems. *Izv. AN SSSR, Otd. Matem. Yest. Nauk.* 4.

———. 1939. On the natural nidality of infectious and parasitic diseases. *Vestn. AN SSSR* 10.

———. 1946. The fundamentals of the study of the natural nidality of transmissible diseases of man. *Zhurn. Obshch. Biol.* 7 (1).

———. *Guidelines of parasitology of human beings with a study of the vectors of transmissible diseases.* Vol. 1, Izd. AN SSSR, M. and L., 1946; vol. 2, Izd. AN SSSR, M. and L., 1948.

———. 1951. *A guide to human parasitology with a study of the vectors of transmissible diseases.* 6th ed. L.

Pavlovsky, Ye. N., G. S. Pervomaysky, and K. P. Chagin. 1951. *The swarm: Its significance and methods of control.* Medgiz, M.

Pearl, R. 1914. The service and importance of statistics to biology. *Quart. Publ. Amer. Stat. Ass.* 40–48.

———. 1930. *The biology of population growth.* New York.

———. 1932. The influence of density of population upon egg production in *Drosophila melanogaster. J. Exp. Zool.* 63.

———. 1937. On biological principles affecting population, human and other. *Amer. Nat.* 71.

Pearse, A. S. 1950. *The emigrations of animals from the sea.* New York.

Pellegrin, I. 1921. Les poissons des eaux douces de l'Afrique et du nord francaise. *Mém. Soc. Sci. Nat. Maroc* 1 (2).

Peredel'sky, A. A. 1957. Fundamentals and tasks of radioecology. *Zhurn. Obshch. Biol.* 18 (1).

Perel', T. S. 1958. The relationship of abundance and species content of rain worms to the type of trees in a forest plantation. *Zool. Zhurn.* 37 (9).

Pereleshin, S. D. 1943. Winter feeding habits of the polar fox in the Yamal okrug. *Zool. Zhurn.* 22 (5).

Petersen, C. G. 1918. The sea bottom and its production of fish food. *Kobenhaven Ber. Biol. Sta.* 20.

Petrishcheva, P. A. 1955. The epidemiological significance of territories on the margins of landscapes. In *Manual: Natural nidality of diseases of man and regional epidemiology.* M.

Petrusevich, K. 1957. Investigations of experimentally induced population growth. *Ekol. Polsk., Ser. A* 9.

———. 1958. Influence of the presence of their own population on the results of fights between male mice. *Bull. Acad. Pol. Sci.* 2 (6) 1.

———. 1958. Differences in male and female quantitative dynamics in a confined population of mice. *Bull. Acad. Pol. Sci.* 2 (6).

———. 1960. On the intraspecies and interspecies relationships. *Zool. Zhurn.* 39 (11).

Pidoplichko, I. G. *On the ice period.* Vol. 2, Izd. Kiev. Gos. Un-Ta, Kiev, 1951; vol. 3, Izd. Kiev. Gos. Un-Ta, Kiev, 1954.

Pielou, D. P., and D. L. Hunn. 1940. The humidity behavior of the mealworm beetle *Tenebrio molitor* L. The reactions to differences of humidity. *J. Exp. Biol.* 17.

Pierce, G. W. 1948. *The songs of insects.* Cambridge.

Pirozhnikov, P. L. 1949. Semi-migratory fishes and river current. *Izv. TINRO* 29.

Plotnikov, V. I. 1927. Locusta (Pachytylus) migratoria *L. and* L. danica *L. as autonomous forms and their derivatives.* Tashkent.

Pogodin, Ye. A., and V. M. Saf'yanova. 1957. Testing of a method of catching bloodsucking Diptera with the aid of a mercury lamp. *Zool. Zhurn.* 36 (6).

Polyakov, I. Ya. 1952. Theoretical bases for prognosing the abundance of mouse-

like rodents and measures for preventing their destructiveness in the European part of the USSR in the Trans-Caucasus. *Avtoref. Diss.*
———. 1954. Toward a theory of prognosing the abundance of small rodents. *Zhurn. Obshch. Biol.* 4 (2).
———. 1958. Biological bases of rodent control. *Tr. VIZR* 12.
Polyakov, I. Ya., and S. G. Pegel'man. 1950. Some age-related properties of the demands of the society vole (*Microtus socialis* Pall.) for temperature conditions. *Zhurn. Obshch. Biol.* 11 (6).
Polyakov, K. L. 1930. Toward a physiology of the olfactory and aural analyzers of the turtle *Emys orbicularis. Russk. Fizio. Zhurn.* 13 (2).
Polyansky, Yu. I. 1957. Temperature adaptations among the infusorians. *Zool. Zhurn.* 36 (11).
Pomerantsev, D. V., and I. A. Shevyrev. 1910. The significance of insectivorous birds in the forest and steppe. *Tr. Lesn. Opyt. Delu Rossii* 24.
Ponomareva, D. A. 1955. Feeding habits and distribution of the euphausids of the Sea of Japan. *Zool. Zhurn.* 34 (1).
Ponugayeva, A. G. 1953. Reflexive regulation of gas exchange in group clusterings of rodents during the nesting period of development. *Tr. in-ta Fiziol. AN SSSR* 2. M. and L.
Ponugayeva, A. G., and A. D. Slonim. 1949. Toward a physiology of acclimatization of the horse in the Tyan'-Shan' Mountains. In *Manual: Experimental study of the regulation of physiological functions.* Izd. AN SSSR, M.
Popov, M. V. 1960. Nutritional conditions and their significance for the abundance dynamics of the belyak hare. In *Manual: A study of the causes and laws of the abundance dynamics of the belyak hare in Yakutia.* Izd. AN SSSR, M.
Popov, T. I. 1914. The origin and development of the basic shrubs within the Voronezh guberniya. *Tr. Dokuch. Pochv. Kom.* 2.
Popov, V. A. 1947. The age makeup, food base, and helminthoses of the ermine as indicators of fluctuations in the abundance of this species. *Tr. O-va Yestyestvois. Kas. Un-te* 57 (3–4). Kazan'.
———. 1949. Materials on the ecology of the mink (*Lutreola vison* Br.) and results of its acclimatization in the Tatar ASSR. *Tr. Kazan. FAN SSSR, Ser. Biol. S.-kh. Nauk.* 2.
———. 1960. *Mammals of the Volga-Kama kray.* Izd. AN SSSR, Kazan'.
Popov, V. G. 1950. The fox as a factor in the abundance of rodents in the Volga-Ural sands. In *Manual: Rodents and their control,* vol. 3. Saratov.
Predtechensky, S. A. 1928. The locust (*Locusta migratoria*) in central Russia. *Izv. Otd. Prikl. Entomol.*
———. 1935. The yearly cycle of the desert locust: Its migrations and periodicity in Persia and adjacent countries of tropical and subtropical Asia. *Tr. Zashch. Rasteniy, Ser. I* 12.
Priklonsky, S. G. 1958. On the death of the eggs of the gray raven from contamination with heavy oils at the mouth of the White River. On the death of waterfowl in fishnets on the steppe lakes in the Novosibirsk oblast'. *Tr. Oksk. Gos. Zap.* 2; *Rab. Oksk. Ornitol. Stantsii* 1.

Prints, Ya. I. 1937. *Pests and diseases of the grape.* Sel'khozgiz, M. and L.

Promptov, A. N. 1940. The species stereotype of behavior and its formation in wild birds. *DAN SSSR* 27 (2).

———. 1940. A study of the daily activity of birds in the nesting period. *Zool. Zhurn.* 19 (1).

———. 1941. *Seasonal migrations of birds.* Izd. AN SSSR, M. and L.

———. 1941. The contemporary state of the study of nesting parasitism in birds. *Usp. Sovr. Biol.* 14 (1).

———. 1946. On the conditional-reflex components in the instinctive behavior of birds. *Fiziol. Zhurn.* 32 (1).

———. 1956. *Notes on the problem of biological adaptation of the behavior of passerine birds.* Izd. AN SSSR, M. and L.

Protasov, V. R., and K. R. Golubtsov. 1960. Some functional properties of the eye of the cod (*Gadus morhua* L.) and the bitterling (*Myxocephalus scorpius* L.). *Tr. in-ta Morf. Zhivot. AN SSSR* 13.

Pshenichny, P. D. 1948. Problems of directed rearing of young farm animals. *Agrobiologiya* 4.

Ptushenko, Ye. S. 1948. Migrations of the lake gull *Larus ridibundus ridibundus. Tr. Tsentr. Byuro Kol'tsev.* 8.

Pütter, A. 1909. *Die Ernährung der Wassertiere und der Stoffhaushalt der Gewässer.* Jena.

Rall', Yu. M. The introduction into ecology of southern gerbils (*Pallasiomys meridianus* Rall.). *Vestn. Mikrob. Epidem. Parazitol.* 17 (3–4), 1938; 18 (1–4), 1939.

———. 1947. *Methodology of field study of rodents and campaigning against them.* Rostov.

———. 1958. *Lectures on the epizootiology of plague.* Stavr. Krayev. Izd., Stavropol'.

Raspopov, M. P. 1935. Toward a biology of the adder. *Byull. Zoopark. Zoosad.* 1–2.

Rayevsky, V. V. 1947. *The life of the kondo-sosvinsk sable.* M.

Réaumur, R. 1734–1742. *Mémoires pour servir à l'histoire des insects.* Paris.

Rensch, B. 1939. Klimatische Auslese vom Grossenvarianten. *Arch. Naturgesch.* 8 (1).

Reykhardt, A. N. 1941. Air transport, insects, and diseases. *Priroda* 1.

Reymers, N. F. 1956. The role of the nutcracker and the mouselike rodents in the cedar forests of the Cis-Baykal. *Byull. MOIP* 61 (2).

Richet, C. 1889. *Chaleur animale.* Paris.

Ricker, W. E. 1954. Stock and recruitment. *J. Fish. Res. Bd. Can.* 11 (5).

Rikhter, G. D. 1948. The role of snow cover in the physico-geographic process. *Tr. in-ta Geog. AN SSSR* 40.

Robinson, W. 1928. Water conservation in insects. *J. Econ. Ent.* 21.

Rodionov, V. M. 1938. Some data on gas exchange in reptiles in the supercooled state. *Byull. MOIP, Otd. Biol.* 47 (2).

Rodyanskaya, I. S. 1957. Toward the question as to the ability of rainworms to tolerate dehydration. *Uch. Zap. Mosk. Gorod. Ped. in-ta Potemkina* 65.

Rogal', I. G. 1955. On the "cold" method of raising young farm animals and its physiological bases. *Zhurn. Obshch. Biol.* 16 (4).
Rol'nik, V. V. 1939. The temperature regime of natural incubation among the nanda. *Vopr. Ekol. o Biotsenol.* 5–6.
Romanov, G. V. 1957. On retardation by flooding of the reproduction of the common vole in the Volga delta. In *Manual: Rodents and their control,* vol. 5. Saratov.
Romanova, Yu. S., and V. A. Lozinsky. 1958. Experiments in the practical application of egg eaters of the annelid silkworm in the forest. *Zool. Zhurn.* 37 (4).
Rotshil'd, Ye. V. 1958. The influence of the mole's digging activities on the development of salt-flat vegetation. *DAN SSSR* 120 (1).
———. 1958. Replacement of the vegetation in colonies of large gerbils in the northern Kyzyl-Kum. *Byull. MOIP, Otd. Biol.* 63 (5).
Rowan, W. 1926. On photoperiodism, reproductive periodicity, and annual migration of birds and certain fishes. *Proc. Boston Soc. Nat. Hist.* 38.
———. 1930. Experiments in bird migration. II. Reversed migration. *Proc. Nat. Acad. Sci.* 16.
———. 1932. Experiments in bird migration. III. The effect of artificial light, castration, and extracts on the autumn movements of the American crow. *Proc. Nat. Acad. Sci.* 18.
———. 1938. Light and seasonal reproduction in animals. *Biol. Rev.* 13.
Rubner, M. 1902. *Die Gesetze des Energieverbrauchs bei der Ernährung.* Berlin and Vienna.
———. 1908. *Das Problem der Lebensdauer und seine Beziehungen zu Wachstum und Ernährung.*
———. 1927. Wärmeregulation. *Handb. Boethe* 6.
Rubtsov, I. A. 1933. Parasites and other causes of death in the Siberian *kubyshka* locusts. *Tr. Zashch. Rasteniy Sibiri* 3.
———. 1935. Principles of development and behavior of locusts in Siberia in connection with climatic factors. *Izv. AN SSSR, Ser. Biol.*
———. 1936. Climatic characteristics of reservations and periods of mass reproduction of the Siberian locusts. *Vopr. Ekol. Biotsenol., Sb.* 5.
———. 1938. Integral climatic indices for purposes of regionalizing and predicting mass reproductions of destructive insects. *Zashch. Rasteniy* 16.
———. 1948. *The biological method of controlling destructive insects.* Sel'khozgiz, M. and L.
———. 1950. On the mass reproduction of midges and its probable explanation. *Priroda* 2.
———. 1952. On the rise and inheritance of food reactions acquired in ontogenesis among insects. *Usp. Sovr. Biol.* 34 (1).
———. 1953. On conditions of mass reproduction of insects (the influence of parasites and predators on fluctuations of host abundance). *Zool. Zhurn.* 32 (3).
———. 1959. The basic questions of study and use of entomophages for biological control. *Vopr. Ekol.* 3.
———. 1960. Variability and selection of entomophages. *Zool. Zhurn.* 39 (5).

Rudnev, D. F. 1959. The role of verdure renewal and certain other anthropogenic factors in the reproduction of forest pests in the Ukraine. *Zool. Zhurn.* 38 (2).

Rul'ye, K. F. 1954. *Selected biological works.* Izd. AN SSSR, M.

Russel, E. S., and S. M. Jonge. 1928. *The seas.* London.

Russel, F. S. 1939. Hydrographical and biological conditions in the North Sea as indicated by plankton organisms. *J. Cons. Int. Explor. Mer* 14.

———. 1947. *The problem of complete fish catching.* Pishchepromizdat, M.

Rustamov, A. K. 1955. Toward the question as to the concept of "life form" in ecology. *Zool. Zhurn.* 34 (4).

———. 1957. On fluctuations in abundance of certain predatory birds and their food specialization. *Tr. Turkmensk. S.-kh. In-Ta* 9.

Ruttenburg, S. O. 1953. Chemical thermoregulation in certain species of rats in connection with their geographic distribution. In *Experimental study of regulation of the physiological functions,* vol. 2. M. and L.

Rykovsky, A. S. 1960. On the helminths of the black cock and their role in lowering the numbers of the hosts. *Zool. Zhurn.* 39 (11).

Safonov, N. D. 1958. Characteristics of the aural analyzer of the horse by the conditioned-reflex method. In *Manual: Questions of the physiology of farm animals.* Izd. AN SSSR, M. and L.

Sakharov, N. L. 1928. Toward a study of cold resistance in insects. *Zhurn. Opytnoy Agron. Yugo-Vost.* 6 (2).

———. 1930. *Destructive moths and their control.* Gos. Izd., Saratov.

Sakhno, I. I. 1959. The influence of agrotechnical measures on the relationship of the sexes and the fecundity of certain mouselike rodents in the fields of the Lugansk oblast'. *Zool. Zhurn.* 38 (12).

Schleidt, W. M. 1952. Reaktionen auf Töne hoher Frequenz bei Nagern. *Naturwissenschaften* 39 (3).

Schmidt, P. 1918. Anabiosis of the earthworm. *J. Exp. Zool.* 27.

Schmidt-Nielsen, B. and K. 1950. Evaporation water loss in desert rodents in their natural habitat. *Ecology* 31.

Schulz, F. N. 1930. Zur Biologie des Mehlwurms (*Tenebrio molitor*). I. Der Wasserhaushalt. *Biochem. Z.* 227.

Schulz, N. 1947. *Die Welt der Seevögel.* Hamburg.

Schüz, E., and H. Weigold. 1931. *Atlas des Vogelzuges, nach den Beringungsergebnissen beipalaärktischen Vögeln.* Berlin.

Schwartzkopf, J. 1955. On the hearing of birds. *Auk* 72 (2).

Schwerdtfeger, F. 1941. Über die Ursachen des Massenwechsels der Insekten. *Z. Angew. Ent.* 28.

Sdobnikov, V. M. 1935. Interrelationships between the northern reindeer and the animal world of the tundra and the forest. *Tr. Arkt. In-Ta* 24.

———. 1940. An experiment in the mass banding of polar foxes. *Problemy Arktiki* 12.

———. 1957. Toward a characterization of the life forms of the Arctic animals. *Zool. Zhurn.* 36 (2).

———. 1958. Comparative ecological analysis of the fauna of the tundra and the taiga. *Zool. Zhurn.* 37 (4).
———. 1959. Biotopes of the northern Taymyr and population density of the animals settled there. *Zool. Zhurn.* 38 (2).
Sechenov, I. M. 1861. Two concluding lectures on the significance of the so-called vegetative acts in animal life. *Med. Vest.* 26.
Segal', A. N. 1956. Motile activity and gas exchange in the blue titmouse during changes in the light regime. *DAN SSSR* 109 (5).
———. 1958. Daily changes in gas exchange in the gray owl. *Zool. Zhurn.* 37 (7).
———. 1958. The influence of the species of bird on gas exchange in the sparrow hawk. *Zool. Zhurn.* 37 (8).
Semenov, N. M., *et al.* 1958. The influence of severe winter conditions in 1955/56 on certain mammals in the Cis-Caspian steppes. *Zool. Zhurn.* 37 (8).
Semënov-Tyan-Shansky, A. A. 1910. Systematic boundaries of a species and its subdivisions. *Zap. AN SSSR,* Ser. 8, 25 (1). SPB.
Semënov-Tyan-Shansky, O. I. 1938. The ecology of the wild game in forests of the Lapland Preserve. *Tr. Laplandsk. Gos. Zap.* 1.
———. 1948. The elk on the Kola Peninsula. Wild northern reindeer on the Kola Peninsula. *Tr. Laplandsk. Gos. Zap.* 2.
———. 1959 (1960). *Ecology of the black cocks and hens.* M.
Semper, K. G. 1881. *Animal life as affected by the natural conditions of existence.* New York.
Serbenyuk, Ts. V., and Yu. B. Manteyfel'. 1958. Some data on the physiology of cutaneous thermoreceptors in the fishes. *Zool. Zhurn.* 37 (12).
Seton, E. T. 1911. *The Arctic prairies.*
Severtsov, A. N., 1922. *Evolution and psychics.* M. and S. Sobashnikovy, M.
———. 1939. *The main directions of the evolutionary process.* Biomedgiz, M.
Severtsov, N. A. 1950. *Periodic phenomena in the life of the mammals, birds, and reptiles of the Voronezh guberniya.* Izd. AN SSSR, M. and L.
———. 1953. *Vertical and horizontal distribution of the animals of Turkestan.* Izd. AN SSSR, M. and L.
Severtsov, S. A. 1930. On the mutual relations between life span and fecundity of the various species of mammals. *Izv. AN SSSR, Otd. Fizmat. Nauk.* 8, 9.
———. 1932. Materials toward the recognition of the biology of reproduction of the Tetraonidae in the Bashkir State Preserve, 1930/31, by the method of quantitative calculation. *Zool. Zhurn.* 11 (3–4).
———. 1935. Toward recognition of the ecology of reproduction. *Zool. Zhurn.* 14 (2).
———. 1936. Morphological progress and the struggle for existence. *Izv. AN SSSR, Ser. Biol.* 5.
———. 1937. Darwinism and ecology. *Zool. Zhurn.* 16 (4).
———. 1941. *Dynamics of population and adaptive evolution of animals.* Izd. AN SSSR, M. and L.
———. 1942. On the quantitative characteristics of adaptability of animals and

types of population dynamics of the higher vertebrates. *Zhurn. Obshch. Biol.* 3 (1–2).
———. 1951. *Problems of the ecology of animals* (unpublished works), vol. 1. Izd. AN SSSR, M.
Shakhbazov, V. G., and M. D. Sirotenko. 1949. Methods of studying the daily activity of the larval stage of the butterflies (Lepidoptera). *DAN SSSR* 65 (4).
Shaldybina, Ye. S. 1956. Vertical migrations of oribatid mites. *Zool. Zhurn.* 35 (4).
Shaposhnikov, L. V. 1958. Acclimatization and design formation among mammals. *Zool. Zhurn.* 37 (9).
Sharets, A. S., *et al.* 1956. The resurgence in numbers of gray marmots after their extermination. *Tr. Sred.-Aziat. N.-I. Protiv. In-Ta* 3.
Sharleman, N. V. 1958. Toward the question as to cases of cannibalism among animals. *Zool. Zhurn.* 37 (11).
Shcheglova, A. I. 1949. Thermoregulation and skin and lung losses of water in the large gerbil (*Rhombomys opimus* Licht.). In *Manual: Experimental study of the regulation of physiological functions.* Izd. AN SSSR, M. and L.
———. 1952. Adaptations of water exchange in some species of susliks to conditions of life. *DAN SSSR* 83.
———. 1953. A change in matter exchange among rodents at low temperatures in the environment. In *Manual: Experimental study of the regulation of physiological functions,* vol. 2. Izd. AN SSSR, M. and L.
Shchegolev, V. N., A. V. Znamensky, and G. Ya. Bey-Biyenko. 1937. *Insects which destroy field crops.* Sel'khozgiz, M.
Shcherbakova, O. P. 1938. Materials toward the recognition of daily periodics of the physiological processes among the higher mammals. *Byull. Eksp. Biol. Med.* 5 (2).
Shcherbinovsky, N. S. 1952. *The desert locust* Cysticerca. Sel'khozgiz, M.
Shefer, M. B. 1948. *Some questions on the population dynamics of fish and regulation of commercial catches.* VNIRO, M.
Shekhanov, M. V. 1952. *Ecology of large gerbils in the northern Cis-Aral.* M.
Shelford, V. E. 1911. Physiological animal geography. *J. Morph.* 22.
———. 1913. The reactions of certain animals to gradients of the evaporating power of air. *Biol. Bull.* 25.
———. 1925. The hydrogen ion concentration of certain western American inland waters. *Ecology* 6.
———. 1929. *Laboratory and field ecology.* Baltimore.
Shepeleva, V. K. 1953. Some functional properties of the visual analyzer of the forest polecat. *DAN SSSR* 93 (3).
———. 1954. Some data on a study of color vision in dogs. *DAN SSSR* 96 (6).
———. 1959. On the comparative distance of the olfactory reception of certain mammals. *Soveshch. Ekol. Fiziol., Tez. Dokl.* 2.
Shevyrev, I. Ya. 1892. Insectivorous birds and the forested steppes. *Selsk. Kh-vo o Lesov.*
———. 1893. *A description of destructive insects of the forest-steppe groves and means of controlling them.* SPB.

Sheynkhaus, E. 1952. *Pathology of insects.* Izd. IL, M.

Shibanov, N. V. 1939. *The life of animals* (after A. E. Brem), vol. 3. M.

Shibanov, S. V. 1951. Abundance dynamics of the polar fox in connection with conditions of reproduction, feeding, and migrations. *Tr. VNIO* 11.

Shigolev, A. A., and A. P. Shimanok. 1949. *Seasonal development of nature.* Geografgiz, M.

Shilov, I. A. 1954. The influence of spring flooding on the movement of abundance of the water rat in various types of floodplains. *Zool. Zhurn.* 33 (6).

———. 1957. Some questions about the establishment of thermoregulation in the ontogenesis of sparrows. *Byull. MOIP, Otd. Biol.* 62 (2).

Shilova, S. A., G. B. Mal'kov, V. I. Chabovsky, and Ye. V. Meshcheryakova. 1956. The influence of depressions of abundance among mouselike forest rodents on the feeding of larvae and nymphs of the tick *Ixodes persulcatus* (P. Sch.) in a nidus of tick encephalitis. *Byull. MOIP* 61 (3).

Shilova, S. A., V. B. Troitsky, G. B. Mal'kov, and V. M. Bel'kovich. 1958. The significance of the mobility of mouselike forest rodents in the distribution of the tick *Ixodes persulcatus* P. Sch. in nidi of spring-summer encephalitis. *Zool. Zhurn.* 37 (6).

Shilova, Ye. S. 1954. On the distribution and stationary displacement of the red-tailed gerbil in the northern Cis-Aral. *Byull. MOIP, Otd. Biol.* 58 (2).

Shkorbatov, G. A. 1953. The ecologo-physiological properties and conditions of existence of closely related forms of freshwater animals. *Zool. Zhurn.* 32 (5).

Shmidt, P. Yu. 1947. *Migrations of fish.* Izd. AN SSSR, M. and L.

Shnitnikov, V. N. 1936. *Mammals of the Seven Rivers.* Izd. AN SSSR, M. and L.

Shorygin, A. A. 1952. *Feeding habits and nutritional relationships of the fishes of the Caspian Sea.* Pishchepromizdat, M.

Shostakovich, V. B. 1924. Forest fires in Siberia in 1915. *Izv. Vostochno-Sib. Otd. RGO* 10.

———. 1928. Periodic fluctuations in geographysical phenomena and sunspots. *Tr. Irkutsk. Meteorol. Observat.* 2–3.

Shpakovsky, S. N. 1937. Notes on the factors causing migrations in birds. *Tr. Novosib. Zoosada.* 1.

Shtefan, M. 1958. A physiological analysis of the interrelationship between gas exchange and flocking behavior in some marine and freshwater fishes. *Zool. Zhurn.* 37 (2).

Shvarts, S. S. 1956. Toward the question as to the development of certain interior features of terrestrial vertebrate animals. *Zool. Zhurn.* 35 (1).

———. 1958. The method of morpho-physiological indices in the ecology of terrestrial vertebrate animals. *Zool. Zhurn.* 37 (2).

———. 1959. Some questions on the problem of species among the terrestrial vertebrates. *Tr. in-ta Biol. Ural'sk. FAN SSSR, Sverdlovsk* 11.

———. 1960. Principles and methods of contemporary animal ecology. *Tr. in-ta Biol. Ural'sk. FAN SSSR, Sverdlovsk* 21.

———. 1960. The role of internally secreting glands in the adaptation of mammals to seasonal changes in the conditions of existence. Age structure of popula-

tions of mammals and their dynamics. *Tr. Ural'sk. Otd. MOIP Sverdlovsk* 2.

Shvarts, S. S., K. I. Kopein, and A. V. Pokrovsky. 1960. Comparative study of some biological properties of the voles *Microtus gregalis gregalis* Pall., *M. g. major* Ogn., and their cross-breeds. *Zool. Zhurn.* 39 (6).

Siivonen, L. 1948. Structure of short-cyclic fluctuations in numbers of mammals and birds in the northern hemisphere. *Pap. Game-Res. Helsinki* 1.

———. 1952. On the reflection of short-term fluctuations in numbers in the reproduction of tetraonids. *Pap. Game-Res. Helsinki* 8.

———. 1956. The correlation between the fluctuations of partridge and European hare populations and the climatic conditions of winter in southwestern Finland during the last 30 years. *Pap. Game-Res. Helsinki* 17.

———. 1957. The problem of short-term fluctuations in numbers of tetraonids in Europe. *Pap. Game-Res. Helsinki* 19.

Siivonen, L., and I. Koskimies. 1955. Population fluctuations and the lunar cycle. *Pap. Game-Res. Helsinki* 14.

Silant'yev, A. A. 1894. *Volume in honor of V. L. Naryshkin.* SPB.

Sinichkina (Popova), A. A. 1950. Notes on the daily activity of the small suslik at various periods of its life. In *Manual: Rodents and their control,* vol. 3. Saratov.

Skadovsky, S. N. 1955. *The ecological physiology of water organisms.* Sov. Nauka, M.

———. 1959. New paths in the struggle against contamination of bodies of water by using hydrobiology. *Usp. Sovr. Biol.* 47 (2).

Skrebitsky, G. A. 1940. The influence of the character of nesting colonies on the behavior of their members. *Nauch.-Metod. Sapiski Gl. Up. o Zapov. M.* 7.

Skryabin, K. I., and R. Shul'ts. 1940. *Fundamentals of general helminthology.* Sel'khozgiz, M.

Slonim, A. D. 1952. *Animal warmth and its regulation in the organism of the mammal.* Izd. AN SSSR, M. and L.

———. 1954. On the ways and means of studying the different analyzers in the food-getting activities of rodents. *Zool. Zhurn.* 33 (5).

———. 1957. On the influence of factors in the external environment on gas exchange. *Probl. Fiziol. o Tsentr. Nervn. Sistemy. Sb. 70 Let Ak. K. Bykova;* Izd. AN SSSR, M.

———. 1959. Chemical thermoregulation in ontogenesis and in the adult animal. *Usp. Sovr. Biol.* 47 (2).

———. 1961. *Fundamentals of general ecological physiology of mammals.* Izd. AN SSSR, M. and L.

Sludsky, A. A. 1953. Jutes in the deserts of Kazakhstan and their influence on the abundance of animals. *Tr. in-ta Zool. AN Kazakh. SSR* 2.

Smirnov, N. A. 1927. The biological relationships of certain pinnipeds to ice. In *Collection in honor of Knipovich.* L.

Smirnov, N. P. 1938. Bioclimate of the USSR. *Izv. Geogr. O-va* 70 (6).

Smirnov, Ye. S., and S. S. Shvarts. 1957. Seasonal changes in the relative weight of the adrenals of mammals under natural conditions. *DAN SSSR* 115 (6).

Soldatov, V. K. 1912. *Investigations of the biology of the Amur salmon.* SPB.

Soldatova, A. N. 1955. On certain properties of the periodic phenomena of life of the small suslik in the southern Trans-Volga area. *Tr. in-ta Geog. AN SSSR* 66; *Mater. Biogeog.* 2.

Spangenberg, Ye. P., and V. V. Leonovich. 1958. The influence of human activity on birds of the eastern coast of the White Sea. *Byull. MOIP, Otd. Biol.* 63 (5).

Spencer, H. A. 1852. A theory of population deduced from the general law of animal fertility. *Westminster Rev.* 57.

Stakhrovsky, V. G. 1932. Toward a study of the biology of the squirrel while running free in a park cage. *Zool. Zhurn.* 11 (1).

Stark, V. N. 1955. Retardation in the development of lumber-eating species of insects and the significance of it for determination of the age makeup and flight times of the population. *Zool. Zhurn.* 34 (3).

Stebayev, I. V. 1958. The animal population of the primeval rock soils and its role in soil formation. *Zool. Zhurn.* 37 (10).

———. 1959. Soil invertebrates of the Salekhard tundras and changes in their groupings under the influence of plowing. *Zool. Zhurn.* 38 (11).

Steggerda, F. R. 1937. Comparative study of water metabolism in amphibians injected with pituitrin. *Proc. Soc. Exp. Biol. Med.* 36.

Stetter, H. 1929. Untersuchungen über den Gehörsinn der Fische, besonders von *Phoxinus laevis* L. und *Ameiurus nebulosus* Raf. *Z. Vergl. Physiol.* 9.

Stier, T., and H. Taylor. 1939. Seasonal variation in behavior of the intact frog heart at high temperature. *J. Cell. Comp. Physiol.* 14 (3).

Stogov, I. I. 1956. On cannibalism among southern and red-tailed gerbils. *Tr. Sred.-Aziat. N.-I. Protiv. In-Ta* 3.

Strel'nikov, I. D. 1934. Light as a factor in the ecology of animals. I. The action of solar radiation on the temperature of the body of certain poikilothermic animals. *Izv. Yest.-Nauch. in-ta Lesgrafta* 17–18.

———. 1935. The action of solar radiation and microclimate on body temperature and behavior of the larvae of the Asian locust. *Tr. ZIN AN SSSR* 2 (4).

———. 1940. The significance of thermoproduction in movements and under the action of solar radiation in the ecology of high-mountain insects. *Zool. Zhurn.* 19 (3).

———. 1940. The significance of thermal exchange in the ecology of the digging rodents. *Izv. AN SSSR, Ser. Biol.* 2.

———. 1944. The significance of solar radiation in the ecology of high-mountain reptiles. *Zool. Zhurn.* 23 (5).

———. 1948. The significance of solar radiation and the mutual actions of physico-chemical factors in the ecology of animals of different landscapes. *Probl. Fizich. Geog.* 13; Izd. AN SSSR, M.

———. 1950. The evolution of warm-bloodedness in connection with circumstances of the environment. The significance of the nest in the heat regime of rodents. In Second Ecological Conference on the problem "Mass reproduction of animals and their prognosis," *Tezisy dokladov,* ch. 2. Kiev.

Stroganova, Ye. V. 1950. Toward the question as to the species and age-related

tolerance of birds to lowering of the barometric pressure. *Fiziol. Zhurn.* 31 (3).

Strogy, A. A. 1920. *Forest fires in the Amur oblast'.*

Su-De-Lon. 1958. Toward the question as to the influence of the temperature factor on the life activities of *Oncomelania nupensis. Zool. Zhurn.* 37 (9).

Sukachev, V. N. 1957. Fundamentals of the theory of biocenology. *Yubil. Sb. AN SSSR* 2.

Sullivan, C. 1954. Temperature reception and responses in fish. *J. Fish. Res. Bd. Can.* 11 (2).

Summerhayes, V. S. 1941. The effect of voles (*Microtus agrestis*) on vegetation. *J. Ecol.* 29.

Sumner, F. B. 1909. Some effects of external conditions upon the white mouse. *J. Exp. Zool.* 7.

———. 1921. Desert- and lava-dwelling mice and the problem of protective coloration in mammals. *J. Mammal.* 2 (3).

———. 1934. Does "protective coloration" protect? Results of some experiments with fishes and birds. *Proc. Nat. Acad. Sci.* 20 (10).

Sun Ju-yung. 1958. The geographic variability of certain ecologo-physiological properties of rust-colored and common voles in the Moscow oblast'. *Avtoref. Diss.*, M.

Suvorov, Ye. K. 1948. *Commercial waters of the USSR.* Izd. LGU, L.

Svetozarov, Ye., and G. Shtraykh. 1940. Light and sexual periodicity among animals. *Usp. Sovr. Biol.* 12 (1).

Sviridenko, P. A. 1934. Reproduction and mortality in mouselike rodents. *Tr. Zashch. Rasteniy* 4 (3).

———. 1936. Forest mice of the Caucasus and Cis-Caucasus. *Byull. NII Zool. MGU* 3.

———. 1940. Feeding habits of the mouselike rodents and their significance in the problem of restoration of forests. *Zool. Zhurn.* 19 (4).

———. 1944. Rodents, the wasters of forest seeds. *Zool. Zhurn.* 23 (4).

———. 1944. Distribution, feeding habits, and epidemiological significance of the field mouse (*Apodemus agrarius* Pall.). *DAN SSSR* 42 (2).

———. 1951. Theoretical disagreements on the role of the olfactory receptor of the rodents and their practical interpretation. *Zool. Zhurn.* 30 (4).

———. 1954. Nonconditioned and conditioned reflexes among wild rodents to food and nonfood supplies. *Vopr. Fiziol. AN SSSR Kiev.* 10.

———. 1957. *Storing of food by animals.* Izd. AN SSSR, Kiev.

Sviridov, N. S. 1958. Feeding habits of the raccoon dog acclimatized in the lower Volga region and in the North Caucasus. *Izv. Irkut. Kh. in-ta Irkutsk* 8.

Syrovatsky, I. Ya. 1953. On the biological role and fish-farming significance of the zander in reservoirs. *Zool. Zhurn.* 32 (6).

Tanasiychuk, V. S. 1951. Computation of the young of commercial fish in the northern Caspian. In *Manual: The fishes of the Caspian Sea. Tr. VNIRO* 18; Izd. MOIP, M.

Tarasov, N. I. 1938. On the classification of aquatic organisms according to the temperature. *Izv. Vs. Geogr. O-va* 3.

Tarasov, P. P. 1944. Biological observations on predatory birds in the southeastern part of the Trans-Baykal. *Izv. Irkut. Gos. Protiv. in-ta Sib. DV Irkutsk* 5.

———. 1956. On certain properties of the anatomy of the northern reindeer as a tundra animal. *Byull. MOIP, Otd. Biol.*

———. 1959. Some properties of the intraspecies relationships among stenotype rodents. *Tr. Sred.-Aziat. N.-I. Protiv. in-ta Alma-Ata* 5.

———. 1960. On the biological significance of the odor glands of mammals. *Zool. Zhurn.* 39 (7).

Taurin'sh, E., and G. Mikhelson. 1950. The influence of artificial nesting grounds on the quantitative and qualitative makeup of the ornithofauna in a pine forest in the Latvian SSR. In Second Ecological Conference on the problem "Mass reproduction of animals and their prognosis," *Tezisy dokladov.* Kiev.

Taylor, W. P. 1934. Significance of extreme and intermittent conditions in distribution of species and management of natural resources, with a restatement of Liebig's law of the minimum. *Ecology* 15.

———. 1936. What is ecology and what good is it? *Ecology* 17.

Teichmann, H. 1959. Über die Leistung des Geruchsinnes beim Aal (*Anguilla anguilla* L.). *Z. Vergl. Physiol.* 42 (3).

Telenga, N. A. 1953. On the role of entomophages in the mass reproduction of insects. *Zool. Zhurn.* 32 (1).

Teplov, V. P. 1952. Computation of the common squirrel. In *Manual: Methods of computing the abundance and geographic distribution of terrestrial vertebrates.* Izd. AN SSSR, M.

———. 1954. Toward the question as to the relationship of the sexes among wild mammals. *Zool. Zhurn.* 33 (1).

———. 1955. Toward a winter ecology of the wolverine in the region of the Pechoro-Ilych Preserve. *Byull. MOIP* 60 (1).

———. 1957. On the significance of predatory mammals in various landscapes. *Vopr. Ekol. Kiev.* 2.

Teplova, Ye. N. 1947. Feeding habits of the fox in the Pechoro-Ilych Preserve. *Tr. Pechoro-Ilych Zap.* 5.

Teplova, Ye. N., and V. P. Teplov. 1947. The significance of winter cover in the biology of the mammals and birds of the Pechoro-Ilych Preserve. *Tr. Pechoro-Ilych Zap.* 5.

Terent'yev, P. V. 1957. On the applicability of the concept of "subspecies" in the study of intraspecies variability. *Vestn. LGU* 21 (4).

Tereza, S. I., and L. G. Shalimov. 1937. Worm-harboring as a possible cause of barrenness in mammals. *Byull. Eksp. Biol. o Med.* 3 (3).

Thienemann, A. 1925. Der See als Lebenseinheit. *Naturwissenschaften* 12.

———. 1931. Der Produktionsbegriff in der Biologie. *Arch. Hydrobiol.* 22.

Thorson, G. 1950. Reproductive and larval ecology of marine-bottom invertebrates. *Biol. Rev.* 25.

Thorson, T., and A. Svihla. 1943. Correlation of the habitats of amphibians with their ability to survive the loss of body water. *Ecology* 24.

Tikhomirov, V. A. 1955. On the influence of animals on the vegetation of the Taymyr tundra. *Byull. MOIP* 60 (5).
Tikhonov, V. N., ed. 1939. *Biology and commercial catching of the Murmansk herring.* Izd. PINRO.
Tikhvinsky, V. I. 1938. On the connections between meteorological factors and fluctuations in abundance. *Tr. O-va Yestest. Kazan. Un-te* 55 (3–4).
Timiryazev, K. A. 1922. The historical method in biology. Russk. Bibliogr. Iz-vo Br. Granat. M.
Timofeyev, V. V. 1958. *The sable of East Siberia.* Irkutsk.
Tinker, I. S. 1940. The epizootiology of plague in susliks. *Rost. Vedom. Izd. Rostov.-D.*
Tischler, W. 1955. *Synökologie der Landtiere.* Stuttgart.
Tkachenko, M. Y. 1939. *General forestry.*
Tothill, I. D., T. N. C. Taylor, and R. W. Payne. *The coconut moth in Fiji.* London.
Treus, V. D. 1957. Seasonal displacement and migrations of the gray duck. Seasonal displacement and migrations of the broadbill. *Tr. Tsentr. Byuro Kol'tsevaniya* 9.
Trippensee, R. E. 1948. *Wildlife management; upland game and general principles.* New York.
Tserevitinov, B. F. 1958. Topographical properties of the hair covering of furbearing animals. *Tr. VNIZhP (VNIO)* 17.
Tupikova, N. V. 1947. Ecology of the house mouse of the middle belt of the USSR. *Mater. Pozn. Fauny Flory SSSR, Otd. Zool.*, Nov. Ser. 8 (26); *Mater. po Gryzunam* 2; Izd. MOIP, M.
Tupikova, N. V., and I. L. Kulik. 1954. The daily activity of mice and its geographic variability. *Zool. Zhurn.* 33 (2).
Turkin, N. V. 1900. A general statistical survey of hunting and trapping animals in Russia. In *Zveri Rossii,* vol. 1. SPB.
Turkin, N. V., and K. A. Satunin. 1900. *Zveri Rossii,* vol. 1. SPB.
Ugolev, A. M. 1950. Conditional salivary reflexes in cats and means of getting food. *Avtoref. Diss.*, L.
———. 1953. Conditioned salivary reflexes in cats. In *Manual: Experimental study of the regulation of the physiological functions,* vol. 2. M. and L.
Unterberger, V. K. 1953. Experimental attraction of wintering birds into a nidus of reproduction of the pine moth *Panolis flammea* Schift. *Zool. Zhurn.* 32 (3).
Ushakov, B. P. 1955. Thermostability of the somatic musculature of amphibians in connection with the conditions of existence of the species. *Zool. Zhurn.* 34 (3).
———. 1959. Thermostability of the tissues—a species feature of the invertebrate animals. *Zool. Zhurn.* 38 (9).
Ushatinskaya, R. S. 1950. *Fundamentals of cold resistance in insects.* Izd. AN SSSR, M.
Van Wijngarden, A. 1957. The rise and disappearance of continental vole plague zones in the Netherlands. *Versl. Landb. Onderz. 's Grav.* 63 (15).

Varming, Ye. 1901. *Ecological geography of plants.* SPB.
Varshavsky, S. N. 1941. Geographic properties of daily activity of the small suslik. *Zool. Zhurn.* 20 (2).
———. 1952. Some results of the application of the method of relative abundance counting of predatory birds in a steppe landscape. In *Manual: Methods of computing the abundance and geographic distribution of terrestrial vertebrates.* M. and L.
Varshavsky, S. N., and M. N. Shilov. 1955. Biological foundations and methodology of predicting changes in the abundance of large gerbils in the desert zone of the northern Cis-Aral. *Tr. Probl. Tem. Soveshchan. ZIN* 5; IV Soveshch. o Izuch. Vredn. o Promysl. Gryzunov. M. and L.
———. 1956. Ecologo-geographic properties of distribution and territorial delimitation of the large gerbil in the northern Cis-Aral. *Tr. Sred.-Aziat. N-I. Protiv. in-ta Alma-Ata* 3.
Vasil'yev, G. A. 1947. Toward the question as to the possible thermoregulatory significance of certain periodic phenomena in the life of the birds. *Izv. AN SSSR, Ser. Biol.* 1.
Vasnetsov, V. V. 1938. Ecological correlations. *Zool. Zhurn.* 17 (4).
———. 1947. The growth of fishes as an adaptation. *Byull. MOIP* 2 (1).
———. 1948. *Morphological properties determining the feeding habits of the bream, zander, and sazan at all stages of development.* Izd. AN SSSR, M. and L.
———. 1953. Principles of development and abundance dynamics of the fishes. *Tr. Vs. Konf. o Vopr. Ryb. Khoz-va.*
———. 1953. Stages of development of the bony fishes. *Ocherki o Obshch. Vopr. Ikhtiol.*
Vavilov, N. I. 1931. *The Linnaean species as a system.* Sel'khozgiz, M.
Vavilov, S. I. 1950. *The eye and the sun.* 2nd ed. Izd. AN SSSR, M.
Vereshchagin, N. K. 1959. *Mammals of the Caucasus.* Izd. AN SSSR, M. and L.
Vernadsky, V. I. 1926. *The biosphere,* vols. 1 and 2. L.
———. 1928. Evolution of species and living matter. *Priroda* 3.
———. 1934. *Outlines of geochemistry.* Gorgeonefteizdat, M.
———. 1939. *Problems of geochemistry.* 2nd ed. Izd. AN SSSR, M. and L.
———. 1940. *Biogeochemical sketches.* Izd. AN SSSR, M. and L.
———. 1944. A few words on the noosphere. *Usp. Sovr. Biol.* 18 (2).
Viktorov, G. A. 1955. Toward the question as to the causes of the mass reproduction of insects. *Zool. Zhurn.* 34 (2).
———. 1956. On the differences in cold resistance between healthy and infected caterpillars of the bean moth *Etiella zinckenella* Tr. *Byull. MOIP* 61 (4).
Vil'yams, V. R. 1948. *Collected works,* vol. 1.
Vinberg, G. G. 1937. Temperature and sizes of biological objects. *Usp. Sovr. Biol.* 6 (1).
———. 1950. Intensity of exchange and sizes of crustaceans. *Zhurn. Obshch. Biol.* 11 (5).
Vinberg, G. G., and Yu. S. Belyatskaya. 1959. The relationship between exchange and body weight in the freshwater gastropod mollusks. *Zool. Zhurn.* 38 (8).

Vinogradov, A. P. 1933. *The geochemistry of living matter.* Izd. AN SSSR, M.
———. 1949. Biochemical provinces. In *Sb. Tr. Yubil. Sessii, Posvyashch. Stol. o Dnya Rozh. V. V. Dokuchayeva.* M. and L.
Vinogradov, B. S. 1937. The jerboas. In *Animals of the USSR: Mammals,* vols. 3 and 4. M. and L.
Vishnyakov, S. V. 1957. Materials on the ecology of the water rat of the central oblast' of the RSFSR. *Mater. po Gryzunam* 5.
Vlasov, Ya. P. 1937. The burrow as a unique biotope in the environs of Ashkhabad. In *Prob. parazitologii i fauny Turkmen., Tr. SOPS, Ser. Turkmen.* 9.
———. 1940. The burrow of the thin-fingered suslik (*Spermophilopsis leptodactylus* Licht.) and the large gerbil (*Rhombomys opimus* Licht.) as a unique biotope in the environs of Ashkhabad. In *Tezisy Mezhr. Zov. Kozh. Leysh. Moskit. Probl., Tr. Turk. Kozhno-Venerolog. In-Ta.* Ashkhabad.
Volchanetskaya, G. I. 1954. Seasonal changes in the reaction of certain species of voles to the influence of environmental temperature. *Uch. Zap. Khar'kov, Gos. Univ.* 52; *Tr. N-I. in-ta Biol. Khar'k. Gos. Univ.* 20.
Volterra, V. 1926. Variazioni e fluttuazioni del numero d'individui in specie animali conviventi. *Mem. Accad. Lincei* 6.
———. Variations and fluctuations of the numbers of individuals in animal species living together. *J. Comm.* 2, 1927; 3, 1928.
Voronov, A. G. 1957. The mutual relationships of the animal and plant worlds. *Priroda* 2.
Voronov, A. G., and L. N. Sobolev. 1960. The content and tasks of biogeography. *Voprosy Geog.* 48.
Voronov, N. P. 1957. Toward a study of the fauna of mole tunnels. *Zool. Zhurn.* 36 (10).
Vorontsov, N. N., O. Yu. Ivanova, and M. F. Shemyakin. 1956. Materials on the winter feeding habits of the sparrow owl. *Zool. Zhurn.* 35 (4).
Voskresensky, K. A. 1948. The filtrate belt as a biological system of the sea. *Tr. Gos. Okeanograf. In-Ta* 6 (18).
Voyeykov, A. I. 1884. *Climates of the globe, that of Russia in particular.* SPB.
———. 1889. Snow cover: Its effect on the soil, climate, and weather. *Zap. PGO Obshch. Geog. SPB* 18.
Vul'f, Ye. V. 1944. *Historical geography of plants.* Izd. AN SSSR, M. and L.
Vysotsky, G. N. 1930. A study of forest pertinence. In *A course in forestry.* Ch. III. Forestry and forest trapping. 1.
Walgren, H. 1954. Energy metabolism of two species of the genus *Emberiza* as correlated with the distribution and migration. *Acta Zool. Fenn. Helsinki* 82–86.
Waloff, N. 1948. Development of *Ephestia elutella* Hb. (Lepidoptera, Phycitidae) on some natural foods. *Bull. Ont. Res.* 39.
Weiss, K. 1957. Zur Gedächtnisleistung von Wespen. *Z. Vergl. Physiol.* 39 (6).
Wellington, W. G. 1940. The effects of temperature and moisture upon the behaviour of the spruce budworm, *Choristoneura fumiferana* Clemens (Lepidoptera: Torticidae). *Sci. Agric.* 29.

———. 1952. Air-mass climatology of Ontario north of Lake Huron and Lake Superior before outbreaks of the spruce budworm *Choristoneura fumiferana* (Clem.) and the forest tent caterpillar *Malacosoma disstria* Hbn. (Lepidoptera, Torticidae, Lasiocampidae). *Can. J. Zool.* 30.

Wellington, W. G., *et al.* 1950. Physical and biological indicators of the development of outbreaks of the spruce worm *Choristoneura fumiferana* (Clem.) (Lepidoptera, Torticidae). *Can. J. Res. D* 28.

Wells, H. G., J. S. Huxley, and G. P. Wells. 1939. *The science of life.* Garden City, N.Y., and L.

West, A. S. 1947. The California flathead borer (*Melanophila californica* van Dyne) in ponderosa pine stands of northeastern California. *Can. J. Res. D* 25.

Wigglesworth, V. B. 1931. A certain effect of desiccation on the bed bug (*Cimex lectularius*). *Proc. R. Ent. Soc. Lond.* 6.

———. 1941. The sensory physiology of the human louse, *Pediculus humanus corporis* de Geer (Anoplura). *Parasitology* 33.

Wisby, W. I., and A. D. Hasler. 1954. The effect of olfactory occlusion on migrating silver salmon *O. kisutch. J. Fish. Res. Bd. Can.* 11 (4).

Wolfson, A. 1945. The role of the pituitary, fat deposition, and body weight in bird migration. *Condor* 47.

Woodbury, A. M. 1954. *Principles of general ecology.* New York.

Wrangel, F. V. 1841. *A journey along the northern shores of Siberia and the Arctic Ocean,* chs. I–II. SPB.

Yanushko, P. A. 1957. The way of life of the Crimean reindeer and their influence on natural forest restoration. *Tr. Krymsk. Gos. Zap. Simferopol'* 4.

———. 1958. Abundance dynamics of the Crimean reindeer. *Zool. Zhurn.* 37 (8).

Yegorov, O. V. 1952. The ecology of the Siberian capricorn. *Avtoref. Diss.*, L.

Yermolayev, M. F. 1950. The influence of factors of the environment on periodicity of the appearance of gamma worms. In Second Ecological Conference on the problem "Mass reproduction of animals and their prognosis," *Tezisy dokladov,* ch. 2. Kiev.

Yershova, I. P. 1952. On certain properties of olfactory reception among rodents. *Zool. Zhurn.* 31 (1).

Yershova, I. P., and B. Yu. Fal'kenshteyn. 1948. On the role of the olfactory receptor in the feeding habits of voles and mice. *Zhurn. Obshch. Biol.* 5.

Yesilevskaya, M. A., L. V. Kirichenko, and Yu. A. Leberman. 1955. Properties of the reaction of the Ukrainian population of the oak bombyx in the Ukraine, Crimea, and Trans-Caucasus. *Zool. Zhurn.* 34 (6).

Yurgenson, P. B. 1939. Typology of stations of the forest marten. *Vopr. Ekol. o Biotsenol.* 4.

———. 1947. On the sexual dimorphism in feeding habits as an ecological adaptation of a species. *Byull. MOIP, Otd. Biol.* 52 (6).

———. 1950. An experiment in comparative ecology of the genus *Martes. Avtoref. Diss.*, M.

———. 1950. Supplementary feeding of valuable fur-bearing animals of the

carnivore order. In *Manual: Reconstruction of the vertebrate fauna of our country*. M.

———. 1954. On the influence of the forest marten on the abundance of the squirrel in the northern taiga. *Zool. Zhurn.* 33 (1).

Yurovitsky, Yu. G. 1958. On the factors determining the abundance of tits in the Rybinsk Reservoir. *Zool. Zhurn.* 37 (12).

Zakharov, L. Z. The basic principles of development of the lower Volga nidi of the Asian locust. *Zool. Zhurn.* 25 (1).

———. 1950. Mass reproductions of the Asian locust and their prognoses. In Second Ecological Conference on the problem "Mass reproduction of animals and their prognosis," *Tezisy dokladov*. Kiev.

———. 1950. The behavior of the Asian locust. *Uch. Zap. Sarat. Gos. Un-ta* 26.

Zavadsky, K. I. 1954. On certain questions of the theory of species and species formation. *Vestn. LGU* 10.

Zavgorodnaya, V. K. 1933. Daily dynamics of flight of the bees (Hymenoptera, Apoidea) on bean plants used as food. *Entomol. Obozrev.* 33.

Zenkevich, L. A. 1937. History of the invertebrate system. In *Handbook of zoology*, vol. 1. M.

———. 1951. Some problems of the biogeography of the sea as a part of general geography. *Voprosy Geog.* 24.

———. *Fauna and biological productivity of the sea.* Vol. 1, 1951; vol. 2, 1947.

———. 1955. Morya SSSR, ikh. fauna i flora. Uchpedgiz, M.

Zenkevich, L. A., and Ya. A. Birshteyn. 1956. Studies of the deep-water fauna and related problems. *Deep Sea Res. Lond.* 4.

Zernov, S. A. 1949. *General hydrobiology*. Izd. AN SSSR, M. and L.

Zhadin, V. I. 1938. Formation of the biological regime of the reservoir. *Usp. Sovr. Biol.* 9.

———. 1940. Fauna of streams and lakes. *Tr. ZIN AN SSSR* 5 (2–3).

———, ed. *The life of the fresh waters of the USSR*. Vol. 1, 1940; vol. 2, 1949. Izd. AN SSSR.

———. 1951. Toward the question as to certain concepts and tasks of hydrobiology. *Tr. Probl. Tem. Soveshch. ZIN* 1.

Zharkhov, I. V. 1938. The ecology and significance of forest mice in the forests of the Caucasus Preserve. *Tr. Kavkazsk. Zap.* 1.

Zhinkin, L. 1931. Materials on the parasitic infection rate of certain wild mammals. *Parazitol. Sborn. ZIN AN SSSR* 2.

Zhitkov, B. M. 1904. *The Yamal Peninsula.*

———. 1934. *Acclimatization of animals.*

Zimina, R. P. 1953. A short sketch of the mammalian fauna and birds of the region of the Tyan'-Shan' physico-geographic station. *Tr. in-ta Geog. AN SSSR* 56.

Zverev, M. D. 1929. The biology of Eversman's susliks and experiments in controlling them with poisoned baits. *Izv. Sibkray. Novosibirsk* 3 (6).

INDEX

N. P. Naumov is one of the USSR's leading ecologists. He is in charge of zoological research at the Gameleya Institute of Epidemiology and Microbiology in Moscow. Among his other professional activities he has published widely in Russian scientific journals, and he participated in the Symposium on Theoretical Questions of Natural Foci of Diseases, held in Prague, Czechoslovakia, 26–29 November 1963. There he met Norman D. Levine, editor of this volume.

Norman D. Levine is professor of veterinary parasitology, veterinary research, and zoology and director of the Center for Human Ecology at the University of Illinois, Urbana. He has published many articles in professional journals and has authored, co-authored, or edited several books, among them two others translated from the Russian: *Natural Nidality of Transmissible Diseases . . .* (1966) and *Natural Nidality of Diseases and Questions of Parasitology* (1968).

Frederick K. Plous, Jr., is a free-lance Russian translator whose major interest is biology. He translated the two works mentioned above, published by the University of Illinois Press. He has also translated many other miscellaneous Russian papers in the fields of ecology, epidemiology, and biology.

This translation of THE ECOLOGY
ANIMALS is pa the program of s
veillance of ific literature
gun a

N IDALITY
OF T DISEASES
CE
OLOGY
S

ne
Jr.